Perspectives in Animal Ecology and Reproduction

The Editors

Dr. Vijay Kumar Gupta, Ph.D., FLS, London (born 1953-) Chief Scientist, CSIR- Indian Institute of Integrative Medicine, Jammu, India. He did his M.Sc. (1975) and Ph.D. (1979) in Zoology both from University of Jammu, Jammu-India. His research capabilities are substantiated by his excellent work on histopathology, ecology and reproductive biology of fishes, turtles, birds and mammals, which has already got recognition in India and abroad.

Dr. Gupta has to his credit more than 100 scientific publications and review articles which have appeared in internationally recognized Indian and foreign journals. Founder fellow, life member and office bearer of many national societies, academies and associations. He has successfully completed a number of research/consultancy projects funded by government, private and multinational agencies. His current areas of interest are histopathology, toxicology, pre-clinical safety pharmacology, reproductive efficacy studies of laboratory animals and biodiversity.

He is the Series Editor of the recently published multi-volume set of books, "**Comprehensive Bioactive Natural Products (Vols. 1-8)"**, published by M/S Studium Press, LLC, USA. He is also Editor-in-Chief of the books, "**Utilisation and Management of Medicinal Plants (Vols. 1-3)"**, "**Medicinal Plants: Phytochemistry, Pharmacology and Therapeutics (Vols.1-4)"**, "**Traditional and Folk Herbal Medicine (Vols. 1-3)"**, "**Natural Products: Research Reviews (Vols. 1-4)"**, "**Bioactive Phytochemicals: Perspectives for Modern Medicine (Vols. 1-3)"**, "**Perspectives in Animal Ecology and Reproduction (Vols. 1-10)**" and "**Animal Diversity, Natural History & Conservation (Vols. 1-5)"**. The Editor-in-chief of the American Biographical Institute, USA, has appointed him as *Consulting Editor* of *The Contemporary Who's Who*. Dr. Gupta also appointed as Nominee for the *Committee for the Purpose of Control and Supervision of Experiments on Animals* (CPCSEA, Govt. of India). The *Linnaean Society of London, U.K.* has awarded fellowship to him in November 2009 in recognition of his contribution towards the cultivation of knowledge in Science of Natural History. Recently, Modern Scientific Press, USA has nominated Dr. Gupta as the Editor of the *International Journal of Traditional and Natural Medicine*.

Dr. Anil K. Verma, Ph.D., M.N.A.Sc., FLS, London (born 1963-) Prof. & Head, Department of Zoology, Govt. (P.G.) Degree College Rajouri, J&K State, did his M.Sc. in Zoology (1986) from University of Jammu, Jammu. He has undergone his M.Phil. (1988) and awarded first rank and Ph.D.(1993) in the field of animal reproduction at the same University. Dr. Verma has published about 60 research papers and review articles in reputed journals and books. He is also a member Editorial Board of the book series "Advances in Fish and Wildlife: Ecology and Biology" a Daya Publishing House, New Delhi. In recognition of his standing in greater scientific community, the Board of Directors of the American Association for the advancement of science (AAAS) New York, Washington, has awarded membership to him. The *Linnaean Society of London, U.K.* has awarded fellowship to him in October 2006 in recognition of his contribution towards the cultivation of knowledge of Science of Natural History.

Dr. Verma was also nominated as Nominee, CPCSEA by Ministry of Environment & Forests, Govt. of India, and also awarded "The Rashtriya Gourav Award" and "The Best Citizen of India" award by the India International Friendship Society in 2009-2010. He is also a nominated member of the Global Academy of Science, India.

Dr. Gurdarshan Singh (born 1962), working as Principal Scientist in the PK-PD-Toxicology Division of Indian Institute of Integrative Medicine (CSIR), Jammu. He did his M.Sc. (1987), M.Phil. (1989) and Ph.D. (1993) in Zoology from University of Jammu, Jammu-India.

Dr. Singh has published 16 research papers, both in national and international journals of high repute and impact factor.

Perspectives in Animal Ecology and Reproduction

— Volume 10 —

— Editor-in-Chief —

Dr. V.K. Gupta

Formerly Chief Scientist
CSIR-Indian Institute of Integrative Medicine,
Canal Road, Jammu – 180 001, India

Editors

Dr. Anil K. Verma

Professor and Head
Department of Zoology, Government (P.G.) Degree College,
Rajouri, Jammu – 180 001, India

Dr. G.D. Singh

Principal Scientist
Pharmacology Division,
CSIR-Indian Institute of Integrative Medicine,
Canal Road, Jammu – 180 001, India

2015

Daya Publishing House®

A Division of

Astral International Pvt. Ltd.

New Delhi – 110 002

Cataloging in Publication Data--DK
    Courtesy: D.K. Agencies (P) Ltd. <docinfo@dkagencies.com>
    Perspectives in animal ecology and reproduction / editor-in-chief,
Dr. V.K. Gupta ; editors, Dr. Anil K. Verma, Dr. G.D. Singh.
        volume 10   cm
    Includes bibliographical references and index.
    ISBN 9789351306627 (International Edition)

    1. Animal ecology--India.   2. Reproduction--India.   I. Gupta, V. K.
(Vijay Kumar), 1953-, editor. II. Verma, Anil K. (Anil Kumar), 1963-, editor.
III. Singh, G. D. (Gur Darshan), 1962-, editor.
        DDC 591.70954   23

Published by   :   **Daya Publishing House®**
                              A Division of
                              Astral International Pvt. Ltd.
                              – ISO 9001:2008 Certified Company –
                              4760-61/23, Ansari Road, Darya Ganj
                              New Delhi-110 002
                              Ph. 011-43549197, 23278134
                              E-mail: info@astralint.com
                              Website: www.astralint.com

Laser Typesetting   :   **Classic Computer Services**, Delhi - 110 035

Printed at   :   **Replika Press Pvt. Ltd.**

PRINTED IN INDIA

Editorial Board

Prof. Tej Kumar Sherstha

Head,
Central Department of Zoology, Tribhuvan University, Kathmandu, Nepal
E-mail: drtks@ccsl.com.np

Dr. Justin Gerlach

Chief Scientist,
The Nature Protection of Seychelles,
133 Cherry Hinton Road, Cambridge CBI 7BX, UK
E-mail: justgerlach@aol.com

Prof. Vanitha Kumari

Head,
Department of Zoology, Bharathiar University, Coimbatore, India
E-mail: regr@bharathiaruni.org

Dr. Sunil Kumar

Deputy Director (Sr. Grade),
Division of Reproductive Toxicology & Histochemistry,
National Institute of Occupational Health,
Meghani Nagar, Ahmedabad – 380 016, India
E-mail: sunilnioh@yahoo.com

Dr. A.K. Pandey

Principal Scientist,
NBFGR (ICAR),
Canal Ring Road, Lucknow – 224 002, U.P. India
E-mail: akpandey_cifa@yahoo.co.in

ISTANBUL UNIVERSITY
Science Faculty
DEPARTMANT OF BIOLOGY
Phone:+90(212) 455 57 00
Fax: +90(212) 519 08 34; +90(212) 522 08 06
34134 Vezneciler-ISTANBUL/TURKEY

Foreword

Life is based on the mutual interactions of numerous organism from one cell life form to most developed living organism known to us. Each species has evolved a different reproductive strategy and the unique strategies which they have are mainly responsible for sustaining generations in future. The existence and conservation of convenient environmental condition is primarily factor for maintaining reproductive strategies without any detrimental affect on them.

Even if we cast aside the uncontrollable events by human-being, I sincerely want to do something for controllable events in which the each of us can play an active role for the future of our planet, shared by all living creatures. It is apparently obvious that all living organism have been directly or indirectly influenced by adverse environmental conditions from man-made sources. Humans must be aware of environmental contaminants generated by primarily antropogenic sources and their negative effects on all living beings and the environment, and this awareness is indispensable for protecting and sustaining all life forms on earth.

Nowadays, a lot of environmentally hazardous chemicals such as; new chemical fertilizers, insecticides and herbicides etc., are producing day-by-day and consequently we contaminate air, water and soil with these potentially toxic compounds. Global warming and ozone layer depletion that caused primarily from our continued addiction to fossil have become the most serious environmental problems of the present times. Taking together all these results, we will go into an irreversible way.

As we well known over many years, a range of environmental contaminants and reagents have detrimental effects on animal reproduction, and many efforts are currently being made to determine the potential effects of these chemicals. In spite of these effort, comparative studies and more comprehensive approaches are needed to assess environmental effects of natural and anthropogenic chemicals from ultrastructural malformation to physiological and biochemical responses, to molecular toxicity. Further information's and examples based on up-to-date research can be found in the chapters of this book.

I would like to congratulate Dr. V.K. Gupta, Dr. A.K. Verma and Dr. G.D. Singh, in the presence of you all, for sharing their current and enlightening research papers with people who interested in this field. I strongly believe that this book is an irreplaceable scientific treasure for undergraduate and graduate students, lecturers and researchers.

Melike Erkan

Preface

All life has a common basis and all life activities such as growth, development, reproduction etc. depend on enzyme mediated bio-chemical reactions. There is a certain range of ecological conditions in which most of these complex chemical reactions occur at their maximum ease and efficiency. This does not mean that all organisms require similar set of conditions for their breeding and optimum growth.

However, it does suggest that within an optimal range a large number and type of organisms are likely to occur, grow and multiply. In habitat which do not provide their optimum conditions, organisms shall have to adopt themselves to the prevailing adverse conditions. The process of natural selection shall automatically exclude the delicate forms restricting the number of species to only those which are able to exists under the adverse set of conditions.

Among the millions of species which constitute the biosphere, man indeed is a marvelous product of organic evolution. He has been gifted with unique mental facilities and a pair of hands. The biosphere without man took many hundred million years to produce the environment as it exists today but man appeared on the planet barely one million years ago and during the first 40-60 thousand years of his existence man lived as scavenger, hunter and gatherer of food material. Modern man has, therefore, built up an extraordinary accumulation of "Human Biomass" in which energy and material flow, in enormous quantity from all other systems. He lives a life of comfort and luxury which require additional resources and his activities have acquired immense significance and the consequence of his actions have a global impact.

The ecology and reproduction have become the focus of attention of the scientists today and all the efforts are underway for a better understanding of the complex interactions among various components of the environment. Science which now dominates every field of our activities and has changed our life style, needs uninterrupted documentation keeping in view the latest advancements and increments in the various disciplines of science and technology.

The present Volume 10 of the book series, *Perspectives in Animal Ecology and Reproduction* has been presented to our esteemed readers in simple, lucid and comprehensive language by contributors who are authority in their respective areas of research and who have always been a source of inspiration for bringing our this humble effort to provide the up-to-date knowledge in the subject.

We hope this book will serve the needs of the biologists, engaged in the field of animal diversity, ecology, breeding biology, natural history, conservation and management strategies. Suggestions and healthy criticism for the improvement of quality of the further volumes will be highly appreciated and thankfully acknowledged. We are grateful to all those contributors who have always been a source of inspiration for bringing out the 10th volume of the book series to provide latest trends and knowledge in the field of animal sciences and for their help and suggestions wherever needed.

Dr. V.K. Gupta, *FLS*

Contents

2015, Perspectives in Animal Ecology and Reproduction, Vol. 10   *Pages 1–11*
Editors: V.K. Gupta, Anil K. Verma and G.D. Singh
Published by: DAYA PUBLISHING HOUSE, NEW DELHI

Chapter 1
Bacterial Biofilms Impact on Aquatic Ecosystems

Carla Mottola, Rui Seixas, Luís Tavares and Manuela Oliveira*

CIISA/Faculty of Veterinary Medicine of the University of Lisbon, Avenida da Universidade Técnica, 1300-477 Lisbon, Portugal

ABSTRACT

In recent years, biofilm formation has shown a tremendous impact in medicine, changing antimicrobial therapeutics or even our comprehension about chronic infections. Despite the intensive research in this topic, biofilm impact on ecosystems and more specifically in aquatic environment has not been given the same importance.

Biofilms has been recognized as the predominant form of microbial life in aquatic ecosystems, since bacterial interactions within this matrix provide to the microorganism a number of benefits. This matrix produced by the bacteria protects the cells from external environment and facilitates bacterial communication through chemical and physical signals, also known as *quorum sensing*, that enables bacteria to remain into a favorable environmental niche. Many ecological advantages can be attributed to biofilms like resistance to nutrient deprivation or to external components. Biofilm is also influenced by external conditions such as the flow speed, light and physico-chemical components, being a dynamic structure with the ability to adapt to its external environment setting. In research, several techniques are applied to study marine biofilm, especially *in situ* hybridization methods which are very successfully to investigate biofilm production and community structure allowing an accurate qualitative and spatial distribution analysis of biofilm formation by single or even polymicrobial communities.

* *Corresponding author.* E-mail: moliveira@fmv.ulisboa.pt

Biofilms in aquatic environment represent a complex matrix with a significant role on biological cycles or even in the dynamic of the marine system. Further research is needed in order to understand the contribution of these microbial consortia and its impact in the largest ecosystem, the marine environment.

Keywords: *Bacterial biofilms, Aquatic ecosystems, Microorganism, Biological cycles.*

Introduction

Biofilm Formation in Marine Environment

Marine ecosystems are among the largest, complex and self-sustaining earth's aquatic ecosystems. They cover two-thirds of the surface of the Earth, including oceans, salt marsh and intertidal ecology, estuaries and lagoons, mangroves, coral reefs, the deep sea and the sea floor (Kaiser *et al.*, 2005). These ecosystems are very important for the health of terrestrial and marine environments, providing food and shelter to the highest levels of marine diversity in the world (Kaiser *et al.*, 2005).

In this aquatic environment, biofilms on solid surfaces are ubiquitous. These highly organized multicellular communities are attached to a surface and enclosure microorganisms of a single or multiple species (Donlan and Costerton, 2002). Bacteria, in most environments are not in a planktonic phenotype but mostly in multicellular structures (Bendaoud *et al.*, 2011).

The transition to a biofilm phenotype is a sequential process that leads to the production of a matrix by the bacterial community (Donlan and Costerton, 2002). This matrix is mainly composed by polymeric compounds, mostly exopolysaccharide (40-95 per cent) and proteins (1-60 per cent), but also includes nucleic acids (1-10 per cent) and lipids (1-40 per cent) (Flemming and Wingender, 2010).These compounds protect the bacterial cells from external environment and facilitate bacterial communication through chemical and physical signals, also known as *quorum sensing*, that enables bacteria to remain into a favorable environmental niche. Many ecological advantages can be attributed to biofilms like withstand nutrient deprivation, pH modifications and resistance against toxins or antimicrobial compounds, more than planktonic bacteria (Dang and Lovell, 2000).

Biofilms formation can occur on a wide variety of abiotic hydrophobic and hydrophilic surfaces, including glass, metals, plastics and in almost any hydrated environment with the required nutrient settings. (Stepanovic *et al.*, 2004). These structures can also form in biotic surfaces which often represent a highly active interface between host and biofilm (Donlan and Costerton, 2002).

Biofilm formation, as mentioned earlier, is a regulated process with several sequential steps required to produce a mature matrix (O'Toole *et al.*, 2000). First, planktonic bacteria in the marine environment interact with the organic and inorganic components of a surface and form an initial and temporary structure.This initial attachment is reversible, however with time it becomes stronger attached to a surface and irreversible (Dang and Lovell, 2000). Bacteria seem to initiate biofilm development in response to specific environmental characteristics, such as nutrient availability (Kotler *et al.,* 1993).

Second, bacteria that first colonized the substratum are accumulated in the biofilm through growth and reproduction which change surface's setting allowing a suitable environment for the colonization by other bacteria. Third, attached and planktonic bacteria communicate with each other, by *quorum sensing* (QS). This communication mechanism plays a vital role in synchronizing gene expression within the bacterial community (Dong and Zhang, 2005). In *quorum sensing,* when the concentration of signal molecules produced by the bacterial community exceeds a determined threshold, the bacterial population starts acting as a single organism and together expresses virulence genes (Boyen *et al.,* 2009). There are several systems in gram-positive and gram-negative bacteria (Boyen *et al.,* 2009). One of the best studied is the *Staphylococcus aureus agr* (accessory gene regulator) *quorum sensing* system which plays a key role during the infection (Kong *et al.,* 2006). Briefly, in the *agr* QS system, modified oligopeptides called auto-inducing polypeptides (AIP) act as QS signal molecules (Lyon and Novick, 2004). Presently, there are four different *agr* groups described in *S. aureus* and the AIP of one *agr* group inhibit the expression of the other *agr* groups (Jarraud *et al.,* 2002). Many virulence traits such as haemolysins, enterotoxins, exfoliative toxins and surface proteins are also regulated by the *agr* system. Therefore, this communication process is essential for the formation of an initial biofilm community (Antunes *et al.,* 2010).

The last step includes the formation of a mature biofilm by a competitive and synergistic interaction between the bacteria within the biofilm and the recruitment of new-colonizing bacteria (Dang and Lovel, 2000).

It is also important to refer that mature biofilms undergo an essential step that is the responsible for high economics loses due to marine biofouling or even the persistence of chronic relapsing infections in both human and veterinary medicine, which is the detachment and dispersal of bacterial biofilm cells. The release of biofilm-associated bacteria into the surrounding environment is a natural phenomenon. It can occur by several pathways, from hydrodynamic shearing to concerted activation of motility and matrix degradation. In a mature biofilm, detachment is the primary process balancing growth by limiting biofilm accumulation, leading to the release of planktonic bacteria. These released bacteria may colonize different substrates and start producing another biofilm matrix (Kon *et al.,* 2006).

Formation of biofilms can have profound positive and negative impact in these environments, and as a consequence, higher costs in terms of both economics and human health (Radjasa *et al.,* 1999).

Ecology and Bacterial Marine Species Involved in Biofouling

The assembly biofilm complex can be made up of many non-bacterial marine species including algae and filamentous fungi. Regarding bacterial biofilms, it is important to note that different communities were detected when different methods were applied, like culture-dependent or independent methods. Therefore, this difference between techniques shows that bacterial species composition within biofilms remains very inconclusive (Lee *et al.*, 2008). Some studies identified Proteobacteria (such as *Pseudoalteromonas agarovorans, Pseudoalteromonas piscicida, Shewanella baltica, Vibrio parahaemolyticus, Vibrio pomeroyi*) as the most prevalent group isolated by plate assays. In contrast, molecular assays reported α-Proteobacteria (*Ochrobactrum anthropi* and *Paracoccus carotinifaciens*) as the most common in marine biofilms (Jones *et al.*, 2007). The bacteria belonging to *Bacteroidetes* division, previously known as the *Cytophaga-Flavobacteria-Bacteroides* (CFB) group, are also isolated (*Cytophaga latercula* and *Tenacibaculum mesophilum*). Others studies reported different genera as the most prevalent, isolated from marine biofilm which includes *Streptomyces* spp., *Bacillus* spp., *Actinopolyspora* spp., and also typical species involved in many human and animal infections like *Staphylococcus* spp. and *Pseudomonas* spp.It is also relevant to note that only a few percent of aquatic microorganisms are culturable, so there are maybe many microorganisms that not have been identified (Kell *et al.*, 1998).

Regarding ecological characteristics of marine bacteria, they can be oligotrophic, psychrotrophic or psychrophilic excluding those that growth in the surface layers in tropical waters (Parsek and Fuqua, 2004). Optimal growth at a salt concentration between 33-35 per cent, with a poorly development if salt is absent. Depending on the habitat, they are barophobe, barotolerantor or barophile, usually displaying pleomorphism, probably determined by oligotrophs and the effects of high hydrostatic pressure. The majority are gram-negative (80 per cent/95 per cent), mobile, aerobe, optionally aerobe and obligatory anaerobe for those living in the depth of the sediments (Hinrichsen, 2011).

At last, this microbial diversity in biofilms may be also affected by the physic-chemical compositions of the aquatic surfaces. The bacterial species involved in the composition of biofilms depends primarily on qualitative and quantitative aspects of the inoculum, light, substratum, nutrient supply, competition and grazing, which are important to determine if the colonized surface will become a dynamic biofilm ecosystem (Rao *et al.*, 2005). Despite this topic still remains unclear, studies regarding biofilm formation under an aquatic environment and the influence of the substratum are essential to understand the diversity of bacterial species enrolled in marine biofilm formation.

Environmental Factors Affecting Marine Biofilm Formation

Considering that biofilm structure is a large diversity community, its structure varies with environmental conditions as temperature, salinity, nutrient availability and light, which can change in time and space. Nutrient availability was found to be one of the major factors affecting biofilm diversity and composition, and to vary with seasons. Studies show that higher enrichment growth conditions leads to higher values of bacterial growth and attachment. Contrariwise, with low temperature and salinities, the marine bacteria have a propensity to adapt to the environment and cell concentration is lower.

In 2006,Chiu *et al.*, studying the planktonic and biofilm bacterial communities in Hong Kong waters, had already shown that the numbers and types of bacteria isolated were different in summer and winter; community compositionsin biofilms were affected by salinity in summer and more by temperature than by salinity in winter.

The nutrient availability also varies with human impact, *i.e.* eutrophication, that occurs naturally over centuries but that has been accelerated by human activities in the last century and consists in excessive nutrients that cause a shift from autotrophic to heterotrophic and to sulphur reducing bacteria as response to decreased light availability and increased load in organic material, whereas the diversity of bacterial biofilms was found to remain unaffected or even to increase (Sawall *et al.,* 2012; Meyer-Reil *et al.,* 2000).

On the other hand, there are many endogenous and exogenous factors that influence the microbial diversity of these communities and their development. Endogenous mechanisms include the interactions between the different bacterial species involved in biofilm formation that allow to a dynamic exchange of metabolites which contribute to the type of ecological niches. Several exogenous factors can consent the predominance of certain species in the biofilm community or the co-existence and balanced nutrient consumption of different species in the same niche, depending of the extensive variety of sources and substrate available in the ecosystem. Examples of exogenous factors are predation or viral lysis of bacteria. For these reasons the bacterial community structure is affected by several environmental effects that are related with season, microhabitat and cross-talk in the fouling communities.

Biological Impact of Marine Biofilm Formation

Marine biofilms constitute complex communities in which the different microorganisms interact vigorously and allow them important advantages, such as more access to nutrients and more eliminated waste matter, protection against toxins and antibiotics, refuge from predation and preservation of enzyme activities (Dang and Lovell, 2000).

Besides these "internal" benefits, the marine biofilms also contribute to the development and metamorphosis of larvae from sessile invertebrates once have been found that they are important in the settlement process of the most marine invertebrates groups including sponges, annelids, echinoderms, tubeworms, cnidarians, algae, bryozoans, phoronids. Several studies have suggested that some soluble bacterial products of the complex protein-lipopolysaccharides as well as extracellular polymers are responsible to help the settlement process because represent signal molecules (Mojica, 2008).

Egan *et al.* (2012), as well as other authors, showed that the development of larvae from sessile invertebrates can be influenced by marine biofilms through different ways, like production of water soluble pigments required for the process or production of rich surfaces that provide necessary information to larvae growth. For example, marine *Pseudoalteromonas* spp. is able to produce bioactive compounds that might afford settlement cues for invertebrate larvae. The G protein-coupled receptors (GPCRs) are abundant and represent one of the best-studied chemoreceptive molecules but never were identified in settlement process, whereas the lectin receptors seem to be involved in ascidian larvae growth (Hadfield *et al.,* 2011).

Another important component of marine biofilm that play an important role in the ecology of fouling communities are the diatoms. They are the earlier colonizers, growth hard surfaces at any time of biofilm development and need of sufficient light; are responsible for most of the energy input in the form of reduced carbon (Cooksey *et al.,* 1984).Diatoms from the biofilms are reported to provide signaling cues to a variety of larval forms of invertebrates such as polychaetes (Lam *et al.,* 2003) and barnacles (Patil and Anil, 2005).

Research in Marine Biofilms: A Phenotypic and Genotypic Approach

Until recently the identification of microorganisms required the isolation of pure cultures followed by physiological and biochemical tests but evidently the main disadvantage of this traditional approach is that only culturable microorganisms could be identified. Over the years more efficient and specific laboratory techniques have been developed. In 1989 Delong *et al.,* introduced Fluorescent *in situ* Hybridization (FISH) method using a 16SrRNA probe for detecting the spatial distribution of a specific bacterial species in a complex microbial community. Nowadays this technique is largely used for taxonomic detection of genus, species or even higher phylogenetic groups as well as for identification or quantification of bacteria in multi-species communities. The advent of molecular microbiology has radically altered our understanding of natural microbial communities. Schmidt *et al.* (1991) used the 16S rRNA approach to identify dominant bacteria in a marine community. The screening of clone libraries with specific 16SrDNA probes followed by sequencing allows discovering novel microbial groups. For example, in 1990 Giovannoni *et al.,* amplified the novel gene SAR 11, but none of the SAR 11 sequences were identical to any cultivated marine bacteria. The introduction of fragment length polymorphism (RFLP) by Delong *et al.,* in 1993 led to found that there was difference in population structure between free-living and macroaggregate-associated bacteria in marine environment. Phylogenetic identification revealed that macroaggregate-associated rRNA clones were related to *Cytophaga, Planctomyces* and *Gammaproteo* bacteria, while the free-living bacterial population was closely related to *Alphaproteo* bacteria, which also reinforced previous findings (Giovannoni *et al.,* 1990; Schmidt *et al.,* 1991). Later, in 1997, in order to profile the complexity of microbial communities, techniques such as denaturing gradient gel electrophoresis (DGGE) or temperature gradient gel electrophoresis (TGGE) have also been used (Ferris *et al.,* 1996; Ferris and Ward, 1997; Muyzer, 1999).

These methods allow differentiating the microorganisms of marine populations and to a better characterization of his complex constituents. Glöckner *et al.* (1999) used FISH to show that *Betaproteobacteria* were dominant in freshwater systems while *Alphaproteobacteria* dominated marine waters. In contrast, Cottrell and Kirchman(2000) have shown that a marine bacterioplankton community analysed by FISH was substantially different to that determined from clone libraries. Eilers *et al.* (2000) combined 16S rDNA cloning with cultivation of North Sea bacterioplankton using different oligotrophic media. The abundance of 16S rDNA clones and cultured isolates was determined using FISH, which showed that none of the readily culturable genera *Pseudoalteromonas, Alteromonas* and *Vibrio* were detected among the 16S rDNA clones, or by FISH. Conversely, SAR 86 clusters and a *Cytophaga-Flavobacteria* cluster were detected by FISH but not by the traditional culture methods.

Consequences to Marine Industry

Biofouling is the term used to define the regular and unwanted deposition of microbial stratum or their waste products on a submerged surface and represent one of the most problems for marine industry. Bacteria as part of the marine microbial population have been considered as the primary colonizers within biofilm and play a role as dominant components. The effects of biofouling formation are usually related with decreased production efficiency trough energy loss, chemical interference, and physical deterioration causing high economic damages and have negative effects on fishery and maritime industry by reducing efficiency in operation of vessels (Lee *et al.*, 2003).

Another costly result of biofouling is the long-term injury of equipment that results from the biocorrosion, a synergistic phenomenon resulted from combination of natural abiotic corrosion and biofilm-generated corrosion. Biofilms that form on ship hulls lead to corrosion and cause "drag," which results in much higher consumption of fuel during transport as well as higher ship hull maintenance costs. It is estimated that industrial biofouling and biocorrosion cost to industry at least $200 billion per year in the United States alone and the damage touches almost every industry in every corner of the globe. The effects of biofilm formation also leads to serious problems in other industries such as oil and gas extraction processes in which cause corrosion and block of pipelines or in agricultural crops carrying to bacterial disease that ruin them. Water contamination can also be a cause of microbial biofilms regarding drinking water, wastewater, desalinization and industrial cooling water (ONR, 2009). For these reasons biofilm research has been primarily focused on their adverse role in industrial biofouling in the last century and the industries actually encounter the impact of microbial biofilms.

Thus, in order to reduce the financial losses, many different antifouling strategies have been developed. Most of these antifouling techniques consist of coating the submerged surface with a chemically enhanced paint, tributyltin (TBT) based antifoulants, which prevents the successful attachment of fouling organisms (Chambers *et al.*, 2006). Unfortunately the best coatings contained heavy metals that had mutagenic effects and bioaccumulation in many marine animals; consequently in 1989 the European Union countries, followed by International Maritime Organization (IMO) and the Maritime Environment Protection Committee (MPEC) restricted the use (Takeuchi *et al.*, 2004).Thenceforth has been carried out the development of alternatives in biofouling battle, including the use of biocides or low energy polymers, but neither of the new methods have been so successful like TBT and have been found that several biocides accumulate in marine environment. On the other hand, some marine natural products seem to be effective but it's very difficult their production for large scale use. The sea-grasses are one of the most prolific coastal eco-system and represent a rich source of secondary metabolites with anti-fouling properties, for example *Cymodocea serrulata* and *Syringodium isoetifolium* were found to inhibit the growth of bacterial marine biofilm.The problem of supply has hampered the development of most secondary metabolites from marine organisms and plants, so it is important to highlight the possible role of marine bacteria associated with sea-grasses as an alternative to the commercial coatings (Marhaeni *et al.*, 2011).

A promising solution may be the creation of synthetic products made of natural products derived from sponge metabolites (Montaser and Luesk, 2011).

Conclusions

The marine biofilm is definitively a natural process starting with the pioneering activity of bacteria that recruits other groups of bacteria and successively many invertebrate fauna and flora. As we have seen in this chapter the marine biofilm formation cause many nuisances in marine activities and industries, so it's very important the right choice of an antifouling strategy that needs to be strongly powerful while preserving the environment.

For this purpose, studies on the dynamics of bacterial communities can provide information particularly useful for formulation of novel cost-effective measures for biofouling prevention, as like as studies about the bacteria composition and influence of substratum properties that can affect the physico-chemical properties of solid surfaces (Jones *et al.*, 2007). Since biofilms are ubiquitous on all submerged surfaces, understanding the basis of the interactions between settling cells, spores and larvae is vital if methods are to be found that interfere with the colonization process.

Due to the microbes living in the biofilm-encased inside a protective polymeric matrix that confers exceptional resilience to removal by disinfectants and mechanical cleaning processes, the pathway to addressing biofouling lies in the ability to proactively prevent formation of this prolific biological polymer.

It's extremely important remember that the microbial biofilms are a remarkable adaptation to challenging environmental conditions and have been ensuring the survival of bacteria for over 3 billion years and the ideal resolution of the biofouling problem would be made preserving the biological process of the nature.

Acknowledgments

Carla Mottola (SFRH/BD/72872/2010) and Rui Seixas (SFRH/BD/75836/2011) hold PhD fellowships from Foundation for Science and Technology (FCT), Portugal.

References

Antunes, L.C.M., Ferreira, R.B.R., Buckner, M.M.C. and Finlay, B.B. 2010. Quorum sensing in bacterial virulence. *Microbiology*, **156**: 2271-2282.

Bendaoud, M., Vinogradov, E., Balashova, N.V., Kadouri, D.E., Kachlany, S.C. and Kaplan, J.B. 2011. Broad-spectrum biolm inhibition by Kingella kingae exopolysaccharide. *Journal of Bacteriology*, **193**: 3879-3886.

Boyen, F., Eeckhaut, V., Van Immerseel, F., Pasmans, F., Ducatelle, R. and Haesebrouck, F. 2009. Quorum sensing in veterinary pathogens: mechanisms, clinical importance and future perspectives. *Veterinary Microbiology*, **135(3-4)**: 187-195.

Chambers, L.D., Stokes, K.R., Walsh, F.C. and Wood, R.J.K. 2006. Modern approaches to marine antifouling coatings. *Surface and Coatings Technology*, **201**: 3642-3652.

Chiu, J.M., Thiyagarajan, Y.V., Tsoi, M.M.Y. and Qian, P.Y. 2006. Qualitative and quantitative changes in marine biofilms as a function of temperature and salinity in summer and winter. *Biofilms*, **2: (3)**: 183-195.

Cooksey, B., Cooksey, K.E., Miller, C.A., Paul, J.H., Rubin, R.W., Webster, D. 1984. The attachment of microfouling diatoms. In: Costlow JD, Tipper RC, editors. Marine biodeterioration, an interdisciplinary study, London, USA.

Dang, H. andLovell, C.R. 2000. Bacterial primary colonization and early succession on surfaces in marine waters as determined by amplified rRNA gene restriction analysis and sequence analysis of 16S rRNA genes. *Applied and Environmental Microbiology*, **66(2)**: 467-475.

Dong, Y. H. and Zhang, L. H. 2005. Quorum-sensing and quorum-quenching enzymes. *Journal of Microbiology*, **43**: 101-109.

Donlan, R.M. and Costerton, J.W. 2002. Relevant Microorganisms Biofilms: Survival Mechanisms of Clinically. *Clinical Microbiology Reviews*, **15(2)**: 167.

Egan, S., Harder, T., Burke, C., Steinberg, P., Kjelleberg, S. andThomas, T. 2012. The seaweed holobiont: understanding seaweed-bacteria interactions. *FEMS Microbiology Reviews*, **37(3)**: 462-76.

Flemming, H.C.andWingender, J. 2010. The biofilm matrix. *Nature Reviews Microbiology*, **8**: 623-633.

Hadfield, M.G. 2011. Biofilms and marine invertebrate larvae: what bacteria produce that larvae use to choose settlement sites. *Annual Review of Marine Science*, **3**: 453-70.

Hinrichsen, E. 2011. Effect of temperature on salinity, the hub for bright minds. *Marine Engineering*, 1-3.

Jarraud, S., Mougel, C., Thioulouse, J., Lina, G., Meugnier, H., Forey, F., Nesme, X., Etienne, J. and Vandenesch, F. 2002. Relationships between *Staphylococcus aureus* genetic background, virulence factors, *agr* groups (alleles), and human disease. *Infection and Immunity*, **70(2)**: 631-641.

Jones, P.R., Cottrell, M.T., Kirchman, D.L. and Dexter, S.C. 2007. Bacterial community structure of biofilms on artificial surfaces in an estuary. *Microbial Ecology*, **53**: 153-162.

Kaiser, M.J., Attrill, M.J., Jennings, S., Thomas, D.N., Barnes, D.K.A., Brierley, A.S., Hiddink, J.G., Kaartokallio, H., Polunin, N.V.C. and Raffaelli, D.G. 2005. Marine Ecology: Processes, Systems, and Impacts. Oxford University Press.

Kell, D.B., Kaprelyants, A.S., Weichart, D.H., Harwood, C.R. and Barer, M.R. 1998. Viability and activity in readily culturable bacteria: a review and discussion of the practical issues. *Antonie Van Leeuwenhoek*, **73(2)**: 169-87.

Kong, K. F., Vuong, C. and Otto, M. 2006. *Staphylococcus* quorum sensing in biofilm formation and infection. *International Journal of Medical Microbiology*, **296(2-3)**:

133-139.

Kotler, R., Siegele, D.A. and Tormo, A. 1993. The stationary phase of the bacterial life cycle. *Annual Review of Microbiology*, **47:** 855-874.

Lam, C., Harder, T. and Qian, P.Y. 2003. Induction of larval settlement in the polychaete *Hydroideselegans* by surface-associated settlement cues of marine benthic diatoms. *Marine Ecology Progress Series*, **263**: 83-92.

Lee, J.W., Nam, J.H., Kim, Y.H, Lee, K.H. andLee, D.H. 2008. Bacterial communities in the initial stage of marine biofilm formation on artificial surfaces. *Journal of Microbiology*, **46(2)**: 174-182.

Lyon, G.J. and Novick, R.P. 2004. Peptide signaling in *Staphylococcus aureus* and other gram-positive bacteria. *Peptides*, **25**: 1389-1403.

Marhaeni, B., Radjasa, O.K., Khoeri, M.M., Sabdono, A., Bengen, D.G. and Sudoyo, H. 2011. Antifouling activity of bacterial symbionts of sea grasses against marine biofilm-forming bacteria. *Journal of Environmental Protection*, **2:** 1245-1249.

Meyer-Reil, L.A. and Köster, M. 2000. Eutrophication of marine waters: effects on benthic microbial communities. *Marine Pollution Bulletin*, **41**: 255-263.

Mojica, K. 2008. Marine Biofilms: Ecology and Impact. OEST 740. Spring 2008.

Moldoveanu, A.M. 2012. Environmental factors influences on bacterial biofilms formation. Vol.XVII. Issue 1.

ONR. 2009. New Hull Coatings for Navy Ships Cut Fuel Use. Protect Environment Media Release.

O'Toole, G., Kaplan, H.B. and Kolter, R. 2000. Biofilm formation as microbial development. *Annual Review of Microbiology*, **54**: 49-79.

Parsek, M.R. and Fuqua, C. 2004. Biofilms 2003: emerging themes and challenges surface associated microbial life. *Journal of Bacteriology*, **186(4)**: 4427-4440.

Patil, J.S. and Anil, A.C. 2005. Biofilm diatom community structure: influence of temporal and substratum variability. *Biofouling*, **21(3-4)**: 189-206.

Radjasa, O.K., Sabdono, A., and Suharsono. 1999. The Growth inhibition of marine biofilm-forming bacteria by the crude extract of soft coral *Sinularia* sp. *The Journal of Coastal Development*, **2(2)**: 329-334.

Rana, M. and Hendrik L. 2011. Marine natural products: a new wave of drugs? *Future Medicinal Chemistry*, **3(12)**: 1475-1489.

Rao, D., Webb, J.S. andKjelleberg, S. 2005. Competitive interactions in mixed-species biofilms containing the marine bacterium *Pseudoalteromonas tunicate*. *Applied and Environmental Microbiology*, **71(4)**: 1729-1736.

Sawall, Y., Richter, C. and Ramette, A. 2012. Effects of Eutrophication, Seasonality and Macrofouling on the Diversity of Bacterial Biofilms in Equatorial Coral Reefs. *PLoS One*, **7(7):** e39951.

Stepanovic, S., Cirkovic, I., Ranin, L. and Svabic-Vlahovic, M. 2004. Biofilm formation by *Salmonella* spp. and *Listeria monocytogenes* on plastic surface. *Letters in Applied Microbiology*, **38(5)**: 428-432.

Stowe, S.D., Richards, J.J., Tucker, A.T., Thompson, R., Melander, C. and Cavanagh, J. 2011. Anti-Biofilm Compounds Derived from Marine Sponges. *Marine Drugs*, **9**: 2010-2035.

Takeuchi, I., Takahashi, S., Tanabe, S., and Miyazaki, N. 2004. Butyltin concentrations along the Japanese coast from 1997 to 1999 monitored by *Caprella* spp. (Crustacea: Amphipoda). *Marine Environmental Research*, **57**: 397-414.

2015, Perspectives in Animal Ecology and Reproduction, Vol. 10   *Pages 13–28*
Editors: V.K. Gupta, Anil K. Verma and G.D. Singh
Published by: DAYA PUBLISHING HOUSE, NEW DELHI

Chapter 2

Effects of Edaphic Factors on Population Dynamics of Soil and Plant Parasitic Nematodes of Banana in West Bengal, India

*Viswa Venkat Gantait**
Zoological Survey of India, M-Block, New Alipur,
Kolkata – 700 053, West Bengal

ABSTRACT

Effects of edaphic factors *i.e.* temperature, moisture, pH and organic carbon content of soil on the dynamics of concomitant populations of 8 soil-free living nematodes belonging to the order Dorylaimida *viz. Dorylaimus innovatus, Prodorylaimus sukuli, Laimydorus siddiqii, Aporcelaimellus subhasi, Thonus garhwaliensis, Oriverutus lobatus, Paractinolaimus aruprus,* and *Promumtazium elongatum* and 6 plant-parasitic nematodes *viz. Hoplolaimus (Basirolaimus) indicus, Helicotylenchus crenacauda, Rotylenchulus reniformis, Pratylenchus coffeae, Meloidogyne incognita* and *Tylenchorhynchus coffeae* were investigated in a banana plantation of West Bengal, India during March 2004 to February 2006. All the species exhibited a bimodal pattern of population fluctuation. Maximum population build-up of each species was observed during monsoon (July/August) but declined during post-monsoon, and maintained a low level at the winter (January). During early spring (March) population again increased but summer (May/June) caused further a decline in number. The present study provides a valuable data base about population fluctuation of soil and plant parasitic nematodes associated with banana

* E-mail: v.gantait@rediffmail.com

with a view to aiding the diagnostics and advisory services in future for adopting the control measures against them. This would ultimately be helpful in improving the production of this valuable fruit crop.

Keywords: *Edaphic factors, Nematode, Population dynamics, Banana plantation, West Bengal.*

Introduction

Banana is an economically important crop, which has been extensively planted in tropical and subtropical countries of the world (Gowen and Queneherve, 1990). India is the largest banana producing country of the world (FAO, 2005). In India, banana is grown in an area of 332.2 thousand hectares with an annual production of 3633 thousand tons (Jonathan and Rajendran, 2003). West Bengal is the second largest banana producing state in the country. Banana plantation covers 25.73 thousand hectares with total production of 502.11 thousand tons per annum in the state.

Nematodes constitute one of the major limiting factors in banana production. Tylenchids, the plant parasitic nematodes have been recognized as a major constraint in banana production and are responsible for serious yield losses (Sundararaju, 2006). The phytoparasitic dorylaims also cause yield losses of banana by root damages. Besides, they are known to serve as casual agents and vectors in transmitting various soil-borne bacterial as well as fungal pathogens to this important crop. Thus the soil and plant parasitic nematodes belonging to the orders Dorylaimida and Tylenchida have considerable direct and indirect effects on this valuable fruit crop. Information on population structure of nematodes is important for the development of effective management schedule and advisory services (Barker and Campbell, 1981; Chawla and Mittal, 1995). Edaphic factors like temperature, moisture, pH and organic carbon content of soil greatly influenced the nematode population (Korthals *et al.*, 1996; Bilgrami *et al.*, 2003). In the above-mentioned context, an investigation regarding temporal variation of dorylaimid and tylenchid nematode population in relation to different edaphic factors in a banana plantation of Paschim Medinipur district, West Bengal, India is being reported hereunder. It provides a valuable data base for the development of effective management schedule and advisory services in future against these hidden enemies of banana which will be helpful to increase the production of this valuable crop.

Materials and Methods

The investigation was carried out in a plot of 15 m x 10 m with banana plantations (*Musa paradisiaca* L. cv. Kanthali) in Sabang block under Paschim Medinipur district (22°57′10″-21°36′35″ N, 88°12′80″-86°35′50″ E) of West Bengal, India. For sampling, 5 spots were fixed in the plot at an equal distance. Rhizospheric soil sample of 250 gm was taken from each spot, at a distance of about 25 cm from the main bole of the banana orchard. Root sample of 5 gm was also taken from the same plant of the same spot. Thus 5 soil and 5 root samples were collected from the plot during each sampling occasion. The samples were collected during first week of every month, preferably on first Sunday, from March 2004 to February 2006.

Nematodes were extracted from soil by 'Cobb's sieving technique' (Cobb, 1918) and from root by 'Mechanical maceration technique' (Reddy, 1983), decanting method followed by 'Modified Baerman's funnel technique' (Christie and Perry, 1951), processed by 'Seinhorst's slow dehydration method' (Seinhorst, 1959). Nematodes were counted by using 'Syracuse counting dish' under a stereoscopic binocular microscopic. Soil temperature was measured by 'Soil thermometer'. Soil pH and moisture were estimated by 'Soil pH and moisture meter', Model DM-15, Takemura Electric Works Ltd., Tokyo, Japan. These three factors were estimated in the field at the time of sampling. Organic carbon content of soil was determined in the laboratory following 'Walkley and Black's rapid titration method' (Walkley and Black, 1947). Dehydrated nematodes were mounted permanently on glass slides and identified following Jairajpuri and Ahmad (1982) for dorylaimids and Siddiqi (2000) for tylenchids. Specimens were deposited in the National Zoological Collections of Zoological Survey of India, Kolkata, West Bengal, India. The soft wares used for statistical analysis were Statistical package and Ecological methodology version 6.1, STATISTICA version 7.0 (StatSoft, Inc., 2007).

Results and Discussion

In the present study, soil temperature ranged from 17°C (in January, 2005) to 36°C (in June, 2005). Between April to October, the temperature was high (>30°C) and between November and March, it was low (<30°C). After October, temperature declined and reached the minimum level during January. Thereafter, temperature increased till May or June. Thus, temperature of the soil showed a unimodal pattern of variation with distinct peak in May/June (Figure 2.1). Moisture content of soil varied from 18.02 per cent (January, 2005) to 35 per cent (August, 2004). It maintained a high level during July/August and then declined till January. Thereafter, it was increased till March and then decline till May. Thus, moisture content of soil exhibited a somewhat bimodal pattern of variation with a distinct peak during July/August and a small peak in March (Figure 2.2). Soil pH ranged from 5.22 (August, 2004) to 7.42 (January,

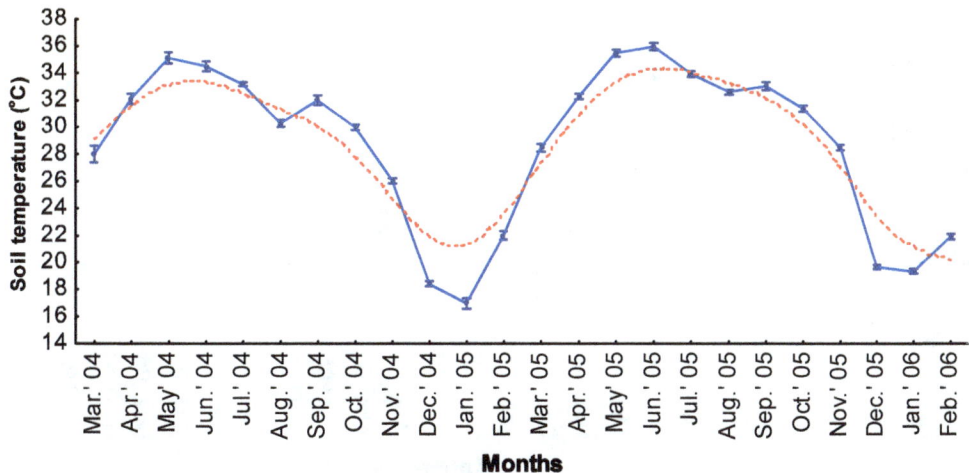

Figure 2.1: Temporal Variation in Soil Temperature.

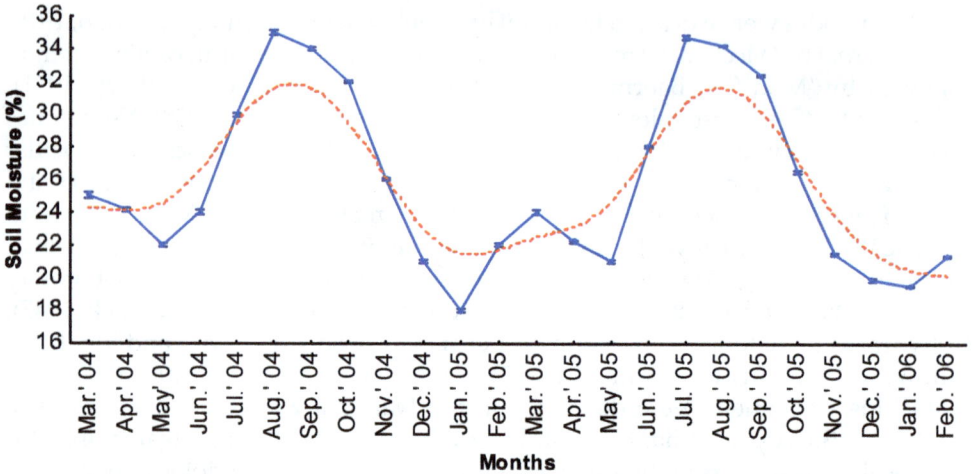

Figure 2.2: Temporal Variation in Soil Moisture.

2005). The soil was mostly acidic in nature (5.22-6.90) during almost entire study period, except in January, when it was alkaline in nature. During monsoon and post-monsoon period, soil was highly acidic, whereas slightly acidic in other months. Soil pH varied with out any distinct trend (Figure 2.3). Organic carbon content of soil varied from 1.82 per cent (June, 2004) to 3.42 per cent (July, 2005). The maximum level was observed during July/August. Thereafter it declined till January followed by an increase till March. Thus, the organic carbon level of soil exhibited a bimodal pattern of variation with distinct peak in July/August and a small peak during March (Figure 2.4).

In the course of the present study, 8 dorylaimid species *viz.*, *Dorylaimus innovatus* Jana and Baqri, 1982, *Prodorylaimus sukuli* Baksi and Baqri, 1985, *Laimydorus siddiqii* Baqri and Jana, 1982, *Aporcelaimellus subhasi* Gantait *et al.*, 2006, *Thonus garhwaliensis*

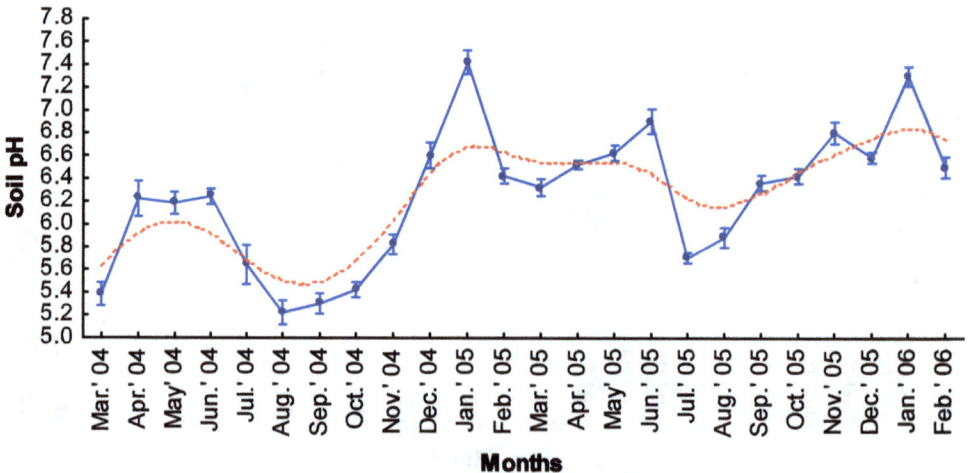

Figure 2.3: Temporal Variation in Soil pH.

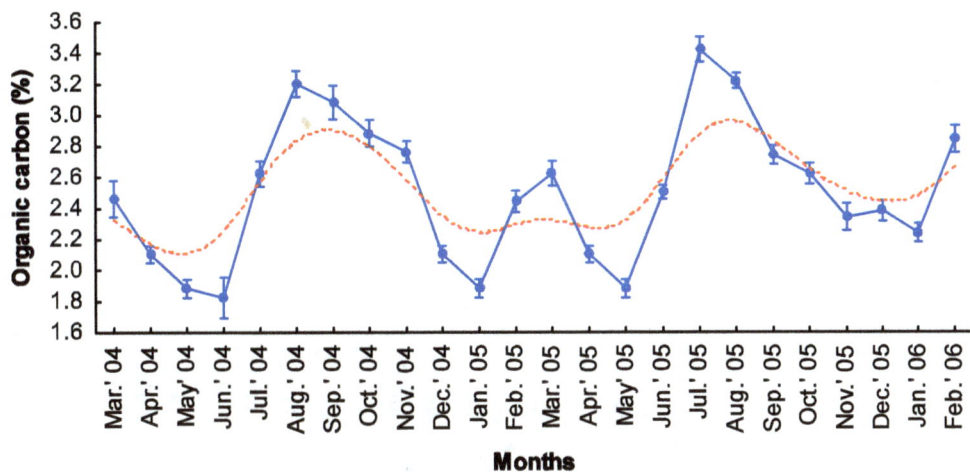

Figure 2.4: Temporal Variation in Organic Carbon Content of Soil.

Ahmad *et al.,* 1986, *Oriverutus lobatus* Siddiqi, 1971, *Paractinolaimus aruprus* Khan *et al.,* 1994 and *Promumtazium elongatum* Ahmad and Jairajpuri, 1984 have been encountered. Six species of plant parasitic nematodes *viz. Hoplolaimus* (*Basirolaimus*) *indicus* Sher, 1963, *Helicotylenchus crenacauda* Sher, 1966, *Rotylenchulus reniformis* Linford and Oliveira, 1940, *Pratylenchus coffeae* (Zimmermann, 1898) Filipjev and Schuurmans Stekhoven, 1941, *Meloidogyne incognita* (Kofoid and White, 1919) Chitwood, 1949 and *Tylenchorhynchus coffeae* Siddiqi and Basir, 1959 belonging to the order Tylenchida have also been encountered.

The number of individuals of *Dorylaimus innovatus* varied from 8 (January, 2005) to 203 (September, 2004). Population declined after September till January. Further it was increased up to March and then declined till May. Thus, the species showed a bimodal pattern of population fluctuation with a prominent peak in September and a low peak during March (Figure 2.5). In *Prodorylaimus sukuli*, maximum number (427) was found during September, 2005 and minimum number (63) was in January, 2005. After September, population declined till January. Thereafter, it was increased till March and then declined till May. Thus, population of this species exhibited a bimodal pattern of variation with a distinct peak in September and a small peak during March (Figure 2.6). In case of *Laimydorus siddiqii*, highest population (21 only) was found in September, 2005. The species showed maximum population in August/ September. Thereafter, it was declined till January when the number was zero. Population was further increased up to February/March and then declined till June when it was nil. Thus, population fluctuation of this species showed a bimodal pattern of variation with prominent peak in August/September and a low peak in February/March (Figure 2.7). Maximum population (376) in *Aporcelaimellus subhasi* was found in August, 2004 and minimum (56) was in May, 2005. The number declined after August till January. Then it was increased up to March and declined further till May. Thus, the species showed a bimodal pattern of population variation with highest peak in August and lowest peak during March (Figure 2.8). In *Thonus garhwaliensis,*

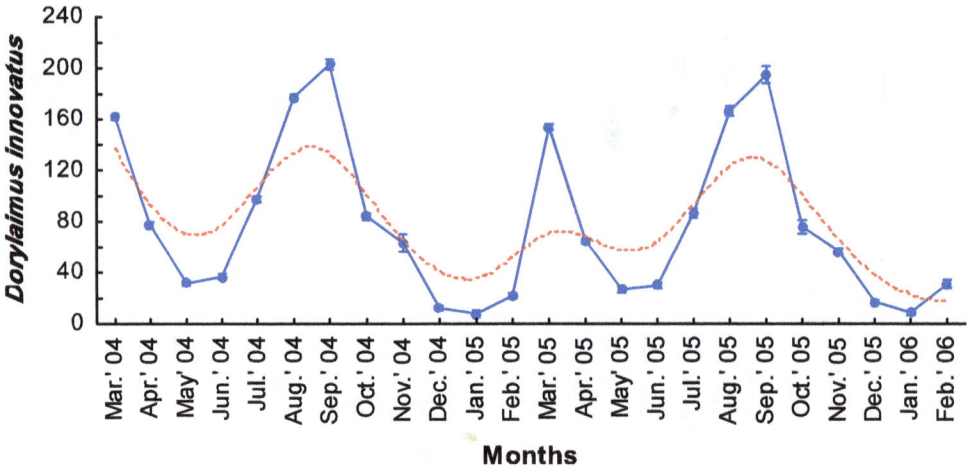

Figure 2.5: Population Fluctuation of *Dorylaimus innovatus.*

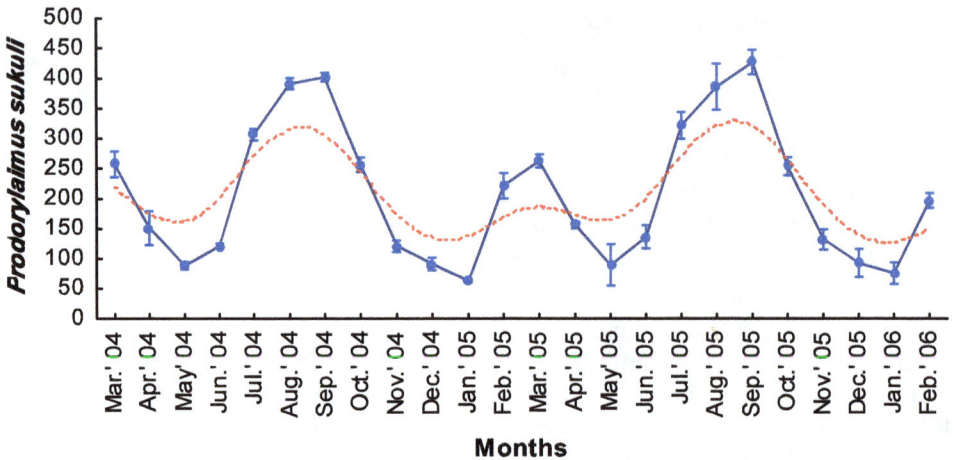

Figure 2.6: Population Fluctuation of *Prodorylaimus sukuli.*

maximum population (14) was found in August, 2004. Thereafter it was declined till January, when the population was nil. Then the number was increased till March and declined again till May. Thus, the species showed a bimodal pattern of population fluctuation with a distinct peak in August and a low peak in March (Figure 2.9). In *Oriverutus lobatus*, maximum population (317) was found in September, 2004 and minimum (43) in May, 2004. Population declined after September till January, then increased up to March. Thereafter, it was declined till May. Thus, a bimodal pattern of population variation was found with a prominent peak in September and a small peak in March (Figure 2.10). The population ranged from 49 (May, 2004) to 352 (August, 2005) in *Paractinolaimus aruprus*. Population declined after August till January. Thereafter it was increased up to March, and then declined till May. Population fluctuation showed a bimodal pattern of variation with a highest peak in August and a low peak in May (Figure 2.11). In *Promumtazium elongatum,* maximum

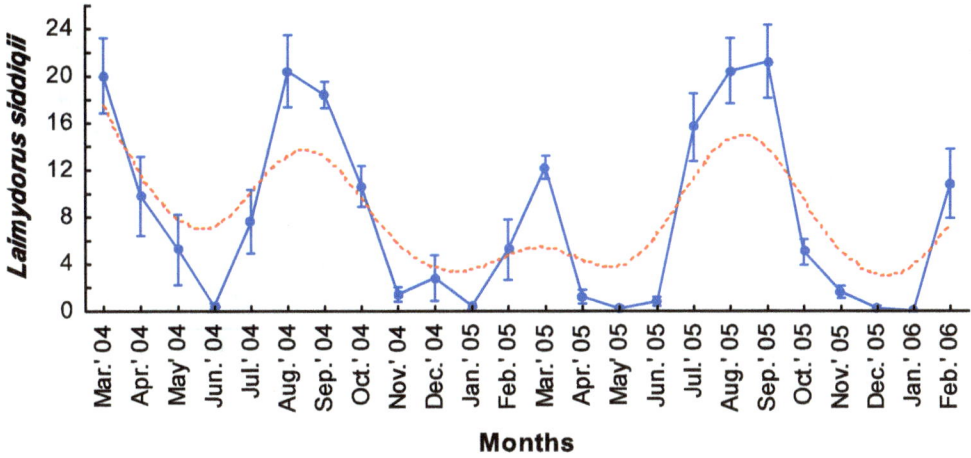

Figure 2.7: Population Fluctuation of *Laimydorus siddiqii*.

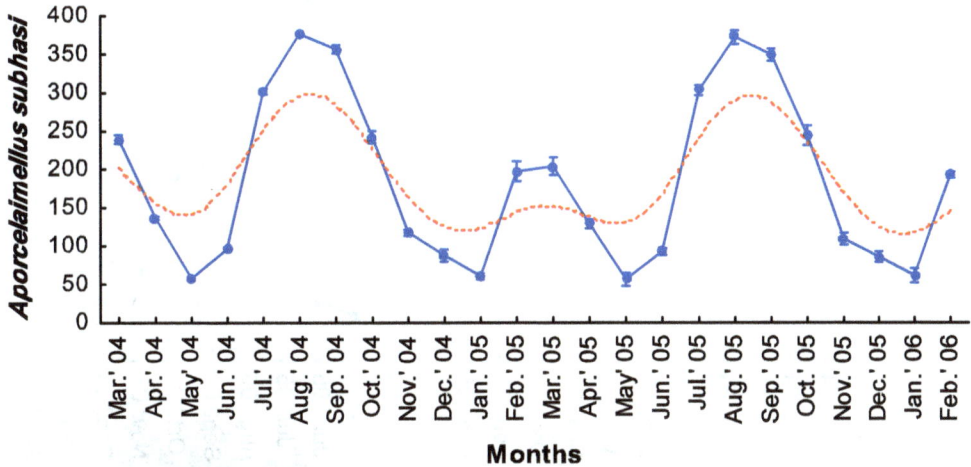

Figure 2.8: Population Fluctuation of *Aporcelaimellus subhasi*.

population was found in September, 2004 (7 only) it was zero in December, 2005. After September, the number was declined till December. Thereafter, it was increased up to February and then declined till May. Thus, the species showed a bimodal patter of population variation with a prominent peak in September and a low peak in February (Figure 2.12). In *Hoplolaimus (Basirolaimus) indicus*, population varied from 30 (January, 2005) to 294 (September, 2005). Maximum population was observed in August/September. It was then declined till January. Furthermore, the population was increased up to March, declined then till June. Thus, a bimodal pattern of population fluctuation was observed in this species with distinct peak during August/September and a very low peak during March (Figure 2.13). In *Helicotylenchus crenacauda*, population varied from 39 (January, 2005) to 311 (September, 2005). The pattern of population variation of this species was more or less similar with *H. (B.)*

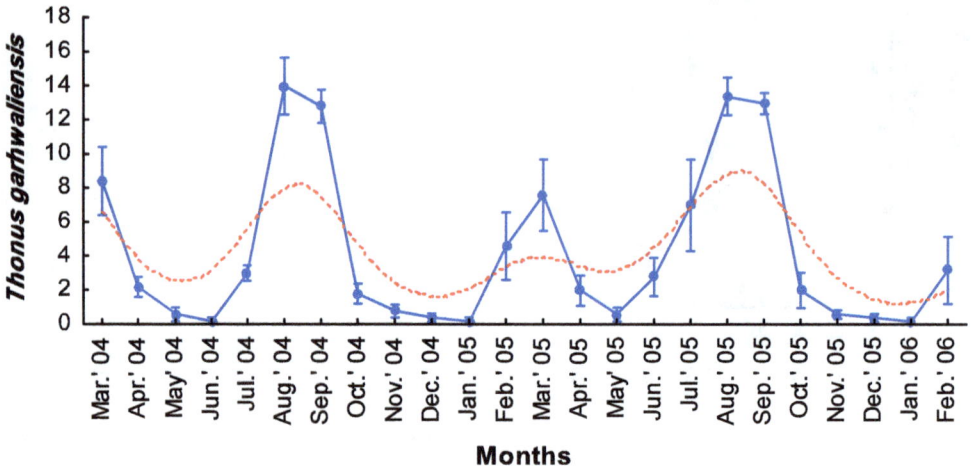

Figure 2.9: Population Fluctuation of *Thonus garhwaliensis.*

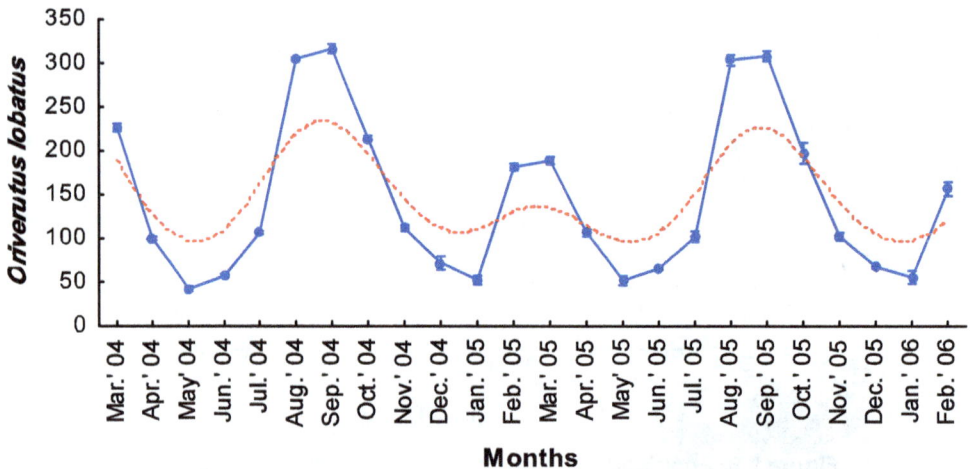

Figure 2.10: Population Fluctuation of *Oriverutus lobatus.*

indicus (Figure 2.14). The population of *Rotylenchulus reniformis* was varied from 89 (June, 2005) to 409 (August, 2004). After August/September, the population was declined till January. Further it was increased till March and then declined till June. Thus, a bimodal pattern of population fluctuation was observed with a prominent peak during August/September and a small peak during March (Figure 2.15). In *Pratylenchus coffeae,* highest population was found in September, 2005 (4 only); the number was zero in January and June, 2005. After September, population declined till January. Then it was increased up to March and declined again till June. Thus, the species exhibited a bimodal pattern of population fluctuation with a prominent peak in September and a low peak in March (Figure 2.16). In *Meloidogyne incognita,* maximum population (24) was found in August, 2004 and it was zero in May, 2005.

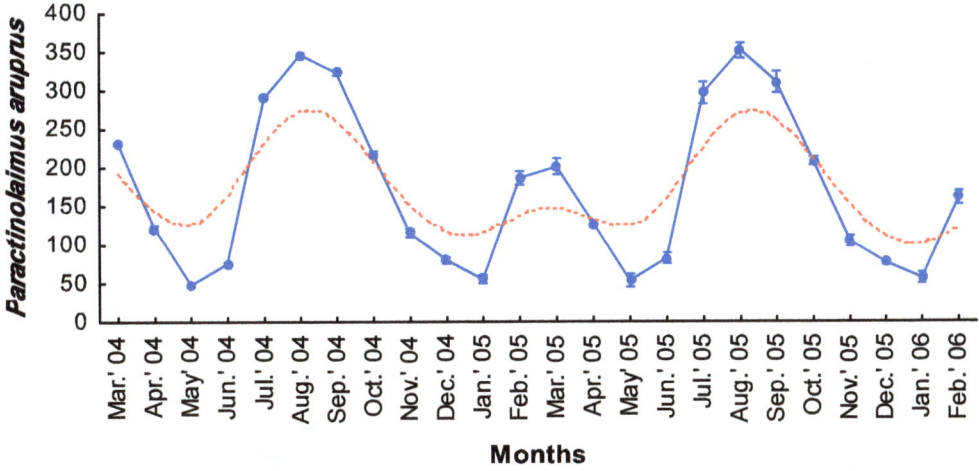

Figure 2.11: Population Fluctuation of *Paractinolaimus aruprus.*

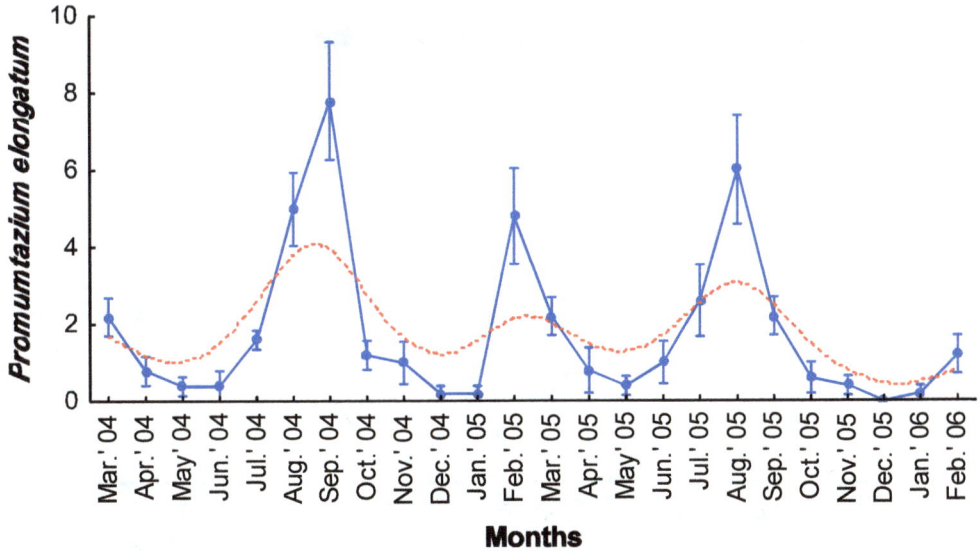

Figure 2.12: Population Fluctuation of *Promumtazium elongatum.*

Population declined after August till January. Further it was increased up to March and then declined till May. Thus, a bimodal pattern of population variation was found in this species with prominent peak in August and a small peak in March (Figure 2.17). The number of individuals of *Tylenchorhynchus coffeae* ranged from 60 (January, 2005) to 316 (September, 2005). The maximum population was found in August/September, which was declined till January. Further, it was increased up to March and then declined during May/June. Thus, the species showed a bimodal population fluctuation pattern with a distinct peak during August/September and a small peak in March (Figure 2.18).

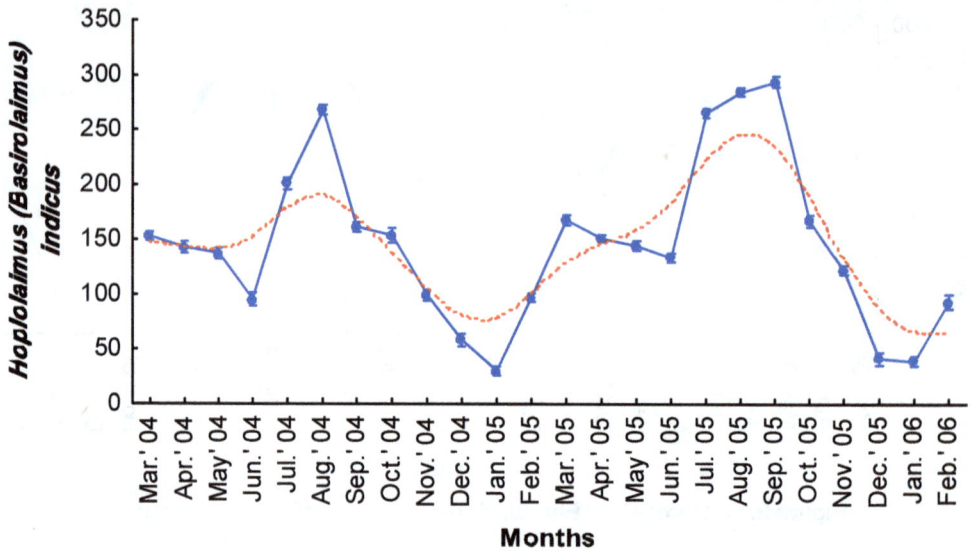

Figure 2.13: Temporal Variation in Population of *Hoplolaimus* (*Basirolaimus*) *indicus.*

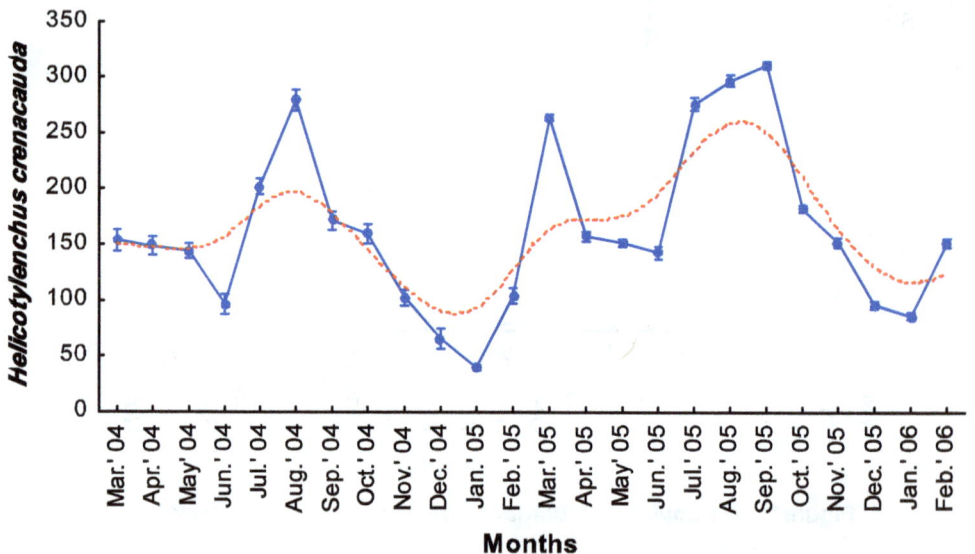

Figure 2.14: Temporal Variation in Population of *Helicotylenchus crenacauda.*

From the above findings it can be summarized that all the nematode species encountered, exhibited a bimodal pattern of population fluctuation with a prominent peak during August or September and a low peak during March every year. Likewise, there were two minima, one during January and again in May or June. The population increased after rainfall but declined during post monsoon period and maintained a low level during winter. During early spring population again increased but summer caused a decline in the number.

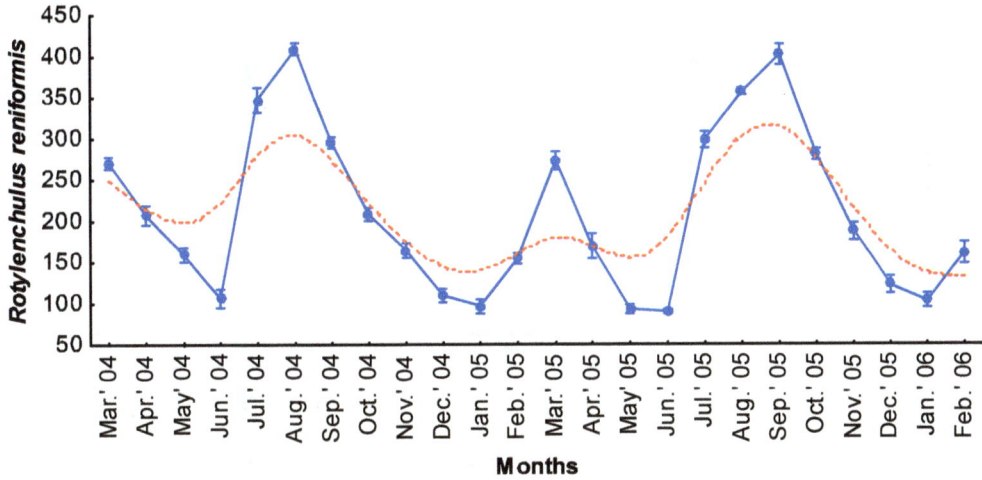

Figure 2.15: Temporal Variation in Population of *Rotylenchulus reniformis*.

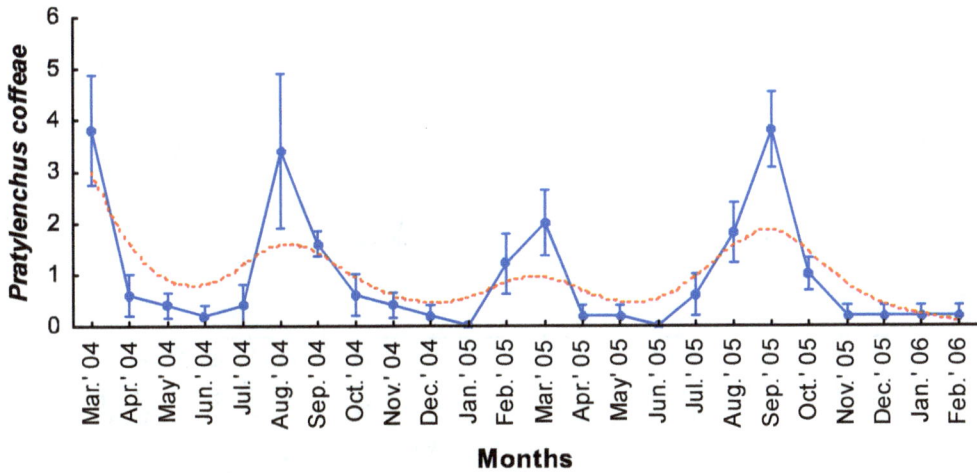

Figure 2.16: Temporal Variation in Population of *Pratylenchus coffeae*.

In any ecosystem, many biotic and abiotic factors are known to influence distribution, dynamics and shift in population densities of nematodes (Dasgupta, 1998). The physio-chemical properties of soil not only varied from one site to another but also exhibited variation from month to month. Bilgrami *et al.* (2003) stated that the ecological factors like temperature, moisture, pH and organic carbon content of soil greatly influenced the nematode community of an ecosystem. Temperature and moisture are important factors determining the population dynamics of nematodes (Azmi, 1995; Bell and Watson, 2001). They also opined that temperature directly affects rates of nematode life processes. Moisture content of soil affects nematode movement (Jones *et al.*, 1969); feeding and reproduction (Baujard and Martiny, 1994). Malik and Jairajpuri (1983) observed soil moisture to be more important factor than soil temperature in determining the population size. Soil pH plays an important role

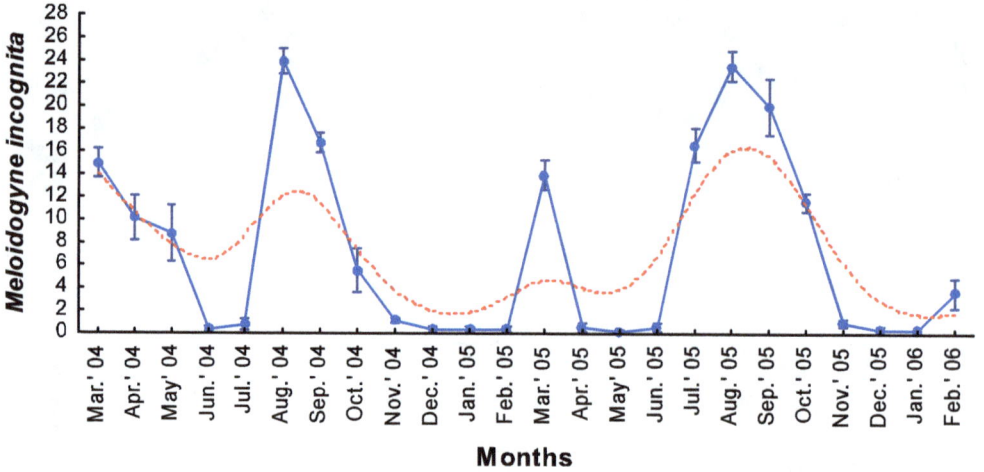

Figure 2.17: Temporal Variation in Population of *Meloidogyne incognita.*

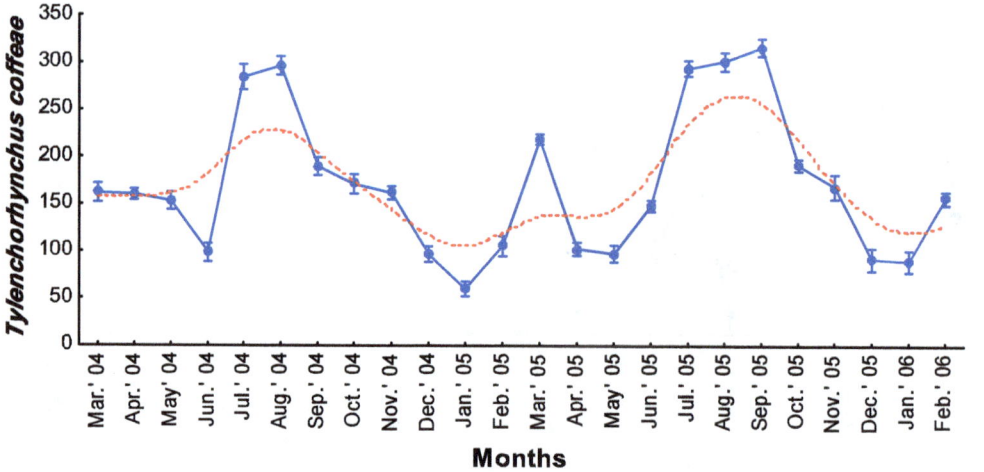

Figure 2.18: Temporal Variation in Number of *Tylenchorhynchus coffeae.*

in managing nematode population (Choudhury and Phukan, 1995; Korthals *et al.,* 1996). Organic carbon content of soil also played an important role in the population dynamics of nematodes (Gantait *et al.,* 2006).

Present investigation revealed that populations of all the nematode species were low in December and January due to low temperature and low moisture content of soil. The low temperature and moisture also might have had adverse effects on the nematode population. It was similar to the findings of Khan *et al.* (1971), Dwivedi and Misra (1990) and Pandey (1999). As the temperature increased in March, the nematode populations tended to increase. Boag (1980) reported that temperature influences feeding rate of nematodes, reproduction and population level also increased. This trend reversed and the abundance of nematodes declined in summer

(May/June). This decline in abundance was associated with high temperature and low moisture content of soil. The result was the agreement with the findings of Griffin and Darling (1964). During rain, the soil moisture gradually increases reducing the temperature of soil. Thus, it resulted in an increase in the nematode population build up. Thereafter, population goes on increasing from July onward reaching peak in August/September. This might be due to optimum moisture conditions of soil and moderate temperature. Thus the present finding is in agreement with that of Azmi (1995). Contrary to these, Robbins and Barker (1974), Singh and Sharma (1995) found that high soil moisture levels exert a negative pressure on the growth and reproduction of nematodes. In the present study, from post-monsoon period onward nematode population decreased. This decline in abundance followed the decline in soil temperature and moisture. Similar observation was also made by Mukherjee and Dasgupta (1983). Thus the present study suggests that, soil temperature and moisture favoured nematode population growth but the extremes of these had an adverse effect on the population. Soil of the study field was mostly acidic in nature having pH in the range of 5.22-6.90, during almost entire study period except in January, when it was slightly alkaline in nature. The maximum abundance of nematodes was found when the pH was moderately high. But in January, the population was minimum. It indicates that population increased in acidic soil and decreased in alkaline soil. Similar findings were also reported by Szczygiel *et al.* (1983). Das *et al.* (1990) also observed increased biotic activities of nematodes in acidic soil and rapid decline in alkaline soil. In other studies Choudhury and Phukan (1995) found that when soil pH increased, the density of most of the nematodes decreased while Bilgrami *et al.* (2003) reported that nematodes preferred less alkaline pH. During monsoon, as the organic carbon content of soil increased, nematode population was also increased. Similar findings were also observed by Szczygiel and Zepp (1983), Bilgrami *et al.* (2003) and Gantait *et al.* (2006).

It can be concluded that in monsoon period, soil temperature was moderate, moisture and organic carbon content of soil were high, and pH was acidic in nature. Besides, there was a consistent increase in the biomass of new feeder roots of banana plantations. In spring, soil temperature and moisture were also moderate. These favoured the population build-up of nematode species. Thus, the edaphic factors appear to interact for the development, multiplication and seasonal behavior of nematodes associated with banana plantations. However, effective control measures against the nematodes should also be adapted during these periods and the production of this valuable fruit crop will be increased significantly.

Acknowledgements

Authors are very much grateful to the Director, Zoological Survey of India, Kolkata, West Bengal, India for providing laboratory and other facilities during this work.

References

Azmi, M.I. 1995. Seasonal population fluctuations behaviour of plant parasitic nematodes in Caribbean Stylo. *Indian Journal of Nematology*, **25** (2): 168-173.

Barker, K.R. and C.L. Campbell 1981. Sampling nematode populations. In: *Plant parasitic nematodes*, (Eds.) B. M. Zuckerman and R. A. Rhode. American Press, New York, USA Vol. III, p. 451-473.

Baujard, P. and Martiny, B. 1994. Ecology and pathogenecity of the nematode *Pratylenchus pernoxius* (Nemata: Tylenchulidae) from Senegal, West Africa. *Afro-Asian Journal of Nematology*, **4**: 7-10.

Bell, N.L. and Watson, R.N. 2001. Population dynamics of *Pratylenchus nanus* in soil under pasture: 1. Aggregation and abiotic factors. *Nematology*, **3** (2): 187-197.

Bilgrami, A.L., W. Liang and Q. Li 2003. Generic diversity, population structure and community ecology of plant and soil nematodes. *International Journal of Nematology* **13** (1): 104-117.

Boag, B. 1980. Effects of temperature on rate of feeding of the plant-parasitic nematodes, *Rotylenchulus robustus, Xiphinema diversicaudatum* and *Hemicycliophora conida*. *Journal of Nematology*, **12**: 193-195.

Chawla, M.L. and Mittal, A. 1995. Nematode population dynamics and modeling. In: *Nematode pest management – An appraisal of eco-friendly approaches,* (Eds.) G. Swarup, D. R. Dasgupta and J. S. Gill. Nematological Society of India, New Delhi, India, p. 285-295.

Chowdhury, B.N. and Phukan, P.N. 1995. Study on the variations of certain plant parasitic nematodes at different levels of soil pH. *Indian Journal of Nematology*, **25** (2): 202-203.

Christie, J.R. and Perry, V.G. 1951. Removing nematodes from soil. *Proceedings of Helminthological Society of Washington,* **18:** 106-108.

Cobb, N.A. 1918. Estimating the nema population of the soil. Agricultural Technology Circular I. Bureau of Plant Industry, Department of Agriculture, United States, 48 pp.

Das, B.K., Sarkar, J., Sarkar, S., Das, N.K., Ray, I. and Sen, S. K. 1990. Correlation between some edaphic factors and *Meloidogyne incognita* infestation on mulberry in Malda, West Benagal. *Indian Journal of Nematology*, **20** (1): 91-94.

Dasgupta, M.K. 1998. *Phytonematology*. Naya Prakash, Calcutta, 846 pp.

Dwivedi, B.K. and Misra, S.L. 1990. Environmental correlates to the population dynamics of plant nematodes around root zones of *Citrus sinensis* at Allahabad. *Current Nematology*, **1**: 25-30.

FAO Statistics-Agricultural data, 2005. http://faistat.fao.org/faostat/serlet/XteServlet.

Gantait, V.V., Bhattacharya, T. and Chatterjee, A. 2006. Fluctuation of nematode population associated with banana plantation in Paschim Medinipur district, West Bengal, India. *Indian Journal of Nematology*, **36** (2): 223-225.

Gowen, S.R. and Queneherve, P. 1990. Nematode parasites of bananas, plantains and abaca. In: *Plant Parasitic Nematodes in Subtropical and Tropical Agriculture.*

(Eds.) M.R. Luc, A. Sikora and J. Bridge. CAB International, Wallingford, p. 431-460.

Griffin, G.D. and Darling, H.M. 1964. An ecological study of *Xiphinema americanum* Cobb in an ornamental spruce nursery. *Nematologica*, **10**: 471-479.

Jairajpuri, M.S. and Ahmad, W. 1992. *Dorylaimida: Free-living, Predaceous and Plant Parasitic Nematodes*. Oxford and IBH Publishing Co. Pvt. Ltd., New Delhi, India, 458 pp.

Jonathan, E.I. and Rajendran, G. 2003. Spatial distribution of root-knot nematode, *Meloidogyne incognita* in Banana, *Musa* sp. *Indian Journal of Nematology*, **33** (1): 47-51.

Jones, F.G.W., Larbey, D.W. and Parrott, D.M. 1969. The influence of soil temperature and moisture on nematodes, especially *Xiphinema, Longidorus, Trichodorus* and *Heterodera* spp. *Soil Biology and Biochemistry*, **1**: 153-165.

Khan, A.M., Adhami, A. and Saxena, S.K. 1971. Population changes of some stylet-bearing nematodes associated with mango (*Mangifera indica* L.). *Indian Journal of Nemalology*, **1**: 99-105.

Korthals, G.W., Alexiev, A.D., Lexmond, T.M., Kammenga, J.E. and Bongers, T. 1996. Long-term effect of copper and pH on the nematode community in an agro ecosystem. *Environ. Toxicol. Chem.*, **15**: 979-985.

Malik, Z. and Jairajpuri, M.S. 1983. Population dynamics and life cycle of *Xiphinema americnum* Cobb, 1913. *Indian Journal of Parasitology*, **7**: 21-28.

Mukherjee, B. and Dasgupta, M.R. 1983. Community analysis of nematodes associated with banana plantations in the Hooghly District, West Bengal, India. *Nematologia Mediterranea*, **11:** 43- 48.

Pandey, R. 1999. Population dynamics of phytonematodes in an arable soil planted with *Costus speciosus* (koeti) Sm. *Indian Journal of Nematology*, **29** (1): 81-82.

Reddy, P.P. 1983. *Plant Nematology*. Agricole Publishing Academy, New Delhi, 287 pp.

Robbins, R.T. and Barker, K.R. 1974. The effects of soil type, particle size, temperature and moisture on reproduction of *Belonolaimus longicaudatus. Journal of Nematology*, **6**: 1-6.

Seinhorst, J.W. 1959. A rapid method for the transfer of nematodes from fixative to anhydrous glycerine. *Nematologica*, **4**: 67- 69.

Siddiqi, M.R. 2000. *Tylenchida. Parasites of Plants and Insects, 2nd Edition*. CABI Publishing, CAB International, Wallingford, UK, 833 pp.

Singh, M. and Sharma, S.B. 1995. Infectivity, development and reproduction of *Heterodera cajani* on pigeonpea, influence of soil moisture and temperature. *Journal of Nematology*, **27**: 370-377.

StatSoft, Inc. 2007. STATISTICA (Data Analysis Software System), Version 7.0. w.w.statsoft.com.

Sundararaju, P. 2006. Community structure of plant parasitic nematodes in banana plantations of Andhra Pradesh, India. *Indian Journal of Nematology,* **36** (2): 226-229.

Szczygiel, A. and Zeep, A. 1983. Effect of organic matter in soil on population and pathogenecity of *Pratylenchus penetrans* and *Longidorus elongatus* to strawberry plants. *Zcszyty Problemowe Postepow Nauk Rolniezych,* **278**: 113-122.

Szczygiel, A., Slowik, K. and Saroka, A. 1983. Effect of soil pH on host-parasite relationship of three plant parasitic nematodes of strawberry plants. *Zeszyty Problemowe Postepow Nauk Rolniezyeh,* **278:** 95-104.

Walkley, A. and Black, I. A. 1947. Chromic acid titration method for determination of soil organic matter. *Soil Science,* **63**: 251-264.

2015, Perspectives in Animal Ecology and Reproduction, Vol. 10   *Pages 29–64*
Editors: V.K. Gupta, Anil K. Verma and G.D. Singh
Published by: DAYA PUBLISHING HOUSE, NEW DELHI

Chapter 3

Arthropod Natural Enemy: Pesticide Interaction

Akhtar Ali Khan*, Mohd. Abas Shah and Somina Majid

Division of Entomology,
Sher-e-Kashmir University of Agricultural Sciences and
Technology of Kashmir, Shalimar, Srinagar – 190 025, J&K

ABSTRACT

The use of pesticides ranks first among the various tactics used to manage insect pests in agriculture. In addition to other problems, pesticides are associated with profound deleterious effects on non-target organisms, the arthropod natural enemies (predators and parasitoids) being the worst affected. The pesticides may cause direct or indirect toxicity to the natural enemies reducing their ecological fitness and intrinsic rate of natural increase. Integration and compatibility of natural enemies with pesticidal control forms the foremost objective of integrated pest management. The two groups differ with respect to modes of pesticide uptake, their metabolism and patterns of pesticide resistance development. The natural enemies are more likely to residual contact from applied pesticides and are less likely to develop resistance under field conditions. Some critical differences in physiological processes are also known among the two groups. Differences in food reserves and their usage, respiratory pattern and some endocrinological aspects are known. The differential aspects may make the natural enemies more susceptible or resistant to the pesticides as compared to the target pests. Many attempts have been made to manipulate such differences for the development of selective pesticides, although with little success.

Keywords: *Cholesterol, Detoxification, Food reserves, Mixed function oxidases, Natural enemies, Residual contact, Resistance.*

* *Corresponding author.* E-mail: akhtaralikhan47@rediffmail.com

Introduction

Modern agriculture has come to rely extensively on synthetic chemical pesticides for pest control. Although these toxins are targeted at plant pests, many of them are broad spectrum biocides that have profound effects on non-target species in agricultural ecosystems. Even the recently developed biorational pesticides, which are based on natural products and are more host specific, can have far-reaching side effects. Arthropod predators and parasitoids, the most important naturally occurring biological control agents of insect and mite pests in most crop ecosystems, happen to be among the most important non-target organisms affected by these pesticides.

Pesticides often disrupt the trophic relationships involving these natural beneficial species. Because of basic physiological similarities between arthropod pests and their natural enemies, pesticides often inflict severe mortality on both groups of organisms. In addition to their direct impact, pesticides often disrupt the trophic relationships by their toxic effects on associated species in the community, including competitors, hyperparasitoids, and alternate host or prey of natural enemies (Khan, 2012). Some of the practical consequences of these disruptions are the outbreaks of primary and secondary pests, increased pest control problems and difficulties in establishing biological controls (Newson *et al.,* 1976; Velasco, 1985). Pesticides affect the biology of natural enemies in even more subtle ways. These species may experience sub-lethal effects on development or behaviour. Fecundity, fertility, rate of development and survivorship may be altered. Behaviours such as host or prey finding and general mobility may also change (Croft, 1977; Desneux *et al.,* 2007).

Because of the negative consequence associated with pesticide use, these toxins and biological control organisms have long been considered incompatible. In time it was realised that problems were associated with either of these classes of pest control methods. Often biological controls were not effective enough and yet pesticides were overly disruptive to non-target species (Shah and Khan, 2013). Thus, the integrated pest management (IPM) *concept* was developed essentially as a response to incompatibility of pesticides and biological controls. From the beginning, its primary emphasis was to integrate these two pest control measures (Ripper, 1956). Modern IPM seeks to exploit the more subtle interaction between biological and chemical pest control agents in the development of selective pesticides (Khan, 2009; Khan and Zaki, 2009; Orr, 2009).

In the following account, various aspects of arthropod natural enemy pesticide interaction have been discussed with emphasis on the differentiating mechanisms between the toxicity of pesticides to pests and arthropod natural enemies. First, the various routes of pesticide exposure to natural enemies have been outlined. It is followed by a comparative account of the physiology and toxicology of pests and natural enemies that results in differential toxicity to the two groups. Some of these aspects are being utilised to develop selective pesticides. In the concluding section, insecticide resistance development in the pests and natural enemies has been examined with an objective to throw light on the fact that why natural enemies are less likely to develop resistance to pesticides under field conditions. The overall objective is to gain an understanding of the various aspects of arthropod natural enemy pesticides interactions so as to devise ways and means to integrate the two more efficiently in insect pest management.

1. Modes of Pesticide Uptake

An arthropod acquires a toxic dose of pesticide by initial contact, followed by transfer of the toxicant to the target site or sites of action within its body. Acquisition of a toxic dose involves processes that are only partially understood for any pest species; much less is known about these events for any natural enemy (Croft, 1990). According to Weiling (1979), intoxication may be discussed in terms of uptake, translocation, activation or degradation and interaction with the target site. Each of these steps occurs internally in the arthropod, except for some aspects of uptake. External pesticide uptake primarily involves contact at the epicuticularl level, through either the exoskeleton or the gut. Types of pesticide exposure may be discussed as direct or immediate contact, residual contact or food chain uptake. All modes of uptake may act collectively to induce mortality or sub-lethal responses in the predator or parasites.

1.1 Pesticide Uptake: A Conceptual Model

The routes of pesticide uptake by natural enemies and associated species are given in Figure 3.1. Pesticide uptake for 1) a predator or free living stage of an entomophagous parasitoid, 2) its arthropod prey or host, 3) the plant host upon

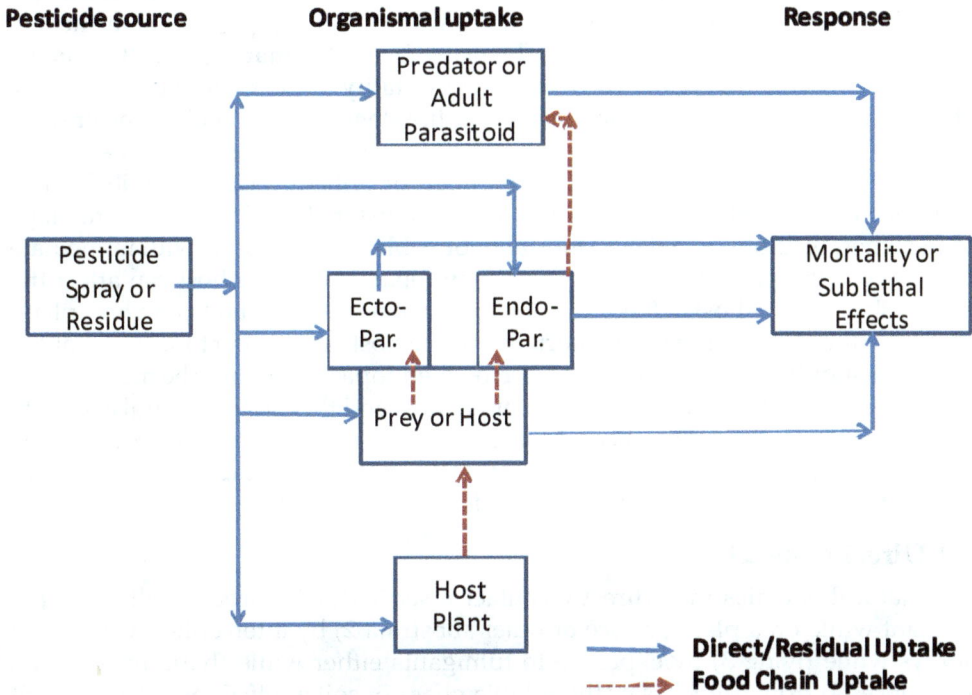

Figure 3.1: A Generalized Conceptual Model of Pesticide Uptake by Natural Enemies.

which the pest arthropod feeds, and 4) an ecto- or 5) an endoparasitoid life stage associated with the plant feeding pest species, is being generalised in the figure.

Direct mortality or sublethal effects of pesticides are caused by 1) direct contact, which includes the immediate exposure resulting from direct interception by the beneficials or associated species; 2) residue uptake, which includes external contact coming after the toxicant has landed on the substratum; 3) food chain transfer, which is contact usually received by natural enemies by feeding on pesticide contaminated food.

Toxicant uptake may kill a free living form of a natural enemy or any exposed ectoparasitoid. An endoparasite larva is often protected from immediate contact or residue uptake within its host. However it may be killed via food chain exposure to the toxicant received from the host. An endoparasite may experience a semidirect type of contact by uptake of residues through the host cuticle and trachea. Pesticide uptake may cause death or sublethal effects to the host which may indirectly affect associate endo-or ectoparasitoids. Indirect effects are not due to the insecticide itself, but due to the death of the host or prey as it affects entomophagous species. Indirect effects are shown only for immature stages of parasitoids. Immobile stages in particular and even adult predators experience indirect effects due to reduction in prey population caused by pesticide. The continuum of species influenced by indirect effects of pesticides include immobile endoparasites that are almost always killed when their hosts are killed to relatively mobile predators that are unable to find sufficient food to survive or reproduce following a pesticide treatment. Parasitism of a host by either an ecto- or endoparasitoid can affect the susceptibility of the host to the toxicant. This effect is termed as influencing factor. Ultimately, this effect on the host can, in turn, feedback to cause indirect mortality of the natural enemy. Toxic effects on predators or free-living parasitoid life stages via pesticide transfer from other organisms can come from either external or internal sources. Pesticides can be taken up by predation and host feeding, plant or pollen feeding, cannibalism or tropholaxis. Latent effects of pesticides involve transfer of pesticides from one stage to the other, resulting in delayed mortality or sublethal effects. As with immediate contact toxicity, parasitism by an ecto- or endoparasitoid of a host can affect the susceptibility of the host to food chain mediated toxicant received from a host plant. This may then contribute to the mortality of the host and indirect mortality of the endoparasitoid. Pesticide effects on an ecto- or endoparasitoid may be mediated by the host insect alone as well as by the plant to host and then to the natural enemy to cause direct or indirect mortality. In some cases food chain transfer from host to the ecto- or endoparasitoid may cause direct mortality of the natural enemy without killing the pest insect (chemotherapeutic effect).

1.2 Direct Contact

Natural enemies may directly contact insecticides 1) through fallout of the toxicant while on a plant surface or other substrata 2) by interception with aerial sprays while flying 3) by exposure to fumigants either while flying or when on substances 4) from drenches or the soluble phase of soil applied insecticides, or 5) systematically via plants and other organisms. Direct contact tends to occur shortly after application, in contrast with residue or food chain uptake that can persist for longer periods after treatment.

a) Direct Fallout

Natural enemies, especially certain sedentary life stages *e.g.*, eggs, larvae, pupae may be directly exposed to and temporarily immersed in a spray that is applied aerially. As with pests, the probability of a natural enemy encountering a pesticide fallout is a function of the spray droplet size, its size distribution and the size of the non-target organism. The relative proportion of a natural enemy population that is protected from a spray in a refuge may vary with species and environment. Hence differences in exposure to direct fallout of pesticides can occur, both among different predators or parasitoids and among different natural enemies and their hosts or preys. Differential exposure can confer ecological selectivity to pesticide application via spatial separation. Active, adult natural enemies may be less affected by direct fallout of sprays than the immatures due to their ability to escape exposure (Bartlett, 1964). Many natural enemies spend considerable time in nontreated habitats outside the crop environment of their principal prey or hosts. They may periodically invade the treated habitat of the pest to search for food or other resources (Tabashnik and Croft, 1985). Exposure to residues as compared to direct fallout sources of pesticides may be more common for highly mobile life stages of natural enemies than for local and sedentary forms of their prey or hosts.

b) Aerial Exposure

Natural enemies are commonly exposed to aerial sprays of pesticides. Natural enemies are often active fliers due to their need to search for food, especially when prey or host density is low to moderate. Under these conditions in a treated area, pests would be less likely to be airborne than natural enemies, and therefore aerial sprays would selectively reduce beneficial populations. Spraying may also differentially predispose one group to exposure to aerially applied pesticides over the other (Bartlett, 1966). However, this exposure is seldom if ever monitored, largely because of the difficulty in studying this short lived act.

c) Vapour Exposure

Volatiles emanating from fumigant pesticides or from rapidly vaporising pesticides can be extremely toxic to natural enemies (Franz, 1974). The size of the organism is an important factor. Smaller organisms are generally more sensitive to pesticides vapours due to their greater surface area to volume ratio (Ripper, 1956). Thus the natural enemies like chrysopids, spiders, ladybeetles may be more tolerant than their prey because of this difference in pesticide uptake. However many predators are of equal size or smaller than their prey. The size discrepancy between parasitoids and their hosts is even greater. These natural enemies usually are much smaller than their hosts and therefore would be discriminated against by pesticide fumes.

d) Soil Uptake

For soil inhabiting natural enemies, soil and its semiliquid environment are a principal source of pesticide uptake. As a source of pesticide uptake, soil at or near saturation is functionally similar to an aquatic habitat. When dry, it is more like a solid substrate. Natural enemies such as dermapterans, carabid beetles, and staphylinid beetles spend much of their time in soil habitat. Their contact with pesticides is affected by their inherent and conditioned behaviours. The factors affecting pesticide uptake and natural enemy mortality in soil habitats have been reasonably well studied. The factors most commonly studied include temperature, moisture, pH and soil structure and other attributes.

e) Systemic and other Types of Exposure

Systemic pesticide transfer through a plant or other organisms to the natural enemy is another common means of direct pesticide uptake. Systemic pesticides can be translocated to surface of the plant or can enter a vapour phase and be toxic to a natural enemy on the plant. (Cate *et al.,* 1972). This exposure is a combination of vapour and residue uptake. The relative toxicity of a toxicant is determined by its translocation and metabolism in the plant and the relative concentration in the plant portion where contact with natural enemy is made (Elliott and Way, 1968; Aveling, 1981).

More specialised means of direct and immediate uptake of pesticides by natural enemies do occur in agricultural systems *e.g.,* with aquatic forms of natural enemies in rice ecosystems etc. and many uptake events are unique to specific natural enemy/ host or prey situations. However more important ones are those cited above.

1.3 Residual Contact

Natural enemies most commonly take up pesticides by residual contact. While residue may be of equal or of somewhat lesser importance than direct fallout immediately following any pesticide application, residues are more persistent phenomena. The greater susceptibility of natural enemies to low concentrations of pesticides often means that during re-entry into treated habits, predators and parasitoids may be subjected to toxic residues longer than pests. This may partially account for the earlier invasion of treated habitat by pests than natural enemies.

The behavioural processes involved in residual uptake of pesticides by natural enemies are complex and dynamic. They include general locomotion, searching for prey or hosts, cleaning and preening and other behaviours. Pesticide deposition, redistribution, weathering, the nature of treated substrata and the dynamics of substrate growth or change influence residual uptake of pesticides. This kind of information can help predict recolonisation rates of natural enemies in treated habitats for IPM of pests (Panis, 1980). The dynamics of weathering can be used to time periodic and inundative releases of natural enemies (Mani and Krishnamoorthy, 1986), besides ecological selectivity (Croft, 1990).

a) Physical and Morphological Factors

Many of the same physical processes that apply to direct exposure to pesticides apply to residual contact as well *e.g.,* distribution, particle size, nature of treated surfaces etc. as with pests, the uptake of residual deposits during walking is affected by particle size; uptake generally increases upto a 20-40 µm range, decreases at higher particle size (Barlow and Hadaway, 1952). Particle pickup depends on the relative adhesion of the particles to thee substrata as compared with body part of the insect contacted. Sites of uptake for residual deposits are mostly on tarsal receptors or setae (Hartley and Graham-Bryce, 1980). Particles often are transferred thereafter to other parts of the body. Gratwick (1957) reported that while walking, the rate of uptake per step is greater for larger insects, but the smaller insects picked up more particles per unit of body weight. However, as noted before, this size relationship between natural enemies and pests is not a common rule.

Crawling or sluglike forms of natural enemies (larval Hymenopteras and Diptera) are exposed to pesticide residues over larger body areas and longer time periods than walking arthropods. Particles in the range of 15-20μm are most toxic to this type of organisms (Hartley and Graham-Bryce, 1980). While these beneficials may contact more area of treated surface, they also move more slowly, which could offset the potential for greater exposure. Generally organisms with such morphology and type of locomotion are some of the most pesticide susceptible natural enemies (Hassan *et al.,* 1983).

b) Behaviour and Residue Uptake

Behaviour of a natural enemy plays an important role in determining the extent of residual exposure. Innate and modified behavioural responses of natural enemies are some of the primary determinants influencing the uptake of pesticide residues (Hartley and Graham-Bryce, 1980). Gratwick (1957) noted the importance of body cleaning movements both in spreading particles picked up on the tarsi and in enabling pests and natural enemies to move pesticides from their bodies. Many authors have commented on the extensive preening and cleaning activities of parasitic wasps. Kuhner *et al.* (1985) observed that by cleaning of its legs, wings and head, adult *Diaretiella rapae* contaminated its whole body with the moderately toxic herbicide Ramrod. Activity rates were increased and normal searching patterns greatly modified by sublethal levels of this pesticide. Thus residual uptake of adult parasitoids is more by this rate than those of the larval stages.

Since natural enemies may take up more pesticide per unit body weight per step and search longer on treated substrata than pests, these factors together may multiply the overall uptake of pesticide residues. Many authors have speculated this is a major explanation for the greater susceptibility of predators and parasitoids over their prey or host (Hoy and Dahlsten, 1984; Waage *et al.,* 1985). Hartley and Graham-Bryce(1980) point out that factors contributing to grater pesticide exposure in natural enemies must be balanced by those causing them to receive lesser exposure. For example, the lesser tendency for natural enemies to be present in the treated habitat than pests should be considered. Within the treated habitat, natural enemies also have a tendency to be in refugia from pesticides more often than pests. In projecting net impact, field toxicity data and other types of evaluations indicate that pesticides having persistent residues generally discriminate against natural enemies as compared with their hosts or prey. This may be the partial explanation as to why short-lived, nonpersistant compounds often show relatively selective properties to natural enemies over the pests, and why more persistent compounds have such devastating effects on these smaller and more active arthropods.

Another behavioural attribute of natural enemies that influences residue uptake is searching activity relative to prey density and distribution. A pesticide sprays often decimates a natural enemy's food supply and alters its distribution in space. A predator or parasite's movements may increase as hunger escalates (Hoy and Dahlsten, 1984). An interesting aside to this discussion of comparative uptake of pesticides by pests and natural enemies is the differentials searching behaviour of primary and secondary parasitoids as impacted by pesticide residues. Horn (1983) noted that *Diaeretiella rapae,* which attacks *Myzus persicae* is parasitized in turn by a complex of secondary parasitoids including *Aphidencyrtus aphidivorus* and *Asaphes lucens.* In fields plots treated with carbaryl, and malathion, he observed higher rates of primary parasitism by *D. rapae,* but lower rates of secondary parasitism by hyperparasitoids. Horn attributed this susceptibility differential to the searching behaviour of the primary and secondary parasitoids. Secondary parasites tend to walk slowly over the leaves searching for suitable hosts, whereas *D. rapae* spend more searching time flying among leaves. Hence the later species was less likely to acquire a lethal dose of insecticides.

c) Pesticide Deposition

Relationships between pesticides distribution a, residue levels and selectivity in ambulatory arthropods are not well understood. Since residual deposits of pesticides are not uniform in the field, insecticide uptake should be a function of the area of insecticide contacted and the accumulate time of exposure. Generally, more mobile the insect, the more vulnerable it should be to a highly dispersed pesticide. However these variables don't act in solitary. For example, there may be conditions where total contact would be greater between a toxicant and a more sedentary pest. A slow species might remain in contact with a pesticide deposit long enough to receive a greater exposure than the larger number of deposits contacted by a fast moving natural enemy. Of course, overall impact must be measured at the population level.

d) Weathering of Residues and Substrate Dynamics

The nature of the treated substrate and the weathering of residues are important variables influencing residue uptake by natural enemies. Reports of the toxicity of weathered deposits to predators and parasites are common in the literature for a wide variety of crops and species (Theiling, 1987). Most studies however are only timed evaluations, with little reference to the factors influencing changes in residues. The most extensive studies of factors affecting the dynamics of pesticide residues on arthropod natural enemies have been done on citrus crops. Citrus grows year-round in arid regions and only sporadically experiences rainfall. Bellows *et al.* (1985) evaluated the residual activity of insecticides on the aphelinid *Aphytis melinus*, the coccinellid *Cryptolaemus montrozieri,* and the phytoseiid mite *Euseius stipulates.* Foliage samples were collected at regular intervals from trees sprayed with acephate, dimethoate, formatenate and sabadilla, and were analysed for residue levels of pesticides which coincided well with the results of bioassay of acute toxicity over a 70 day interval.

e) Exposure to Residues via other Organisms and Sources

Contact via external residues on other organism and on sources other than plant surfaces is another component of pesticide uptake by natural enemies. Residual uptake occurs via ingestion of foods, water or other substances. Sources of external residues may be other plant parts, arthropod excrement, prey or hosts of natural enemies, members of their own species or other arthropods laden with pesticides.

With arthropod sources, there is probably little difference between uptake of external residues and internal food chain sources of pesticides insofar as toxicity is concerned. One difference might be in the extent of metabolism of the pesticide, when coming from an internal versus an external source. Pesticides via food chain sources would be more likely to have been metabolised than from external sources. Metabolic transformation could render a toxicant more or less toxic to a beneficial, depending on the compound and host or prey involved. Some natural enemies may primarily feed on internal fluids of their host or prey (piercing and sucking natural enemies), such as hemipterans, mites and many parasitoid adults and would therefore be exposed primarily to pesticide via food chain. Conversely, chewing natural enemies usually consume whole organisms, and would be exposed to both external and internal pesticide deposits in their hosts or prey.

There are several ways by which external residues can be transferred to natural enemy. Adult parasitoids and predators may obtain sustenance from consumption of plant nectar, pollen, sap, free water, insect excrement such as the honeydew and exposed body fluids of their host. Many studies have demonstrated the toxicity of pesticides administered via these sources (Brettell and Burgess, 1973; Hegazi *et al.*, 1982). Other studies have documented residual uptake of pesticides via predation (Teotia and Tiwari, 1972; Nakashima and Croft, 1974). Both external and internal food chain residues are usually involved. Cannibalism is yet another means of uptake of external or internal pesticide residues by natural enemies. May predaceous species and immature stages of parasitoids are cannibalistic (Clausen, 1962). Cannibalism among different life stages can serve as a source of residue uptake to a free living natural enemy, as can feeding on fragments such as cast skins or the egg chorion (*e.g.*, among certain coccinellids; Zeleny, 1965).

1.4 Food Chain Uptake and Transfer

Certain toxicants administered via plant to a pest often will kill the pest, but not cause mortality to a natural enemy which subsequently feeds upon the pest. Many studies have demonstrated the selectivity of systemic pesticides; however, systemic pesticides are always selective. There have probably been more studies published which demonstrate significant mortality to natural enemies by feeding on the pesticide laden tissues of their prey or hosts (Croft, 1990). In addition, natural enemies' take up pesticides from food chain sources in pest honey dews or plant nectars, pollen or other internal sources of toxicants (Croft and Brown, 1975). Important ones among these are:

a) Plants as Source of Food Chain Uptake

Many natural enemies feed on plant materials at some stage in their life cycles to obtain nutrients or water (Brown and Shanks, 1976). Studies with predaceous mites have shown that certain species directly take up systemic toxicants administered to plants. Following foliar treatment with a systemic poison which decimates their prey, predators may survive initial fallout, residues and food chain uptakes. However they may again be exposed in a more susceptible physiological state of semistarvation by taking up water or nutrients from the treated plant (Porres *et al.,* 1975; Congdon and Tanigoshi, 1983). Similar uptake of pesticides from facultative herbivorous/ entomophagous hemipterans is probably a frequent event.

Studies demonstrating pesticide uptake from plants by parasitoids have been uncommon, probably because phytophagy is rare among them (Clausen, 1962). In most cases, parasitoids take up pesticides while consuming pollen or sugars and nutrients from nectars.

b) Uptake via Predation

Documentation of pesticide transfer through predator-prey food chain system is extensive (Croft and Brown, 1975; Croft, 1977). A few generalisations can be made regarding the uptake of systemic pesticides from prey by predator. Toxic effects can apparently be conferred from consumption of live, dead or morbid prey. Pesticide uptake by a predator from its prey is analogous to the uptake of a systemic pesticide from a treated plant by a pest. The extent of the effect on natural enemies likely depends on the behaviour and feeding habits of the prey (localisation. Concentration and metabolism of the toxicant), the feeding habits of the predator (amount and rate of uptake), and the metabolic and detoxicative abilities of the predator.

Metabolism and transfer of a toxicant from plant o herbivore versus from herbivore to natural enemy may differ appreciably. The physiological systems involved in each case are quite different from each other. Since a toxicant is selected for tits lack of phytotoxicity, it's unlikely to metabolise the toxicant, however physiological systems of the pest and natural enemy are more similar. A pest is more likely to metabolise the toxicant. Subsequent effects on the beneficial would depend on how toxic the bye-product was to the natural enemy.

c) Intraspecific Transfer

Food chain transfer of a toxicant can occur from one intraspecific organism to another. Cannibalism is one such source of intraspecific food chain transfer for both predators and internal parasitoids. In some species cannibalism occurs almost independent of prey density like the coccinellids and chrysopid larvae. At low prey density and high natural enemy density, many natural enemy species will engage in cannibalism and in the process receive a toxic dose of pesticide.

Another mechanism of intra specific transfer of pesticides is tropholaxis. Pesticides in toxicant laden food supplies may be transferred among ants and other social entomophagous insects. Dmitrienko (1979) administered chlordane and several organophosphate insecticides tom adults of predaceous ant colonies. He observed high rate of mortality in the immature life stages which were subsequently fed these pesticide laden food materials.

d) Uptake via Honeydew and Host Feeding

It has long been known that certain pesticides, which move systemically through plant-herbivore food chain via honeydew from homopteran species, can be toxic when fed upon by predators or parasitoids (Doutt and Hagen, 1950; Bartlett, 1964). While honey dews may be repellent to natural enemies, toxicants taken up through feeding on honey dews can be extremely toxic to natural enemies of many groups (Bartlett, 1964, 1966) on the other hand, little has been published about the systemic effects of pesticides on adult parasitoids via oviposition and host feeding or on immature endo- or ectoparasitoid stages (excluding chemotherapeutic effect).

e) Food Chain Accumulation

Pesticide movement in the food chain may result in metabolic concentration of the toxicant at the predator – parasitoid level. Such pesticide biomagnification has been most commonly documented for persistent pesticides, especially among predators in aquatic and soil environments. Wilkes and Weiss (1971) measured a 2700 fold accumulation of DDT in dragonfly nymphs due to feeding on pesticide laden prey and direct uptake from water. A relatively detailed study of food chain toxicity in plant-herbivore-natural enemy system was conducted by Kiritani and Kawahara (1973). They examined the fate and effect of BHC passing from irrigated soil to rice, to the green leafhopper and finally to the predaceous spider *Lycosa pseudoannulata*. After feeding on rice for 2 days, the leaf hoppers contained three times as much BHC as the rice plants. BHC applied to the rice plants contained 13 per cent of the gamma isomer. The gamma isomer was found to increase and accumulate successively in plants and then in the leafhoppers. The proportion of the gamma isomer reached 35 per cent in the leafhopper. Biological concentration of gamma BHC was particularly significant since the spider was much more susceptible to this isomer than its leafhopper prey. Similarly in model studies of the metabolism and fate of DDT, aldrin, dieldrin and lindane in paddy ecosystems, Hsu and Hsu (1980) observed high levels of biomagnification of and low biodegradation indices in the arthropod pests *Oxyahyla intricata* and *Nilaparvata lugens* and the predaceous spider *Lycosa*.

f) Food Chain Uptake by Ecto- and Endoparasitoids

The interaction between a parasitoid and its host relative to toxicant transfer is complex and involves numerous feedback loops. Toxicant transfer is mediated by the physiology of the host and is initially determined by the extent of penetration through the parasitoid cuticle. The insecticide can cause death of the host, which may or may not kill the parasitoid depending on its stage of development (MacDonald and Webb, 1963; Lingappa *et al.*, 1972). If the host survives it may detoxify the pesticide or convert it to a more toxic metabolite which can subsequently affect the internal parasitoid.

Of the most detailed studies of host-parasitoid-pesticide interactions was reported by Novozhilov *et al.* (1973). They conducted detailed studies with *Trissoleus grandis*, a scelenoid egg parasite of the scutllerid *Eurygaster integriceps* in the USSR. After pesticide treatment eggs were washed and pulverised to measure homogenate levels of the organophosphate trichlorphon. The chorion of the host egg absorbed from 95-99 per cent of the toxicant. The amount of the toxicant in the homogenate of parasitized eggs was much less (0.7-4.43 per cent) and varied according to the developmental stages of the egg parasitoid at the time of treatment. To some extent, the increasing rate of penetration was attributed to the aging of the host egg and difference in gas exchange. Differences in penetration were suggested to be due to the disruption of embryonic tissues of the host as the parasite developed. Data suggested that host embryos were to some degree involved in the active exclusion of the toxicant. Emergence of adult parasites for different treatments was highest when host eggs were treated while the parasite was in its egg stage. Emergence was lowest when host eggs were treated just prior to adult parasite emergence. Similarly, mortality of parasite adults was highest when treatments were applied to host eggs containing parasitoid pupae. Adults emerged shortly thereafter and contacted residue on host eggs.

g) Chemotherapeutic Effect on Endo- and Ectoparasitoids

Another very specific interaction between toxicants, parasites and their hosts is what one might term a chemotherapeutic effect (Croft and Brown, 1975). In these cases, a pesticide does not cause host mortality, but will kill an ecto- or endoparasitoid and may free the host from its natural enemy. This phenomenon has been noted primarily among relatively large lepidopteran hosts and their hymenopteran parasitoids (Sewall and Croft, 1987). Detailed documentation of the chemotherapeutic effect was first reported by Teague *et al.* (1985). When incorporated into the artificial diet of *Heliothis zea* parasitized by the braconid *Microplitis croceipes*, the fungicide benomyl caused no effect on *H. zea*, nor did it affect the parasitism of the host by adult wasps. However subsequent emergence of parasites from the host was depressed by more than 90 per cent at higher concentrations.

h) Latent Transfers

Latent effects of pesticides are those expressed by a life stage of a natural enemy subsequent to the one initially exposed. Tamashiro and Sherman (1955) first described the delayed mortality from pesticides observed in adult *Opius oophilus*, a parasite of the oriental fruit fly *Dacus dorsalis*. When larval hosts containing parasitoid larvae were treated with parathion and six organochlorine compounds, emerging adults exhibited symptoms of pesticide poisoning and died shortly after eclosion. Parasites exposed to aldrin, dieldrin, endrin and chlordane exhibited latent mortality ranging from 50 per cent to 54 per cent of the total emerging population at the highest dosage tested.

The physiological or toxicological basis for latent effects of pesticides is not well understood. Increased mobilisation of fat body containing high pesticide residues in later life stages is a plausible explanation. Changes in detoxification enzyme levels in subsequent life stages are another possibility.

i) Parasitism Effects on Host/Parasitoid Uptake or Transfer

Parasitism has numerous effects on both host physiology and behaviour (Slansky, 1986), and many of these effects can result in changes in uptake and metabolism of a pesticide by a host and parasitoid. Physiological changes leading to increased susceptibility in hosts, and indirectly in parasitoid. Behavioural changes may alter the amount of toxicant taken up by the host. This in turn feeds back to affect the physiology of the parasitoid.

2. Entomophage versus Herbivore Physiology

2.1 Nutrition

Herbivorous and carnivorous arthropods consume foods that are fundamentally different in terms of nutrition; foods of arthropod origin are higher in protein and lower in carbohydrate than are plant based diets. Besides, animal derived diets usually provide a balanced complement of amino acids, fatty acids and vitamins for entomophagous insects while as plant derived diets lack some of these. Because of these nutritional deficiencies, entomophages have almost double the assimilation efficiency as compared to herbivores (Slansky and Scriber, 1982). As nutrition determines the basic enzyme kinetics and ultimate biochemical composition of body tissues; the contrasting dietary intake of herbivores and their natural enemies may contribute to differences in pesticide susceptibility and resistance potential between these groups. Some of the promising aspects are:

a) Fats and Fatty Acid Composition

A general observation relating to lipid constituents is that herbivores tend to have higher fat body contents than entomophagous insects (Fast, 1964). Higher lipid content can be important in the sequestration of the toxicants and can confer tolerance to lipophilic compounds such as the synthetic insecticides. Fatty acid composition of many dipteran and hymenopteran parasitoids is a close mimic of their hosts (Thompson and Barlow, 1974). Most of them exhibit fatty acids almost identical to their hosts while some, for instance, *Exerrites comstockii* duplicate their fatty acid profile of its hosts, regardless of their nature. Initially it was thought these parasitoids lack enzymes for de novo synthesis of fatty acids. However later studies confirmed the presence of enzyme pathways for both fatty acid synthesis and turnover in these species (Thompson and Barlow, 1974; Jones *et al.*, 1982). It seems that many parasitoids are deficient in fatty acid transferaces or that specificity of these enzymes has been altered relative to their hosts (Jones *et al.*, 1982). These enzymes could secondarily confer differential susceptibility to pesticides in pests versus beneficial species by the interactions between lipophilic toxicants and lipid sinks used for sequestration.

b) Sterols

All arthropods require external sources of sterols for growth and development. Cholesterol meets the sterol needs of most insects, but the predominant sterols in in plants (*e.g.*, sitosterol, stigmasterol and campesterol) must be converted to cholesterol to satisfy this need. Most phytophagous insects remove the methyl and ethyl groups at C-24 in plant sterols and convert them to cholesterol. However carnivores seem to lack this capability and require cholesterol or a similar analogue in their diets (Svoboda *et al.*, 1978). Several entomophagous hemipterans and hymenopterans that are unable to dealkylate 24-alkylsterols and have insufficient cholesterol in their diets will utilize

24-alkylecdesteroids as moulting hormone (Svoboda and Lusby, 1986). This differential capability to dealkylate sterols occurs within the coccinellidae, which includes both phytophagous and carnivorous members. This aspect and other such approaches have been used to manipulate insecticidal activity among herbivorous insects (Prestwich *et al.,* 1983) but has not been exploited to achieve selectivity to entomophages.

c) Lipid Content to Carbohydrate Ratio

Dietary triglycerides are of limited value as energy sources for some hymenopteran parasitoids larvae. Interestingly, Diptera and Hymenoptera, which comprise over 90 per cent of entomophagous parasitoids (Price, 1981), rely on carbohydrates such as trehalose as energy source for flight. Many phytophagous pests, however, use lipids for flight energy (Fast, 1970). Therefore, entomophagous insects retain three to four fold higher levels than herbivores of key regulatory enzymes for carbohydrate catabolism, including hexokinase, glycogen phosphorylase and phosphofructokinase (Crabtree and Newsholme, 1975). Conceivably, chemicals that interfere with carbohydrate mobilisation could have greater effect on entomophages than herbivores. Conversely the compounds selectively inhibiting the lipid metabolism could favour beneficial species over their host or prey due to disruption of nutrient utilisation. However mammalian safety and the ability of either arthropod group to metabolically compensate for these critically blocked enzyme pathways are major roadblocks.

2.2 Respiration

Predaceous insects, especially adults and larvae, have higher respiratory rates than their prey, presumably due to the increased metabolic demands associated with searching behaviour. Oxygen consumption in a carnivorous mite (Thurling, 1980) and a coccinellid predator (Tanaka and Ito, 1982) were about twice that of herbivores from the same taxonomic groups, yet carnivores survived starvation better by markedly reducing respiration. Higher rates of locomotion and other respiratory-associated activities such as energy metabolism may confer faster pesticide uptake, activation and metabolism by natural enemies, which could render them more susceptible than their prey or hosts. In contrast, entomophages could have distinct advantage over phytophages by having greater respiratory reserves to sustain them during food stress.

2.3 Endocrinology

Endocrinological factors may suggest other means of achieving selectivity between pest and beneficials. For example, the juvenile hormones (JHs) and ecdysteroids that regulate various developmental processes in insects have no comparable structures in vertebrates (Kircher, 1982). Insect bioregulators having JH action have received extensive study recently because of their safety to non-targets other than arthropods. However, much less is known about differences in endocrine physiology and metabolism between herbivores and their natural enemies. The selectivity of juvenoids to beneficial and injurious arthropods has been mixed. Certain diaryl juvenoids are inactive against hymenopteran parasitoids. While hydroprene has little activity on some dipteran and hymenopteran parasites, it is deleterious to others at dosages used for pest control (Ascerno *et al.,* 1980). While endocrinological impediments of parasitoids by the direct action of JH mimics in the host is expected, it is gratifying that some JH mimics have favourable selectivity to some life stages of insects (Bull *et al.,* 1973) and acarine predators (El-Banhawy, 1980).

One major difference in JH biochemistry that may be exploited to achieve selectivity between lepidopterans and their parasitoiuds (hymenopteran and dipteran) is that the parasotoid orders appear to biosynthesise only JH III, whereas major lepidopterans pests synthesise the higher homologs, JH II, JH I, JH 0, and 4-methyl-JH I instead of or in addition to JH III. Thus inhibitors of JH synthesis for the higher homologs of JH III might be more selective to the beneficials.

3. Arthropod versus Herbivore Toxicology

The strategies that provide basis for physiological selectivity or differential pesticide effects from toxicological point of view are; 1) presenting a toxicological barrier, such as an integument or internal lipid barriers 2) avoidance by sequestering the chemical in insensitive storage tissues such as in fat or the integument 3) rapid excretion of the toxicant 4) metabolic detoxification of the toxicant 5) developing an insensitive target site for the toxicant. These strategies are in general used by any animal to avoid toxicosis.

3.1 Penetration

That the integument can confer some selectivity is depicted by the increased toxicity upon direct injection of pesticide into an organism. A number of cases are known where penetration barriers in natural enemies to specific insecticides have provided some basis for selectivity over their host or prey. However, no patterns are present in these cases to suggest that either group is favoured by having greater penetration barriers to exogenous toxins.

Bull and Ridgway (1969) examined comparative rates of penetration of P[32]-labelled trichlorfon among adults of the herbivores *Heliothis virescens* and *Lygus hesperus,* and predaceous larvae of the lacewing *Chrysoperla carnea. C. carnea,* which is extremely tolerant to this insecticide showed poor penetration of trichlorfon over a 4 hour period, where the mirid showed a high rate of penetration and the tobacco budworm was intermediate. Comparative studies with the C[14] labelled organophosphate sulprophos to third instar larvae of *H. virescence,* the adult boll weevil *Anthonomus grandis* and lady beetle *Hippodamai convergens* showed extremely rapid penetration and loss of external dose in the highly susceptible predator. The more tolerant phytophages showed slower rates of absorption of topically applied toxicant. Several studies have focused on the differential penetration of pof pesticides through the cuticle of host insects versus the less sclerotized epicuticular membranes of their endoparasitoids.

3.2 Sequestration

Lipophilic sinks comprised of fatty tissues and intra and inter cellular lipoproteins that serve in chemical transport allow insects to sequester dietary toxicants and defend against predation (Duffey, 1980, Blum, 1981). The greater non-specific accumulation of lipophilic toxicants in insects has been attribute to higher ingestion rates per unit of body weight, a higher ratio of lipid to body water content of insects compared to non-targets (Duffey, 1980). It appears likely that storage of toxicants in lipid reserves can confer some pesticide tolerance and resistance in arthropods in general (Hollingworth, 1976, Duffey, 1980). In relation to sequestration of toxicants, the most extensive work with natural enemies has been associated with the study of fat body lipid reserves as buffers to toxicosis. Takeda *et al.* (1965) demonstrated in *Coccinella septempunctata* a direct correlation between low fat content in early season and the severity of malathion toxicity. In late season, summer population which had fed extensively and had presumably obtained higher fat body reserves, showed a lower degree of effects from the organophosphate insecticide (Table 3.1).

Table 3.1: Toxicity of Malathion to Adult *Coccinella septempunctata* in Relation to Fat Body Content (After Takeda *et al.*, 1965)

Date	Per cent Mortality	Per cent Fat on Dry Weight Basis
September 20-30	66	17.2
October 16	75	17.8
October 23	60	21.7
October 29	33	25.3
November 8	25	24.9

Although the fat body metabolism vis-a-vis insecticide toxicology is receiving greater emphasis these days, several authors have inferred causal relationships between pesticide side effects on natural enemies and the role of these organ systems in sequestration and pesticide dynamics (Croft, 1990).

3.3 Excretion

The excretory apparatus is primarily responsible for the transfer and elimination of harmful metabolites from internal to external fluids, where adsorption to waste solids, may occur prior to defecation. Important selectivities among insects in excretory capacities are known, particularly for polar insecticides. For example ingested nicotine is rapidly excreted in tobacco hornworm (90 per cent in 4 hours), whereas houseflies only excrete 10 per cent of an applied dose in 18 hour (O'Brien, 1967). With regard to pesticide excretion in natural enemies, the most work has been done with DDT and metabolites in entomophagous predators. Attalah and Newson (1966) studied metabolism and excretion in the coccinellid *Coleomegilla maculata* and observed the tolerance to DDT was due, in part, to rapid metabolism of the compound to DDE and then the excretion of both the metabolised compounds in faces. Dempster (1968) studied conversion of DDT to DDE in the DDT tolerant carabid *Harpalus rufripes* and noted rapid conversion and elimination of DDE as a nontoxic metabolite.

Although little is known about excretion in parasitic Diptera and Hymenoptera, endoparasitic larvae can store nitrogenous wastes internally and then deposit them in a meconium just prior to pupation (Vinson and Iwantsch, 1980). This may confer differential sensitivity to the parasite over its host to nitrogenous wastes, and perhaps nitrogenous pesticides. In summary, excretory mechanisms have been associated with pesticide metabolism and may even account for some tolerance to pesticides in natural enemy species. However, no distinctive excretory differences that might confer differential selectivity to entomophages over their prey or hosts have been noted.

3.4 Target Site Selectivity

Binding of a toxicant with its target site is the ultimate event prior to harmful effect, and occurs only after penetration, alternative binding, excretory and metabolic barriers have been circumvented. The nerve toxicants that are inhibitors of acetyl cholinesterase (AChE) including organophosphates (OP) and carbamates or the acetyl choline receptors such as nicotine in insects must penetrate a penultimate barrier prior to reaching the target site. This is necessary since the cholinergic junctions are limited the central nervous system (CNS) sand the CNS is ensheathed in by the neural lamella and perineurium. Comparative studies of cholinergic target site differences between herbivorous insects and their entomophagous enemies have been limited. Substrate preferences for cholinesterase, the target for OP and carbamate pesticides showed no distinction in substrate-activity profiles between herbivorous, omnivorous and carnivorous insects including a syrphid, blowfly and tiger beetle (Metcalf *et al.*, 1955).

Using studies of choline esterase (AChE) activity in whole body homogenates of herbivores and entomophagous insects, Singh and Rai (1976) found much higher levels of Ach hydrolysis in *Cocccinella septumpunctata* (252 mg hydrolysed/g tissue/hr) and *Cicindella sexpunctata* (220 mg) than in hemipteran plant bugs *Leptocorisa acuta* (30 mg) and *Dysdercus koenigii* (63 mg) and the aphid *Lipaphis erysimi* (6 mg). Similar results were obtained with AChE activity on a per insect basis. Phosphamidon, an insecticide selectively toxic to these phytophagous insects showed, showed a higher inhibition rate with AChE from *L. erysimi* and *L. acuta* than with AChE from their predators *C. septumpunctata* and *C. sexpunctata*, respectively. This was due to higher affinity between enzyme and inhibitor rather than to the rate of phosphorylation. The authors suggest that these differences in properties of AChE between the two groups might be exploited to develop selective pesticides.

More detailed comparative work on cholinesterase activity has been done with silkworm moths and their tachinid parasitoids (Bai Shang *et al.*, 1981). Enzyme activity from the head and nerve cord of the larvae of the silkworm, *Anthraea pernyi* and from individual larvae and head and thorax of the adult tachinid *Blepharia tibialis* were measured in relation to the toxicity of the organophosphate insecticide dimethoate and the related compound omethoate. The bimolecular rate constants K_i for omethoate to cholinesterase of the tachinid fly were from 12 to 80 greater than those of the silkworm. These results for *in vitro* cholinesterase inhibition were in close agreement *in vivo* toxicity tests with dimethoate. They concluded that the mechanism of selective toxicity favouring the silkworm over its natural enemy was principally due to difference in selective inhibition of the AChE. Hollingworth (1976) concluded that the successful design of selective neurotoxicants will largely depend in a continued polyfactorial approach to the dynamics of toxicant target site interaction.

3.5 Detoxification Enzymes

In consideration of the detoxification capacities of pests versus natural enemies, the discussion must begin by reviewing background about pest pre-adaptations to plant foods. In this regard, feeding experiences as it relates to the degree of polyphagy is thought to alter pesticide susceptibility in herbivorous arthropods. It has been suggested that elevated levels of detoxification enzymes due to prolonged exposure to plant toxicants are often responsible for altered susceptibility- the allelochemical preadaptation hypothesis (Gordon, 1961). Early biochemical evidence for this hypothesis was provided by Krieger *et al.* (1871) who found increasing aldrin epoxidase (Mixed Function Oxidase, MFO) levels in lepidopteran larvae capable of consuming many plant hosts. While herbivores may retain many preadaptive detoxification abilities for pesticides because they must detoxify plant allelochemicals, entomophages may be less exposed to dietary toxicants and thus have lower detoxification capabilities. This hypothesis was used to elaborate the general higher susceptibility of natural enemies than herbivorous pests to pesticides (Croft and Morse, 1979). It was suggested that a natural enemy may be able to express its capacity to adapt to a natural plant toxicant only after the pest had exploited its genetic plasticity in resisting the toxicant (Croft and Brown, 1975). Expression of enhanced detoxification may occur within a generation through induction to influence the tolerance to resistance mechanisms.

Evidence to evaluate detoxification differences between natural enemies and pests initially came from the data reported by Brattsten and Metcalf (1970, 1973), Croft and Morse (1979) and Mullin and Croft (1985). They tested the susceptibility of a wide variety of arthropod species to carbaryl in the presence and absence of the MFO inhibitor piperonyl butoxide. Piperonyl butoxide synergizes pesticide potency by inhibiting a major detoxification system; it is well known that carbaryl is primarily detoxified by oxidative metabolism (Brattsten and Metcalf, 1970). Thus a synergistic ratio provides an estimate of the MFO available for detoxification in the test animal. Although a high degree of variation was apparent in each trophic group, herbivores had a 10 fold higher mean tolerance to carbaryl and two fold higher mean synergistic ratio than carnivores. Although predators and parasitoids may lack well developed MFO's, they may be just as adapted as herbivores in relation to nonoxidative detoxification mechanisms (Plapp, 1981). Several researchers observed that oxidatively detoxified insecticides were more toxic to entomophagous arthropods than to herbivores, whereas insecticides detoxified primarily hydrolytically were safer to carnivores (Plapp, 1981; Bashir and Crowder, 1983; Brown and Casida, 1983).

Mullin *et al.* (1982) compared detoxification enzymes in the polyphagous spider mite *Tetranychous urticae* and its major acarine predator *Amblysius fallacis*. These species were chosen as an herbivore-predator model because of their similar biologies and morphologies. Oxidative (aldrin epoxidase, MFO), hydrolytic (esterases and epoxide hydrolases), and conjugating (glutathione-S-transferase) enzymes that typify the metabolism of lipophilic toxicant to excretable products were measured from whole body homogenates (Table 3.2). Major enzyme differences were found that were closely associated with patterns of susceptibility observed among strains. Comparing pesticide-susceptible strains, the herbivore had five fold higher MFO and six fold higher trans-epoxide hydrolase levels than the carnivore. Since both elevated MFO and trans-epoxide hydrolase have been ascribed to previous encounters with plant defensive chemicals, these results support the allelochemical preadaptation hypothesis. Also the herbivore and carnivore had similar esterase activities for 1-naphthyl acetate. These data support the hypothesis that both herbivores and entomophages should have similarly developed hydrolytic detoxification pathways. Esteratic pathways are important in both detoxification and basal metabolism and carboxylesterase, lipase, amidase, cholinesterase and thioesterase activities may all be incorporated into this general esterase measurement (Heymann, 1980). A carnivore can have an advantage over an herbivore for specific enzymes in potential detoxification capability. Thus *A. fallacis* had eleven fold higher glutathione-S-transferase and four fold higher *cis*-epoxide hydrolase activities than the prey mite. These enhanced levels may be due to the higher respiration rate of the predator, which can lead to autooxidative metabolites that require detoxification by these enzyme systems (Mullin *et al.*, 1982).

These findings with a predator herbivore system led to the examination of a representative parasitoid herbivore system, that of polyphagous tortricid *Argyrotaenia citrana* and its braconid ectoparasitoid *Onchophanes americanus* (Craft and Millin, 1984). MFO comparisons between the whole body levels of the parasitoid versus gut levels of the lepidopteran larvae indicated that the host had substantially higher levels of oxidative enzymes than did the natural enemy, even when calculated on a whole body weight basis. Again, *trans*-epoxide hydrolase (EH) activity was elevated in the herbivore relative to the entomophage, whereas hydrolytic esterase activity was similar in the two species (Table 3.2) which supports the hypothesis relating detoxification potential with feeding ecology. In contrast to the predator prey model, the glutathione-S-transferase activity was similar between the parasitoids and the host. This may be related to the sessile ectoparasitoid *O. americanus*, hence the parasitoids might require less of a protective enzyme such as glutathione-S-transferase that prevents oxygen toxicity. Similar results were found for the Mexican bean beetle *Epilachna varivestris* associated with its eulophid endoparasitoid *Pediobius foveolatus* (Mullin, 1895; Table 3.2).

Table 3.2: Detoxification Enzyme Levels in Three Natural Enemies and their Respective Prey/Hosts (Modified from Croft and Mullin, 1984, and Mullin, 1985)

Species	Enzyme Activity (pmol/min-mg-protein)					
	Aldrin epoxidase	trans-EH	cis-EH	trans/ cis-EH	Esterase $(x10^{-3})$	Glutathione transferase
Amblyseius fallacis	0.27	310	431	0.72	318	1095
Tetranychus urticae	1.44[b]	1710[b]	117[b]	14.6[b]	389	102[b]
Onchophanes americanus	0.85	407	727	0.5	307	135
Argyrotaenia citrana	26.8[b]	1340[b]	536	62.5[b]	593	288
Pediobius foveolatus	0.67	198	415	0.48	83	–
Epilachna varivestris	4.16[b]	780[b]	352	2.22[b]	93	–

b: Interspecific differences between significant values for the herbivore versus entomophage at P < 0.01 level.

A useful; comparative index that emerged from the detoxification enzyme studies with natural enemies and herbivores was the ratio of trans to cis-EH. Herbivorous prey consistently had a higher trans/cis ratio than associated entomophages. One explanation for this phenomenon may be the distribution of epoxicides within plant and animal tissues. Plants synthesise trans and higher substitute olefins, including fatty acids, phenolics, alkaloids and terpenoids that are rare or absent in animals. Many are allelochemical defences against herbivory. In contrast, cis-olefins usually have constitutive and homeostatic functions in both plants and animals (Mullin and Croft, 1985).

Epoxidation of olefins largely by MFO can produce harmful epoxicides which are subsequently detoxified by the EH of appropriate selectivity. Thus, herbivores in general might be expected to exhibit a high trans- relative to cis-EH level to evade plant defences, while entomophages, due to their feeding specialisation on animal tissues, would be expected to have a lower ratio. This tendency has been confirmed for 20 species of herbivorous pests from the orders Acari, Coleoptera, Lepidoptera and Diptera and for seven species of natural enemies from the orders Acari, Neuroptera, Coleoptera and Hymenoptera (Mullin and Croft, 1984,1985; Mullin, 1985). Indeed the herbivores surveyed had, on an average 21-fold higher trans-EH activity and 10 fold total EH activity than the entomophages. These highly significant biochemical differences suggest that EH may be an appropriate biochemical target for development of a broad-spectrum insecticide against herbivorous pests that will have less impact p0n entomophages (Croft, 1981a). Presumably, crop pests should be more susceptible than their natural enemies to inhibitors of trans-EH.

More extensive studies of the biochemical defence capacities of an entomophage versus several phytophagous species in relation to both secondary plant compounds and pesticides have been published by Yu (1987, 1988). He examined selectivity relationships between the pentatomid spined soldier bug *Podisus maculiventris* and its lepidopteran prey, the velvet caterpillar *Anticarsia gemmatalis,* the tobacco budworm *Heliothis virescens,* the fall armyworm *Spodoptera frugiperds* and the corn earworm *Heliothis zea* using six bioassays for MFO, three for glutathione-S-transferase, and three for esterase detoxification systems. The selectivity of thirteen compounds from three major insecticide classes (organophosphates, carbamates and pyrethroids) to the hemipteran predator and each prey were examined, as well as the ability of the phytophages oxidatively metabolise a number of secondary plant compounds (Table 3.3).

The detoxification survey of Yu (1987) indicated that the predator showed more enzyme activity than its prey in only four of 24 MFO assays, 3 of twelve glutathione transferase assays and three of 12 assays for esterases. Bioassay results showed that the predator was I general more susceptible to organophosphate and carbamate insecticides than the prey. Yu (1988) concluded that these differences probably were due to lesser detoxification capabilities of the beneficial species. It was not known why the predator was less susceptible to several pyrethroids s compared to its prey. Although pyrethroids are known to be metabolised by MFO and esterases in other insects, the spines soldier bug had lower activities of these enzymes than the prey species.

Table 3.3: Detoxification Enzyme Activities in Spined Soldier Bug and Four Species of its Lepidopteran prey (from Yu, 1987)

Detoxifying Enzymes	Ratio of Relative Enzyme Activity among Species				
	Spined Soldier Bug	Velvet Caterpillar	Tobacco Budworm	Fall Armyworm	Corn Earworm
MICROSOMAL OXIDASES					
Aldrin epoxidation	1[a]	3.8	12	2	5.7
Biphenyl hydroxylation	1	2.5	2.5	1.5	6.7
PCMA *N*-demethylation	1	3.3	1.5	1.3	1.4
PNA *O*-demethylation	1	1.9	3	1.6	4.1
Parathion desulfuration	1	0.2	0.02	0.2	0.1
Phorate sulphoxidation	1	29.1	4.6	44.6	79.8
GLUTATHIONE TRANSFERSES					
DCNB conjugation	1	14	7.3	15	2.7
CDNG conjugation	1	0.05	1.7	0.6	0.8
PNPA conjugation	1	5.4	7.2	5.2	3.7
HYDROLASES					
General esterases	1	2.8	4.8	4.2	4
Carboxylesterases	1	1.8	2.5	0.9	2.2
Epoxide hydrolases	1	1.5	2.6	1.9	0.8

a: Eatio of pmol/nin mg protein of SSb/similar value for lepidopteran prey species.

In this study the predator examined might be least likely to exhibit substantial differences in detoxification enzyme profiles as compared to its prey. More specialised entomophages such as certain parasitoids would more likely exhibit differential detoxification enzyme profile than their hosts. The data of Yu (1987, 1988) tend to support this hypothesis, although the detoxification enzyme activity in the pentatomid predator compared to tits lepidopteran prey is perhaps lower than expected.

From these studies, it is clear that a detailed perspective of natural enemy/pest detoxification is emerging. As with pets, natural enemies show many exceptions to the broad generalisations proposed initially to explain the patterns of susceptibility of secondary plant compounds and pesticides observed between pests and natural enemies (Croft and Morse, 1979; Plapp, 1981). In predicting trends in detoxification and susceptibility, one has to consider the different roles individual detoxification enzyme systems in a carnivorous species, the diversity of secondary plant compound content of the arthropod prey or hosts food, the feeding specificity of the natural enemy, and its evolutionary background and degree of specialisation as an entomophage.

4. Pesticide Resistance in Natural Enemies

Resistance to pesticides in agricultural pests and vectors of human diseases is an urgent worldwide problem (Roush and Tabashnik, 1990). In contrast, documented

Table 3.4: Documented Case of Resistance to Pesticides in Arthropod Natural Enemies

Species (Family)	Location	Habitat	Pesticides	Maximum Resistance Ratio at LC_{50}
Amblyseius fallacis (Phytoseiidae)	Michigan, NY	Apple, Soybean	Permethrin	15
			Fluvalinate	5
			Carbaryl	8
			Methoxychlor	13
			DDT	>43
Amblyseius hibisci (Phytoseiidae)	California	Citrus	Dimethoate	22
Metaseiulus occidentalis (Phytoseiidae)	California	Grape	Dimethoate	11
			Methomyl	3
Typhlodromus pyri (Phytoseiidae)	England, The Netherlands,	Apple	Azinphosmethyl	6
	New Zealand		Carbaryl	>62
			Permethrin	2
Xylocoris flavipes	South Carolina	Stored grain	Malathion	33
Aphidoletes aphidimyza (Cecidomyiidae)	Michigan	Apple	Azinphiosmethyl	3
Aphytis africanus (Aphelinidae)	South Africa	Citrus	Methidathion	6
A. holoxanthus (Aphelinidae)	Israel	Citrus	Malathion	3
A. malinus (Aphelinidae)	California	Citrus	Carbaryl	2
			Chlorpyrifos	2
			Dimethoate	3
			Malathion	8
Bracon hebator (Braconidae)	Georgia, South Carolina	Stored Grain	Malathion	8
Comperiella bifaciata (Encyrtidae)	South Africa	Citrus	Methidathion	66

Contd...

Table 3.4–*Contd...*

Species (Family)	Location	Habitat	Pesticides	Maximum Resistance Ratio at LC_{50}
Diaretiella rapae(Aphidiidae)	California	Cauliflower, wild mustard	Methomyl	2
Diglyphus begini (Eulophidae)	Hawaii, California	Bean, Tomato	Fenvelerate	17
			Methomyl	21
			Permethrin	13
Ganaspidium utilis (Ecucoilidae)	Hawaii, California	Bean , Tomato	Fenvelerate	3
			Malathion	9
			Oxamyl	2
			Permethrin	12
Pholetesor ornigis (Braconidae)	Ontario	Apple	Fenvelerate	2
			Methomyl	4
			Permethrin	3
Trioxys pallidus (Aphidiidae)	California	Walnut	Azinphosmethyl	5
Chrysoperla carnea (Chrysopidae)	California	Alfalfa	Carbaryl	3
			Diazinon	2
			Fenvelerate	2
			Permethrin	5
			Phosmet	2
Chrysoperla carnea (Chrysopidae)	Ontario, Canada	Apple	Cypermethrin	9*
			Delatamethrin	31
			Parathionm methyl	19
			Malathion	5

Contd...

Table 3.4—*Contd...*

Species (Family)	Location	Habitat	Pesticides	Maximum Resistance Ratio at LC_{50}
Encarsia formosa (Eulophidae)	UK	Greenhouses	Lindane	3*
			Permethrin	15
			DDT	12
Coloeomegilla maculata (Coccinellidae)	USA	Cotton	DDT	6*
			Toaxaphene	4
			Methyl parathion	10-35
			Monocrotophos	7-12
Stethorus punctum (Coccinellidae)	USA	Apple	Azinphosmethyl	R*
Stethorus puncticillum (Coccinellidae)	Italy	Apple	Azinphosmethyl	R*
Trichogramma evanescens (Trichogrammitae)	Poland	–	Oxydemeton- methyl	22*

Modified after Tabashnik and Johnson, 2005, and Croft (1990). * = fold resistance.

examples of pesticide resistance in field populations of natural enemies are relatively rare. Beneficial organisms account for fewer than 3 per cent of the 447 species of insects and mites reported as resistant by Georghiou (1986). Croft (1990) reported 31 cases of resistance in predators or parasitoids (Table 3.4).

The hypothesis proposed to explain the scarcity of the documented cases of resistance in natural enemies can be put into three general categories: 1) bias in documentation of resistance, 2) differential preadaptation to pesticides, and 3) differences in population ecology (Table 3.5).

Table 3.5: Hypotheses Proposed to Explain the Scarcity of Pesticide Resistance in Natural Enemies

Category	Specific Hypothesis	References
Documentation bias		Georghiou, 1972; Croft and Brown, 1975
Preadaptation	Detoxification enzymes	Croft and Morse, 1979; Croft and Strickler, 1983
	Intrinsic tolerance	Rosenheim and Hoy, 1986
	Genetic variation	Georghiou, 1972
	Fitness cost	Tabashnik and Johnson, 2005
Population ecology	Food limitation	Huffaker, 1971; Georghiou, 1972
	Life history	Croft, 1982; Tabashnik and Croft, 1985
	Exposure	Croft and Brown, 1975
	Genetic systems	Tabashnik and Johnson, 2005

4.1 Bias in Documentation

Resistant pest populations can attract immediate attention when pesticide treatments fail to control them. Resistance in natural enemies, however, does not create problems and may go unnoticed. Thus, the bias in documentation hypothesis states that pesticide resistance is more likely to be documented in pests than in natural enemies. Croft and Brown (1975) noted that if resistance in natural enemies appears rare due to inadequate documentation, then systemic testing of samples of natural enemies in heavily treated ecosystems should detect more cases of resistance. Data available on pesticide impact on natural enemies more than doubled from 1970 to 1984 (Theiling and Croft, 1988) and the number of natural enemy species reported as resistant to one or more pesticides also doubled during the same period. Thus, the data indicates that documentation of the cases of natural enemy resistance remained consistent through time. Tabashnik and Johnson (2005) reviewed the cases of natural enemy resistance and concluded that available evidence is not sufficient to conclusively refute or support the documentation bias hypothesis and bioassays of several pest and natural enemy species from two or more populations representing variation in insecticide exposure for a particular crop and region could help to assess the extent of bias in documentation.

4.2 Differential Preadaptation

The differential preadaptation hypothesis state that resistance to pesticides evolves more readily in pests than in natural enemies due to differences in responses to pesticides between pests and natural enemies that existed before selection for pesticide resistance. Several specific hypothesis fall under the general category of differential preadaptation; these state that pest evolve resistance more readily because

a. They have better intrinsic detoxification capabilities than natural enemies

b. They have greater intrinsic tolerance to pesticides than natural enemies

c. They have more genetic variation in tolerance to pesticides than natural enemies

d. The fitness cost associate with resistance is lower for pests than for natural enemies.

a) Detoxification Enzymes (Croft and Morse, 1979; Croft and Strickler, 1983)

The detoxification enzymes found in pests are also present in natural enemies. Data from *in vivo* synergism tests show that MFO enzyme levels in natural enemies are not consistently lower than those in pest species; evidence to date shows the opposite trend. Although *in vitro* enzyme assays have indicated many cases in which pasts have higher levels of detoxification enzymes than natural enemies have, the reverse has been reported with nearly equal frequency. Thus the fact that pests have better intrinsic detoxification capabilities than natural enemies is not supported. In those cases where detoxification enzyme levels are higher in pests than in natural enemies, there is little or no evidence indicating that this difference contributes to an ability to evolve resistance more rapidly.

b) Intrinsic Tolerance (Rosenheim and Hoy, 1986)

Differences in intrinsic pesticide tolerance between pets and natural enemies are difficult to assay. Surveys suggest that intrinsic pesticide tolerance is not consistently higher for pests than for natural enemies, but comparative data are limited and may be biased. Higher intrinsic tolerance in pests could account in part for more documented cases of resistance in pests, particularly of the criteria for resistance include the ability to survive the field applications of pesticides.

c) Genetic Variation (Georghiou, 1972)

Lack of genetic variation for pesticide tolerance in natural enemies could retard their evolution of resistance. This facet of the preadaptation hypothesis needs investigation.

d) Fitness Cost (Tabashnik and Johnson, 2005)

It is often presumed that in the absence of pesticides, a resistant individual is less fit than a susceptible individual. If this fitness cost resistance were substantially greater for natural enemies than pests, it might retard evolution of resistance in natural enemies. Reviews of the data available for pests suggest that the fitness cost is not generally high, but may depend on the nature of resistance mechanism (Roush and McKenzie, 1987). Studies of the predators *M. occidentalis* and *C. carnea* and the parasitoid *Abisopteromalus calandrae* show little or no fitness costs associated with resistance (Baker, 1998). Likewise, Baker (1995) found that in two hymenopteran parasitoids, resistance to malathion were stable in the absence of exposure to insecticides. However, more data is needed to determiner categorically whether fitness costs associated with pesticide resistance varies appreciably among the groups.

4.3 Population Ecology

The concept underlying the population ecology hypothesis is that pesticide resistance evolves more readily in pests than in natural enemies because of the differences in population ecology. There are several specific hypotheses under the general category of differences in population ecology and state that pets evolve resistance more readily because:

a. Natural enemies suffer from food limitation following insecticide treatments because of the differences in life history traits.

b. They are more exposed to pesticides than natural enemies because of differences between pests and natural enemies in their genetic systems.

a) Food Limitation (Huffaker, 1971; Georghiou, 1972)

The food limitation hypothesis is based on the population dynamics of interactions between natural enemies and pests. The idea is that the few resistant pests surviving an initial pesticide treatment will have an abundant food supply, the crop. In contrast, resistant natural enemies surviving treatment will find their food supply severely reduced. Thus resistance evolves more slowly in natural enemies because they starve, emigrate or have reduced reproduction following treatments that eliminate much of their food supply.

Indirect support for the food limitation hypothesis is provided by the success of laboratory selection programmes in which natural enemies are provided abundant food, by then general trend that natural enemies evolve resistance in the field only when their prey or hosts are resistant or more protected from pesticide exposure, and by theoretical models. In simulation studies that included the evolutionary potential for resistance in both predator and prey, as well as coupled predator-prey population dynamics, the predator and prey were assumed to have equal intrinsic tolerance and equal genetic potential for evolving resistance. The key assumption of these simulations was that low prey density reduced the predators' rate of consumption, survival and fecundity. Intensive pesticide use caused rapid resistance development in the prey but suppressed the resistance evolution or caused local extinction of the predator (Tabashnik, 1986). These theoretical results imply that food limitation is sufficient to account for pests' ability to evolve resistance readily than natural enemies evolve resistance. Tabashnik and Johnson (2005) concluded that food limitation due to reduction in host or prey population by pesticides is major factor influencing natural enemy populations. If this is true there are some important implications for management. First, many natural enemies should evolve resistance when provided abundant food in artificial selection programs. Second that intensive pesticide use may disturb biological control, even if the natural enemies are not susceptible-either because of natural tolerance or resistance. Thus to maintain effective biological control, use of selective pesticides should be sparing and judicious.

b) Life History Traits (Croft, 1982; Tabashnik and Croft, 1985)

Life history traits include number of generations per year, rate and timing of reproduction, survival, developmental rate and sex ratio. Theoretical background and historical patterns suggest that the rate of resistance evolution increases the reproductive capacity, particularly the number of generations per year, increases. However the median number of generations per year is less for pests than natural enemies (Stiling, 1990). Further, analysis of extensive database and re-evaluation of relevant theory suggests that there is no simple relationship between generations per year and the rate of resistance evolution.

Differences between pests and natural enemies in other life history traits may affect resistance evolution. For example pests generally have higher fecundity, and may maintain larger populations in treated habitats as compared with natural enemies.

c) Exposure (Croft and Brown, 1975)

The rate of resistance evolution is function of selection intensity, which is determined in part by extent of exposure to pesticides. Croft and Brown (1975) speculated that natural enemies are often less intensively selected than pest are because insecticides are directed at pest,; natural enemies contact pesticides only insofar as they occupy the same habitat as their prey or host. However mobile predators and parasites might pick up more toxicant and thus suffer greater mortality from residual deposits than would sedentary pests in the same habitat. Evaluation of these hypotheses requires detailed information about exposure to pesticides in the field. Because most of the information is not available, much empirical; work is needed to assess the effects of pesticides exposure on resistance evolution in pests versus natural enemies (Tabashnik and Johnson, 2005).

d) Genetic Systems (Tabashnik and Johnson, 2005)

Most evolutionary considerations about pesticide resistance are based on the assumption that organisms are diploid and sexually reproducing, but a variety of genetic systems between pests and natural enemies could influence their relative ability to evolve resistance.

Hymenopteran insects are haplodiploid, in one modelling study resistance evolved faster under haplodiploidy (Horn and Wadleigh, 1988) which suggests that haplodiploidy does not slow evolution of pesticide resistance in hymenopteran parasitoids. Phytoseiid mites have various genetic systems including thelytoky, parahaploidy and perhaps true arrhenotoky. Parahaploid phytoseiid such as *Typhlodromus occidentalis* and *Phytoseiules persimilis* may have some advantages of haploidy and diploidy (Hoy, 1985). This could help to explain why phytoseiids evolve resistance to pesticides more readily than do other natural enemies, but it does not support the idea that genetic systems of natural enemies retard the resistance development.

Related factors that could affect the evolution of resistance in pest and natural enemies are the extent of sexual versus asexual reproduction, inbreeding and sociality. Social insects would not be expected to evolve resistance readily because they have small effective population size (few reproductives) and their reproducing individuals usually have limited exposure to pesticides. These factors may explain the paucity of documented cases of resistance in social Hymenoptera and Isoptera insects (Georghiou, 1981). Thus, genetic systems and related factors may influence resistance development in pest and natural enemies. It does not however appear that there are consistent differences in these traits that would favour resistance development in pests compared with that of the natural enemies.

References

Ascerno, M.E., Smilowitz, Z., and Hower, A.A., 1980. Effects of the insect growth regulator hydroprene on diapausing *Microtonus aethiopoides*, a parasite of the alfalfa weevil. *Environmental Entomology*, **9**: 262-264.

Atallah, Y.H., and Newson, L.D., 1966. Ecological and nutritional studies on *Coleomagilla maculate* (Col: Coccinellidae). III. The effect of DDT, toxaphene and endrin on the reproductive and survival potentials. *Journal of Economic Entomology*, **59**: 1181-1187.

Aveling, C., 1981. Action of mephosfolan on anthocorid predators of *Phorodon humuli*. *Annals Applied Biology*, **97**: 155-164.

Bai-Shiang, T., Lin, H., and Hsu, T.S., 1981. Studies on the cholinesterases from the Tussah silkworm *Anthereae pernyi* and its tachnid parasite *Blepharipa tibialis*. *Acta Entomologia Sinica*, **24**: 1-8.

Baker, J.E., 1995. Stability of malathion resistance in two hymenopterous parasitoids. *Journal of Economic Entomology*, **88**: 232-236.

Baker, J.E., Parez-Mendoza, J., Beeman, R.W., and Thorne, J.E., 1998. Fotness of a malathion resistant strain of parasitoid *Anisopteromalus calandrae* (Hym: Pteromalidae*). Journal of Economic Entomology*, **91**: 50-55.

Barlow, F., and Hadaway, A.B., 1952. Studies on aqeous suspension of insectcicdes. Part II: Quantitaive determinations of weights of DDT picked up and retained. *Bulletin of Entomological Research*, **42**: 769-776.

Bartlett, B.R., 1964. Integration of biological and chemical control, pp. 489-514. In: *Biological Control of Insect Pests and Weeds*. (Ed.) P. DeBach, Reinhold Press, New York, p. 844.

Bartlett, B.R., 1966. Toxicity and acceptance of some pesticides fed to parasitic Hymenoptera and predatory coccinellids. *Journal of Economic Entomology*, **56**: 1142-1149.

Bashir, N.H., and Crowder, L.A., 1983. Mechanism of permethrin tolerance in the common green lacewing (Neuroptera: Chrysopidae). *Journal of Economic Entomology*, **76**: 407-409.

Bellows, T.S., Morse, J.G., Hadjidemetriou, D.G., and Iwata, Y., 1985. Residual toxicity of four insecticides used for control of citrus thrips (Thysanoptera: Thripidae) on three beneficial species in an citrus agroecosystem. *Journal of Economic Entomology*, **78**: 681-688.

Blum, M.S., 1981. *Chemical Defenses of arthropods*. Academic Press, New York, 562 pp.

Brattesten, L.B., and Metcalf, R.L., 1970. The synergistic ratio of carbaryl with piperonyl butoxide as an indicator of the distribution of multifunction oxidases in the Insecta. *Journal of Economic Entomology*, **36**: 101-104.

Brattesten, L.B., and Metcalf, R.L., 1973. Synergism of carbaryl toxicity to natural insect populations. *Journal of Economic Entomology*, **66**: 1347-1348.

Brettell, J.H., and Burgess, W.M., 1973. A prelinary assessment of the effect of some insecticides on predators of cotton pests. *Rhod. Agric. Res.*, **20**: 77-84.

Brown, G.C., and Shanks, C.H., 1976. Mortality of two-spotted spioder mite predators caused by the systemic insecticide, carbofuran. *Environmental Entomology*, **5**: 1155-1159.

Brown, M.A., and Casida, J.E., 1983. Oxime ether analogs of pyrethoids help determine the contribution of pyrethroide esterases toward selective toxicity. *185th Nat. Mfg. Amer. Chem. Soc.*, Seattle, Wash.

Bull, D.L., and Ridgway, R.L., 1969. Metabolism of trichlorfon in animals and plants. *Journal of Agriculture and Food Chemistry*, **17**: 837-841.

Bull, D.L., Ridgway, R.L., Buxkemper, W.E., Schawarz, M., McGovern, T.P., and Sarmiento, R., 1973. Effect of synthetic juvenile hormone analogous on certain injurious and beneficial arthropods associated with cotton. *Journal of Economic Entomology*, **66**: 623-626.

Cate, J.R., Ridgway, R.L., and Lingren, P.D., 1972. Effects of systemic insecticides applied to cotton on adults of an ichneumonid parasite, *Campoletis perdistinctus*. *Journal of Economic Entomology*, **65**: 484-488.

Clausen, C.P., 1962. *Entomophagous Insects*. Hafner, New York, 688 pp.

Congdon, B.D., and Tanigoshi, L.K., 1983. Indirect toxicity of dimethoate to the predaceous mite *Euseius hibisci* (Chant) (Acari: Phytoseidae). *Environmental Entomology*, **12**: 933-935.

Crabtree, B., and Newsholme, E.A., 1975. Comparative aspects of fuel utilization and metabolism by muscle. In: *Insect Muscle*. (Eds.) P.N.R. Usherwood, Academic Press, New York. p. 405-500

Croft, B.A., 1977. Susceptibility surveillance to pesticides among arthropod natural enemies: modes of uptake and basic responses. *Z. Pflkrankh. Pfl;schutz.*, **84**: 140-157.

Croft, B.A., 1982. Developed resistance to insecticdes in apple arthropods: A key to pest control failures and successes in North America. *Entomologia Exerimentalis et Applicata*, **31**: 88-110.

Croft, B.A., 1990. *Arthropod Biological Control Agents and Pesticides*. New York: Wiley. pp. 723.

Croft, B.A., and Brown, A.W.A., 1975. Responses of arthropod natural enemies to insecticides. *Annual Review of Entomology*, **20**: 285-335.

Croft, B.A., and Morse, J.G., 1979. Recent advances on pesticide resistance in natural enemies. *Entomophaga*, **24**: 3-11.

Croft, B.A., and Mullin, C.A., 1984. Comparison of detoxification enzyme systems in *Argyrotaenia citrana* (Lep: Tortricidae) and the ectoparasite, *Oncophanes americanus* (Hym: Braconidae). *Environmenral Entomology*, **13**: 1330-1335.

Croft, B.A., and Strickler, K., 1983. Natural enemy resistance to pesticides: Documentation, characterization, theory and application. In: *Pest Resistance to Pesticdes*. (Eds.) G.P.Georghiou and T.Saito. New York, Plenum Press, p. 669-702.

Dempster, J.P., 1968. The sublethal effect of DDT on the rate of feeding by the ground beetle *Harpalus rufipes*. *Entomologia Expermientais et Applicata*, **11**: 51-54.

Desneux, N., Decourtye, A., and Delpuech, J., 2007. The sublethal effects of pesticides on beneficial Arthropods. *Annual Review Entomology*, **52**: 81–106.

Dmitrienko, V.K., 1979. The effect of pesticides containing chlorine and phosphorus on ants. *Ekologiya*, **1**: 53-60.

Doutt, R.L., and Hagan, K.S., 1950. Biological control measures against *Pseudococcus maritimus* on pears. *Journal of Economic Entomology*, **43**: 94-96.

Duffey, S.S., 1980. Sequestration of plat natural products by insects. *Annual Review of Entomology*, **25**: 447-477.

El-Banhawy, W.M., 1980. Comparison between the response of the predaceous mite Amblyseius brazilli and its prey *Tetranychous desertorum* to the different IGRs methoprene and Dimilin (Acari: Phytoseiidae, Tetranychidae). *Acarologia*, **21**: 221-227.

Elliot, W.M., and Way, M.J., 1968. The action of some systemic aphicides on the eggs of *Anthocoris neumorum* (L.) and A. confuses Reut. *Annals of Applied Biology*, **62**: 215-226.

Fast, P.B., 1964. Insect lipids: a review. *Memoirs of Entomological society of Canada*, 37, 50 pp.

Fast, P.B., 1970. Insect lipids. *Progressive Chemistry-Fats and other Lipids*, **11**: 181-242.

Franz, J.M., 1974. Testing of side-effects of pesticides on beneficial arthropods in the laboratory-a review. *Z. Pflkrankh. Pflschutz.*, **81**: 141-174.

Georghiou, G.P., 1972. The evolution of resistance to pesticides. *Annual Review of Ecology and Systemics*, **3**: 133-168.

Georghiou, G.P., 1981. *The Occurrence of Resistance to Pesticides in Arthropods: An Index of Cases Reported Through 1980*. Rome: FAO.

Georghiou, G.P., 1986. _Pesticide Resistance: Strategies and Tactics for Management_ Washington, DC, National Academy, pp. 14-43.

Gordon, H.T., 1961. Nutritional factors in insect resistance to chemicals. _Annual Review of Entomology_, **6:** 27-54.1

Gratwick, M., 1957. The contamination of insects of different species exposed to dust deposits. _Bulletin Entomological Research_, **48**: 741-753.

Hartley, G.S., and Graham-Bryce, I.J., 1980. _Physical Principles of Pesticide Behaviour._ Academic Press, New York. 1024 pp.

Hassan, S.A., Bigler, F., Bogenschutz, H., Brown, J.U., and Firth, S.I., _et al.,_ 1983. Results of the second joint pesticide testing programme by the IOBC/WPRS Working group "Pesticides and Beneficial Arthrpods". _Z. Angew. Entomol._, **95**: 151-158.

Hegazi, E.M., Rawash, I.A., El-Gayar, F.H., and Kares, E.A., 1982. Effect of parasitism by _Microplitis rufiventris_ Kok. on the susceotibility of _Spodoptera littoralis_ (Boisd.) larvae to insecticides. _Acta Phytopathol. Acad. Sci. Hungar._, **17**: 115-121.

Heyman, E., 1980. Carboxylestearses and amidases. In: _Enzymatic Basis of Detoxification._ (Ed.) W.B. Jakoby, Academic Press, New York, Vol. 2. p. 291-323.

Hollingworth, R.M., 1976. The biochemical and physiological basis of selective toxicity. In: _Insecticide Biochemistry and Physiology._ (Ed.) C.F. Wilkinson. Plenum Press, New York. p. 431-506.

Horn, D.J., 1983. Selective morta;ity of parasitoids and praedators of _Myzus persicae_ on collards treated with malathion, carbaryl or _Bacillus thuriengiensis. Entomologia Experimentalis et Applicata_, **34**: 2-08-211.

Horn, D.J., and Wadleigh, R.W., 1988. Resistance of aphid natural enemies to insecticdes. In: _Aphids, their Biology, Natural Enemies and Control._ (Eds.) P.Harrewinjin and A.K. Minks. Amsterdam, Elseweir. p. 337-347.

Hoy, J.B., and Dahlsten, D.L., 1984. Effects of malathion and Staley's bait on the behavior and survival of parasitic hymenoptera. _Environmental Entomology_, **13**: 1483-1486.

Hoy, M.A., 1985. Recent advances in genetics and genetic improvement of the Phytoseiidae. _Annual Review of Entomology_, **30**: 345-370.

Hsu, E.L., and Hsu, S.L., 1980. Studies on metabolism and fate of organochlorinated insecticides DDT, aldrin, dieldrin and y-HCH in the rice paddy model ecosystem. _NTU Phytopathology and Entomology_, **7**: 44-65.

Huffakar, C.B., 1971. The ecology of pesticide inferences with insect populations. In: _Agricultural Chemicals-Harmony or Discord for Food-People-Environment._ (Ed.) J.W. Swift. University of California, Division of Agricultural Science, Berkeley. p. 92-107.

Jones, D., Barlow, J.S., and Thompson, S.N., 1982. _Exeristes, Itoplectis, Apaereta, Brachymeria,_ and _Hyposotor_ species: _in vitro_ glyceride synthesis and regulation of fatty acid composition. _Experimental parasitology_, **54**: 340-351.

Khan, A.A., 2009. Relative toxicity of pesticides for safety of predatory coccinellids. *Annals of Plant Protection Sciences,* **17**: 307-311.

Khan, A.A., 2012. Comparison of spider diversity in relation to pesticide use in apple orchards of Kashmir. *Journal of Biological Control,* **26:** 1-10.

Khan, A.A., and Zaki, F.A., 2009. Effect of spray oils on ladybird beetles (Coccienellidae: Coleoptera). *Journal of Insect Science,* **22**: 59-62.

Kircher, H.W., 1982. Sterols and insects. In: *Cholesterol Systems in Insects and Animals.* (Eds.) J. Dupont, CRC Press, Boca Raton, Fla. p. 1-50.

Kiritani, K., and Kawahara, S., 1973. Food-chain toxicity of granular formulations of insectcicides to a predator, *Lycosa pseudoannulata,* of *Nephotettix cincticeps. Botyu-Kagaku,* **38**: 69-75.

Krieger, R.I., Feeny, P.P., and Wilkinson, C.F., 1971. Detoxification enzymes in the guts of caterpillars: an evolutionary answer to plant defences. *Science,* **172:** 579-581.

Kuhner, C. Klingaulf, F., and Hassan, S.A., 1985. Development of laboratory and semi-field methods to test the side effect of pesticides on *Diaretiells rapae* (Hymenoptera: Aphidiidae). *Meded. Facult. Landbouww., Rijks. Gent,* **50:** 531-538.

Lingappa, S.S., Starks, K.J., and Eikenbary, R.D., 1972. Insecticidal effect on *Lysiphlebus testaceipes,* a peasite of the greenbug at three developmental stages. *Environmental Entomology,* **1**: 520-521.

MacDonald, D.R., and Webb, F.E., 1963. Insectiocides and the spruce budworm. *Memoirs of Entomological Society of Canada,* **31**: 288-385.

Mani, M., and Krishnamoorthy, A., 1986. Susceptibility of *Telenomus remus* Nixon, an exotic parasitoid of *Spodoptera litura* to some pesticides. *Tropical Pest Management,* **32**: 49-51.

Metcalf, R.L., Mrach, R.B., and Maxon, M.G., 1955. Substrate preferences of insect cholinesterases. *Annals of Entomological Society of America,* **48**: 222-228.

Mullin, C.A., 1985. Detoxification enzyme relationships in arthropods of differeing feeding stratesies. In: *Bioregulators for Pest Control.* (Ed.) P.A. Hedin, Symposium Series 276, American Chemistry Society. p. 267-278.

Mullin, C.A., and Croft, B.A., 1985. An update on development of selective pesticides favouring arthropod natural enemies. In: *Biological Control of Agricultural Integrated Pest Management Systems.* (Eds.) M.A. Hoy and D.C. Herzog. Academic Press, New York, p. 123-150.

Mullin, C.A., Croft, B.A., Strickler, K., Matsumara, F., and Miller, J.R., 1982. Detoxification enzyme differences between a herbivorous and predatory mite. *Science,* **217**: 1270-1272.

Nakashima, M.J., and Croft, B.A., 1974. Toxicity of benomyl to the life stages of *Amblyseius fallacies. Journal of Economic Entomology,* **67**: 657-677.

Newson, L.D., Smith, R.F., and Whitcomb, W.H., 1976. Selective pesticides and selective use of pesticides. In: *Theory and Practice of Biological Control*. (Eds.) C.B. Huffakar and P.S. Messenger, Academic Press, New York, p. 565-591.

Novozhilov, K.V., Kamenkova, K.V., and Smirnova, I.M., 1973. Development of the parasite *Trissolcus grandis* Thoms. (Hymenoptera: Scelionidae) where organophosphorus insecticides are in use against *Eurygaster interriceps* Put. (Hemiptera: Scutelleridae). *Entomol. Obozr.*, **52**: 20-28.

Orr, D., 2009. Biological Control and Integrated Pest Management. In: *Integrated Pest Management: Innovation-Development Process*. (Eds.) R. Peshin and A.K. Dhawan. Netherlands, Springer Science. p. 207-240.

Panis, A., 1980. Damage caused by the Coccidae and Pseudococcidae (Homoptera, Coccoidae of citrus in France and special effects of some pesticides on the orchard entomocoenosis. *Fruits*, **35**: 779-782.

Plapp, F.W., 1981. The nature, modes of action and toxicity of insecticides. In: *Handbook of Pest Management in Agriculture*. (Ed.) D. Pimental, CRC Press, Boca Raton, Fla., vol. 3, p. 3-16.

Porres, M.A., McMurthy, J.A., and March, R.B., 1975. Investigations on leaf sap feeding by three species of phytoseiid mites by labeling with radioactive phosphoric acid ($H_2{}^{32}PO_4$). *Annals of Entomological Society of America*, **68**: 871-872.

Prestwich, G.D., Gayen, A.K., Phirwa, S., and Kline, T.B., 1983. 29-Flourophytosterols: novel pro-insecticides which cause death by dealkylation. *Biotechnology*, **1**: 62-65.

Price, P.W., 1981. Semiochemicals in evolutionary time. In: *Semiochemicals: Their Role in Pest Management*. (Eds.) W. Luckmann and R.L. Metcalf, Wiley Interscience, New York. p. 251-279.

Ripper, W.E., 1956. Effect of pesticides on the balance of arthropod populations. *Annual Review of Entomology*, **1**: 403-438.

Rosenheim, J.A., and Hoy, M.A., 1986. Intraspecific variation in levels of pesticide resistance in field populations of a parasitoid, *Aphytis malinus* (Hym: Aphelinidae): The role of past selection pressures. *Journal of Economic Entomology*, **79**: 1161-1173.

Roush, R.T., and McKenzie, J.A., 1987. Ecological genetics of insecticide and acaricide resistance. *Annual review of Entomology*, **32**: 361-380.

Roush, R.T., and Tabashnik, B.E. (Eds). 1990. *Pesticide Resistance in Arthropods*. New York, Chapman and Hall.

Sewall, D., and Croft, B.A., 1987. Chemotheuraputic and non-target side-effects of benomyl to the orange tortrix, *Argyrotaenia citrana* (Lepidoptera: Tortricidae) and braconid endoparasite *Apanteles aristoteliae* (Hymenoptera: Braconidae). *Environmental Entomology*, **16**: 507-512.

Shah, M.A., and Khan, A.A., 2013. Functional response-a function of predator and prey species. *The Bioscan*, **8**: 751-758.

Singh, D.S., and Rai, L., 1976. Properties of cholinesterasein herbivorous and carnivorous insects and its implication in developing specific insecticides. *Indian Journal of Entomology*, **51**-204-205.

Slansky, F., 1986. Nutritional ecology of endopaerasitic insects and their hosts: an overview. *Journal of Insect Physiology*, **32**: 255-261.

Slansky, F., and scriber, J.M., 1982. Selected bibliography and summary of quantitative food utilization by immature insects. *Bulletin of Entomological Society of America*, **28**: 93-99.

Stilig, P., 1990. Calculating the establishment rates of parasitoids in classical biological control. *American Entomologist*, **36**: 225-229.

Svoboda, J.A., and Lusby, W.R., 1986. Sterols of phytophagous and omnivorous species of hymenoptera. *Archives of Insect Biochemistry and Physiology*, **3**: 13-18.

Svoboda, J.A., Thompson, M.R., Robbins, W.E., and Kaplanis, J.N., 1978. Insect steroid metabolism. *Lipids*, **13**: 742-753.

Tabashnik, B.E., 1986. Evolution of pesticide resistance in predator-prey systems. *Bulletin of Entomological Society of America*, **32**: 156-161.

Tabashnik, B.E., and Croft, B.A., 1985. Evolution of pesticide resistance in apple pests and their natural enemies. *Entomophaga*, **30**: 37-49.

Tabashnik, B.E., and Johnson, M.W., 2005. Evolution of pesticide resistance in natural enemies. In: *Handbook of Biological Control: Principles and Applications of Biological Control*. (Eds). T.S. Bellows and T.W. Fischer. Academic Press, CA, USA, p. 673-689.

Takeda, S., Hukusima, S., and Yamada, H., 1965. Some physiological aspects of coccinellid beetles in relation to their tolerance against pesticide treatments. *Food and Agriculture Bullitin*, GIFU University, pp. 83-91.

Tamashiro, M., and Sherman, M., 1955. Direct toxicity of insecticides to oriental fruit fly larvae and their internal parasites. *Journal of Economic Entomology*, **48**: 75-79.

Tanaka, K., and Ito, Y., 1982. Different response in respiration between predaceous and phytophagous lady beetles (Coleopteran: Coccinellidae) to starvation. *Res. Popul. Ecol.*, **24**: 132-141.

Teargue, T.G., Horton, D.L., Yearian, W.C., and Phillips, J.R., 1985. Benomyl inhibition of Cotesia marginiventris ib four lepiodopterous hosts. *Journal of Entomological Science*, **20**: 76-81.

Teotia, T.P.S., and Tiwari, G.C.D., 1972. Toxicity of some important insecticides to the coccinellid predator *Coccinella septempunctata* Linn. *Labdev Journal of Science and Technology*, **10**: 17-18.

Theiling, K.M., 1987. The SELCTV database: the susceptibility of arthropod natural enem, ies of agricultural pests to pesticides. MS Thesis, Oregon State University, Corvallis, 170 pp.

Theiling, K.M., and Croft, B.A., 1988. Pesticide side-effects on arthropod natural enemies: A database summary. *Agriculture Ecosystems and Environment*, **21**: 191-218.

Thompson, S.N., and Barlow, J.S., 1974. The fatty acid composition of parasitic hymenoptera and its possible biological significance. *Annals of Entomological Society of America*, **67**: 627-632.

Thurling, D.J., 1980. Metabolic rate and life stage of the mite *Tetranychus cinnabarinus* from its host, the tobacco hornworm. *Annals of Entomological Society of America*, **65**: 547-550.

Velasco, L.R.I., 1985. Field parasitisation of *Apanteles plutellae* Kurdj. (Braconidae, Hymenoptera) on the diamondback moth of cabbage. *Phillipine Entotmol.*, **6**: 539-553.

Vinson, S.B., and Iwantsch, G.F., 1980. Host regulation by insect parasitoids. *Quarterly Review of Biology*, **55**: 143-165.

Waage, J.K., Hassal, M.P., and Godfray, H.C.J., 1985. The dynamics of pest-parasitoid-insecticide interactions. *Journal of Applied Ecology*, **22**: 825-838.

Welling, W., 1979. Dynamic aspects of insect-insecticide interactions. *Annual Review of Entomology*, **22**: 53-78.

Wilkes, F.G., and Weiss, C.M., 1971. The accumulation of DDT by the dragonfly nymph, Tetragoneuria. *Transactions of American Fisheries Society*, **100**: 222-236.

Yu, S.J., 1987. Biochemical defense capacity in the spined soldier bug (*Podisus maculiventris*) and its lepidopterous prey. *Insect Physiology and Biochemistry*, **28**: 216-223.

Yu, S.J., 1988. Selectivity of insecticides to the spined soldier bug (Hemiptera: Pentatomodae) and its lepidopterous prey. *Journal of Economic Entomology*, **81**: 119-122.

Zeleny, J., 1965. The effects of insecticides (fosfotion, Intration, Soldep) on some predators and parasites of aphids (*Aphis craccivora* Koch, *Aphis fabae* Scop.). *Rozpr Col. Acad. Ved.*, **75**: 3-73.

2015, Perspectives in Animal Ecology and Reproduction, Vol. 10   *Pages 65–71*
Editors: V.K. Gupta, Anil K. Verma and G.D. Singh
Published by: DAYA PUBLISHING HOUSE, NEW DELHI

Chapter 4

Sensory Structures on Caudal Lamellae and Tergites of Damsel Fly Larva *Ischnura senegalensis* (Zygoptera : Odonata) as Revealed by Scanning Electron Microscopy

Susmita Gupta*

Department of Ecology and Environmental Science,
Assam University, Silchar – 788 011, Assam, India

ABSTRACT

Cuticle dominates the arthropod ways of life and cuticular structures such as different types of sensilla provide a basis for understanding the relationship between their morphology and function. This paper describes the cuticular morphology of the 8th,9th and 10th tergites and caudal lamellae of a damselfly species *Ischnura senegalensis* (Charpentier) (Zygoptera : Odonata) as revealed by Scanning Electron Microscopy. The study revealed that the caudal gill is formed of a tough, chitinous cuticle. The mid rib of the caudal gill is armed with series of stout sensory structures sensilla basiconica (Sb1). The dorsal and ventral edges are equipped with spine like and hooklike sensilla basiconica (Sb 2). In the central portion of the lamellae there are a few very long pliable sensilla trichoidea (St 1). Among the last three tergites only 10th tergite is having spiny margin and posterior extremities of the three tergites are provided with several short *Sensilla basiconica* (Sb 1). In the lateral side of the 9th tergite a prominent hair plate is seen with

* *Corresponding author.* E-mail: susmita.au@gmail.com, susmita_au@rediffmail.com

sensilla trichoidea (St 2). Spoon shaped sensilla trichoidea (St.3) are also found in the joint of the caudal lamellae with the last abdominal segment. The role of these sensory structures has been discussed in the paper.

Keywords*: Cuticular morphology, Ischnura senegalensis, Scanning electron microscopy, Sensilla basiconica, Chemosensors.*

Introduction

Odonates are primarily aquatic insects and their life history is closely linked to specific aquatic habitats. This habitat specificity makes them a good indicator of wetland health. About 6,000 extant species are distributed all over the world. India is highly diverse with more than 500 known species. The order Odonata are divided into three groups, *viz.* damselflies (Zygoptera), Anisozygoptera and dragonflies (Anisoptera). The suborder Anisozygoptera is a living fossil with two species of which *Epiophlebia laidlawi* is known from Darjeeling. Three conspicuous leaf like structure at the posterior end of the abdomen distinguish larval damselflies from dragonflies and other aquatic insects. These leaf like structures are caudal gills or caudal lamellae and serve most important function like respiration. Damselfly larvae are important predators and prey in many freshwater communities. Beside respiration caudal lamellae play a role in swimming by allowing a larva to take advantage of its body size. Lamellae provide the primary propulsive surface area for swimming especially during escape, a supplemental surface area for respiration and perhaps a means of signalling conspecifics (Bose and Robinson, 2012).

Similar to aerobic animals, insects must obtain oxygen from their environment and eliminate carbon dioxide respired by their cells. This is gas exchange through series of gas filled tubes providing surface area for gaseous exchange (Respiration strictly refers to oxygen-consuming, cellular metabolic processes). Gas exchange occurs by means of internal air-filled tracheae. These tubes branch and ramify through the body. The finest branches called tracheloe contact all internal organs and tissues and are numerous in tissues with high oxygen requirements. Insects with closed tracheal systems have no functional spiracles, so gas exchange must occur by diffusion through the cuticle (Chapman *et al.,* 2004). This group includes two major respiratory categories: cutaneous respiration and tracheal gills (Eriksen *et al.,* 1996). Mayfly, dragonfly, and damselfly larvae have developed tracheal gills that can be important sites for gas exchange, but when external, may also serve other functions including locomotion (Burnside and Robinson, 1995) and predator escape (Robinson *et al.,* 1991). Cuticle dominates the arthropod ways of life (Neville, 1975). Cuticular structures such as different types of sensilla, their position and orientation provide a basis for understanding the relationship between their morphology and function. In India studies on the sensory structures on aquatic insect larvae are very scanty (Kapoor and Zacharia, 1983; Gupta *et al.,* 2000, 2002; Gupta and Gupta, 1996, 1998,1999; Gupta, 1998 a, b). There is only one study on scanning electron microscopy of different cuticular structures of Odonata of North East India (Gupta and Gupta,1999). In India there is no record of Scanning Electron Microscope study of caudal lamellae of Zygoptera. This paper describes the cuticular morphology of the 8th 9th and 10th tergites and caudal lamellae of a damselfly species *Ischnura senegalensis* (Charpentier) as revealed by Scanning Electron Microscopy and also discusses the ecological role.

Materials and Methods

Larva of *Ischnura senegalensis* (Charpentier) was collected from water bodies of Keibul Lamjao National Park, Manipur, North-east India. The insect was fixed for 2hrs. in 2.5 per cent glutaraldehyde and post fixed for 2hrs. in 2 per cent osmium tetroxide, both fixatives buffered with 0.1M cacodylate. It was dehydrated in graded concentration of acetone, dried in a critical point drying apparatus, mounted on brass stubs and coated with gold in a fine coat ion sputter JFC 1100 (Gupta *et al.,* 2000). The specimens were examined and photographed in Scanning Electron Microscope, JEOL JSM6360.

Results and Discussion

The species selected for study is *Ischnura senegalensis* (Charpentier) (Figure 4.1). The three important characters of the species are

a) Premental setae 4+4; palpal setae 5 and 5, a few spiniform setae present on the sides of prementum.

b) Movable hook medium sized.

c) Caudal lamellae duplex lamellae, tapering at both ends, light brown with darker tracheae.

The caudal gills are three in number, *viz.*, a single unpaired *median gill,* placed dorsally, and a pair of *lateral gills,* placed latero-ventrally, to right and left of the median gill respectively. The median gill is bilaterally symmetrical. The *lateral gills* are formed from the two *cerci* of the larva. Morphologically, each caudal gill is a hollow outgrowth of the body-wall, lined by a tough, chitinous integument or *cuticle,* beneath which lies a continuous *layer* of *hypoderm cells.* The space enclosed by the gill-walls always carries one or more large *longitudinal trachea,* from which more or less numerous branch trachea pass outwards to all points of the gill-wall. The trachea and blood-canals are supported in a more or less strongly developed meshwork of *alveolar tissue,* which fills up all the rest of the space in the interior of the gill (Tillyard, 1917 a).

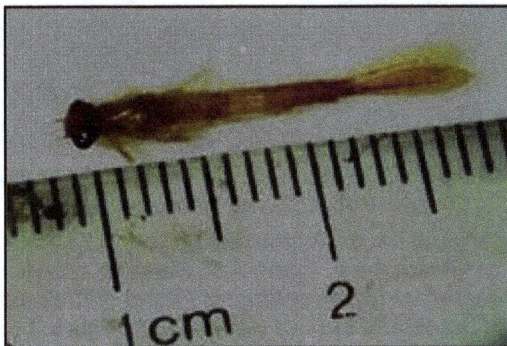

Figure 4.1: Image of *Ischnura senegalensis.*

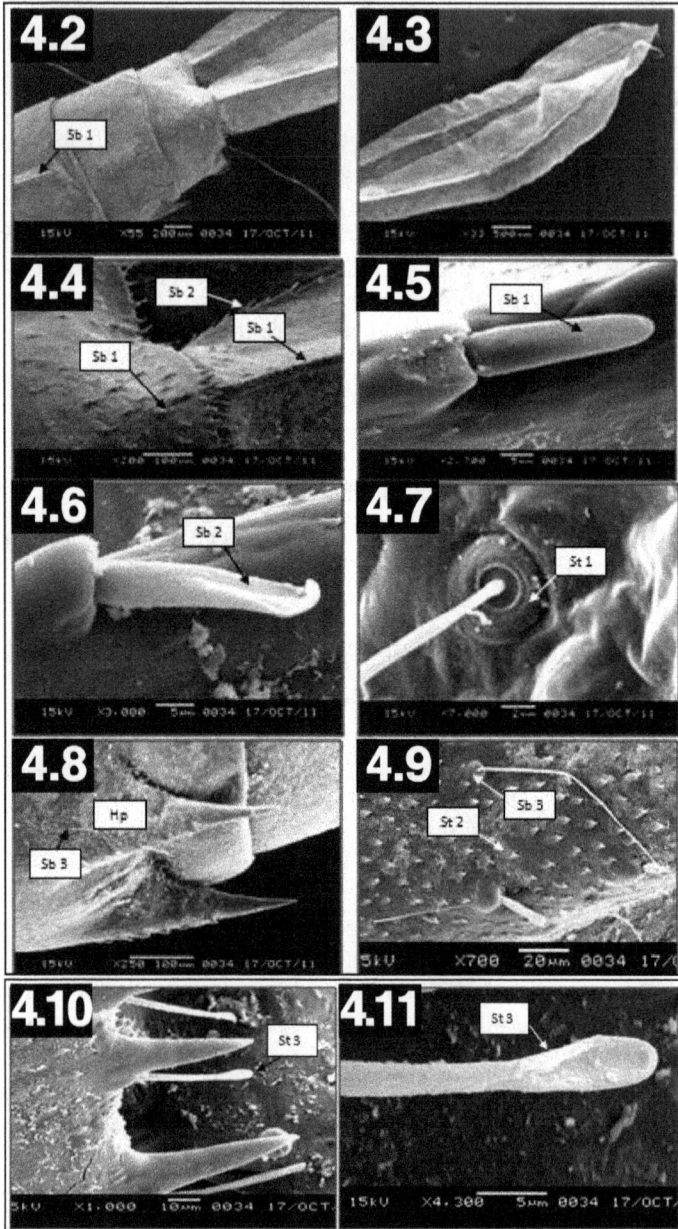

Figures 4.2–4.11: SEM Micrographs of Caudal Lamellae and Tergites of *Ischnura senegalensis*.

Figure 4.2: Part of tergites and caudal lamellae; Figure 4.3: Caudal lamellae; Figure 4.4: Joint of tergite and caudal lamellae; Figure 4.5: *Sensilla basiconica* (Sb 1) enlarged; Figure 4.6: *Sensilla basiconica* (Sb 2) enlarged; Figure 4.7: *Sensilla trichoidea* (St 1); Figure 4.8: Hairplate in the lateral side of tergite (Hp); Figure 4.9: Hairplate enlarged showing St 2 and Sb 3; Figure 4.10: Joint of tergite and caudal lamellae enlarged; Figure 4.11: *Sensilla trichoidea* (St 3) enlarged.

Figure 4.2 shows the cuticular surface structure of the 8th, 9th and 10th tergites and base of the caudal lamellae. Figure 4.3 is showing the leaf like caudal lamellae having tough hard anterior portion and fragile posterior portion with long pointed tip. The caudal gill of *Ischnura senegalensis* is formed of a tough, chitinous cuticle. In the anterior portion of the gill, the central area is clearly thickened while outer portion becomes thin and blade like. The mid rib of the caudal gills of *Ischnura senegalensis* is armed with series of stout sensory structures sensilla basiconica (Sb 1) (Figures 4.4 and 4.5). Similarly dorsal and ventral edges of the blade are equipped with some spine like and some hook like sensilla basiconica (Sb2) (Figures 4.4 and 4.6). This marks the position of the nodes along the outer edge of the gill. According to Tillyard (1917 a and b) the cuticle of gill varies in thickness from 5μ to as much as 15 μ, according to the type of gill and lamellar gills are stoutest and strongest. In the central portion of the lamellae there are a few very long pliable sensilla trichoidea (St 1) emerging from circular sockets. Sensilla trichoidea (St1) (Figure 4.7) with their long pliable setae and sockets that allow movement in all the directions are likely to be mechano- or contact chemosensilla (Frazier, 1985; Crouau and Crouau Roy,1991) and may be used for cleaning detrital or other debris from the cuticular surface (Honegger, 1977; Zack and Bacon, 1981).

Among the three tergites only 10th tergite is having spiny margin and posterior extremities of the three tergites are provided with several short sensilla basiconica (Sb 1) with bulbous base (Figure 4.4). In the tergite 8, similar to caudal lamellae, mid rib with series of sensilla basiconica (Sb 1) are found (Figure 4.2). In the lateral side of the 9th tergite a prominent hair plate near the lateral spine with unsocketed, swollen based, short, pointed sensilla trichoidea (St 2) is seen. Strewn among them are found a few socketed bulbous based long sensilla basiconica (Sb 3) (Figures 4.8 and 4.9). The trichoid sensilla (St 2) forming hairplate explain its function as position detectors providing information about the load and relative angles between body parts and joints (Graham, 1985; Schmidt and Smith, 1987). The spines found in the edge of the lateral spines are used as defence by the insect for escaping from the predator. Figure 4.10 shows the joint of the caudal lamellae with the last abdominal segment where along with the spinous margin spoon shaped sensilla trichoidea (St 3) are seen (Figures 4.10 and 4.11).

Conclusions

Different types of sensilla basiconica present in the caudal lamellae of *Ischnura senegalensis* are chemosensors and they are used for getting chemical cues of their immediate environment *i.e,* water. According to Crespo, (2007) damselfly nymphs have been shown to use infochemicals from a variety of predators and even learn to associate them with the presence of predators.

Acknowledgements

Author is thankful to the Head, Sophisticated Analytical Instrument Facility (SAIF), North Eastern Hill University, Shillong, Meghalaya for providing SEM facility. Thanks also go to Miss Kiranbala Takhelmayum for collecting the specimen.

References

Bose, A.P.H., and Robinson, B.W., 2012. Invertebrate predation predicts variation in an autotomy-related trait in larval damselfly. *Evolutionary Ecology,* (Published on line May 2012). DOI 10.1007/s10682-012-9581-3.

Burnside, C. A., and Robinson, J. V., 1995. The functional morphology of caudal lamellae in coenagrionid (Odonata: Zygoptera) damselfly larvae. *Biological Journal of the Linnean Society,* **114:** 155–171.

Chapman, L.G., Schneider, K.R., Apodaca, C., and Chapman, C. A., 2004. Respiratory Ecology of Macroinvertebrates in a Swamp–River System of East Africa. *Biotropica,* **36 (4):** 572-585.

Crespo, J.G., 2011. A review of chemosensation and related behaviour in aquatic insects. *Journal of Insect Science,* **11**: 62.

Crouau, Y., and Crouau-Roy, B., 1991. Antennal sensory organ of a troglobitic Coleoptera, Speonomus hydrophilus Jeannel, (Catopidae): an ultrastructural study by chemical fixation and cryofixation. *International Journal of Insect Morphology and Embryology,* **20**: 169-184.

Eriksen, C.H., Resh, V. H., and Lamberti, G.A., 1996. Aquatic insect respiration. *In* R. W. Merritt and K. W. Cummins (Eds.). An introduction to the aquatic insects of North America, 3rd edition, pp. 29–40. Kendall/Hunt Publishing Co., Dubuque, Iowa.

Frazier, J.L., 1985. Nervous system; sensory system- In: blum, M. S. (ed.) *Fundamentals of Insect Physiology:* 287-356. John Wiley and Sons, New York, 400 pp.

Graham, D., 1985. Pattern and control of walking in insects. *Advances in Insect Physiology,* **18**: 31-140.

Gupta, A., and Gupta, S., 1999. Sensilla on the antenna and leg of the larvae of *Crocothemis servilia* (Drury) (Anisoptera: Libellulidae). *Fraseria,* **5**: 29-32.

Gupta, S., and Gupta, A., 1996. Antennal sensilla of the subimago of *Cloeon* sp. (Ephemeroptera: Baetidae). *Journal of Animal Morphology and Physiology,* **43**: 137-138.

Gupta, S., and Gupta, A., 1998. Cuticular sensory structures on the cerci of the nymphs and adults of mayfly *Cloeon* sp. (Ephemeroptera: Baetidae). *Proceedings of the National Academy of Sciences, India,* **68 (B)** III and IV: 319-321.

Gupta, S., and Gupta, A., 1999. Morphology of the mouthparts and its relation to feeding of *Cloeon* sp. *Geobios New Records,* **18**: 66-68.

Gupta, S., 1998a. External morphology of the antennal sensilla of the imago of *Cloeon* sp. (Ephemeroptera: Baetidae) by Scanning Electron Microscopy. *Journal of Animal Morphology and Physiology,* **45 (1 and 2):** 142-144.

Gupta, S., 1998b. External morphology of the interommatidial hairs of *Cloeon* sp. (Ephemeroptera: Baetidae) as revealed by SEM. *Geobios New Reports,* **17**: 63-67.

Gupta, S., Gupta, A., and Meyer-Rochow, V.B., 2000. Post-embryonic development of the lateral eye of *Cloeon* sp. (Ephemeroptera: Baetidae) as revealed by Scanning Electron Microscopy. *Entomologica Fennica,* **11**: 89-96.

Gupta, S., Michael, R.G., and Gupta, A., 2002. Scanning electron microscopic studies on the post embryonic development of the dorsal Eye of *Cloeon* sp. (Ephemeroptera: Baetidae). *Entomon,* **27:** 447-453.

Honegger, H. W., 1977. Interommmatidial hair receptor axons extending into the ventral nerve cord in the cricket *Gryllus campestris. Journal of Comparative Physiology,* **130**: 49-62.

Kapoor, N. N., and Zachariah, K., 1983. Ultrastructure of the sensilla of the stone fly nymph *Thaumatoperla alpina* Burns and Neboiss (Plecoptera: Eustheniidae). *International Journal of Insect Morphology and Embryology,* **12 (213)**: 157-168.

Neville, A.C. 1975. Biology of arthropod cuticle.- Springer, Berlin.

Robinson, J.W., Hayworth, D. A., and Harvey, M.B., 1991. The effect of caudal lamellae loss on swimming speed of the damselfly *Argia moesta* (Hagen) (Odonata: Coenagrionidae). *American Midland Naturalist,* **125**: 240–244.

Roy, B., and Gupta, S., 2002. Impact of paper mill waste on *Channa punctatus. Journal of Industrial Pollution Control,* **18 (2):** 231-235.

Schmidt, J.M., and Smith, J.J.B., 1987. The external sensory morphology of the legs and hairplate system of female *Trichogama minutum* Riley (Hymenoptera: Trichogrammatidae). *Proceedings of the Royal Society of London,* B, **232**: 323-366.

Tillyard, R.J., 1917 a. On the morphology of the caudal gills of the larva of zygopterid dragonflies. *Proceedings of the Linnean Society of New South Wales, 1917, Vol. xlii., Part 3.*

Tillyard, R. J., 1917 b. The biology of dragonflies. Cambridge Univ. Press, Cambridge. 396 pp.

Zack, S., and Bacon, J., 1981. Interommatidial sensilla of the praying mantis: their central neural projections and role in head- cleaning behaviour. *Journal of Neurobiology,* **12**: 55-65.

2015, Perspectives in Animal Ecology and Reproduction, Vol. 10   Pages *73–82*
Editors: **V.K. Gupta, Anil K. Verma and G.D. Singh**
Published by: **DAYA PUBLISHING HOUSE, NEW DELHI**

Chapter 5

Synergistic Effect of NPV, Azadirachtin and Titanium Dioxide (Nanoparticale) on the Larvicidal and Pupicidal Effect of *Helicoverpa armigera*

T. Nataraj*[1,5], K. Murugan[1], P. Madhiyazhagan[1],
A. Nareshkumar, J. Subramaniam[1], P. Mahesh kumar[1]
Jiang-Shiou Hwang[3], R. Chandrasekar[4],
P. Sakthivel[5] and O.P. Sharma[5]

[1]*Division of Entomology, Department of Zoology,*
Bharathiar University, Coimbatore – 641 046, Tamil Nadu, India
[2]*Division of Entomology, Department of Zoology,*
Periyar University, Salem, India
[3]*Institute of Marine Biology, National Taiwan Ocean University,*
Keelung 202-24, Taiwan
[4]*Department of Biochemistry and Molecular Biophysics,*
238 Burt Hall, Kansas State University, Manhattan 66506, KS, USA
[5]*National Institute of Plant Health Management,*
Ministry of Agriculture, Hyderabad, India

ABSTRACT

The microbial Insecticides of Nucleopolyhedroviruses (NPVs) is Insect – Specific baculoviruses provide to control of *Helicoverpa armigera*. The NPVs had 2 major disadvantages were slow to kill the target insect and degradation by the ultraviolet, aim of this study to require this disadvantages of NPVs. In the present study has use Titanium dioxide (TiO_2) and Azadirachtin (AZA) for improve NPVs

* *Corresponding author.* E-mail: bionataraj@gmail.com

activity. The Photo stabilized TiO_2 reflects ultraviolet light and so could be expected to protect the OBs of NPVs, from degradation by sunlight and AZA bitter compound was control insect food utilization. The result was carried individual and combined treatment of NPV, TiO_2 and AZA Insecticides was evident replace the NPVs disadvantages to show 100 per cent mortality in I, II, III instar larvae and pupa of *Helicoverpa armigera* in lower concentration.

Keywords: *Nucleopolyhedroviruses, Helicoverpa armigera, Titanium dioxide, Azadirachtin.*

Introduction

The cotton bollworm, *Helicoverpa armigera* (Hubner), is a pest of major importance in India in most agroecological zones ranging from Andaman Nicobar Islands to Jammu and Kashmir (Singh *et al.,* 2002). Crop losses of 75–100 per cent in chickpea (Lal, 1996) and 57–80 per cent in cotton (Gupta, 1999) have been recorded. The estimated monetary loss in Tamil Nadu was Rs.20.12 million USD on different crops (Jayaraj, 1990). In Punjab, Haryana and Rajasthan the damage due to the pest on cotton was estimated at Rs.296.93 million USD (Harish, 2002). The ability of this pest to adopt transient habitats in a short span of time accelerated the excessive use of pesticides resulting in development of resistance to various classes of insecticides (Armes *et al.,* 1992) and outbreaks in several areas since 1983–1984 (Singh *et al.,* 2002). This necessitated the search for ecofriendly alternatives, especially microbial insecticides in view of their high specificity, potential activity and environmental safety. Biopesticides based on the baculovirus group, the nucleopolyhedrovirus (NPV), offer a great scope against *H. armigera*. Successful utilisation of *H. armigera* NPV (HaNPV) under field conditions was reported on chickpea (Rabindra *et al.,* 1989) and cotton (Sathiah and Rabindra, 2001).

Their Major disadvantage is that they are relatively slow to kill the target insect and particularly the susceptibility of their occlusion bodies (OBs) to degradation by the ultraviolet (UV) portion of sunlight (jaques, 1977). The viruses are particularly sensitive so quickly decreased at temperatures $> 60°C$. Thus extensive research has use Titanium dioxide (TiO_2) and Azadirachtin (AZA).

Neem (*Azadirachta indica* A. Jusieu) is an evergreen, multipurpose tree found in the Indian subcontinent and southeast Asian countries Every part of the tree is useful in one way or the other (Nutan Kaushik *et al.,* 2007). Azadirachtin is a chemical compound belonging to the limonoids. It is a secondary metabolite present in the neem Tree seeds. The molecular formula is $C_{35}H_{44}O_{16}$. AZA is a highly oxidized tetranortriterpenoid which boasts a plethora of oxygen functionality, comprising an enol ether, acetal, hemiacetal and tetra substituted oxirane as wellas a variety of carboxylic esters. It is classified among the plant secondary metabolites.

Titanium dioxide (100nm) was first produced commercially in 1923. It is a multifaceted compound. TiO_2 reflects UV, It also potent photo catalyst that can break down almost any organic compound and a number of companies are seeking to capitalize on TiO_2's reactivity by developing a wide range of environmentally beneficial products (Apr. 2001, Environmental health perspectives). Hence, in the present study status of *H. armigera* on cotton an attempt has been made to study the impact of NPV, AZA and TiO_2 on mortality of *H. armigera*.

Materials and Methods

Laboratory Mass Culture of *H. armigera*

Helicoverpa armigera larvae were collected from the cotton and sunflower fields in and around Coimbatore and were cultured in the laboratory and fed with *Ricinius communis* leaves *ad libitum* at 27±1—C; 10: 14 L.D: 75 per cent Rh. Pre-pupa of *H. armigera* were separated and provided with vermiculate clay which is a good medium for pupation. Pupae of *H. armigera* were kept on cotton in Petri dishes inside an adult emergence cage. The emerging moths were fed with 10 per cent sucrose solution fortified with a few drops of vitamin mixture (MULTDEC drops) to enhance oviposition.

Moths in the ratio of one male to two females were allowed inside oviposition cages containing the adult food mentioned above. The egg cage of *H. armigera* was covered with white muslin cloth for egg laying. The egg clothes were removed daily and surface sterilized using 10 per cent formaldehyde solution to prevent virus infection. The egg clothes were moistened and kept in a plastic container for the eggs to hatch. This process facilitated uninterrupted supply of test insects.

Preparation of NPV, TiO_2 and Azadirachtin

Preparation of NPV

NPV : HaNPV containing a suspension concentrate containing atleast 1 LE/ml of the viral particles was initially obtained from Project Directorate of Biological Control (ICAR), Bangalore, India. Virus concentrations were determined by three Independent counts of polyhedral occlusion bodies (POB/ml) made with a hemocytometer under a phase-contrast microscope (400X). Various concentration of viral suspensions were made by diluting with distilled water. For isolation of OBs, cadavers were treated with 0.1 per cent sodium dodeycl sulphate (SDS) (1 ml per cadaver) for one night at 4°C and filtered through five layers of cheesecloth. OBs were pelleted by centrifugation at 3600 g for 10 minutes at room temperature in 50 ml centrifuge tubes. The pellet was resuspended in 0.5 per cent SDS and centrifugation and resuspension repeated with 0.3 M NaCl before final resuspension of OBs in distilled water (O'Reilly *et al.*, 1992).

Preparation of Azadirachtin

5.265 mg of Azadirachtin was dissolved in 50 ml of distilled water to obtain stock solution of 100 ppm concentration. Desired concentrations ranging from 0.01 to 12 ppm were prepared from the stock solutions using distilled water. Combined treatments were prepared by mixing equal volume of dilutions of the two treatments.

Preparation of Titanium Dioxide

Photostabilized TiO_2 (Nano Tek-titanium dioxide) was obtained from Manufacturer of industrial chemicals (Delhi). This material is water dispersible, and was obtained as a mixture of water and coated TiO_2 particles (50 per cent water by weight). Nonphotostabilized TiO_2 (dry powders) was also obtained from Nanophase Manufacturer of industrial chemicals (Delhi). Photostabilized TiO_2 was also obtained from Indian Institute of Chemical Technology (Hyderabad, India). This material was obtained dry, and is not dispersible in water without added wetting agents.

Treatments

The Treated leaves were air dried in shade and fed to the larvae for 24h. Fresh uncontaminated leaves were supplied daily thereafter. Another set of 30 larvae of each instar were fed with leaves treated with sterile water which served as control. Each treatment was replicated five times. The biological parameters such as larval, pupal and adult durations were recorded. In addition, the fecundity was assessed by counting the number of eggs laid during the life span. Food utilization efficiency measures were calculated by the method of (Waldbauer, 1968; Murugan and George, 1992).

Larval Mortality

Larval mortality was recorded and the percentage mortality was calculated by counting the number of dead larvae and compared to the number of larvae introduced. NPV infected cadavers exhibited oozing of the body contents, transparency and stretching of the larval body. The larvae that were unable to move and feed were confirmed dead. Data were corrected for control mortality Abbott (1925) using the formula: The lethal concentration of 50 per cent and 90 per cent (Lc_{50} and Lc_{90}) were used to measure difference between test samples.

Observed mortality in treatment-Observed

$$\text{Corrected mortality} = \frac{\text{Mortality in control}}{100 - \text{Control mortality}} \times 100$$

$$\text{Percentage of mortality} = \frac{\text{Number of dead larvae}}{\text{Number of larvae introduced}} \times 100$$

Lc_{50}, Lc_{90} were calculated from toxicity data by using probit analysis (Finney, 1971).

Statistical Analysis

Data analysis all replicates should be pooled for analysis. LC_{50} and LC_{90} values were calculated from a log dosage-probit mortality using computer software programs; the 95 per cent confidential limit was calculated to the test concentrations and also DMRT (Duncan Multiple Range Test) was calculated (Alder and Roessler, 1964).

Results

Table 5.1 reveals the percentage of larvicidal and papicidal activity of NPV at various concentration against *H. armigera*. At 1 ppm of NPV treatment. The Larval mortality observed for different larval instars were 49 per cent in I instars, 46 per cent in II instars, 40 per cent in III instars, 33 per cent in fourth instars, 28 per cent in V instars, 20 per cent in VI instars and 43 per cent of pupa. At 2 ppm The Larval mortality at different instars were increased 79 per cent in I instars, 72 per cent in II instars, 70 per cent in III instars, 62 per cent in IV instars, 52 per cent in V instars, 49 per cent in VI and 75 per cent in pupa. At 3 ppm The larval mortality at different instars were highly increased 100 per cent in I instars,100 per cent in II instars, 96 per cent in III instars, 94 per cent in IV instars, 94 per cent in V instars, 90 per cent in VI and 96 per cent in pupa.

Table 5.1: Larvicidal and Pupicidal Activity of NPV at Various Concentrations against *H. armigera*

Larval and Pupal Stages	Per cent of Larval and Pupal Mortality			LC_{50} and LC_{90} (per cent)	Regression Equation	95 per cent Confidence Limit		Chi Square Value
	Concentration of NPV (ppm)					LC	UCL	
						LLC_{50}	LC_{50}	
	1	2	3			(LC_{90})	(LC_{90})	
I	49	79	100	0.26532 (0.61201)	3.69567	0.18885 (0.54006)	0.3178 (0.73685)	0.584
II	46	72	100	0.34519 (0.65577)	4.12639	0.29491 (0.58489)	0.38884 (0.77245)	0.595
III	40	70	96	0.3293 (0.80278)	2.70668	0.24894 (0.71240)	0.3926 (0.94182)	0.399
IV	33	62	94	0.40287 (0.88315)	2.66831	0.33278 (0.78857)	0.46381 (1.02492)	0.230
V	28	52	94	0.46668 (0.91699)	2.84598	0.40645 (0.82494)	0.52421 (1.05123)	0.226
VI	20	49	90	0.53066 (0.98634)	2.81239	0.47201 (0.89162)	0.58955 (1.12210)	0.422
Pupa	43	75	96	1.26532 (2.68491)	0.90276	0.99898 (1.94717)	2.05551 (5.13200)	0.200

Table 5.2 Reveals The percentage of larvicidal and papicidal activity of NPV+AZA at various concentration against *H. armigera*. At 1 +0.25 ppm of NPV+AZA treatment. The Larval mortality observed for different larval instars were 52 per cent in I instars, 50 per cent in II instars, 49 per cent in III instars, 47 per cent in fourth instars, 47 per cent in V instars, 45 per cent in VI instars and 50 per cent of pupa. At 2 + 0.5 ppm the larval mortality at different instars were increased 85 per cent in I instars, 82 per cent in II instars, 79 per cent in III instars, 76 per cent in IV instars, 75 per cent in V instars, 70 per cent in VI and 78 per cent in pupa. At 4 + 1.0 ppm the larval mortality at different instars were highly increased 100 per cent in I instars,100 per cent in II instars, 100 per cent in III instars, 99 per cent in IV instars, 98 per cent in V instars, 96 per cent in VI and 100 per cent in pupa.

Table 5. 3 shows the synergistic effect of larvicidal and pupicidal activity of NPV + TIO_2 + AZA at various concentration against *H. armigera*. At 1 + 0.5 + 0.25 ppm of NPV + TIO_2 + AZA treatment. The larval mortality observed for different larval instars were 51 per cent in I instars, 47 per cent in II instars, 40 per cent in III instars, 34 per cent in fourth instars, 29 per cent in V instars, 25 per cent in VI instars and 47 per cent

Table 5.2: Larvicidal and Pupicidal Activity of NPV and AZA at Various Concentrations against *H. armigera*

Larval and Pupal Stages	Per cent of Larval and Pupal Mortality Concentration of NPV+AZA (ppm+ppm)			LC_{50} and LC_{90} (per cent) (LC_{90})	Regression Equation	95 per cent Confidence Limit LC LLC$_{50}$ (LC_{90})	UCL LC$_{50}$ (LC_{90})	Chi Square Value
	1.00+0.25	2.00+0.5	4.00+1.0					
I	52	85	100	1.20116 (2.77200)	0.81548	0.79449 (2.44730)	1.45788 (3.35457)	0.127
II	50	82	100	1.27409 (2.92075)	0.77827	0.88557 (2.57755)	1.53128 (3.52634)	0.28
III	49	79	100	1.32659 (3.06044)	0.73913	0.94426 (2.70012)	1.58899 (3.68427)	0.584
IV	47	76	99	1.37594 (3.39062)	0.63611	0.96948 (2.99510)	1.66943 (4.03284)	0.009
V	47	75	98	1.48327 (4.03693)	0.57324	0.91181 (3.17304)	1.68337 (4.26965)	0.046
VI	45	70	96	1.48327 (4.03693)	0.50185	1.00921 (3.55917)	1.83386 (4.78965)	0.02
Pupa	50	78	100	1.30646 (3.10955)	0.71075	0.90548 (2.74008)	1.57977 (3.74589)	0.838

Table 5.3: Larvicidal and Pupicidal Activity of NPV, AZA and TiO_2 at Various Concentrations against *H. armigera*

Larval and Pupal Stages	Per cent of Larval and Pupal Mortality Concentration of NPV+ TiO_2+ AZA (ppm+ppm+ppm)			LC_{50} and LC_{90} (per cent) (LC_{90})	Regression Equation	95 per cent Confidence Limit LC LLC$_{50}$ (LC_{90})	UCL LC$_{50}$ (LC_{90})	Chi Square Value
	1+ 0.5+ 0.25	1+0.5+ 0.25	1+0.5+ 0.25					
I	51	82	100	1.51045 (3.44569)	0.66222	1.0616 (3.06613)	1.35512 (3.04247)	0.836
II	47	79	100	1.67188 (3.58791)	0.66886	1.27965 (3.20391)	1.53977 (3.18309)	1.267
III	40	76	100	1.90895 (3.64866)	0.71606	1.59477 (3.32316)	1.69362 (3.51828)	0.027
IV	34	70	98	2.1614 (4.23396)	0.61834	1.83787 (3.82137)	1.93045 (3.84460)	0.115
V	29	61	97	2.46705 (4.59897)	0.60113	2.1638 (4.16970)	2.10604 (3.87871)	0.033
VI	25	54	97	2.69082 (4.74417)	0.62413	2.41031 (4.32035)	2.24426 (3.98467)	0.127
Pupa	47	79	100	1.69188 (3.58791)	0.66889	1.27965 (3.20391)	1.47081 (3.10646)	0.892

of pupa. At 2 + 1.0 + 0.5 ppm. The larval mortality at different instars were increased 82 per cent in I instars, 79 per cent in II instars, 76 per cent in III instars, 70 per cent in IV instars, 61 per cent in V instars, 54 per cent in VI and 79 per cent in pupa. At 4 + 1.5 + 1.0 ppm. The larval mortality at different instars were highly increased 100 per cent in I instars, 100 per cent in II instars, 100 per cent in III instars, 98 per cent in IV instars, 97 per cent in V instars, 97 per cent in VI and 100 per cent in pupa.

Discussion

The consumption of neem and NPV resulted in retarded growth and development of the larvae, similar to the results obtained using whole neem extracts (Murugan *et al.,* 1999). Similarly Senthil Nathan and Kalaivani (2005) reported the feeding (CI) and growth (RGR) were affected more by the combined treatment of AZA and NPV than individual treatment.

It may be concluded from this study that extended larval period is coupled with lower consumption and utilization, which is more likely due to longer retention of food in the gut for maximization of AD to meet the increased demand of nutrients by multiplying virus and the host for tissue growth (Rath *et al.,* 2003; Reynolds *et al.,* 1985). A similar result was reported by Shapiro *et al.* (1994) using aqueous neem extract in combination with NPV. Food consumption by *H. armigera* larvae infected with NPV at the beginning of their fifth instar was greatly inhibited. Our findings agree with the result of Beach and Todd (1988) who showed reduction in consumption of soybean foliage by NPV-infected *Pseudoplusia includes* (Walker) and *Anticarsia gemmatalis* (Hubner). Decreased values of food consumption and utilization in NPV-infected *H. armigera* have been reported by Subrahmanyam and Ramakrishnan (1981).

The NPV infection was not observed in tissues other than the midgut epithelium until 20 h. It is thus unlikely that infectious parental nucleocapsid pass the midgut epithelium and basal membrane to infect other larval tissues (Tanada *et al.,* 1975; Flipsen *et al.,* 1995). In the early stage of virus multiplication, biosynthesis of virus proteins and nucleic acids would likely to cause changes in the metabolic activity of the cells. A large number of nucleocapsids in the cylinder cells of the midgut could be made 30 h after NPV infection. Infected viruses multiply and produce many polyhedra in the midgut and finally the cells are broken and the polyhedra are freed into the intestinal tract (Xeros *et al.,*1952; Martignoni *et al.,*1969).

Our results show that NPV affected the gut physiology of *H. armigera* at several doses. The NPV tested in the present study are reported to have less environmental impact than most commercial chemical insecticides. In addition, chemical insecticides are toxic to beneficial insects like natural parasitoids (Mumuni *et al.,* 2001). The larval mortality due to the disease was more rapid in older instars than in early first- and second-instar larvae (Narayanan, 1985). Patil *et al.* (1989) found that chlorpyrifos (0.1 per cent) produced 96.7 per cent mortality of the fourth instar *H. armigera* larvae, while our study similarly mortality results also showed in Tables 5.1 and 5.2. So our results were also found between the limits reported previously.

While TiO_2 reflects UV, it also has a property that would be undesirable in a formulation of an NPV: it is a photocatalyst. In the presence of water and light, TiO_2 catalyzes the formation of hydrogen peroxide (H_2O_2) (Hoffman *et al.,* 1995). Reactive oxygen species, such as peroxides, are known to inactivate OBs of NPVs (Ignoffo and Garcia, 1994; Hunter-Fujita *et al.*, 1998), though the role of these chemicals in inactivation of OBs on plant surfaces in the field has not been determined. Thus, TiO_2 has the potential to reduce the activity of OBs even though it can reflect UV. The catalytic activity of TiO_2 can be eliminated, however, by "photostabilizing," or coating of the TiO_2 particles with materials such as silica or stearate (Fairhurst and Mitchnick, 1997). Herein, we report the results of tests of photostabilized TiO_2 as a UV protectant for OBs of the NPV of the *H. armigera* (HaNPV). Table 5.3 shows combined treatment of NPV+AZA+TiO_2 was good evident of TiO_2 to protect OBs of the NPV and increase NPVs life span for kill the target organism.

Thus, based on the results of the present investigation, it can be inferred that use of NPV, TiO_2 As well as neem has good scope in the management as well as IPM strategy against *H. armigera*. Use of NPV and neem products will not only be economically viable but will also be environmentally sustainable in the long run.

Conclusion

The present study it is also proved that the biodegradable property of NPV is highly reduced and it extends its property for long periods. Form this we conclude that the combined properties of NPV and TiO_2 reduce the repeated application of biopesticides and reduce the cost and time taken for the farmers.

References

Abbott, W.S. 1925. A method of computing the effectiveness of an insecticides. *Journal of Economic Entomology*, **18**: 265-267.

Alder, H.L., E.B. Roessler 1964. Introduction to probability and statistics, 3rd Ed., W.H. Freeman and Company. p82-84.

Armes N.J., Jadhav D.R., Bond G.S., and King A.B.S. 1992. Insecticide resistance in *Helicoverpa armigera* in Southern India. *Pesticide Science*, **34**: 355–364.

Beach, R.M., 1988. Todd, Discrete and interactive effects of plant resistance and nuclear polyhedrosis viruses for suppression of soybean looper and velvet bean caterpillar (Lepidoptera: Noctuidae) on soybean. *J. Econ. Entomol.*, **81:** 684–691.

Fairhurst, D., and M.A. Mitchnick 1997. Particulate sun blocks: general principles. In: N.J. Lowe, N.A. Shaath and M.A. Pathak (eds), *Sunscreens: development, evaluation, and regulatory aspects* (2nd edn., revised and expanded). Cosmetic Science and Technology Series 15. Marcel Dekker, New York. pp. 313–352.

Finney, D.J. 1971. Probit Analysis, Cambridge University Press. p. 333.

Flipsen, J.T.M., J.W.M. Martens, M.M. Van Oers, J.M. Vlak, and J.W.M. Van Lent 1955. Passage of *Autographa californica* nuclear polyhedrosis virus through the midgut epithelium of *Spodoptera exigua* larvae. *Virology*, **208** 328–335.

Gupta G.P. 1999. Use of safe chemicals in cotton IPM system: an overview. *Cotton Research Development*, **13**: 56.

Harish D. 2002. Cotton crop loss seen at Rs.1, 364 cr. Available at http://www.kisanwatch.org/eng/news/nov/11200/nws2.htm

Hoffman, M.R., S.T. Martin, W. Choi and D.W. Bahnemann 1995. Environmental applications of semiconductor photocatalysis. *Chem. Rev.,* **95**: 69–96.

Hunter-Fujita, F.R., P.F. Entwistle, H.F. Evans and N.E. Cook 1998. *Insect viruses and pest management.* Wiley, Chichester. xii + 620 pp.

Ignoffo, C.M., D.L. Hostetter, P.P. Sikorowski, G. Sutter and W.M. Brooks 1977. Inactivation of representative species of entomopathogenic viruses, a bacterium, fungus, and protozoan by an ultraviolet light source. *Environ Entomol.,* **6**: 411–415.

Jaques, R.P. 1977. Stability of entomopathogenic viruses. In: C.M. Ignoffo and D.L. Hostetter (eds), *Environmental stability of microbial insecticides.*Misc. Publ. Entomol. Soc. Am., 10. pp. 99–116.

Jayaraj S. 1990. The problem of *Heliothis* in India and its integrated management. In: Heliothis Management. Proceedings of National Workshop. Tamil Nadu Agricultural University, Coimbatore: 1–16.

Lal O.P. 1996. An outbreak of pod borer *Heliothis armigera* (Hubner) on chickpea in eastern Uttar Pradesh (India). *Journal of Entomological Research,* **20**: 179–181.

Martignoni, M.E., P.J. Iwai, K.M. Hughes, and R.B. Addison 1969. A cytopasmic polyhedrosis of *Hemerocampa pseudotsugata. J. Invert. Pathol.,* **13**: 15–18.

Mumuni, A., B.M. Shepard, and P.L. Mitchel 2001. Parasitism and predation on eggs of *Leptoglossus phyllopus* (L.) (Hemiptera: Coreidae) in cowpea: impact of endosulfan sprays. *J. Agric. Urban Entomol.,* **18**: 105–115.

Murugan, K., and Ancy George 1992. Feeding and Nutritional Influence on Growth and Reproduction of *Daphnis nerii* (Linn.) (Lepidoptera: Sphingidae). *J. Insect Physiol.,* **38**: 961-967.

Murugan, K., S. Sivaramakrishnan, N. Senthil Kumar, D. Jeyabalan, and S. Senthil Nathan 1999. Potentiating effect of neem seed kernel extract and neem oil on nuclear polyhedrosis virus (NPV) on *Spodoptera litura* Fab. (Lepidoptera: Noctuidae). *Insect Sci. Appl.,* **19**: 229–235.

Narayanan, K. 1985. Susceptibility of *Spodoptea litura* (F.) to a granulosis virus. *Curr. Sci. India,* **54**: 1288–1289.

Nutan Kaushik, B., Gurudev Singh, U. K., Tomar, S. N., Naik Satya Vir, S. S., Bisla, K. K., Sharma, S. K., Banerjee and Pramilla Thakkar. Regional and habitat variability in azadirachtin content of Indian neem (*Azadirachta indica* A. Jusieu).

O'Reilly, D.R. 1997. Auxiliary genes of baculoviruses pp. 267-301 in: (Ed.) The Baculoviruses, LK Miller, Plenum Press, New York.

Patil, R.S., S.D. Bhole, S.P. and Patil 1958. Laboratory evaluation of some chemicals for control of swarming caterpillar (*Spodoptera litura*) infesting rice (*Oryza sativa*) in Konkan region of Maharashtra. *Indian J. Agr. Sci.,* **59**: 381–383.

Rath, S.S., B.C. Prasad, and B.P.R.P. Sinha 2003. Food utilization efficiency in fifth instar larvae of *Antheraea mylitta* (Lepidoptera: Saturniidae) infected with *Nosema* sp. and its effect on reproductive potential and silk production. *J. Invertebr. Pathol.,* **83**: 1–9.

Reynolds, S.E., S.F. Nottingham, and A.E. Stephens 1985. Food and water economy and its relation to growth in fifth instar larvae of tobacco hornworm, *Manduca sexta. J. Insect Physiol.,* **31**: 119–127.

Sathiah N., and Rabindra R.J. 2001. Field efficacy of certain improved formulations of the nuclear polyhedrosis virus against the bollworm, *Helicoverpa armigera* (Hubner) on cotton. In: Proceedings of Symposium on Biocontrol Based Pest Management for Quality Crop Protection in the Current Millenium. Punjab Agricultural University, Ludhiana, India: 82.

Senthil Nathan, and S., K. Kalaivani 2005. Efficacy of nucleopolyhedrovirus and azadirachtin on *Spodoptera litura* Fabricius (Lepidoptera: Noctuidae). *Biological Control,* **34**: 93–98.

Shapiro, M., J.L. and Robertson, R.E. 1994. Webb, Effect of neem seed extract upon the gypsy moth (Lepidoptera: Lymantriidae) and its nuclear polyhedrosis virus. *J. Econ. Entomol.,* **87**: 56–360.

Singh S.P., Ballal C.R., and Poorani J. 2002. Old world bollworm: *Helicoverpa armigera* associated Heliothinae and their natural enemies. Project Directorate of Biological Control, Bangalore, India, Technical Bulletin, No. 31.

Subrahmanyam, B., and N. Ramakrishnan 1981. Influence of a baculovirus infection on molting and food consumption by *Spodoptera litura. J. Invertebr. Pathol.,* **38**: 161–168.

Tanada, Y., and R.T. Hess 1975. Omi, Invasion of a nuclear polyhedrosis virus in the midgut of the armyworm, *Pseudaletia uipuncta,* and the enhancement of a synergistic enzyme. *J. Invert. Pathol.,* **26**: 99–104.

Waldbauer, G.P. 1968. The consumption and utilization of food by insects. *Adv. Insect physiol.,* **5**: 229-288.

Xeros, N. 1952. Cytoplasmic polyhedral virus disease. *Nature,* **170**: 1073.

2015, Perspectives in Animal Ecology and Reproduction, Vol. 10   *Pages 83–102*
Editors: **V.K. Gupta, Anil K. Verma and G.D. Singh**
Published by: **DAYA PUBLISHING HOUSE, NEW DELHI**

Chapter 6

Present Status and Prospects of Black Tiger Shrimp Farming: A Case Study in Maritime State of West Bengal, India

Basudev Mandal* and Sourabh Kumar Dubey

Department of Aquaculture Management and Technology, Vidyasagar University, Midnapore, Paschim Medinipur – 721 102, West Bengal, India

ABSTRACT

Among the Asian countries, India is one of the major shrimp producing country, possesses a huge brackish water resources of over 1.2 million hectare suitable for farming. A total of 1, 35,778 metric ton of shrimps were produced from an area of 1, 15,342 ha from India in the year 2011-12. The state of West Bengal in India has an immense potential for commercial shrimp farming and presently three coastal districts *viz.*, North 24 Parganas, South 24 Parganas and East Midnapore having brackish water area of about 2.10 lakh ha are suitable for coastal farming. The dominant species under culture in West Bengal is *Penaeus Monodon* due to its high unit value realisation and over expanding export demand. The production of *Penaeus monodon* through aquaculture in West Bengal was about 46,000 metric ton from an area of 48,558 ha. This state significantly shares in export production and contributes nearly 34 per cent of total production of farmed shrimp and prawn in the country, being at 2nd position after Andhra Pradesh. Although rapid expansion of shrimp aquaculture in West Bengal considerably contribute in rural livelihood and economy, but it has been accompanied by various management problems coupled with environmental sustainability issues. Considering the back drop, this present paper reviews the

---

* *Corresponding author.* E-mail: bmandalamtvu@gmail.com, bdmandal@yahoo.co.in

present status, culture matrix, key issues, constrains, challenges and opportunities of sustainable shrimp culture in West Bengal. The lessons learned from the review are considered in the context of recommendations to encompass a socially equitable and ecologically viable coastal aquaculture.

Keywords*: Black tiger shrimp production, Farming system, Present status, Prospect, Constraints, Maritime state of West Bengal.*

Introduction

World aquaculture is overwhelmingly concentrating in Asia. Aquaculture productions in Asia increased from 30.3 million tons in 2001 to 55.5 million tons in 2011 contributing 88.5 per cent to the global aquaculture. India is second dominant aquaculture producing country after China with a production of 4,57,3,465 tons in 2011 (FAO, 2013). In last decades, shrimp significantly contributes in global aquaculture and production has increased 631471 tons in 2001 to 781582 tons in 2010 (FAO, 2012). With respects of both quantity and value, about 80 per cent of global shrimp production originates from Asia. Over the last decade, Indian shrimp farming industry has transformed from a traditional shrimp trapping system to a capital oriented semi-intensive farming system. India is blessed with a coastline of over 8,129 km and 2.02 million sq km of Exclusive Economic Zone (EEZ). The country possesses huge brackish water resources, more than 1.2 million hectare (ha) suitable for shrimp farming of which around 15 per cent is utilized for farming (Abraham and Sasmal, 2009), producing about 1.04 lakh tons of shrimps and prawns which is going for marine export basket (MPEDA, 2009). During the financial year 2011-12, for the first time in the history of Indian marine product exports, earning have crossed USD 3.5 billion accounting a growth of 6.02 per cent in quantity, 28.65 per cent in Rupee and 22.81 per cent growth in USD earning compared to previous year. Frozen shrimp continued to be the major export value item accounting for 49.63 per cent of the total USD earnings in India (MPEDA, 2011). West Bengal (Lat 21°38N´ – 27°10N´, Long 85°38E´ – 89°50E´), a maritime state of India, historically has vast potential for commercial farming of marine shrimps as well as freshwater prawns besides several species of commercially important fishes. Nearly 4.05 lakh ha of coastal area located in and around backwaters, estuaries and other brackish water impoundments in West Bengal provide potential sites for brackish water shrimp farming. Development in coastal aquaculture in West Bengal is concentrated on shrimp farming throughout the 158 km coastline of the state. The dominant species under shrimp culture is black tiger shrimp, *Penaeus monodon* due to its unique test, high unit value and ever-expanding export demand in global market. Scientific culture of tiger shrimp started in West Bengal during mid 1980's and more than 39,000 hectares of area was brought under culture. Presently, three coastal districts of West Bengal *viz.*, North 24 Parganas, South 24 Parganas and East Midnapore having brackish water area of about 2.10 lakh ha are suitable for coastal farming (Abraham *et al.,* 2013). The area under shrimp culture in the North 24 Parganas and South 24 Parganas districts including mangrove based Sundarbans eco-regions are around 46500 ha. In North 24 Parganas, out of 1, 10,000 ha potential brackish water area, 35,000 ha extensively used for shrimp farming.

In South 24 Parganas district, nearly 11,500 ha area are under shrimp aquaculture out of 70, 000 ha brackish water area. In East Midnapore district, out of 5,618.22 ha brackish water area, 1950 ha area potentially suitable for shrimp farming (Anon, 2007; Anon, 2010). Traditional/improved traditional farming is normally practiced at North 24 Parganas and South 24 Parganas districts and productivity is very low (200-300 kg/ha), while extensive/improved extensive (Scientific) farming is mainly concentrated at East Midnapore district and hold prestigious position in West Bengal. With proper management practices, high quality of peletized feed and artificial aeration system the farms are capable to achieve 1-1.5 ton/ha production per year.

However, rapid development of shrimp culture has been accompanied by many man-made and environmental hazards. Failures in shrimp production due to various reasons like, post larvae (PL) quality, feed, water and soil quality and disease outbreak, but in most cases origin of problem are poor management practices. Disease outbreaks is a common phenomena in the shrimp industry, especially in 1994-96 caused massive drop in World shrimp production. In 2004, the shrimp farmers of East Midnapore District, West Bengal noticed significant mortalities of *Penaeus monodon*, which occurred between 45 and 70 days of stocking (Abraham and Sasmal, 2008). According to The Marine Products Export Development Authority statistics, Shrimp production in West Bengal was showed decreasing trends from 2007 to 2010 than the previous years (MPEDA, 2010). Moreover, the sustainability of shrimp culture does not only refer to ecological and disease sensitivity but also the economic sustainability as high yielding intensive and semi-intensive farming are subjected to huge financial risks.

Considering the back drop, this present attempts try to explore the present status of shrimp aquaculture and also to understand the different culture matrix, constrains, challenges, and opportunities of shrimp culture practices in West Bengal.

Production Trends and Area under Black Tiger Shrimp Culture in the Country

In Indian scenario, according to the Marine Products Export Development Authority statistics (2011), during the year 2011-12 a total of 1, 35,778 metric ton of shrimps were produced from an area of 1, 15,342 ha. State wise details of tiger shrimp farming are depicted in Table 6.1. During the year 2009-10, shrimp production was estimated to be around 95,918.89 metric ton from 1, 02,259.98 ha area compared to the 2008-09 year production which was 75,996.54 metric ton from 1, 08,788.68 ha area. This trend showed that in 2009-10, area got reduced by around 6 per cent over the previous year but the production has remarkably increased by around 26.21 per cent due to the factors like better management practices adopted by the aqua farmers. Compared to the production of 1, 18,575 metric ton in the year 2010-11, an increase in production by 14.51 per cent in 2011-12 was recorded. The estimated value of shrimp produced during the year 2011-12 was INR 4,073 crore, registered an increase of INR 993 crore over the previous year.

The aquaculture production of shrimp in West Bengal increased from 12,500 metric ton during the year 1990-1991 to approximately 46000 metric ton in 2011-12. The production from North 24 Pgs, South 24 Pgs and East Midnapore district estimated was around 25104, 11,072 and 9823 metric ton respectively in 2011-12 (Tables 6.2–6.4).

Table 6.1: State-wise Details of Tiger Shrimp Farming in India (2011-12)

State	Estimated Potential Brackish Water Area (ha)	Area Utilized (ha)	Production (MT)	Productivity (MT/Ha/year)
West Bengal	4,05,000	48,558	45,999	0.95
Odisha	31,600	8,597	10,901	1.27
Andhra Pradesh	1,50,000	35, 274	51,081	1.45
Tamil Nadu	56,000	6,322	12,409	1.96
Kerala	65,000	12,809	8,138	0.64
Karnataka	8,000	650	609	0.94
Goa	18,500	53	51	0.96
Maharashtra	80,000	1,098	1,721	1.57
Gujarat	3,76,000	1,971	4,869	2.47
Total	11,89,200	1,15,342	1,35,778	1.18

Source. MPEDA, 2011.

Table 6.2: Year-wise Production Details on Shrimp Farming in West Bengal

Year	2003-04	2004-05	2005-06	2006-07	2007-08	2008-09	2009-10	2010-11	2011-12
Area utilized (ha)	49,955	50,215	50,474	49,715	48,236	47,488	47,488	47,588	48,558
Estimated production (MT)	29,714	35,432	42,336	42,006	27,668	27,418	33,685	40,725	45,999

Source. MPEDA, 2011.

Table 6.3: District-wise Production of Shrimp Farming in West Bengal (09-10)

District	2009-10	2010-11	2011-12
North 24 Parganas	21,643	24,372	25104
South 24 Parganas	5,792	8,725	11,072
East Midnapore	6,250	7,628	9823
Total	33,685	40,725	45999

Source. MPEDA, 2011.

Table 5.4: Year-wise Details of Shrimp Farming in East Midnapore District

Year	Area Under Culture (Ha)	Production (MT)	Productivity (MT/Ha)
2007-08	1,200	4,310	3592
2008-09	1,488	5,500	3696
2009-10	1,488	6,250	4200
2010-11	1,588	7,628	4803
2011-12	1,950	9,823	5037

Source: MPEDA, 2011.

Shrimp Farming Practices in West Bengal

Traditional versus Scientific Shrimp Farming Systems

Production techniques in shrimp farming are conventionally classified into four main categories based on the level of stocking density, area, management practices and yield. These are traditional, extensive, semi-intensive and intensive while based on a review of techniques of shrimp culture it split into five categories namely, extensive, improved extensive or traditional, semi-intensive, intensive and super-intensive (Primavera, 1991; Raux and Bailly, 2002). However basic characteristics of traditional and scientific farming are summarised in Tables 6.5 and 6.6.

Table 6.5: Basic Characterises of Traditional versus Scientific Shrimp Farming Systems

Parameters	Traditional Shrimp Farming	Scientific Shrimp Farming
Water management	Based on tidal influences and water intake through sluice gate	Water filling and exchange through pumping
Salinity	Varies according to monsoon regime	Salinity relatively stable
Mode of stocking	Seed mixed with other species by auto stocking and additional stocking of natural seeds	Selective stocking with hatchery seeds
Feeding	Based on natural food, often farm made feed	Usage of high nutritive feed
Harvesting	Periodic harvesting during full and new moon periods, collection at sluice gate by traps and bag nets	Harvesting at the end of one crop, normally in 120 days
Production	Average production 200-350kg/ha	Average production 2000-2200 kg/ha

Traditional Systems: South and North 24 Parganas Perspectives

The traditional shrimp farming techniques is dominated in mainly South and North 24 Parganas district in West Bengal. It is based on tidal influence, and hence practiced in the low lying coastal areas. Annual production of shrimp from these fields is very low, 200 – 350 kg/ha (Alagarswami, 1995). Brackish water from the adjacent water body *e.g.* river, creek, backwater, lake, etc. is let into by gravity flow to

Table 6.6: Technical Consideration in Shrimp Farming

Parameters	Traditional	Semi-intensive	Intensive
Stocking density	< 2 m²	4-6 m²	30–80/m²
Pellet feed	No	Yes	Yes
Feed conversion ratio	–	1-1.5	>1-1.5
Liming	No	Yes	Yes
Fertilizer	No	100-150kg/ha	150-200kg/ha
Pesticides	Yes	Yes	Yes
Aerator	No	<10hour/day	>20hour/day

a large and shallow area of 10 to 100 ha, enclosed by constructed earthen bunds on its periphery. Entry of water is normally regulated by means of a wooden sluice made of planks and bamboo poles. Screen made of bamboo is used for filtration of the water. Although seed of different fish and shrimp species also find entry into these areas along with the tidal water (auto stocking). Such farming area is locally called as *bheri*. Stocking and harvesting is periodic which is according to the lunar phase. Harvesting is normally done during the full moon and new moon period using traps made of bamboo and partially by cast netting. In most of the *bheries* a single sluice is used for both letting in and draining out the water. Water depth of the rearing areas varies between 1 – 3 feet. Partial water exchange is done during the lunar phases taking advantage of the tidal amplitude of spring and neap tides. Usually no supplementary feed is used and shrimp grow on the natural productivity. Natural growth of filamentous algae, *hydrilla* and other aquatic micro/macrophyte in these large and shallow rearing areas provide an excellent natural environment for the growth of tiger shrimp. Depending upon salinity variation in the water body, these areas are used exclusively for shrimp farming or together with paddy cultivation. In areas experiencing higher fluctuation in the salinity in the range of 10 – 25 ppt during the year, only shrimp farming is practiced. In the improved traditional system, farmers usually stock natural or hatchery produced seed of *Penaeus monodon* for better yield with stocking density of 2 PL/m² and with farm made feed. However, in few cases farmers also apply lime during the culture period. In such areas, farming starts during January/February months and harvesting is completed by October/November.

Intensification of Culture System: East Midnapore Perspectives

Extensive Systems

Depending upon type of management practice adopted and site conditions, extensive/improved extensive (scientific) shrimp farms are utilized for a comparatively higher production in East Midnapore. Extensive systems apply monoculture with stocking density of 5 PL/m² and usually supply water through pumping from canals, creeks or from the sea. Farmers use locally prepared feeds and, under good management, are able to harvest up to some 700 kg/hectare per crop with one or two crops per year. In modified extensive systems, ponds are prepared with tilling, liming and fertilization, which enables the application of higher stocking densities (up to 10 PL/m^2) and increases the potential yield to some 1 000 kg/ha per crop. Small and marginal farmers holding farm area of below 1.0 ha per farmer dominate the scene.

Semi-intensive farms are the more scientific farming systems with 0.25 – 4.0 ha culture area where stocking density is high. Stocking is generally done with hatchery seeds and 15-25 PL/m^2 stocking was maintained. Average annual yields of semi-intensive farms in India are about 2200 kg/hectare with an average of 1.2 to 1.5 crops a year (ADB/NACA, 1998).

Intensive Systems

The ponds are 0.25 to 0.5 ha in size with four aerators per pond and a central drainage system to remove accumulated sludge. Feeding with pelletized food takes place a number of times per day and the stocking density increases to 30 – 80/m^2. Yields of over 8 000 kg per hectare are possible, but the actual average annual yield in India is about 4500 kg/hectare in 1.6 crops per year (ADB/NACA, 1998). Although this system is very common in Thailand and Taiwan Province of China, it is not frequently used in India.

Factors Influencing the Black Tiger Shrimp Farming

Site Selection

Selection of favourable site is the important criteria for successful shrimp farming. Among the ecological factors, land elevation, soil quality, climate, water, etc is most important. Climatic variables like temperature, rainfall, wind, evaporation, sunshine, humidity etc have to be thoroughly studied before selecting potential site. Healthy and unpolluted water supply from estuary, creek, and backwater is needed to be assured. Water quality is one of the crucial factors that affect shrimp farming. Successful shrimp farming depends on maintaining salinity, pH, dissolved oxygen and other water and soil quality parameters in appropriate ranges. Temperature is an important component, which greatly influences and directly affects the pond dynamics, controls the effect on the rates of both food consumption and metabolism and the growth (Wootton, 1996). The favourable range of physico-chemical parameters for ideal shrimp farming was presented in Table 6.7. Clay loam type soil (particle size less than 0.002 mm) which has low permeability is better for construction of shrimp pond.

Pond Preparation and Management

Individual nursery and grow out ponds of a size of 500 m^2 to 10,000m^2 with a depth of 80 and 100 cm respectively are ideal. For semi-intensive farming, a smaller size with depth of 150-200 cm is preferred. Square or rectangular shaped ponds with width-length ratio of 1:2 are preferable for effective water exchange. Presence of inlet and out let with provision of sluice gate is necessary to facilitate water exchange during tide.

Table 6.7: Favourable Physico-chemical Parameters for Shrimp Farming Pond

Parameters	Range
Temperature (ºC)	25-33
Salinity (ppt)	15-35
Transparency (cm)	25-60
Dissolved Oxygen (ppm)	>5
pH	7.5-8.5
Total alkalinity (ppm)	< 20
Total ammonia (ppm)	< 1.0
Free ammonia (ppm)	< 0.25
Nitrite (ppm)	< 0.25
Hydrogen sulphide (ppm)	< 0.25
Soil pH	7 and a little above
Organic carbon content	1.5-2.5 per cent
C: N ratio	10: 15

Modified from *Handbook on Shrimp Farming*, MPEDA, 1996.

Successful pond management requires proper maintenance of the following aspects before stocking *viz.*, complete sun drying of the pond bottom until the soil cracks to a depth of 1-2 cm; installation of nylon net (40-60 mesh/inch) in the inlet and out let channels; eradicating the ponds to eliminate predators which have hibernation and aestivation habits; lime application for improving pH as per requirement and to fertilize for benthic algae and phytoplankton by adding organic manure and inorganic fertilizers.

If pond is to be sterilized to get free from pathogens, hydrated or burnt lime at the rate of 1000-2000kg/ha can be applied. Application of potassium permanganate ($KMnO_4$) at the rate of 1-2 ppm is another method of sterilization. When ponds cannot be dewatered completely, any one of the following namely lime, ammonia, mahua oil cake, derris root, or tea seed cake can be used for eradicating of the predators. Quick lime (CaO) of 100-600 kg/ha can be uniformly applied. Mahua oil cake (4-6 per cent saponin) application at the rate of 200 ppm is effective to eradication process. Application of ammonium sulphate [$(CNH_4)_2SO_4$] in combination with lime [$Ca(OH)_2$] at the rate of 5-10: 1 ratio (1000 kg of lime with 100-200 kg ammonium sulphate per ha) which liberated free ammonia that highly toxic to the weed fish and other shrimp. Application of this combination is beneficial to the soil fertility and conditioning of pond soil. After eradication process fertilization should be started at least 15 days before stocking. The post-larvae (PL) depend on natural food available in the pond, and to achieve a sustainable production of the plankters it is essential to addition organic and inorganic manure. Ponds having a productivity of 2000mg/c/m^3 need no fertilization. Quick lime (400-500 kg/ha), urea (50-100 kg/ha), cowdung (1000-

4000kg/ha) and poultry litter (500-1000kg/ha) are commonly used for enhancing phytoplankton production. Zeolite (porous alminosilicate) is applied to the ponds at a level of 200kg/ha once in two weeks to maintain pond bottom and water in a good condition and also metabolites removal (Anon, 1996; Kurian and Sebastian, 2002).

Stocking

Selections of good quality of seeds, scientific transport, acclimatization of seeds etc before stocking are critical criteria to achieve a sustainable production. Hatchery reared shrimp seed or wild collection are generally used for stocking in the shrimp farms. Selection of disease free and healthy seeds (post-larvae) solves most of the disease problems that are encountered by farmers. Uniform sized post larvae of about 18 mm length are appropriate for stocking. Differences in size indicates different age levels reflected throughout the grow out cycle and may result in cannibalism. The quality of seed is characterized by active movement against current in the container, without bright coloration and adaptability to stress conditions of reduced salinity and formalin content. For short duration of transport, 2-2.5 lakh seed (PL 15-20) can be transported when duration of transport is 1-2.5 hours. 25000 seed (PL 15-20) in water can be transported in 60 litre bags or bins provided with continuous aeration or oxygen packing when the duration of transport is within two hours. For long distance transport, one third of the bag is filled with water and seed at the rate of 250/ l is recommended. Shrimp seed have to be acclimatized to the temperature and salinity of the farm before stocking. Instead of direct transferring shrimp seed to ponds, better survival can be ensured by rearing them in nurseries for short period. Depending on the species stocked, type of system, quality of water, natural fertility of pond and management practices adopted, stocking density ranged from 50,000 – 5 lakhs/ha. Oxygenation has to be done in grow out culture by aerators and aspirators when high stocking density culture is practised. (Anon, 1996; Kurian and Sebastian, 2002; Ravichandran, 2006).

Feed Management

Feeding and feed management is one of the important operational functions in *Penaeus monodon* farming. In traditional system, a complex of natural food plays a pivotal role. These are mainly blue green algae which form a complex with associated zooplankters. The natural feeds thus developed are called *'Lab Lab'* or *'Lumut'*. *Lab Lab* is benthic blue green algae, diatoms any many others form of plants. *Lumut* is composed of filamentous green algae and many others form of life. As the traditional shrimp farming has shifted to extensive or and semi-intensive systems, formulated feed (dry pellets) containing adequate protein levels with minimum water stability should be required in successful, economically viable and sustainable shrimp farming. In artificial shrimp feed, crude protein 30-35 per cent, non protein nitrogen < 0.2 per cent, lipids 6-8 per cent, crude fibre 3-5 per cent, fatty acids 0.5 per cent, moisture 11 per cent and digestible energy 3200-3600 kcl/kg is recommended. Feeding methods may be of check tray feeding, broadcasting or in some cases bag feeding. Artificial feed has to be environmentally viable and should not contain any antibiotics. The daily requirement of feed can be split into smaller rations and feed 4-6 times a day with major percentage during evening and night feeding.

Health Management

The intensification coupled with unscientific farming practices has led to serious environmental and shrimp health related problems in most of the farms (Abraham *et al.*, 2009). The important pre-disposing factors leading to disease outbreaks in shrimp culture are adverse environment, high stocking density, nutritional deficiency, inadequate aeration, insufficient water exchange, heavy algal blooms, physical injury and presence of high numbers of virulent pathogens (Alavandi *et al.*, 1995). Commonly occurring diseases in farmed shrimps can be infectious or non-infectious and sometimes caused by nutritional and environmental deficiencies. However, some common disease encountered in *Penaeus monodon* farming, causative pathogens, gross signs and their prophylactic measures were depicted in Table 6.8 in a concise manner (Lio-Po, 1988; Lio-Po *et al.*, 1989; Lightner, 1993; Parado-Estepa, 1994; Flagel, 1997, Karunasagar *et al.*, 1997; kongokeo, 2005; Abraham and Sasmal, 2008; Abraham *et al.*, 2009, 2013). Better Management Practice (BMP) is of utmost importance for disease prevention and control. Judicious use of chemotherapeutics may be required in some cases. In view of the risks associated with shrimp trading in international market with tissue containing residual antibiotics, MPEDA banned some antibiotics and pharmacologically active substances for use in aquaculture (Table 6.9). However, use of immunostimulants to induce non-specific immunity and application of probiotics/prebiotics is being considered as alternate method for prevention of diseases. Infected seed has been identified as the major source of pathogen outbreak. In that view, farmer have been advised to stock disease-free or Specific Pathogen Free (SPF) seed ensured through PCR testing to good crop.

Bio-security Management

Bio-security management for brackish water shrimp farming is very much significant to achieve highest fish production (Mandal *et al.*, 2009). Management of bio- security in brackish water prawn farm throughout the culture period to achieve extra advantage against antibiotic treatment and ensure maximum production in per unit water area. Probiotic farming and bio-security management improve non pathogenic beneficial bacterial population over pathogenic micro-organism and it also prevent spread in or spread out of infection from/to shrimp farm environment. The outcome of an experiment had showed the probiotic has no obvious effect on water salinity, hardness, NO^+_3, NO^+_2, alkalinity and plankton density. But probiotic treated and un-treaded farm unit has show comparatively high pH value than the other culture year. After 30 days of culture period the random sampling of cultured shrimp from both the farm unit come with significant difference. The shrimp production, survivability and quality has increased by this two farming concept without any negative impact on production, sustainability and their surrounding environment.

Table 6.8: Diseases of Black Tiger Shrimp: Symptoms and their Preventive Measures

Disease	Symptoms	Prophylactic Measures/Treatment
I. Viral diseases		
Monodon Baculovirus (MBV) Disease (*P. monodon-type Baculovirus*)	Affected shrimps exhibit pale bluish-gray to dark blue-black coloration, sluggish and inactive swimming movements, loss of appetite and retarded growth.	Destruction of affected shrimp and Better Management Practice (BMP)
Infectious Hypodermal and Hematopoietic Necrosis Virus (IHHNV) Disease (*Picornavirus*)	Shrimps demonstrated erratic swimming behaviour, rising slowly to the water surface, hanging and rolling over until the ventral side is up and become weak and lose their appetite for food. They repeat the process of rising to the surface and sinking until they die usually within 4-12 hours.	Destruction of affected shrimp and Better Management Practice (BMP)
Hepatopancreatic Parvo - like Virus (HPV) Disease (Parvo virus)	Shrimp showed retardation in growth, loss of appetite and occasional white opaque areas on the abdominal muscles.Moreover benthic diatoms, protozoans such as *Zoothamnium* sp. and filamentous bacteria may cause fouling on the exoskeleton.	Destruction of infected stock and Better Management Practice (BMP)
White Spot Syndrome Virus (WSSV)	White spots or patches under the carapace and over the exoskeleton. This may be associated with red discoloration. Non specific signs of ill-health, including damaged appendages and external fouling. Initially, the shrimp are often observed near the surface of the water and stop feeding. The typical whites spots appear soon after the first signs of ill-health and up to 100 per cent mortality can occur in less than 7 days.	Stocking of PCR tested seed; Treat water with calcium hypochlorite at 15 – 20 ppm and Better Management Practice (BMP)
II. Bacterial diseases		
Luminous Bacterial Disease (*Vibrio harveyi, V. splendidus*)	PL becomes weak and opaque-white. Heavily infected larvae exhibit a continuous greenish luminescence when observed in total darkness.	Water change must be 80 to 90 per cent replacement daily. Use of Terramycin 3-5 g/kg fortified with feed
Shell Disease, Brown/ Black Spot, Black Rot/ Erosion, Blisters, Necrosis of Appendages. (Shell-degrading bacteria belonging to *Vibrio, Aeromonas*, and *Pseudomonas* groups.)	Appearance of brownish to black erosion of the carapace, abdominal segments, rostrum, tail, gills, and appendages. Blister containing cyanotic gelatinous fluid may develop on the carapace and abdominal segment. The blister may extend to the underside of the ventrolateral section of the carapace creating a bulge on the underside.	Water quality monitoring, minimize handling and avoid overcrowding. Avoid injuries to the exoskeleton of the shrimps to prevent the development of primary portals of entry.

Contd...

Table 6.8—*Contd...*

Disease	Symptoms	Prophylactic Measures/Treatment
Filamentous Bacterial Disease (*Leucothrix mucor*)	Presence of fine, colour less, thread-like growth on the body surface and gills. In larger shrimps, filamentous bacteria on the gills and other body surfaces may result in respiratory distress at the point of attachment.	Maintenance of good water quality with optimum dissolved oxygen level; Cutrine-Plus® (0.2-0.5 ppm copper/l) in 4-to 6-h static treatments.
III. Fungal diseases		
Larval Mycosis (*Lagenidium* sp. *Haliphthoros* sp. and *Sirolpidium* sp.)	Infected post larvae appear whitish, become weak, and may eventually die. Signs are readily apparent when the disease is already widespread.	Reduced stocking density; application of Treflan or Malachite green oxylate or trifluralin at 0.2 ppm for 24 hours.
Black Gill Disease (*Fusarium* sp.)	Blackish gill observed and shrimp eventually die	Maintenance of soil and water quality
IV. Protozoan diseases		
Vorticella, Epistylis, Zoothamnium, Acineta and *Ephelota*	Fuzzy mat on shell and gills of heavily infected juveniles and adults; reddish to brownish gills; mortality due to oxygen deficiency	50-150ppm formalin treatment for 30 minutes; Better management Practice (BMP); water exchange by draining pond and tank bottom daily to remove excess feeds, faecal matter, and other organic wastes.
Microsporidiosis (*Microsporadium* sp.)	Muscle tissues are affected, infection may result in sterility of spawners with white ovaries	100-200 ppm formalin treatment; Better management Practice (BMP)
Gregarine disease (*Gregarines*)	Affected in digestive tract causing lower appetite; molluscs is an intermediate host	Eliminate the molluscan intermediate host
V. Nutritional, toxic and environmental diseases		
Chronic Soft Shell syndrome	Nutritional deficiency; pesticide contamination; poor pond water and soil condition; high soil pH (>6), low water phosphate (<1 ppm) and low organic matter content (< 7 per cent); affected shrimps are soft-shelled, grow slowly, and eventually die. Shrimps become more susceptible to wounding, cannibalism, and surface fouling by *Zoothamnium* and other epicommensals.	Improvement of pond management practice; application of supplementary feed, (mussel meat at 8-14 per cent of the body weight daily for 2-4 weeks, or a diet containing a 1: 1 ratio of calcium and phosphorus.

Contd...

Table 6.8– *Contd...*

Disease	Symptoms	Prophylactic Measures/Treatment
Muscle nencrosis	Deterioration of water quality and stressful environmental conditions like low oxygen levels, temperature or salinity shock, overcrowding, and severe gill fouling. Shrimps characterized with opaque white areas on the abdomen, blackening on edges of the uropod followed by erosion. Shrimp becomes weak and eventually die.	Improvement of environmental status, maintenance of good water and soil quality, avoidance of over stocking and over feeding
Red disease	Presence of aflatoxin (produce by the fungi *Aspergillus* spp.) in feeds; high input of lime (2-6 tons/ha) followed by high initial pH, prolonged exposure to low salinity (6-15 ppt). Signs associated is sudden drop in feed consumption, shrimps become lethargic and show general body weakness as, yellowish and eventually reddish discoloration of the body and appendages, red, short streaks on gills, reddish colour of faecal matter, Increased fluid in the cephalothorax emitting foul odour.	Improvement of feeding management practice; selection of appropriate dose of inputs while pond preparation
Acid Sulphate Disease Syndrome	Low water and soil pH is the main cause. Shrimp signed with decreased moulting frequency, yellow to orange to brown discoloration of the gill and appendage surfaces.	Improvement of low pH soil condition by liming and flushing of pond bottom before stocking. Broadcasting of lime on pond dike surfaces or hang lime bag as the need arises.

Table 6.9: List of Antibiotics and Pharmacologically Active Substances Banned for Use in Aquaculture

Sl.No.	Banned Antibiotics and Pharmacologically Active Substances
1.	Chloramphenicol
2.	Nitrofurans including Furazolidone, Nitrofurazone, Furaltadone, Nitrofurantoin, Furylfuramide, Nituratel, Nifursoxime, Nifurprazine and all their derivatives.
3.	Neomycin
4.	Nalidixic Acid
5.	Sulphamethoxazole
6.	*Aristolochia* spp and preparations thereof
7.	Chloroform
8.	Chlorprpmazine
9.	Colchicine
10.	Dapsone
11.	Dimetridazole
12.	Metronidazole
13.	Ronidazole
14.	Ipronidazole
15.	Other nitroimidazoles
16.	Clenbuterol
17.	Diethylstilbestrol (DES)
18.	Sulfonamide (except approved sulfadimethoxine, sulfabromomethazine and sulfaethoxyrphyidazine)
19.	Floroquinolones
20.	Glycopeptides

Harvesting and Quality Assurance

Harvesting is given importance because when the shrimps fresh and handled without stress the quality used to be good after processing. Harvesting of shrimps in night gives better catch. The ideal time for harvest is morning and evening. Harvesting in full moon and new moon period may be avoided as there will be more soft shell fetching low price. Harvesting is generally relies on partial and total harvesting. As per the market demand and availability of freezing facilities, the farmer may resort either to single complete harvest or multiple partial harvests. Draining, seining, cast netting and trapping are the usual practiced methods for harvesting. Draining is the most efficient method followed in extensive and semi-intensive systems by keeping the screens in position at the sluice and draining the water completely so that shrimps gather in the harvest pit near the sluice gate. The screens are then removed and shrimps are swept into the collection net fixed at sluice gate. In traditional culture, harvesting in vast area is quite impossible at a time. Therefore, for farms of 50-100 ha, half area is used for harvesting near the inlet and drain channel. During full moon and new moon or the time of water exchange, harvesting takes place. Seining is practiced in un-drainable ponds. In case of large ponds, seining can be done using two nets starting from the centre portion of the ponds and pulling the nets to opposite sides. Cast nets and drag nets are also used to capture culture shrimps. Hand picking are also supplement the usual shrimp harvesting by draining and netting operations. (Anon, 1996, Kurian and Sebastian, 2002).

Shrimp continues to represent one of the safest forms of muscle protein consumed in the world. Moreover, there are a lots of quality relevant issues, such as backward machinery and equipment to catch and maintain raw materials, unskilled work force, polluted field, lack of knowledge regarding the use of antibiotics, pesticides, and so on. These lead to low quality of raw materials, of which shrimp is a typical product and the final product form is frozen, shell-on shrimp with or without heads. To ensure total quality assurance, it is mandatory to involve a pre-planning procedure to fully outline an operation, noting potential hazards that could occur and identifying control points to prevent, minimize or correct hazards.

Constraints and Conflicts in Progress of Shrimp Farming in West Bengal

Rapid expansion of shrimp aquaculture has been accompanied by various management problems coupled with environmental sustainability issues. The first and foremost problem being faced by shrimp farmers and entrepreneurs of West Bengal is non availability of good quality seeds for stocking at appropriate time into their farms. There is no shrimp hatchery in West Bengal. This has resulted supply of poor quality seeds, diseased seed by the hatchery or operating outside the state. This seeds are procured by owner of the kati centre (Seed Bank) reared in unhealthy condition and being supplied to the shrimp farmers. In this juncture there is every chance for spread of disease through infected shrimp seed.

In West Bengal, majority of the farms are traditional type. Those farms are fed by tidal water through some river or creek. However, several such water sources have badly silted resulting in less water flow and causing almost freshwater condition. Furthermore, those traditional farms are in operation since several decades without drying and any bottom soil management. Settlement of dead floating algae and other vegetal matters at the pond bottom and decay of other organic matters has deteriorated their ecology very badly. Bottom soil of several such bheries has turned to black in colour, creating unfavourable condition for aquaculture. There productivity has been badly hampered and there are incident of outbreak of diseases in few cases.

Shrimp farmers of West Bengal have been facing difficulty in realising optimum price for the farmed shrimp for in majority of cases they can't sale it directly to the processors. The produce is initially sold to local auctioneers. From there, it is procured by the middlemen who beheads it and keeps it in ice cold water for 3 to 6 days for weight gaining. After beheading and chilling, it is resold to processing plant agents for supply to processing companies. By that time its quality is deteriorated resulting in lowering the unit price of the shrimp.

Among various sustainability issues in West Bengal, in particular, mangrove destruction is a serious ecological offence which has far reaching impacts on physical, chemical and biological properties of ecosystem especially in Sundarbans mangrove eco-region of global concern.

Conversion of agriculture land is a burning issue, which is increasing day by day especially in East Midnapore district of West Bengal. In certain parts of West Bengal, agriculture lands have been converted into shrimp farming due to mono-cropping pattern of land and high profit associated with this farming resulting many fold increase in coastal land. This has led to some social conflicts among local villagers and shrimp farm owners. Moreover, salinization of agriculture land and freshwater aquifers due horizontal and vertical seepage of saline water from shrimp farms which led to social conflicts in some part of this state. Employment opportunity both direct and indirect in shrimp farming is increasing due to possible two-cropping pattern than agriculture which also led to livelihood conflict in some region.

However several technological, environmental and social issues need to be addressed to ensure sustainability of shrimp farming and widely acceptable to all classes of farmers (Table 6.9).

Prospects of Black Tiger Shrimp Farming in West Bengal

At the global level, the model predicts that aquaculture production in 2030 would expand from 93.2 million tons under the baseline case to 101.2 million tons under this scenario and shrimp in 2030 would be higher by more than 10 percent (Anon, 2013). West Bengal significantly shares in export production and contributes nearly 34 per cent of total production of farmed shrimp and prawn in the country, being at 2nd position after Andhra Pradesh. Being a pioneer in shrimp production, shrimp farming industry of West Bengal need to have more sustainable and scientific in future to compete with the global demand. Various ways to make shrimp aquaculture environmentally and socio-economically sustainable have been suggested (Macintosh and Phillips, 1992; FAO/NACA, 1995; Barraclough and Finger-Stich, 1996; Primavera, 1998; Troell *et al.,* 1999; Kautsky *et al.,* 2000). Achieving the potential the shrimp farming sector of West Bengal needs to pay attention to many aspects, some of which are depicted below:

Achieving the Ultimate Potential

i. Careful selection of sites and resources

ii. Use of specific pathogen-free (SPF) seeds

iii. Improvement of grow out management; wide acceptance of Better management practice (BMP)

iv. Advancement in post-harvest technology for ensuring good quality of product for both domestic and export market

v. Enhancement of research, particularly health management, water quality and processing and policy research

vi. Adoption of public-private partnership (PPP) model

vii. Ensuring environmental and social responsibility

Table 6.10: Constraints, Concerns and Conflict of Shrimp Aquaculture in West Bengal

Constraints	Concerns	Conflicts
Non-availability of good quality seeds	Ecological footprint	Producer versus middleman
Dearth knowledge of Better Management Practice (BMP)	Ground water salinity and soil salinization	National guideline and formulation of policies
Long term unsatisfactory water quality	Collection of wild seeds from river and sea	Competition of natural resources
Quality of feed ingredients	Mangrove degradation	Agriculture land encroachment
Emergence of disease attack and lack of quarantine	Water pollution and nutrient loading in source water	Social disruption and disempowerment
Realisation of optimum price	Introduction of alien species	Freshwater irrigation
Transport	Use of banned antibiotics	Privatisation of common land
High price of equipments and consumables	Research and training	Marginalisation of coastal communities
Labour	Climate change and extreme weather events	'Aquaculture colonisation'

Conclusion and Suggestions

The commercialisation of shrimp culture has been driven by lucrative profits from export markets and fuelled by governmental support, private sector investment, and external assistance. The formulation of policies directed to assure the sustainability of shrimp aquaculture should involve many different actors, and include consumers, aquaculture entrepreneurs, local communities and government representatives. Mangrove restoration programs should be initiated in areas where shrimp aquaculture development has caused significant damage to this ecosystem. Moreover, the decline in rice cultivation and the increase in the role of aquaculture represent an important structural change in the economy in the rural areas of the region in recent years. Therefore, more consideration and support is expected to lead to development of small-scale aquaculture.

References

Abraham, T. J., and Sasmal, D., 2009. Influence of salinity and management practices on the shrimp (*penaeus monodon*) production and bacterial counts of modified extensive brackishwater ponds. *Turkish Journal of Fisheries and Aquatic Sciences,* **9**: 91-98

Abraham, T. J., Sasmal, D., Das, G., Nagesh, T.S., Das, Mukhopahyayay, S.K., and Ganguly, S., 2013. Epizootology and pathology of bacterial infections in cultured shrimp *Penaeus monodon* Fabricius 1798 in West Bengal, India. *Indian J. Fish.,* **60(2)**: 167-171.

Abraham, T. J., and Sasmal, D., 2008. Incidence of different disease conditions in shrimp culture systems of West Bengal with special reference to white spot syndrome virus infection. *J. Inland Fish. Soc. India,* **40 (2)**: 1-6.

ADB/NACA., 1998. *Aquaculture sustainability and the environment.* Report on a regional study and workshop on aquaculture sustainability and the environment. Asian Development Bank and Network of Aquaculture Centres in Asia and the Pacific Bangkok, Thailand.

Alagarswami, K., 1995. India country case study. In: *Regional study and workshop on the environmental assessment and management of aquaculture development (TCP/RAS/2253).* NACA Environment and Aquaculture Development Series No. 1. Network of Aquaculture Centres in Asia and the Pacific. Bangkok, Thailand.

Alavandi, S.V., Vijayan, K.K., and Rajendran, K.V., 1995. Shrimp diseases, their prevention and control. CIBA Bulletin No. 3, p. 1-17.

Annonymous., 2013. Fish to 2030: prospects for fisheries and aquaculture. Agriculture and environmental services discussion paper 03. The World Bank report (number 83177-GLB), New York, Washington, 102pp.

Anonymous., 1996. A Manual of on shrimp farming. The marine products export development authority, 171pp.

Anonymous., 2007. Handbook on Fishery Statistics of West Bengal 2002–2007. Department of Fisheries, Aquaculture, Aquatic Resources and Fishing Harbours. Government of West Bengal, Writer's Building, Kolkata, India, 112 pp.

Anonymous., 2010. Annual Report, 2009-10, Department of Fisheries, Aquaculture, Aquatic Resources and Fishing Harbours. Government of West Bengal, Writer's Building, Kolkata, India, 68 pp.

Barraclough, S., and Finger-Stich, A., 1996. Some ecological and social implications of commercial shrimp farming in Asia. Discussion Paper. United Nations Research Institute for Social Development and World Wildlife Fund.

FAO., 2013. Global aquaculture production statistics for the year 2011. Fisheries and aquaculturedepartment, FAO, Rome [online]. ftp://ftp.fao.org/FI/news/Global Aquaculture Production Statistics 2011.pdf

FAO., 2012. Yearbook: Fishery and aquaculture statistics 2010. FAO, Rome. 78 pp.

FAO/NACA., 1995. Report on a regional study and workshop on the environmental assessment and management of aquaculture development. Network of Aquaculture Centres in Asia-Pacific, Bangkok, Thailand.

Karunasagar, I., Otta, S. K., and Karunasagar, I., 1997. Histopathological and bacteriological study of white spot syndrome in *Penaeus monodon* along the west coast of India. *Aquaculture,* **153**: 9-13.

Kautsky, N., Rönnbäck, P., Tedengren, M., and Troell, M., 2000. Ecosystem perspectives on management of disease in shrimp pond farming. *Aquaculture,* **191**: 145-161.

Kongkeo, H., 2005. Cultured Aquatic Species Information Programme. *Penaeus monodon.* FAO Fisheries and Aquaculture Department [online]. Rome[Cited 6 February 2014].http://www.fao.org/fishery/culturedspecies/ Penaeus_monodon/en

Kurian, C.V., and Sebastian, V. O., 2002. Prawns and prawn fisheries of India. Hindustan publishing cooperation, New Delhi, India, 307 pp.

Lightner, D. V., 1993. Disease of cultured penaeid shrimp. In: *CRC handbook of mariculture, 2nd edn., Vol.1, Crustacean Aquaculture.* (Eds.) J. P. McVey. CRC Press Inc. Boca Rough, FL, p. 393-486.

Lio-Po, Gilda., 1988. Prawn health in aquaculture. In: Technical Consideration for the Management of Intensive Prawn Farms, (Eds.) Y.N.Chiu, L.M. Santos and R.O. Juliano. Iloilo City, Philippines: University of the Philippines Aquaculture Society. pp. 130-133.

Lio-Po, G. D., Fernandez, R. D., Cruz, E. R., Baticados, M. C. L., and Llobrera, A. T., 1989. Recommended practices for disease prevention in prawn and shrimp hatcheries. Aquaculture Department, Southeast Asian Fisheries Development Center, Iloilo, Philippines.

Macintosh, D.J., and Phillips, M.J., 1992. Environmental considerations in shrimp farming. In: *Proceedings of the Third Global Conference on the Shimp Industry,* (Eds.) H de Haan and T Singh, INFOFISH, Kuala Lumpur, pp. 118-145.

Mandal, B., Ganguli, D., and Chand B. K., 2009. Probiotic farming and biosecurity management for sustainable brackish water shrimp farming. Proceedings of the International Seminar. Published by PG Deptt. of Zoology, Raja N.L. Khan Womens' College, Midnapore, West Bengal. pp. 62-65.

MPEDA., 2009. Annual Report 2009-10. The marine products export development authority, Cochin.

MPEDA., 2010. Annual Report 20011-12. The marine products export development authority, Cochin.

Parado-Estepa, F.D., 1994. Shrimps. In: Proceedings of the Seminar-Workshop on Aquaculture Development in Southeast Asia and Prospects for Seafarming and Searanching, (Eds.) F. Lacanilao, R.M. Coloso and G.F. Quinitio, Aquaculture Department, Southeast Asian Fisheries Development Center, Iloilo, Philippines. pp. 32-39.

Primavera, J.H., 1991. Intensive prawn farming in the philippines: ecological, social and economic implications. *Ambio,* **20(1)**: 28-33

Primavera, J.H., 1998. Tropical shrimp farming and its sustainability. In: *Tropical Mariculture,* (Ed.), S. de Silva. Academic Press, London, pp. 257-289.

Rabichandran P., 2006. Shrimp farming. In: *Handbook of fisheries and aquaculture,* (Eds.) S. Ayyappan. Indian Council of Agricultural Research, New Delhi, p. 392-403

Raux, P., and Bailly, D., 2002. Literature review on world shrimp farming, individual partner report for the project: policy research for sustainable shrimp farming in Asia. European Commission INCODEV Project, CEMARE, University of Portsmouth UK and CEDEM, Brest, France.

Troell, M., Rönnbäck, P., Kautsky, N., Halling, C., and Buschmann, A., 1999. Ecological engineering in aquaculture. The use of seaweeds for removing nutrients from intensive mariculture. *J. Appl. Phycology,* **11**: 89-97.

Wootton, R.J., 1996. Ecology of teleost fishes. Chapman and Hall, NY., 404 pp.

2015, Perspectives in Animal Ecology and Reproduction, Vol. 10   *Pages 103–121*
Editors: **V.K. Gupta, Anil K. Verma and G.D. Singh**
Published by: **DAYA PUBLISHING HOUSE, NEW DELHI**

Chapter 7

Breeding, Larval Biology and Seed Production of Indian River Prawn, *Macrobrachium malcolmsonii* (H. Milne Edwards)

Rubu Mukherjee[1], D.R. Kanaujia[1]* and A.K. Pandey[2]*

[1]Central Institute of Freshwater Aquaculture,
Kausalyaganga, Bhubaneswar – 751 002, Odisha, India
[2]National Bureau of Fish Genetic Resources,
Canal Ring Road, Lucknow – 226 002, India

ABSTRACT

Breeding, larval biology and seed production of the freshwater Indian river prawn, *Macrobrachium malcolmsonii,* were studied in captivity as well as under hatchery conditions. This species is being reared and maintained in ponds of Central Institute of Freshwater Aquaculture, Bhubaneswar. Physico-chemical parameters and biological spectrum of the three selected ponds were included for the study during two successive years. The adults and berried prawns were collected from the three ponds and maintained in FRP tanks whereas breeding and larval biology including stages and individual characteristics, growth and metamorphosis, moulting frequencies, food and feeding at different stages, water quality parameters of the medium during entire seed production cycle as well as the post-larvae (PL) production were undertaken under hatchery conditions. Three trials of larval rearing and seed production were conducted during each year. The larval development was not uniform and varied considerably with progress of the cycle. In different production trials, 1st post-larvae appeared within 49 days and 16,264 PL (post-larvae) were harvested on day 63 of the cycle in the

* *Corresponding author.* E-mail: kanaujia.dr@gmail.com

three different production trials and the average PL/l range was 17.87-18.45 in the trails during first year. The corresponding survival of the larvae from zoea I to PL was about 18 per cent during the 63 days cycle. Similar trend was also recorded during second year in the three trials with total PL production of 16,149, slightly less as compared with that of the first year.

Keywords: *Breeding, Larval rearing, Post-larvae production, Macrobrachium malcolmsonii.*

Introduction

The freshwater prawn farming has become a major contributor to global aquaculture, in terms of quantity and value during the recents years (Ardrill *et al.,* 1973; New and Singholka, 1985; New, 1990, 1994; 2003; New and Valenti, 2000; Valenti and Daniels, 2000). Initially, the freshwater prawn production was 5 per cent of the whole prawn production in the world (New, 1995). However, world production of freshwater prawn between 1990 and 2000 increased from 21,000 to 118,500 tonnes per year which corresponds to an increase of about 500 per cent (FAO, 2000 - cited by New and Valenti 2000). Although these data refers only to *Macrobrachium rosenbergii,* the most cultivated species in the world (New and Valenti, 2000), another two species *M. malcolmsonii* (India) and *M. nipponense* (China) are becoming increasingly more popular in some Asian countries (Kutty *et al.,* 2000; New *et al.,* 2010; New and Nair, 2012; Nair and Salin, 2012). The freshwater prawn culture is simpler and inexpensive than brackishwater prawn farming since the ponds are built in small, medium and large size, located on the coastal as well as inland areas. Freshwater prawns are decapods crustaceans belonging to Family Palaemonidae. Though they are called prawns as their brackishwater counterparts, they are closer to lobster regarding their evolution, thus presenting many similarities mainly with respect to reproduction habits. The vast majority of freshwater prawns species of commercial interest belong to Genus *Macrobrachium* are found in the tropical and subtropical regions of the world (New, 2003; Kanaujia, 2003).

About 125 freshwater prawn species of Genus *Macrobrachium* are distributed in inland waters throughout the world belong to the Family Palaemonidae (Holthuis, 1980). Out of these, 49 species are important to the fisheries point of view (Holthuis, 1980, 2000). About 15 species are being considered for aquaculture (Ling and Costello, 1979). Over 30 species of this genus have been recorded from the freshwater of the Indian subcontinent. They are distributed in different open water bodies such as rivers, reservoirs and estuaries. Among them, the common species are *Macrobrachium rosenbergii, M. malcolmsonii, M. gangeticum* synonym *M. birmanicum choprai, M. villosimanus, M. equidens, M. javanicum, M. scabriculum, M. idella, M. rude, M. mirabile* and *M. lamarrei.* However, the first three species are considered most suitable for aquaculture due to their faster growth (Tiwari, 1975; Towari and Holthuis, 1996; Kanaujia *et al.,* 2001; Kurian and Sebastian, 2002). The commercial culture of *M. rosenbergii* has been well-established throughout the world while the second larger freshwater prawn, *M. malcolmsonii,* is distributed in the Indian subcontinent, Bangladesh, Sri Lanka, Pakistan and Myanmar. This species is available in river Ganga, Hooghly, Brahmputra, Mahanadi, Godavary, Krishna, Cauvery, Narmada,

Tapti, and Sindh *etc* draining into Bay of Bengal and west coast and also in Chilka Lake in Odisha and Kolleru Lake in Andhra Pradesh (Tiwari, 1949, 1955, 1975; Rajyalakshmi and Randhir, 1969; Rajyalakshmi, 1980; Kanaujia, 1989; Kanaujia and Mohanty, 1992; Kurian and Sebastian, 2002 and Jhingran, 2003). This species migrates up to a distance of 1,400 km from the estuary to upper stretches of different riverine ecosystems (Tiwari, 1949, 1955; Jhingran, 1956; Ibrahim, 1962; Kewalramani *et al.*, 1971), an attribute significant for culture of *M. malcolmsonii* in inland freshwater bodies (Rajyalakshmi, 1984; Ahemad, 1984, 1993; Kanaujia and Mohanty, 1996 and Kanaujia, 2003; Radheyshyam, 2009; New *et al.*, 2010; Mishra *et al.*, 2014). Fishery and biology of *M. malcolmsonii* have been studied by Ibrahim (1962a), Subrahmanyam (1975), Rajyalakshmi (1980), Rao (1986) and Kanaujia (1989). Efforts have also been made to develop hatchery technology for large-scale seed production of *M. rosenbergii* and *M. malcolmsonii.* Kanaujia and Mohanty (1992) made major breakthrough in developing hatchery technology for commercial seed production of *M. malcolmsonii* using the broodstock of the river Mahanadi that provided greater scope for its commercial seed production in order to promote the same in the freshwater prawn farming.

Since *M. malcolmsonii* is found in the upper and middle stretches of river Mahanadi (Odisha), they are being reared and maintained in ponds of Central Institute of Freshwater Aquaculture, Bhubaneswar (Kanaujia and Mohanty, 1992, 1996; Kanaujia, 2003). It is being cultured in the freshwater ponds of rural Odisha on experimental basis (Radheyshyam, 2009; Mishra *et al.*, 2014). Since non-availability of sufficient quantity of quality seed being the major constraint experienced in the adoption of the freshwater prawn farming, breeding, larval biology and seed production of the Indian river prawn were studied in captivity as well as under the hatchery conditions for two years. Physico-chemical parameters and biological spectrum of the three selected ponds were also included in the study.

Materials and Methods

The present study was carried out at Central Institute of Freshwater Aquaculture (CIFA), Kausalyaganga (latitude 20°1'6"-20°11' 45" N; longitude 85°50' 52"-85°51' 35" E), Bhubaneswar (India) for a period of two years (January 2002-December 2003). The location of CIFA enjoys a tropical climate of three distinct seasons *i.e.* winter (November-February), summer (March-June) and monsoon (July-October). *M. malcolmsonii* were reared under captive condition in three selected earthen rearing ponds. The adults and berried Indian river prawns were collected from these ponds and maintained in FRP tanks whereas breeding and larval biology studies including pre-mating moult, mating, spawning, larval incubation, hatching, rearing, feeding, growth, harvesting, acclimatization *etc* were carried out under the hatchery conditions. Three trials of larval rearing and seed production were conducted whereas physico-chemical parameters such as temperature, transparency, pH, dissolved oxygen, ammonia, nitrate, salinity, total alkalinity and chloride content as well as live food spectrum (zoo- and phytoplankton) of the three ponds were also recorded during the study period.

Results and Discussion

Physico-Chemical Parameters of the Water

Physico-chemical parameters of the ponds water recorded during two years exhibited significant variations. Maximum water temperature (30.9°C) was in May and minimum (20.6°C) in December during 1st year and slightly lower (30.7°C) in May whereas the minimum water temperature (20.4°C) in December during the 2nd year. The maximum transparency in both the years was recorded in February whereas minimum in August. The pH value of the pond water was within the alkaline range exhibited narrow range of fluctuation ranged between 7.0-7.8 and 7.1-7.7 during first year and second year, respectively, which was lowest in June during both the years. Dissolved oxygen ranged between 5.0-6.7 mg/l during both the years. The seasonal oxygen value was found to be higher during summer and lower in winter during both the years. The total alkalinity ranged from 108 to 127 mg/l and 107 to 128 mg/l during first and second year. The maximum total hardness was recorded in October and September during first year and second year, respectively and minimum in January during both the years. The chloride contents ranged between 14.2-16.7 mg/l during the two years. The seasonal fluctuation indicated higher chloride concentration during summer and lower in winter.

The plankton population exhibited marked monthly and seasonal fluctuations throughout the study period of the successive two years. The average plankton count during first year ranged between 1677-4190, 1302-4127 and 795-1889 u/l in Pond 1, 2 and 3, respectively and between 1385-3246, 1158-2362 and 792-1393 u/l in the respective three ponds during the second year. The greater phytoplankton density was recorded during summer months and was dominant over zooplankton throughout the study period. As it is well-known that the zooplankton are the primary consumers which was found much lower in annual and seasonal population during the two years. The zooplankton density ranged from 202-658 u/l during the study and was represented by the three major groups – Rotifera, Copepoda and Cladocera.

The average water temperature varied between 27.1-27.9 and 27.3-28.2°C in all the three larval rearing trials during the first and second year, respectively. The pH value of the larval rearing medium ranged between 7.3-7.5 in first and second year. The average total alkalinity of the larval rearing medium varied between 82.7-83.4 mg/l in first year and between 82.4-83.35 mg/l during second year. The average total hardness of the larval rearing medium ranged between 2,082-2,100 mg/l in first year and between 2,088.5-2,138.0 mg/l in the second year. The average dissolved oxygen level of the larval rearing medium varied between 7.3-7.4 mg/l in the first year and 7.3-7.4 mg/l in second year. The average ammonia level of the larval rearing medium was observed 0.1 mg/l during both the years. The nitrate level of the larval rearing medium was observed to be 0.1 mg/l in first year and also the same in the second year.

Maturation

The identification of sex was based on secondary sexual characters in males. The appendix masculina was clearly visible in prawns of size 75 mm and above that also possessed bulging propodus. Gonadal development was observed from April onwards. The berried females with yellow eggs were noticed during May till the end of October. A small percentage of mature females with eggs were observed in May and a gradual increase in their number was observed during July-September. In the later phase, most of the females were either with fully matured gonads or in berried condition. Egg-bearing females were also found with mature gonads. Among the size group of 30-45 g, the number of females were considerably higher. However, the number of berried females increased during monsoon and dominated over the males till the end of October. Females in the size range of 75-168 mm were found with ripe ovary of 1.54-15.56 g in weight. The weight of ovary was found to increase with the size of the female. The body weight was taken along with eggs and then without eggs and the total number of eggs were estimated. The females with body length ranging between 75-168 mm and body weight 5.75-56.34 g were found with 1.54-15.56 g egg mass, however, a lower number of eggs *i.e.* 8,400 were recorded in a female with 75 mm length. The details of study regarding eggs carried by the females were correlated with body length to weight, weight of egg mass and total number of eggs carried by the females and number of eggs/gm of body weight (Figures 7.1–7.3).

Maturation

The maturity stages, breeding and occurrence of berried females were observed during the experiment conducted under captive condition. Breeding started with the onset of breeding season, the ovary of the matured female passed through four distinct phases like (i) immature, (ii) early maturing, (iii) maturing and (iv) ripe. Once the ovary attains ripe stage, then the females undertook pre-mating moult that followed mating and spawning. The detail observations were made on maturity stages and occurrences of berried females in three ponds during first and second years are presented in Figure 7.4. Gonadal maturation and occurrence of berried prawns started from May. The number of berried prawns recovered during 2 years in three ponds ranged between 5-82 per cent from May to October with peak being September. The occurrence of berried females started from May and continued till October with peak during September, thereafter ceasing completely from November to April. The females registered comparatively slower and retarded growth with the onset of breeding period whereas males attained maximum growth during this period.

However, female *M. malcolmsonii* has the following maturity stages *i.e.* (i) immature (ii) early maturing, (iii) maturing (iv) ripe, (v) berried and (vi) spent. For attaining the different stages of maturity, the individual female takes different time. Thus the time interval between maturity stages in different females varied. Ovary of *M. malcolmsonii* took 2-22 days with an average of about 8 days for developing into early maturing stage. Development and duration of next stage of ovary maturation followed the similar trend and duration with the previous on, taking an average 4-6 days for attaining ripe stage. Thus, it took about 15 days for developing immature ovary to ripe stage.

Characteristics of Mature Females

The females were observed with fully developed ovaries which was a large yellow coloured mass occupying a large portion of the dorsal and lateral parts of the

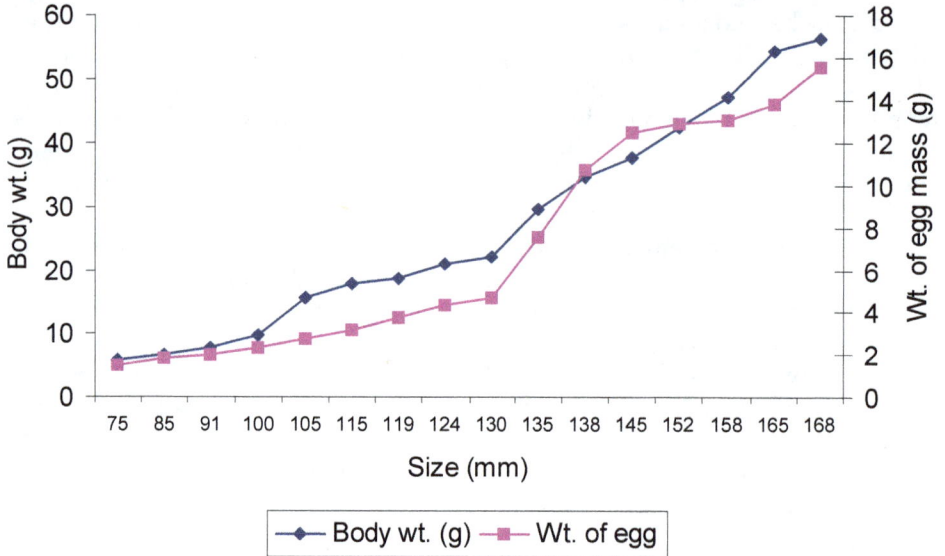

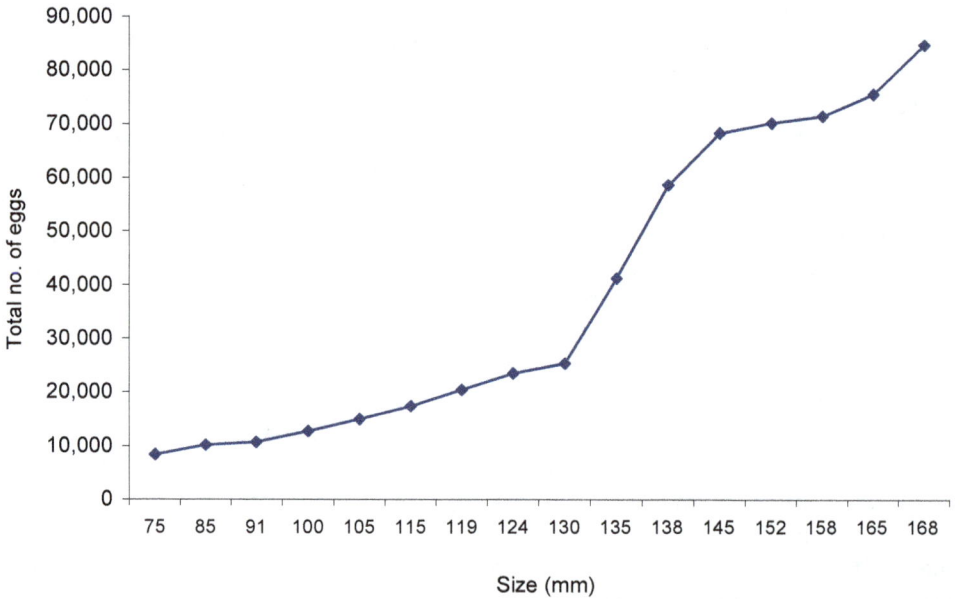

Figure 7.1: Number of Eggs and Body Weight and Weight of Egg Mass in Relation to Size of *M. malcolmsonii.*

cephalothorax. The abdominal pleura bent slightly inwards, the pleopods become slightly distended and arched outward to provide the space in between them and the abdomen enlarged which together with distended pleopods formed a "brood chamber" to accommodate the large number of eggs.

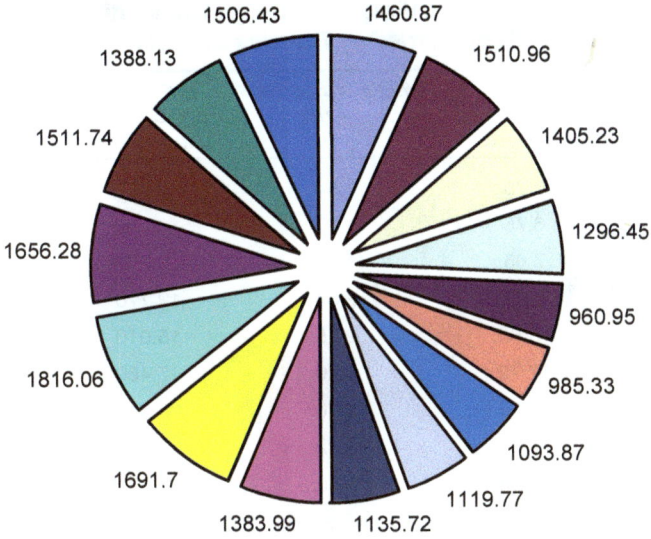

Figure 7.2: Number of Egg/g Body Weight at different Sizes of *M. malcolmsonii*.

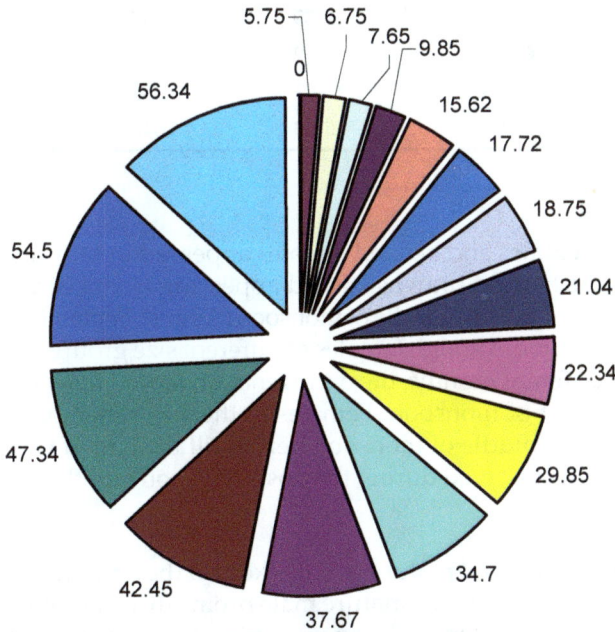

Figure 7.3: Number of Eggs in Relation to Body Weight of *M. malcolmsonii*.

Fecundity

The fecundity of female *M. malcolmsonii* (estimated by the total number of eggs carried) ranged from 8,400-84,872 at the size range of 75-168 mm (Table 7.1). The maximum fecundity 84,872 eggs were recorded with 56.34 g body weight and the minimum 8,400 eggs with 5.75 g body weight.

Table 7.1: Body Weight, Weight of the Egg Mass with Respect to Number of Eggs Carried by Females of *M. malcolmsonii*

Size (mm)	Body Weight (g)	Weight of Eggmass (g)	Total Number Eggs Carried by the Females	Number of Eggs per gram of Body Weight
75	5.75	1.54	8,400	1,460.87
85	6.75	1.87	10,199	1,510.96
91	7.65	1.97	10,750	1,405.23
100	9.85	2.34	12,770	1,296.45
105	15.62	2.75	15,010	960.95
115	17.72	3.20	17,460	985.33
119	18.75	3.76	20,510	1,093.87
124	21.04	4.32	23,560	1,119.77
130	22.34	4.65	25,372	1,135.72
135	29.85	7.56	41,312	1,383.99
138	34.70	10.76	58,702	1,691.70
145	37.67	12.54	68,411	1,816.06
152	42.45	12.89	70,309	1,656.28
158	47.34	13.12	71,566	1,511.74
165	54.50	13.87	75,653	1,388.13
168	56.34	15.56	84,872	1,506.43

Breeding

Adult males, females and berried females appeared in May and continued till end of October. The berried prawns started appearing from last week of May and their number increased during middle of monsoon (August-September) corresponding the peak breeding period. Though females of different size groups (75-168 mm) were observed during different months, the occurrence of berried females was very much related with the onset of monsoon. Females in the size range of 95-160 mm were dominant during the middle of monsoon period till the end of October, the larger sized berried females observed during the post-monsoon months too.

Mating

Though the pre-moulted mature females kept in the aquarium for mating were liable to be attacked by the others, mature male promptly protected and cared them. The female prawn secreted a kind of hormone during pre-mating moult which strongly attracted the males. Sexually mature male was able to mate with the female just after completing its pre-mating moult. Mating was induced under controlled condition in glass aquarium by introducing a mature male with sexually ripe female after completing pre-mating moult. The male after being introduced usually took few minutes to get acquainted with its counterpart. Soon after pre-mating moult, male tried to catch the female and started courtship by displaying masculine grace and

Pond 1

Pond 2

Pond 3

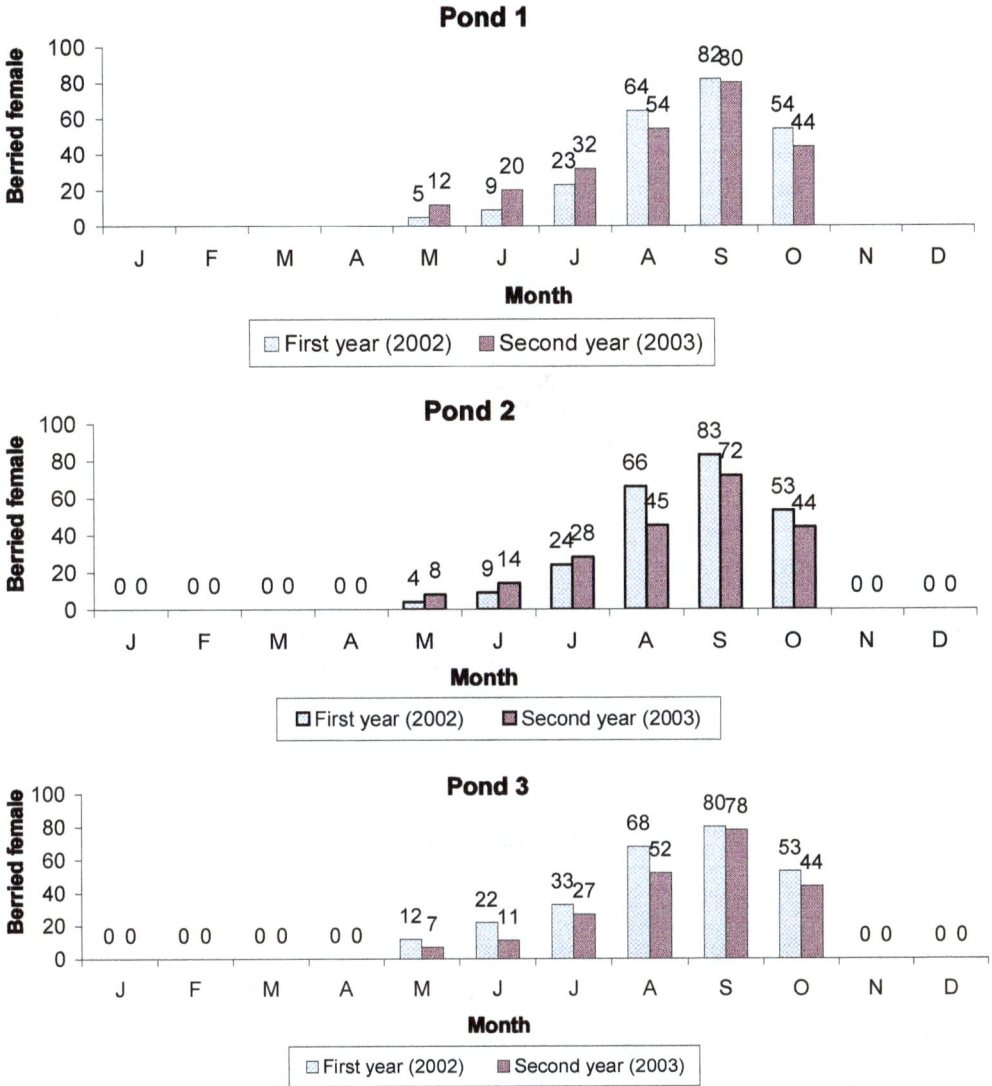

Figure 7.4: Occurrence of Berried Females *M. malcolmsonii* in the Three Rearing Ponds during First and Second Year.

strength by lifting its head, raising body, waving feeler and raising gesture accompanied by intermittent jerking movements. This act was continued for about few minutes to half an hour before the female was successfully won over. The victorious male then chased the female and kept arresting it within the range of its long chelipeds. The courtship, mounting and copulation continued for few minutes and in the mean time, the male actively started clearing the ventral portion of female thoracic shell with other legs. It takes about few minutes to complete this act and the final mating lasted only for a few seconds. Sperms ejected from the male were deposited as a gelatinous mass on the female thoracic region between the base of 3ʳᵈ thoracic legs.

Spawning

Few hours after mating, eggs were ejected and deposited in the brood chamber and get coated with the layer of gelatinous substance for protection. During the release of eggs, the body of the female bends forward to keep contact with the ventral thoracic region in 'U' shape so that the eggs may be extruded through female genital pore directly into the brood chamber. The eggs are hold in bundles like grapes by extremely thin and elastic membranous substance and adhered tightly to the fine ovigerous setae of the 1ˢᵗ four pairs of pleopods. The release of eggs from the ovary to the brood chamber was observed in a tubular form which becomes rounded after leaving the genital pore. The colour of the eggs was yellow which became dark and grey just before hatching of zoea stage I. The eggs were slightly oval in shape measuring about 0.4-1.0 mm depending upon the size of the female.

Incubation

Berried females carried their eggs and cared them till hatching. During the incubation period, the temperature ranged between 28-30°C. The pleopods were observed to beat back and forth intermittently to provide aeration to the eggs during the incubation period of 10-15 days (Table 7.2). Dead eggs and foreign materials were carefully removed at regular intervals by the sensitive and versatile first pair of thoracic legs. Starting from the first day of incubation, the yellow colour of eggs gradually become lighter but the colour changed to slate grey when the larvae inside the eggs were fully developed.

Table 7.2: Instars Duration and Progressive Increase in Size during Larval Development of *M. malcolmsonii*

Larval Stages	Instars Duration (Day)	Total Length (mm)	Growth Increment (mm)
I	1	1.88	–
II	2-1	1.97	0.09
III	4-2	2.10	0.13
IV	10- 6	2.19	0.09
V	18-8	2.56	0.37
VI	26-8	3.78	1.22
VII	31-5	4.44	0.66
VIII	32-1	5.23	0.79
IX	34-2	6.87	1.64
X	37 - 3	8.57	1.7
XI	39 - 2	10.76	2.19
PL	42 - 3	11.64	0.88

Hatching

Once the first stage zoea larvae were fully developed, they were ready to come out of the eggshell to start active life. The process of hatching was slow with continuous vibrations of the mouth parts of the larva accompanied by the stretching of its body forcing the eggshell to elongate gradually. Vibrations of the mouth parts become more and more vigorous accompanied by further stretching of the body. About an hour later, the thoracic appendages started to vibrate vigorously but intermittently for about few minutes with increasing length of periopods vibration which become very vigorous and continuous. The body continued to stretch and telson which was held like a mask, covering and protecting the eyes and head, started pushing outward. Suddenly, the eggshell was broken and the telson thrashed out followed by the head and with a forceful flex and stretch of the body, the hatched zoea larvae started swimming actively.

Larval Rearing

Three trials of larval rearing and seed production experiments were conducted to study the larval biology of *M. malcolmsonii* under laboratory conditions. The detailed data on larval rearing and seed production experiments of the three trails were recorded. During the experimental period, detailed studies on larval stages and their individual characteristics, larval growth and metamorphosis, moulting frequencies, food and feeding efficiency at different stages, water quality parameters of the rearing medium during entire seed production cycle and the post-larvae (PL) production were made. The different larval stages were observed to be planktonic in nature and were attracted by light. They swim actively all the times at an oblique angle by keeping their tail up and head downward. The three larval rearing trials have been undertaken during 2002- 2003. The first stage zoeae larvae completed eleven larval stages and metamorphosed into post-larvae within 40-63 days with 80 per cent metamorphosis.

Larval Growth and Metamorphosis

The larval growth from zoea stage I to IX zoea and subsequently post-larval (PL) stage were found not synchronous in all the stages. The zoea I attained stage IV after 9 days and reach to stage IX in about 34 days after hatching. The first few post-larvae appeared on 49 days of hatching. The growth pattern of the larval stages I to PL observed during the study period have been presented in Table 7.2. The initial size of zoea stage I was 1.88 mm which attained PL stage at 11.64 mm size. The larvae took 4 days to attain stage III and to attain stage IV, it took another 6 days. Larvae took another eight days to attain stage V. Subsequently stage V larvae took 8, 5, 1, 2, 3, 2 and 3 days for attaining VI, VII, VIII, IX, X, XI and PL stages, respectively (Table 7.3).

Different Larval Stages

Stage I

Size 1.88 mm, body transparent, carapace without spine. Telson triangular, posterior edge broad and concave. Abdomen with six somites, the last is not separated from telson. Eyes large, sessile transparent on outer portion light greenish and the lateral side of the black spot on eye chromatophore. Antennular peduncle slightly bluish and reddish ting tip of the antenna bluish with a red dot. Green, red chromatophore present on cephalothorax with blue margin. Blue red tinge on mid-dorsal of body. Chromatophore on the base of walking legs.

Table 7.3: Duration between the Two Larval Stages of
M. malcolmsonii in Rearing Tanks

Larval Stages	Duration (Days)		
	Range	Mean	±SD
I-II	2-4	3.0	1.41
II-III	2-6	4.0	2.83
III-IV	1-3	2.0	1.41
IV-V	1-6	3.0	3.54
V-VI	22-33	29.0	7.78
VI-VII	3-7	5.0	2.83
VII-VIII	2-6	3.0	2.83
VIII-IX	5-9	7.0	2.83
IX-X	7	7.0	0.00
X -XI	6-7	6.0	0.71
XI-PL	6-7	6.0	0.71

Stage II

Size 1.97 mm, body transparent, white tip of the antennule bluish with red dots. Small red dots on ventral side of the abdomen. Light bluish region observed on mid-dorsal prominent red region on carapace, frontal part of the red region is blue. Red chromatophore present on eye. A pair of supra-orbital and branchiostegal spines present. Pleura of abdominal somite developed 1st, 2nd, 3rd, periopod biramous, 5th periopod uniramous. Eyes large stalked, red big point present at lower portion.

Stage III

Size 2.10 mm, uropod develops one rostrum, teeth appear, 6th abdominal somite separated from telson. Bluish or reddish colour at mid-dorsal. Blue tinge at the tip of antennae. Greenish red chromatophore at the base of eye. Base of walking legs reddish and body light yellowish.

Stage IV

Size 2.19 mm, rostrum with 2 teeth, uropod biramous. Body yellowish white becomes transparent at lower portion. At the base of eyestalk, red mid-dorsal blue red. Blue tinge at the tip of antennules.

Stage V

Size 2.56 mm, red chromatophore at the base of 4th pleopod. The telson rectangular shape. Exopod of uropod rounded and endopod pointed, two rostral teeth. Other characters were same as previous.

Stage VI

Size 3.78 mm, body yellowish white, become transparent at lower portion, blue tinge on tip of antennules. Yellowish and small red tinge at the anus. Deep red on 4th pleopods extending toward 3rd, blue yellowish green red tinge present on mid-dorsal. Red tinge on the base of antennae and antennules. Red tinge and yellowish chromatophore present at bases of 1st, 2nd, 3rd pereopods and maxilipods. Red region on posterior lateral side of the carapace just above the 5th periopod from lateral view. Pleopod buds appear 2nd, 3rd, 4th more developed, 1st smallest and 2nd largest. The chromatophore on ventral eyestalk with yellowish green little reddish portion observed. Frontal portion of eyestalk toward antennae is filled with light black dots.

Stage VII

Size 4.44 mm, pleopod buds biramus, chromatophore at the base of 4th pleopod extending towards 4th pleopod extending towards 3rd and dorsal side, chromatophore at the base of 5th walking leg. Red chromatophores on ventral side of the eyestalk. Other characters same as previous.

Stage VIII

Size 5.23 mm, rostrum with 2 teeth, 3-4 spine on 2nd teeth. Eye large stalk dorso-frontal portion blackish ventral and posterior portion is yellowish green or red. Pleopod buds with ciliated structure except 5th is smallest. Body yellowish straw colour, 6th segment transparent. Blue dot at the tip of the antennae disappeared. Dorsal chromatophore disappeared. Red chromatophore at the base of antenna, antennules and mandible. Red chromatophore at anus, 4th, 3rd peleopod and above the 5th walking leg on carapace.

Stage-IX

Size 6.87 mm, appendix interna develop of the endopod of 2nd pleopod. Mid-dorsal pinkish red chromatophore increases on body. The other characters same as previous.

Stage-X

Size 8.57 mm, rostrum with 2 big and 4-5 small teeth and 4 spine on 2nd rostrum teeth. The chromatophore increases on the body. Red chromatophore at ventral side covering the 4th, 3rd, and 2nd pleopod. Red colour observed on the lateral side of the body segment near anus. Red chromatophore at the base of 3rd maxillae, no other portion of walking legs.

Stage XI

Size 10.76 mm, rostrum with 2 big and 10-11 small teeth, lower portion of 1st and 2nd teeth serrated 2nd teeth with 5 spine (4 big one small). Eye large, posterior portion of eye but greenish and anterior blackish, ventral side some other with red tinge chromatophore, body yellowish, last segment transparent. No chromatophore at mid-dorsal pinkish, red chromatophore at the base of 4th, 3rd and 2nd pleopod. Red chromatophore at anus base of the eye, antenna, antennul red, greenish at the back of the eyeball. Red chromatophore at the base of maxilliped at mandible base of 5th walking leg on the posterior lateral side of the carapace just above the 5th walking legs. Red dots present on body segment, chromatophore on mandible.

PL

Size 11.64 mm, ventral sides with red chromatophore, red chromatophore at anus. Red dots at the tip of the telson. Filler of antenna reddish. Sometime base of the walking leg red. All teeth on dorsal side of rostrum equal size, on the dorsal side of the rostrum 10-11 teeth gap, then another one tooth sometime teeth present at the tip of rostrum. 1st teeth with 4-5 spine, 2nd teeth with 3 spine then 2 spine on each teeth, frontal teeth devoid of spine.

Larval Food and Feeding

During larval rearing, the feed items were used at different stages were freshly hatched nauplii of brine shrimp (*Artemia salina*), prepared feed like egg custard and soft tissue of freshwater mussels (*Lamellidens* spp.) were given. In the initial larval stages - *i.e.* from stage I to stage III, larvae were primarily fed with *Artemia* nauplii twice a day during early morning (06-07 hrs) and night (22-23 hrs). On attaining stage III, the larvae were fed with *Artemia* nauplii four times a day during morning (06-07 hrs), noon (11-12 hrs), evening (17-18 hrs) and night (22-23hrs). The larvae generally preferred *Artemia* nauplii since it is marine animal. Once the larvae attained stage V, *Artemia* nauplii feed was supplemented with prepared feed *i.e.* egg custard during morning and noon and mussel meat during evening and *Artemia* was provided only in night hours. This feeding practice was continued till the larvae completely metamorphosed to post-larval stage. Larvae at all the stages of development showed positive feeding response to all the three types of feed used.

Post-larval Production

The newly-metamorphosed post-larvae were harvested by using shell-strings and acclimatized in freshwater condition. They were counted daily and the metamorphosis of post-larvae assessed. The larval development was not uniform and varied considerably with progress of the cycle. In different production trials, 1st post-larvae appeared within 49 days and 16,264 PL (post-larvae) were harvested on day 63 of the cycle in the three different production trials and the average PL/l range was 17.87-18.45 in the trails during first year. The corresponding survival of the larvae from zoea I to PL was about 18 per cent during 63 days of cycle. Similar trend was also recorded during second year in the three trials where total PL production was 16,149 which was slightly less than that of the first year.

Indian aquaculture has been evolving from the level of subsistence activity to that of an industry and prawns enjoy superiority over fishes owing to its universal appeal, unique taste and low fat content. It is fast becoming a popular food item among the young and olds, especially in Japan, United Kingdom, United States, Hong Kong, Singapore and other countries (New and Valenti, 2000). Of the many other species of Sub-Order Caridea, Family Palaemonidae, *Macrobrachium rosenbergii* appears to be the largest and the most desired of the known species. Due to the high export value, the giant freshwater prawn, popularly known as 'scampi', enjoys immense potential for culture in India. About 4 million ha impounded freshwater bodies in the various states of India offer great potential for freshwater prawn culture. Scampi can be cultivated for export through monoculture in existing as well as new ponds or with compatible freshwater fishes in existing ponds. Another species, generally smaller in size but with great potential of aquaculture in inland waters, *M. malcolmsonii*, has also been reported from India, Pakistan and Bangladesh (Ahemad, 1957; Kanaujia, 1989; 2003; Yakoob, 1994; Hossain *et al.,* 2012; Ahamad *et al.,* 2014).

As in Bangladesh, two major species of Genus *Macrobrachium i.e. M. rosenbergii* and *M. malcolmsonii*, are reported to have commercial importance in many estuarine areas of India. In Kerala, it is estimated that 10,000 metric tonnes of shrimps mainly species of Family Penaeidae are harvested annually, small amount of which consisted of *M. rosenbergii*. With the water development works in many estuarine areas, the natural fishery of *Macrobrachium* has been declined. *M. malcolmsonii* is an important fishery in East Godavari district of Andhra Pradesh. It is essentially an inhabitant of flowing waters such as rivers and estuarine areas. Like *Macrobrachium rosenbergii*, it requires brackishwater for spawning and nursing of the post-larval stages. In the Andhra Pradesh, a managed fishery of this species occurs when juveniles brought into flooded ponds during the monsoon season, they grow into marketable sizes when the water recedes after the monsoon. This species also occurs in many other river systems in India such as Mahanadi, Krishna and tributaries, Hooghly estuary system and in Lake Chilka and Kolleru. These river systems usually account for heavy production of juveniles during certain periods of the year which may open up the possibilities of supplying stock for rearing purposes. This species has simple feeding habits utilizing organic matter and detritus as food which simplifies its requirements for feeding. The natural habitat of this species that requires flowing water has to be overcome, as culture has to be done in stagnant impoundments (Ibrahim, 1962a; Rajlakshmi, 1980; Kanaujia, 1989; Kanaujia and Mohanty, 1992).

Although, extensive work has been done on lotic and lentic ecosystem of in India, much attention has not been paid on the study of the ecology and biology of the prawn species available in riverine ecosystems (Ray *et al.,* 1966; Bilgrami and Dutta Munshi, 1979). Studies of riverine ecosystem in India has been confined and limited to some stretches of the few rivers which provide some information about prawn population and fishery. Some aspects of plankton and fish ecology of the river Ganga and Yamuna have been studied at Allahabad suggesting it to sustain a diverse biota (Ray *et al.,* 1966). Bilgrami and Dutta Munshi (1979) made limnological studies with the impact of the human activities on flora and fauna of the river Ganga in a stretch of Patna to Farakka. The distribution, migration, growth, maturation and breeding biology of the prawns have been studied in the Ganges, Hooghly, Mahanadi and Godavari, Krishna, Cauvery and Ojat (Gujurat) to develop hatchery technology for their seed production (Tiwari, 1955; Jhingran, 1956; Ibrahim, 1962a; Subrahmanyam, 1975; Rajyalakshmi, 1980; Ahemad, 1984; Rao, 1986 and Kanaujia, 1989). Ibrahim (1962b) recorded the embryonic development of *M. malcolmsonii* and Kewalaramani *et al.* (1971) studied on larval cycle in brackishwater to produce post-larvae using indigenous feed such as *Tubifix* worm, zooplankton *etc.* Sankolli *et al.* (1984) reported the larval biology and produced few post-larvae following water renewal system. Attempt made by Central Inland Fishery Research Institute, Barrackpore during 1982-

86 resulted in the production of few post-larvae. Kanaujia and Mohanty (1992) achieved success and developed hatchery technology for commercial seed production of *M. malcolmsonii* using broodstock collected from river Mahanadi. In the present study, maturation and breeding were observed under the captive (pond) environment and post-larvae (PL, seed) production under hatchery conditions pave the way for culture of the highly promising species in ponds of the inland freshwater for diversification of aquaculture in this country (Subramanyam, 1974a, b; Ahamed and Naushin, 1990; Kutty, 2001; New and Nair, 2012; Nair and Salin, 2012).

Acknowledgements

We are grateful to the Director, Central Institute of Freshwater Aquaculture (ICAR), Bhubaneswar for providing the necessary facilities to carry out the work. This work was supported by NATP fund of ICAR, New Delhi.

References

Ahamad, F., Fulanda, B., Siddik, M.A.B., Hossain, M.Y., Mondol, M.M.R., Faruque, Z. and Ohtomi, J., 2014. An overview of freshwater prawn fishery in Bangladesh: present status and future prospects. *J. Coast. Life Med.,* **2**: 580-588.

Ahemad, N., 1957. Prawn and prawn fishery of East Pakistan. *Govt. of Pakistan, Directorate of Fisheries,* 31 p.

Ahemad. S.M.J., 1984. Fishery of freshwater prawn, *Macrobrachium malcolmsonii* (H. M. Edwards), in Sambalpur district and prospects of its culture in western Orissa. In: *Souvenir of the Seminar on Freshwater Fisheries and Rural Development. Section II* (06-07 April, 1984). pp. 1-7. Rourkela, Orissa.

Ahemad, S.M.J., 1993. A case for stocking our reservoirs with *Macrobrachium malcolmsonii. Fishing Chimes,* **12**(12): 43-44.

Ahemad, S.K.J. and Naushin, A.S., 1990. Role of river prawn, *M. malcolmsonii* (H. M. Edwards), in aquaculture. In: *Proceedings of the National Seminar on Aquaculture Devlopment in India: Problems and Prospects* (Eds. Natarajan, P. and Jayaprakash, V.). pp. 15-20. Kerala University, Thiruvananthapuram.

Ardrill, A., Jenson, R. and Thompson, R.K., 1973. The introduction of the freshwater prawn, *Macrobrachium rosenbergii,* into Mauritius. *Rev. Agric. Sour. I'Ie Maurica,* **52**: 6-11.

Bilgrami, K.S. and Datta Munshi, J.S., 1979. *Limnological Survey and Impact of Human Activities on the River Ganges: Barauni to Farakka.* Technical Report M.A.B. Project, UNESCO, Paris

Holthuis, L.B., 1980. *FAO Species. Catalogue. Vol. 1. Shrimps and Prawns of the World.* F.A.O. Fish Synop. No. 125. FAO, Rome.

Holthuis, L.B., 2000. Nomenclature and taxonomy. In: *Freshwater Prawn Culture: The Farming of Macrobrachium rosenbergii* (Eds. New, M.B. and Valenti, W.C.). pp. 12-17. Blackwell Science Ltd., Oxford and London.

Hossain, M.Y., Ohtom, J., Jaman, A., Saleah, J. and Robert, L.V.J., 2012. Life-history traits of the monsoon river prawn, *Macrobrachium malcolmsonii* (H. Milne

Edwards, 1844) (Palaemonidae), in the Ganges (Padma) river, northwestern Bangladesh. *J. Freshw. Ecol.* **27**, 131-142.

Ibrahim, K.H., 1962a. Observations on the fishery and biology of the freshwater prawn, *Macrobrachium malcolmsonii* (H. M. Edwards), of river Godavari. *Indian J. Fish.,* 9: 433- 467.

Ibrahim, K.H., 1962b. On the early embryonic development of *M. malcolmsonii* (H. M. Edwards) and *M. scabriculus* (Heller) from river Godavari. *Sci. & Cult.,* **28**: 232-233.

Jhingran, V.G., 1956. The capture fishery of river Ganga at Buxar (Bihar, India) in the years 1952-1954. *Indian J. Fish.,* **3**: 197-215.

Jhingran V.G., 2003. *Fish and Fisheries of India.* Hindustan Pub. Corp., Delhi.

Kanaujia D.R., 1989. Biology of freshwater prawn, *Macrobrachium malcolmsonii*, of river Ganga, Buxar, Bihar. In: *Proceedings of the National Seminar of Freshwater Aquaculture.* pp. 51-54. CIFA, Kaushalyaganga, Bhubaneswar.

Kanaujia, D.R., 2003. Scope of prawn aquaculture in Uttar Pradesh. In: *Brain Storming Session on Production Status and Potential of Fisheries in Uttar Pradesh - Constraints and Opportunity* (22 January 2003). pp. 77-93. Uttar Pradesh Council of Agricultural Research, Lucknow.

Kanaujia, D.R. and Mohanty, A.N., 1992. Breeding and large-scale seed production of the Indian river prawn, *Macrobrachium malcolmsonii* (H. M. Edwards). *J. Aqua.,* **2**: 7- 16.

Kanaujia, D. R. and Mohanty, A. N., 1996. Prospects of both mono- and mixed-culture of *Macrobrachium malcolmsonii. Fishing Chimes,* **15**(12): 33-35.

Kanaujia, D.R. and Mohanty, A.N., 2001. Effect of salinity on the survival and growth of the juveniles of Indian river prawn, *Macrobrachium malcolmsonii* (H.M. Edwards). *J. Adv. Zool.,* **22**: 31- 36.

Kanaujia, D.R., Mohanty, A.N. and Soni, S., 2001. Breakthrough in seed production of Ganga river prawn *Macrobrachium gangeticum* (Bate, 1868): a milestone in Aqua-farming. *Fishing Chimes,* **21**(1): 28-30.

Kewalramani, H.G., Sankolli, K.N. and Shenoy, S.S., 1971. On the larval life-history of *Macrobrachium malcolmsonii* (H. M. Edwards) in captivity. *J. Indian Fish Assoc.,* **1**(1): 1-25.

Kurian, C.V. and Sebastian, V., 2002. *Prawn and Prawn Fisheries of India.* Hindustan Pub. Corp., Delhi. 310 p.

Kutty, M.N., 2001. Diversification of aquaculture. In: *Sustainable Indian Fisheries* (Ed. Pandian, T.J.). pp. 189-212. National Academy of Agricultural Sciences, New Delhi.

Kutty, M.N., Herman, F. and Menn, H.L., 2000. Culture of other prawn species. In: *Freshwater Prawn Culture: The Farming of Macrobrachium rosenbergii* (Eds. New, M.B. and Valenti, W.C.). pp. 349-410. Blackwell Science Ltd., Oxford & London.

Ling, S.W. and Costello, T.J., 1979. The culture of freshwater prawns: a review. In: *Advances in Aquaculture* (Eds. Pillay, T.V.R. and Dill, W.A.). pp. 299-304. Fishing News Books Ltd., London.

Mishra, P., Patra, B. C., Pandey, A.K. and Kanaujia, D. R., 2014. Growth and production of Indian river prawn, *Macrobrachium malcolmsonii*, in monoculture system. *J. Exp. Zool. India,* **17**: 525-530.

Nair, C.M. and Salin, K.R., 2012. Current status and prospects of farming the gaint river prawn, *Macrobrachium rosenbergii* (de Man) and monsoon river prawn, *Macrobrachium malcolmsonii* (H. Milne Edwards), in India. *Aquacult. Res.,* **43**: 999-1014.

New, M.B., 1990. Freshwater prawn culture: a review. *Aqauculture,* **88**: 99-143.

New, M.B., 1995. Status of freshwater prawn farming: a review. *Aquacult. Res.,* **26**: 1-54.

New, M.B., 1994. Freshwater prawn farming: a review of current research, global status, opportunities and constraints. In: *Freshwater Prawn Farming in India* (Eds. Thakur, N.K., Tiwari, R. and Joseph, M.M.). pp. 1-27. Special Pub. No. 10. Asian Fisheries Society: Indian Branch, Mangalore.

New, M.B., 2003. The role of freshwater prawn in sustainable aquaculture. In: *Souvenir International Symposium on Freshwater Prawns* (2003). pp. 10-13. College of Fisheries, Kerala Agricultural University, Kochi.

New, M.B. and Singholka, S., 1985. Freshwater Prawn Farming: A Manual for the Culture of Macrobrachium rosenbergii. FAO Fish. Tech. Pap. No. 225. pp. 1-124.

New, M.B. and Valenti, W.C., 2000. Freshwater Prawn Culture. The Farming of *Macrobrachium rosenbergii.* Blackwell Science Ltd., Oxford & London. 443 p.

New, M.B. and Nair, C.M., 2012. Global scale of freshwater prawn farming. *Aquacult. Res.,* **43**: 960-969.

New, M.B., Valenti, W.C., Tidwell, J.H., D'Abramo, L.R. and Kutty, M.N. (2010). *Frweshwater Prawns: Biology and Farming.* Wiley-Blackwell, Oxford, United Kingdom.

Radheyshyam, 2009. Farming the freshwater prawn, *Macrobrachium malcolmsonii. Aqua. Asia,* **14**(1): 29-32.

Rajyalakshmi, T., 1980. Comparative study of the biology of the freshwater prawn, *Macrobrachium malcolmsonii* of Godavari and Hooghly river system. *Proc. Indian Natl. Sci. Acad.,* **46B**: 72- 89.

Rajyalakshmi, T., 1984. Introduction of *Macrobrachium* group of prawns in rural aquacultures schemes in northern Orissa. In: *Souvenir of the Seminar on Freshwater Fisheries and Rural Development* (6-7 April 1984). pp. 55-62. Rourkela, Orissa.

Rajyalakshmi, T. and Randhir, M., 1969. The commercial prawn *Macrobrachium malcolmsonii* (H. Milne Edwards) of the river Godavari, a discussion on the trend and characteristics of the population during 1963-1966. *FAO Fish Rep.* **3**(57): 903-920.

Rao, K.J., 1986. Studies on maturation, breeding, fecundity and sex-ratio in *Macrobrachium malcolmsonii* (H. Milne Edwards) from Kolleru lake. *J. Aqua. Biol.,* **4**(2): 62-72.

Ray, P., Singh, S.B. and Sehgal, K.L., 1966. A study of some aspects of ecology of the river Yamuna and Ganga at Allahabad (U.P.) in 1958-59. *Proc. Nat. Acad. Sci. India,* **36B**: 235-372.

Sankolli, K.N., Shenoy, S., Jalihol, D.R. and Almekar, G.B., 1984. Egg-to egg mass culture technique of the Godavary prawn, *Macrobrachium malcolmsonii* (H. Milne Edwards), in a static culture system. In. *Souvenir of the Seminar on Freshwater Fisheries and Rural Development* (6-7 April, 1984). pp. 15-18. Rourkela, Orissa.

Subrahmanyam, M., 1974a. Farming potentials of the Indian river prawn, *Macrobrachium malcolmsonii. Seafood Expot J.,* **6**(3): 1- 4.

Subrahmanyam, M., 1974b. New prawns for farming and export from India. *Seafood Export J.,* **6**(12): 1- 3.

Subrahmanyam, M., 1975. Induced breeding of prawns. *Seafood Export J.,* **7**(1): 4-8.

Tiwari, K.K., 1949. On a new species of *Palaemon* from Banaras with a note *Palaemon lanchestri* (de Man). *Rec. Indian Mus.,* **45**: 333- 345.

Tiwari, K.K., 1955. Distribution of Indo-Burmese freshwater prawns of the genus *Palaemon* Fabricius, its bearing on the Satpura hypothesis. *Bull. Nat. Int. Sci. India,* **7**: 230-239.

Tiwari, K.K., 1975. Culture species of prawns and shrimps of India and principles of their selection. In: *Summer Institute of Intensive Freshwater Fish Culture* (15 June-14 July, 1975). pp. 1-11 Central Inland Fisheries Research Institute (ICAR), Barrackpore.

Tiwari, K.K. and Holthuis, L.B., 1996. The identity of *Macrobrachium gangeticum* (Bate), 1868 (Decapoda, Caridea, Palemonidae). *Crustaceana,* **69**: 922-925.

Valenti W.C. and Daniels, W., 2000. Recirculation hatchery systems and management. In: *Freshwater Prawn Culture: The Farming of Macrobrachium rosenbergii* (Eds. New, M.B. and Valenti, W.C.). pp. 69 -90. Blackwell Science Ltd., Oxford & London.

Yakoob, M., 1994. The aquaculture of freshwater prawn *M. malcolmsonii* (H. Milne Edwards), at farmers level in Pakistan. *Agric. Fish. Manage.,* **25**: 355-361.

2015, Perspectives in Animal Ecology and Reproduction, Vol. 10   *Pages 123–136*
Editors: V.K. Gupta, Anil K. Verma and G.D. Singh
Published by: DAYA PUBLISHING HOUSE, NEW DELHI

Chapter 8

Shift in Foot Protrusion Frequency, Cholinergic Impairment in Pedal Ganglia and Histopathology of Labial Palp in Freshwater Mussel Exposed to Cypermethrin

Krishnendu Das, Mitali Ray and Sajal Ray*

Aquatic Toxicology Laboratory, Department of Zoology,
University of Calcutta, 35 Ballygunge Circular Road,
Kolkata – 700 019, West Bengal, India

ABSTRACT

Cypermethrin, a non polar, thermostable, neurotoxic pyrethroid pesticide being widely used by the Indian farmers and also in the other parts of the world to control the agricultural pests and plant diseases. Cypermethrin is readily adsorbed into the soil surface and bound there. Thus sedentary suspension feeders like *Lamellidens marginalis* (Mollusca: Bivalvia) is subjected to exposure of cypermethrin contaminated sediments in freshwater bodies. Reports on cypermethrin toxicity in this economically, nutritionally and medicinally important freshwater bivalve is limited. In this study, toxicity of cypermethrin is screened in *L.marginalis* in respect to foot protrusion behaviour, acetylcholinesterase (AChE) activity in pedal ganglia and histopathological analysis of labial palp. The species was exposed to 0.03ppm, 0.05ppm, 0.07ppm and 0.09ppm of cypermethrin for 1, 2, 3, 4, 7, 15 and 30 days of exposure. In this study, foot protrusion was inhibited in a dose dependent manner. Initial rise in AChE activity was followed by steady decline with the increase of dose and time. Histopathological study on labial palp revealed structural damage in the densely ciliated dorsal fold, labial palp ridge

_____

* *Corresponding author.* E-mail: raysnailmail@rediffmail.com

and oral groove. Prolonged sublethal exposure of cypermethrin may affect normal suspension feeding habit of *L.marginalis* and lead to decline of this important bioresource in its natural habitat. This study is aimed to establish an effective biomarker of cypermethrin toxicity in freshwater ecosystem.

Keywords*: Pyrethroid, Cypermethrin, Lamellidens marginalis, Acetylcholinesterase, Labial palp, Foot protrusion, Biomarker.*

Introduction

Wetland of India, estimated to be only 4.63 per cent of total geographic area is important repositories of aquatic biodiversity (Deepa and Ramchandra, 1999; Prasad *et al.,* 2002). For the last few decades toxic contamination of wetlands by various xenobiotics including pesticides through agricultural runoff has been of concern. The possible consequences of this toxic contamination are gradual loss of biodiversity, dwindling of biofilter species, gradual loss of traditional food items of human, poultry and fish. Pyrethroid insecticides have been introduced over the past two decades in both agricultural and domestic applications. Environmental protection agency (EPA), USA. assessed the risks to non target organisms for ten synthetic pyrethroids *i.e.* bifenthrin, cyfluthrin, cypermethrin, deltamethrin, fenpropathrin, fenvalerate, cyhalothrin, tefluthrin, tralomethrin and permethrin. Compared to other pyrethroids, cypermethrin is relatively stable, with a half life of 8-16 days in direct sunlight. In India, cypermethrin, the reported neurotoxin pesticide is being widely used to control pest for tea, paddy, cotton, cabbage, brinjal, sugarcane, wheat etc. In soil the half-life of cypermethrin is to be as long as 8 weeks and in water as long as 100 days (ETN Cypermethrin., 1996). Cypermethrin bears adequate stability in air and light (Kaufman *et al.,* 1981). Hydrolysis of the ester linkage is the principal degradation route and leads to the formation of 3-phenoxybenzoic acid (PBA) and cyclopropanecarboxylic acid derivatives (Sakata *et al.,* 1986), principally, 3-(2,2-dichlorovinyl)-2,2-dimethyl cyclopropanecarboxylic acid (DCVA) (Kaufman *et al.,* 1981). Cypermethrin undergo photodegradation rapidly on soil surfaces into various byproducts, with half-lives of 8-16 days (Walker and Keith., 1992). Cypermethrin is a reported neurotoxin for aquatic and terrestrial life forms (Curtis *et al.,* 1995). Cypermethrin may leads to malfunctioning of neurotransmitters, followed by hyper and hypoopercular activity, imbalanced swimming activity in fish (Marigoudar *et al.,* 2009). Cypermethrin can produce toxic effects in mammals (Nair *et al.,* 2011; Wang *et al.,* 2010; Grewall *et al.,* 2010) including human (Saxena, 2010). Soils and sediments act as main environmental reservoirs of cypermethrin (Bacci *et al.,* 1987). During monsoon, the freshwater aquatic fields, the natural habitat of *Lamellidens marginalis* are contaminated by agricultural runoff containing pesticides including cypermethrin. Thus, the survival of the species is under threat due to cypermethrin toxicity (Beth *et al.,* 2004; Subramaninan *et al.,* 1999; Das *et al.,* 2012). Cypermethrin is a non-polar pesticide and readily adsorbed into the soil surface and bound there (Kaufman *et al.,* 1981). Thus this sedentary suspension feeder subjected to huge cypermethrin contaminated sediments in freshwater bodies. *L. marginalis* can accumulate a large deposit of cadmium, nickel in its different tissues from its environment and the rate of accumulation was found to be dose and time dependent (Das and Jana, 2003). Cypermethrin induced significant alteration in dynamics of hemocyte density and impairment of gills and digestive glands in *L. marginalis* were reported by Das *et al.*(2012, 2013). This study is aimed to save the economically, nutritionally and medicinally important freshwater bivalve *L.marginalis* and to establish an inexpensive biomarker to carry out continuous screening of cypermethrin toxicity. Burrowing behaviour largely depends on foot protrusion (Allen., 2009). Any significant alteration in foot protrusion frequency under exposure of cypermethrin may affect their feeding. The enzyme acetylcholinesterase (AChE) plays a major role in the neurotransmission (Cajaraville *et al.*, 2000; Rickwood *et al.*, 2004). AChE activity has been established as an efficient biomarker of neurotoxicity in bivalves (Najimi *et al.*, 1997; Radenac *et al.*, 1998). Combined monitoring of behaviour and neurological function helped to estimate toxicological significance of exposure of neurotoxin in target and non target organisms (Beauvais *et al.*, 2000). Information of cypermethrin mediated inhibition of AChE activity in nerve tissue is very limited for *L.marginalis*. Feeding mechanism in *L.marginalis* is highly dependent on selection and rejection mechanism of labial palp (Garrido *et al.*, 2012). Labial palp plays an important site for ingestion volume control and capable of effecting particle selection. Labial palp between the gills and mouth select, transport and convey particulate to be ingested as food or rejected as pseudofaeces (Foster *et al.*, 1978; Kiorboe *et al.*, 1981). As the inner and outer palp of each pair fused along with the dorsal margins forming the oral groove and the dense ciliation on dorsal fold correlates with the particle transport function of this surface. So any structural damage in that labial palp region may affect the food particle transport mechanism. In this present study foot protrusion frequency, AChE activity of pedal ganglia and histopathological survey of labial palp in *L.marginalis* has been investigated thoroughly under exposure of multiple sublethal concentrations of cypermethrin. All data were analyzed to establish its suitability as a biomarker of cypermethrin in aquatic ecosystem.

Materials and Methods

Fresh specimens of *L.marginalis* were collected from some selected unpolluted freshwater ponds of the district of 24 parganas (South) of West Bengal. They were brought to laboratory immediately after collection in water taken from the specific habitat. They were kept in well aerated fiber-glass tanks for 5days in room temperature to acclimate and care was taken to avoid crowding. Continuous flow of tap water was maintained so that oxygen did not act as stress factor. Along with some modification here in this experiment the rearing and maintenance of *L.marginalis* were carried out according to Raut (1991). Aqueous solutions of cypermethrin (United Phosphorus Limited, India; 10 per cent EC; CAS number 52315-07-8) were prepared in Borosilicate glass containers at concentrations of 0.03, 0.05, 0.07 and 0.09 ppm. The pH of the solution was maintained at 7.2. Each experiment set consisted of 10 animals of the same size. They were exposed to a volume of 5 liter of cypermethrin solution. For control, a set of animals were kept in cypermethrin free analytical grade water. The experiments were carried out in static water environment and fresh solutions of cypermethrin were replenished at every 12 hours. The temperature of the water was maintained between 24-26°C. Foot protrusion in *L.marginalis* of uniform size was studied for 30days at room temperature in static water for both control and cypermethrin treated animals respectively. The number of animals showing foot protrusion were examined at every 1 hour a day from 9 A. M. to 9 P. M. for 30 days (Bayne and Widdows., 1978). Acetylcholinesterase [(AChE), E. C. 3. 1. 1. 7] activity of (pedal ganglia was measured after Hestrin, S. (1949) with little modification and the corresponding protein content was estimated after Lowery *et al.* (1951). The samples were allowed to react with alkaline hydroxylamine reagent (freshly prepared v/v solution of 2M hydroxylamine hydrochloride and 3.5 N sodium hydroxide) for 1 minute. Addition of 1.0 ml concentrated hydrochloride acid (sp. Gr. 1.18, diluted with 2 parts by volume of water) and 1.0 ml of the iron solution (0.28 M Ferric chloride, hexahydrate in hydrochloroid acid (0.1 N) were made to brought the pH to 1.2 ± 2.0. Spectrophotometrical (CECIL- CE 4002) estimation of purple-brown coloured solution repeated for 5 times. A standard curve of acetylcholine chloride was used to estimate the optical density of the sample under study. Aseptically dissected labial palp of *L.marginalis* was collected with fine, sharp and sterile scalpel. Removal of mucus, blood from small pieces of collected tissue samples were carried out with chilled sterile snail saline (SSS, 5mM HEPES, 3.7mM NaOH, 36 mM NaCl, 2mM KCL,2mM $MgCl_2.2H_2O$ and 4mM $CaCl_2.2H_2O$ at pH 7.8) (Chakraborty *et al.*,2008). Immediately thereafter the tissues were fixed in Bouin's solution (saturated picric acid : 75parts,37-40 per cent formaldehyde : 25 parts and glacial acetic acid : 5 parts). Then those fixed tissue samples were passed through standard alcoholic gradation. After 5 days of immersion in cedar wood oil the tissue samples were embedded in paraffin and blocks were prepared. Tissue sections with an average thickness of 5-6 µ were made by using a microtome. Harris haematoxylin and eosin stains were used to stain those tissue sections. Stained sections were studied under camera attached digital microscope (Olympus BX 51,Germany)(Chakraborty *et al.,* 2010).

Results

Foot protrusion frequency was inhibited under different sublethal concentration of cypermethrin (Figures 8.1 and 8.2). The maximum foot protrusion frequency was recorded as 81.45 ± 3.67 at 24 h of 0.03 ppm cypermethrin exposure against a control value of 85.29 ± 3.94.The lowest value recorded was 38.53 ± 3.98 for the specimens exposed to 0.09 ppm cypermethrin for a time span of 30 days (Figure 8.3). Highest AChE activity was recorded as 0.71 ± 0.065 µM/mg protein/minute against 0.03 ppm of cypermethrin for 30 days in pedal ganglion lysates against the control value

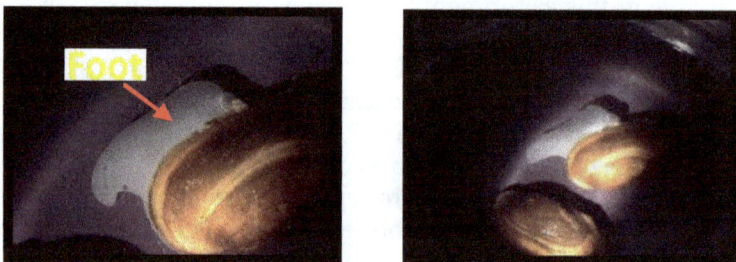

Figure 8.1: Foot Protrusion Response of *L.marginalis* in Control Species.

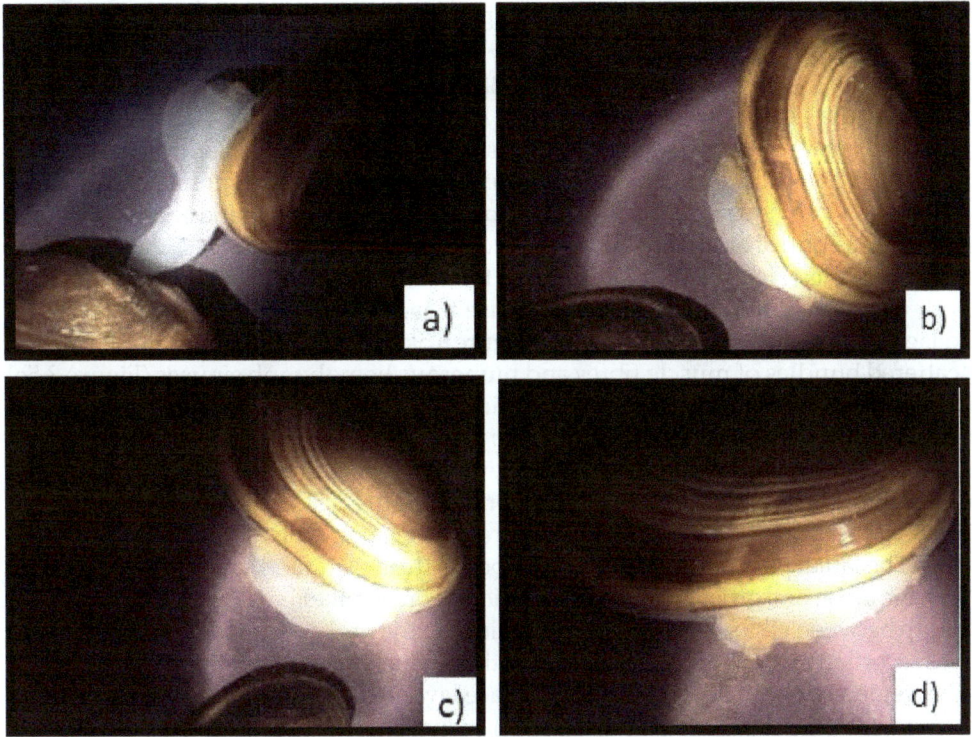

Figure 8.2: Foot Protrusion Response of *L. marginalis* at 15 Days of Exposure to (a) 0.03ppm, (b) 0.05ppm, (c) 0.07ppm and (d) 0.09ppm Cypermethrin.

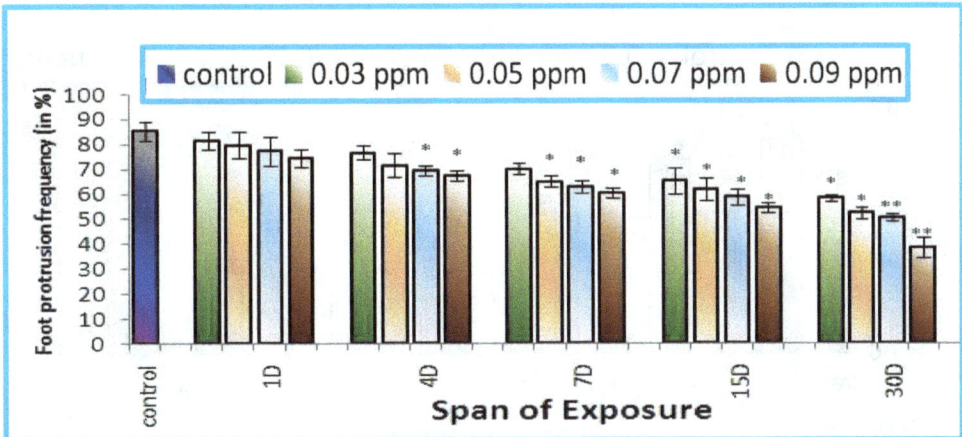

Figure 8.3: Foot Protrusion Frequency of *L. marginalis* at 1, 4, 7, 15 and 30 Days of Exposure to 0.03, 0.05, 0.07 and 0.09 ppm of Cypermethrin. Data represented is the mean ± standard deviation (n = 10). The asterics (*) indicate the values that are significantly different (*P<0.05,* *P<0.01) from the control value.

of 0.59 ± 0.081 µM/mg protein/minute. The lowest AChE activity was recorded as 0.046 ± 0.015 µM/mg protein/minute for the pedal ganglion lysates of *L.marginalis* exposed to 0.09 ppm of cypermethrin for 30 days. No trace of mortality was observed in spite of the significant steady decline in AChE activity (Figure 8.4). Prolonged shell gaps with protruded foot suggested possible paralytic effect on adductor and foot muscles. The study of histopathology of the labial palp of cypermethrin treated *L.marginalis* provides a clear view on damage and destruction of the organ. The labial palp of the control animal bear dorsal fold (Figure 8.5a) consists of densely ciliated epithelial cells (Figure 8.5k) extends over the entire dorsal fold. The troughs present with ciliated columnar epithelium (Figure 8.5a,m). The palps are composed of a collagenous connective tissue network (Figure 8.5k), within which there are distributed scattered bundles of muscle fibres and numerous hemolymph sinuses (Figure 8.5e) also noticed. Dorsal surfaces of each pair of labial palps were deeply ridged and densely ciliated. The ridges led down to the ciliated oral groove. The inner ridge surfaces of the palp are folded into deep ridges and grooves (Figure 8.5c). The epithelium rests on thin basement membrane (Figure 8.5k) consisting of connective tissue fibers and muscles. Water channel remain clear and lined by ciliated epithelial cells. The labial palp tissue exposed to cypermethrin exhibited severe morphological damage in the oral groove, palp crest and ridged oral epithelium (Figure 8.5d). Increase in gap between ciliated columnar epithelium in the oral groove region (Figure 8.5h) was recorded along with hyper vacuolization (Figure 8.5l) in the degenerated muscle fiber region between inner and outer surface layer of labial palp. Scattered infiltrated hemocytes were recorded (Figure 8.5h) in palp tissue. Densely packed ciliated epithelium in the dorsal fold region apparently lost their tightness (Figure 8.5r) and structural integrity.

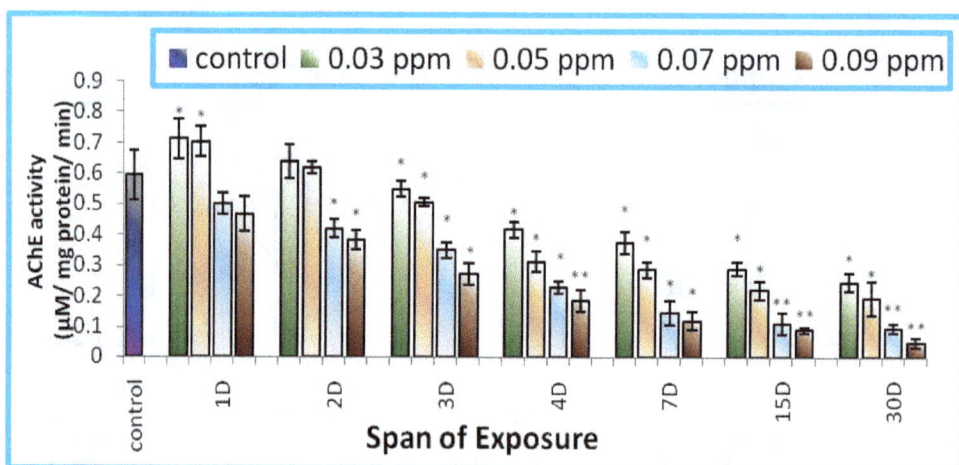

Figure 8.4: AChE Activity in Pedal Ganglion Lysate of *L. marginalis* at 1, 2, 3, 4, 7, 15 and 30 Days of Exposure to 0.03, 0.05, 0.07 and 0.09 ppm of Cypermethrin. Data represented is the mean ± standard deviation (n = 10). The asterics (*) indicate the values that are significantly different (*P<0.05, **P<0.01) from the control value.

Figure 8.5: Transverse Section of the Labial Palps of *L .marginalis.*

(a), (c), (e), (g), (i), (k), (m), (o), (q) normal labial palp exhibiting densely ciliated epithelium in the dorsal fold, trough with ciliated columnar epithelium, collagenous connective tissue network with scattered bundles of muscle fibers, hemolymph sinus, and epithelium rest on thin basal membrane.; (b), (d), (f), (h), (j), (l), (n), (p), (r) cypermethrin exposed (0.09ppm/30days) labial palp exhibiting severely damaged oral groove, palp crest, ciliary columnar epithelium, muscle fiber and with variable number of infiltrated hemocytes.

ASR: Anterior slope of ridge; BM: Basement membrane; C: Cilia; CCE: Ciliated columnar epithelium; DF: Dorsal fold; DGCE: Degenerated ciliated epithelium; DROE: Damaged ridged oral epithelium; DCE: Densely packed ciliated epithelium; DVCT: Degenerated vascular connective tissue with few infiltrated hemocytes; FS: Foot side; GCCE: Gap between ciliated columnar epithelium; HS: Hemolymph sinus; IH: Infiltrated hemocytes; ILP: Inner labial palp; ME: Mouth end; M: Muscle fiber; MS: Mantle side; NIH: No trace of infiltrating hemocytes; NROE: Normal ridged oral epithelium; OG: Oral groove; OLP: Outer labial palp; PC: Palp crest; PSR: Posterior slope of ridge; RPC: Ruptured palp crest; TCCE: Tightly packed ciliated columnar epithelium; VCT: Vesicular connective tissue; VCTV: Vesicular connective tissue with vacuoles; VCTIH: Vesicular connective tissue with infiltrated hemocytes.

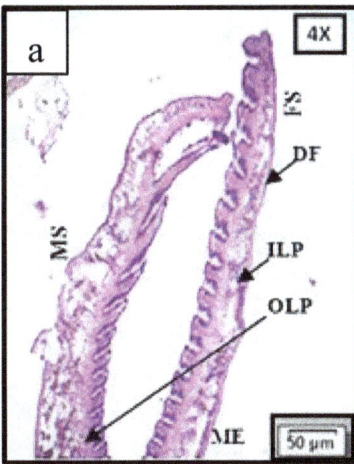

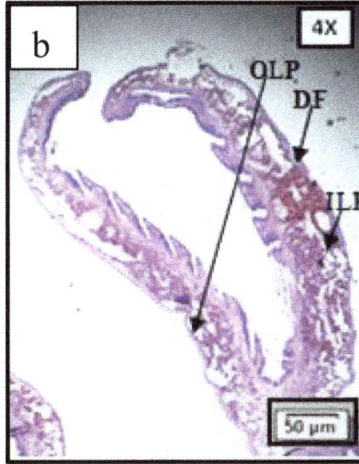

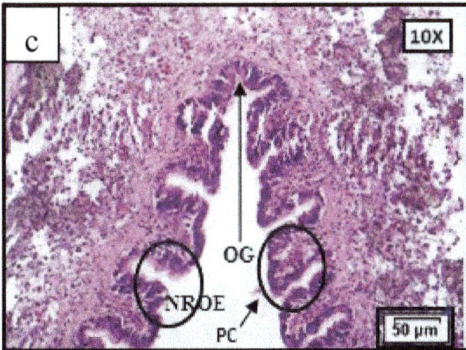

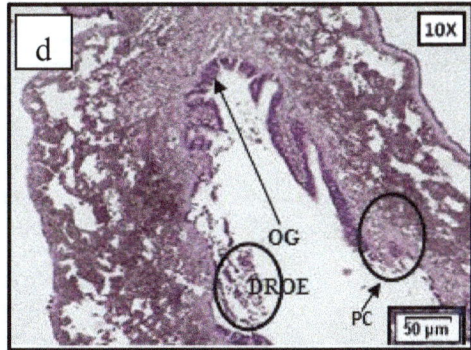

Contd...

Figure 8.5– *Contd...*

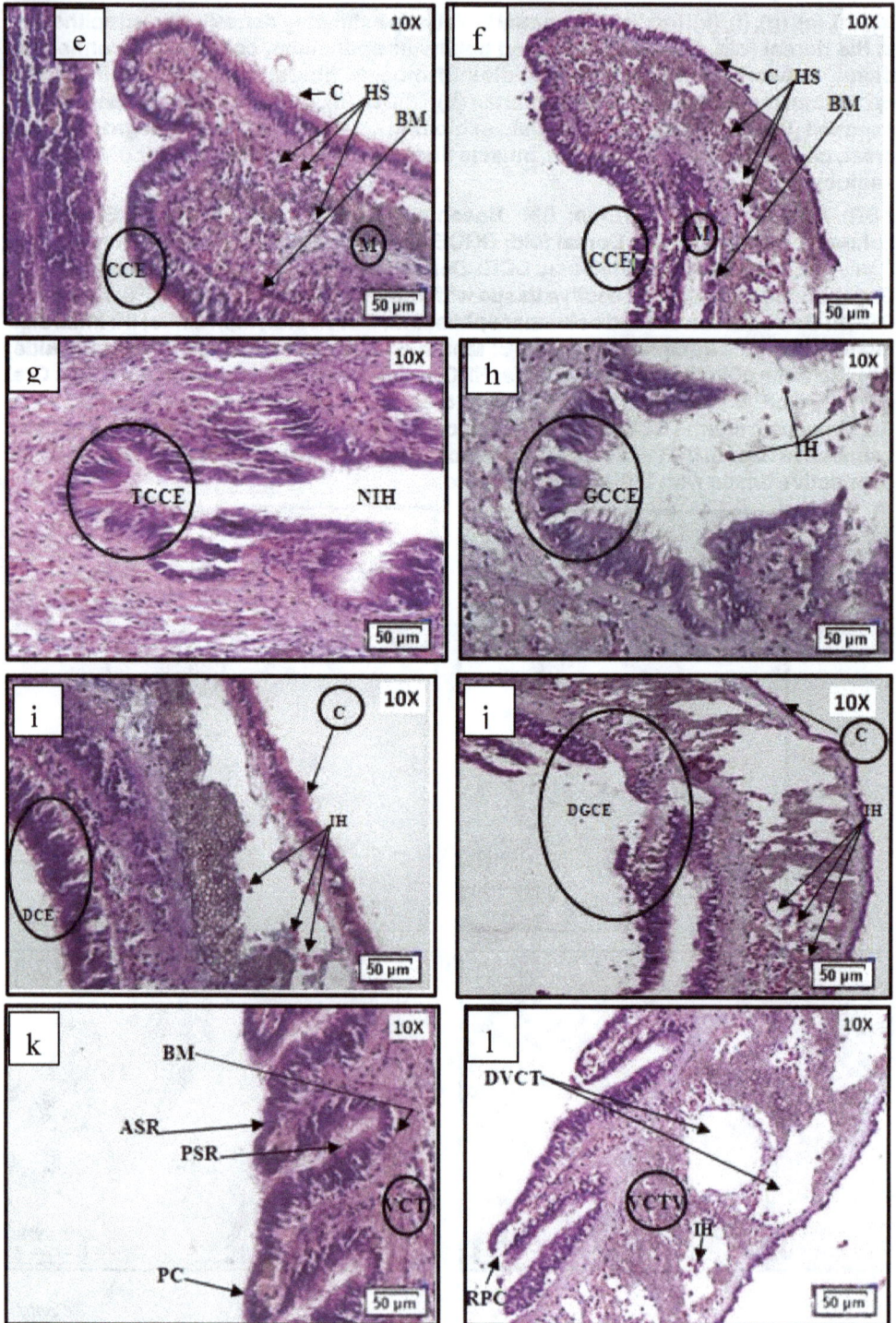

Contd...

Figure 8.5–*Contd...*

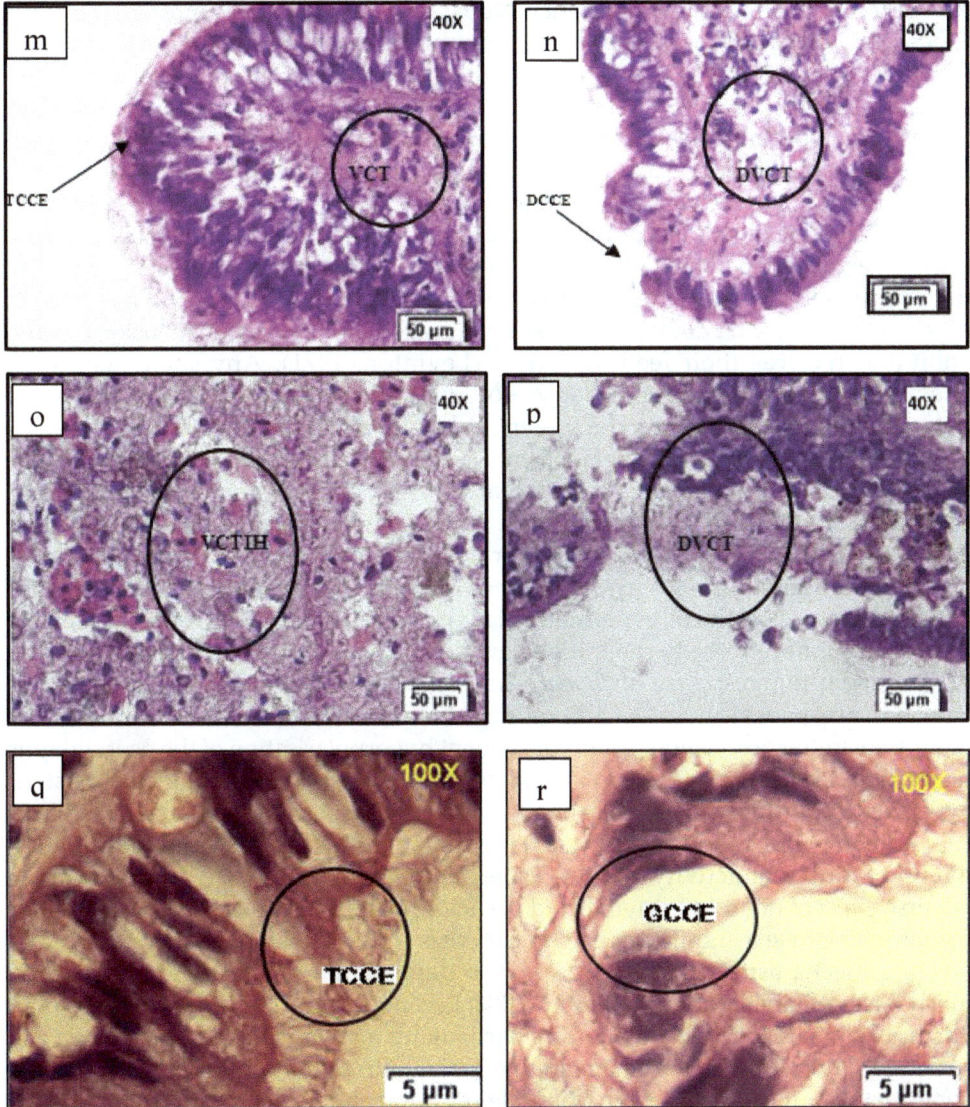

Discussion

Pyrethroids are widely used neurotoxic pesticides and can enter body of an organism to exert their effects. Pyrethroids are estimated at 23 per cent of the insecticides in world market, with more than 3500 registered formulations and are widely used in agriculture, residential areas, public health and food preparation (Casida and Qaistad., 1998). Cypermethrin one of the neurotoxic pyrethroid pesticide is widely used by the farmers in India. Information on cypermethrin toxicity in *L.marginalis* is scanty. This study is aimed to correlate among behavioural response, neurotoxicity and histopathological changes of food sorting organ and to establish the possible reason behind their nutrition deficiency. Nutritional deficiency may affect the survival fitness of *L.marginalis* in cypermethrin contaminated natural habitat. In the present study behavioural response as foot protrusion, neurotoxic response in AChE activity and histopathology of food sorting organ, labial palp of cypermethrin exposed *L.marginalis* a possible decline in their suspension feeding. Unusual gaping of shell valves in response to pollutant was reported by several authors (Hietanen, 1988; Yasmeen, 2012). Inhibitory role of neurotoxic pesticide cypermethrin on AChE activity was reported by several authors (Khan, 2005; Tiwari, 2012) and this may lead to a reduced level of neuromuscular action on foot muscle, adductor muscle and finally to opening and closing pattern of bivalve. Declined in foot protruding response in cypermethrin exposed *L.marginalis* possibly disabled their normal suspension feeding habit and reduces their feeding radius (Levinton, 1971). Any such significant alteration in foot protruding response would interfere in deposit feeding mediated sediment mixing, dissolve oxygen content, habitat of epiphytic and epizoic organisms (Vaughn, 2001) and microbial metabolism (Dame, 1996). The decrease in AChE activity in mollusc has been considered as a biomarker of exposure to neurotoxic compound (Bocquene *et al.,* 1997; Chakraborty *et al.,* 2010). Cypermethrin induced decrease in AChE activity was in report by several authors (Khan, 2005; Tiwari, 2012). Hyperexcitation in nerves due to prolonged opening of sodium channel in response to the neurotoxic pesticide, cypermethrin may cause death to the affected organisms (Baffi., 2005). In this study the initial elevation of AChE activity was indicative to possible adaptive response of *L. marginalis* against cypermethrin toxicity. But the gradual suppression in AChE generation may lead to accumulation of neurotransmitter acetylcholine and subsequent hyperpolarisation of the post synaptic membrane (Singer, 1960; Galloway *et al.,* 2002; Yaqin, 2008) against persistent cypermethrin induced stress. Declined level of AChE activity under cypermethrin exposure in pedal ganglia suggested a possible inhibition in nervous coordination with foot muscular activity during suspension feeding of *L. marginalis*. Due to the adverse effect of neurotoxic pyrethroid pesticide, a gradual loss of such efficient filter feeder may cause a significant ecological crisis in the freshwater habitat of India. Feeding mechanism in *L.marginalis* is also dependent on selection and rejection mechanism of labial palp (Ward *et al.,* 2003; Beninger *et al.,* 1995). Labial palps play an important site for ingestion of food particulates and capable of effecting particle selection. Thus structural damage in that labial palp region under the exposure of cypermethrin may affect the food particle selection and transport mechanism. This present study provide important information regarding the shift in foot protrusion frequency, cholinergic impairment of pedal ganglia and histopathology of labial palp in *L.marginalis* under the exposure of cypermethrin. This study validates the use of behavioural response, neurotoxicity and histopathological alterations as the potential biomarker to assess the degree of cypermethrin mediated aquatic toxicity in freshwater ecosystem of India.

Acknowledgements

Grant support received from the Department of Science and Technology, Government of India, to MR is thankfully acknowledged. UGC SAP DRS II programme at Zoology Department is also acknowledged too.

References

Allen, D. C. and Vaughn, C. C. 2009. Burrowing behaviour of freshwater mussels in experimentally manipulated communities. *J. N. Am. Bent. Soc.*, **28**(1): 93-100.

Bacci, E.D. Calamari, C.Gaggi and M.Vighi. 1987. An approach for the prediction of environmental distribution and fate of cypermethrin. *Chemosphere*, **16**(7): 1373-1380.

Baffi, A. B., Perira, C. D., Souza, G. R. L., Bonetti, A. M., Caron, C. R. and Gourlart, L. 2005. Esterase profile in a pyrethroid-resistant Brazilian strain of the cattle tick *Boophilus microplus* (Acari, Ixodidae). *Gen. Mol. Biol.*, **28**(4): 749-753.

Bayne, B. L. and Widdows, J. 1978. The physiological ecology of two populations of *Mytilus edulis* L. *Oecologia* (Berlin), **37**: 137-162.

Beauvais, S. H., Jones, S. B., Brewer, S. K. and Little, E. E. 2000. Physiological measures of neurotoxicity of diazinon and malathion to larval rainbow trout *Onchorhynchus mykiss* and their correlation with behavioural measures. *Environ. Toxicol. Chem.*, **19**: 1875-1880.

Beninger, P. G., Jean, S. D. and Poussart, Y. 1995. Labial palps of the blue mussel *Mytilus edulis* (Bivalvia: Mytilidae). *Mar. Biol.*, **123**: 293-303.

Beth, C., Michael, S., Mac, L., Robin, O. and Jay, L. 2004. Evaluation of potential health risks to Eastern Elliptio (*Elliptio complanata*) (Mollusca: Bivalvia: Unionida: Unionidae) and implications for sympatric endangered freshwater mussel species. *Jou. Aqu. Eco. Str. Rec.*, **90**: 35-42.

Bocquene, G., Roig, A., and Fournier, D. 1997. Cholinesterases from the common oyster (*Crassostrea gigas*): evidence for the presence of a soluble acetylcholinesterase insensitive to organophosphate and carbamate inhibitors, *Febs. Lett.*, **407**(3): 261-266.

Cajaraville, M.P., Bebianno, M.J., Blasco, J., Porte, C., Sarasquete, C. and Viarengo, A. (2000). The use of biomarkers to assess the impact of pollution in coastal environment of the Iberian Peninsula: a practical approach. *Sci. Tot. Env.*, **247**: 295–311.

Casida, J. E. and Qaistad, G. B. 1998. Golden age of insecticide research: past, present or future. *Ann. Rev. Ento.*, **43**: 1-16.

Chakraborty, S., Ray, M. and Ray, S.2008. Sodium Arsenite Induced Alteration of Hemocyte Density of *Lamellidens marginalis* – An Edible Mollusk from India. *Clean.*, **36** (2): 195 – 200.

Chakraborty, S., Ray, M. and Ray, S. 2010. Toxicity of sodium arsenite in the gill of an economically important mollusk of India. *Fish and Shellfish*, **29**(1): 136-48.

Curtis, J. E. and Horne, P. A. 1995. Effect of Chlorpyrifos and Cypermethrin Applications on Non-Target Invertebrates in a Conservation-Tillage Crop. *J. Aust. Ent. Soc.*, **34**: 229-231.

Dame, R. F. 1996. Ecology of Marine Bivalves: an Ecosystem Approach. 2nd ed. CRC press, New York.

Das, K., Ray, M. and Ray, S. 2012. Cypemethrin-induced dynamics of hemocyte density of Indian mollusc *Lamellidens marginalis*. *Ani. Bio. Jour.*, **3**(1): 39-49.

Das, K., Ray, M., and Ray, S. 2013. Structural impairment of gill and digestive gland of *Lamellidens marginalis* exposed to cypermethrin. *Ani. Bio. Jour.,* **3 (3):** 117-126.

Das, S. and Jana, B. B. (2003). In situ cadmium reclamation by freshwater bivalve *Lamellidens marginalis* from an industrial pollutant-fed river canal. *Chemosphere*, **52**(1), 161-173.

Deepa, R. S. and Ramachandra, T. V. 1999. Impact of urbanization in the interconnectivity of wetlands. Paper presented at the National Symposium on Remote Sensing Applications for Natural Resources. *Wat. Res. Dev.*, **16**(4): 639-650.

Extension Toxicology Network (ETN). 1996. Pesticide information Profiles. http:// ace.orst.edu/cgi-bin/mfs/01/pips/resmethr.htm.

Foster, S. R. L. 1978. The function of the pallial organs of bivalves in controlling ingestion. *J. Mollus. Stu.*, **44**: 83-89.

Galloway, T. S., Millward, N., Browne, M. A. and Depledge, M. H. 2002. Rapid assessment of organophosphorous/carbamate exposure in the bivalve mollusk *Mytillus edulis* using combined esterase activities as biomarkers. *Aqu. Toxicol.*, **61**: 169-180.

Garrido, M. V., Chaparro, O. R., Thompson, R. J., Garrido, O. and Navarro, J. M. 2012. Particle sorting and formation and elimination of pseudofaeces in the bivalves *Mulinia edulis* (siphonate) and *Mytilus chilensis* (asiphonate). *Mar. Bio.*, **159**(5): 987-1000.

Grewall, K.K., Brar, R.S *et al.,* 2010. Toxic impacts of cypermethrin on behavior and histology of certain tissues of albino rats. *Toxicological International,* **17**(2): 94-98.

Hestrin, S. 1949. The reaction of acetylcholine and other carboxylic acid derivatives with hydroxylamine, and its analytical application. *S. J. Biol. Chem.*, **180**: 249.

Hietanen, B., Sunila, I. and Kristoffersson, R. 1988. Toxic effects of zinc on the common mussel *Mytilus edulis* L. (Bivalvia) in brakish water. I. Physiological and histopathological studies. *Ann. Zool. Fenn.*, **25**: 341-347.

Holopainen, I. J. and Lopez, G. R. 1989. Functional anatomy and histology of the digestive tract of fingernail clams (Sphaeriidae, Bivalvia). *Ann. Zool. Fenn.*, **26**: 61-72.

Kaufman, D. D., Russell, B. A., Helling, C. S. and Kayser, A. J. 1981. Movement of cypermethrin, decamethrin, permethrin and their degradation products in soil. *J. Agri. Food. Chem. Soc.*, Washington D.C. 239-245.

Khan, M. Z. 2005. Effects of agro pesticides cypermethrin and malathion on cholinesterase activity in liver and kidney of *Calotes versicolor* Daudin (Agamidae: Reptilia). *Turk. J. Zool.*, **29**: 77-81.

Kiorboe, T. and Mohlenberg, F. 1981. Particle selection in suspension-feeding bivalves. *Mar. Eco. Prog. Ser.*, **5**: 291-296.

Levinton, J. S. 1971. Control of *Tellinacean* (Mollusca: Bivalvia) feeding behaviour by predation. *Limnology and Oceanography*, **V16**: 660-662.

Lowry, H. W., Rosebrough, N. J., Farr, A. L. and Randall, R. J. 1951. Protein measurement with folin phenol reagent. *J. Biol. Chem.*, **193**: 265-275.

Marigoudar, S.R., Ahmed, R.N. and David, M. 2009. Impact of Cypermethrin on Behavioural Responses in the Freshwater Teleost, *Labeo rohita* (Hamilton) *World Journal of Zoology,* **4** (1): 19-23.

Nair, R., Abraham, M.J. *et al.,* 2011. A pathomorphological study of the sublethal toxicity of cypermethrin in Sprague Dawley rats. *Int. jour.Nut. pharm.neu. dis.*, **2**(1): 179-183.

Najimi, S., Bauhaimi, A., Daubeze, M., Sekhniri, P., Ekkerin, J., Narbonne, J. F. and Moukrim, A. 1997. Use of Acetylcholinesterase in *Perna perna* and *Mytilus galloprovincialis* as a biomarker of pollution in Agadir marine Bay (South of Morocco). *Bull. Environ. Contam. Toxicol.*, **58**: 901-908.

Prasad, S. N., Ramachandra, T. V., Ahalya, N., Sengupta, T., Kumar, A., Tiwari, A. K., Vijayan, V. S. and Vijayan, L. 2002. Conservation of wetlands of India. *A. Rev. Trop. Ecol.*, **43**(1): 173-186.

Radenac, G., Bocquene, G., Fichat, D. and Mirimand, P. 1998. Contamination of a dredged material disposal site (La Rochelle Bay, France). The use of the acetylcholinesterase activity of *Mytilus edulis* (L) as a biomarker of pesticides: the need for a critical approach. *Biomarkers*, **3**: 305-315.

Raut, S.K. 1991. Laboratory rearing of Medically and Environmentally Important Molluscks. In: *snails, Flukes and Man.* (Jairajpuri, M. S. Ed.) Zoolocial Survey of India Publication. pp. 79-83.

Rickwood, C. J. and Galloway, T. S. 2004. Acetylcholinesterase inhibition as a biomarker of adverse effect: A study of *Mytilus edulis* exposed to the priority pollutant chlorfenvinphos. *Aqu. Toxicol.*, **67**: 45-56.

Sakata, S., Nobuyoshi, M., Matsuda, T. and Miyamoto, J. 1986. Degradation and leaching behaviour of the pyrethroid insecticide cypermethrin in soils. *J. Pesticide. Sci.*, **11**: 71-79.

Saxena, P. and Saxena, A.K.2010. Cypermethrin Induced Biochemical Alterations in the Blood of Albino Rats. *Jor. Jour. Biol. Sci.*, 3(3): 111-114.

Subramanian, A.N., Lal Mohan, R.S., Karunagaran, V.M. and Babu Rajendran, R.1999. Concentrations of HCHs and DDTs in the tissues of River dolphins *Platanista gangetica. Chemical Ecology*, **16**: 143–150.

Tiwari, S., Tiwari, R. and Singh, A. 2012. Impact of cypermethrin on Fingerlings of Common Edible Carp (*Labeo rohita*). *The. Sci. Worl. Journ.*, **2012**: 291-395.

Vaughn, C. C. and Hakenkamp, C. C. 2001. The functional role of burrowing bivalves in freshwater ecosystem. *Freshwater Biology*, **46**: 1431-1446.

Walker, M. H. and Keith, L. H. 1992. EPA's Pesticide Fact Database. Lewis Publishers, Chelsea, MI.

Wang, H., Wang, Q. *et al.,* 2010. Cypermethrin exposure during puberty disrupts testosterone synthesis via downregulating StAR in mouse testes. *Arch Toxicol.,* **84**: 53–61.

Ward, J. E., Levinton, J. S. and Shumway, S. E. 2003. Influence of diet on pre-ingestive particle processing in bivalves I: Transport velocities on the ctenidium. *Jour. Exp. Mar. Bio. Eco.,* **293**(2003): 129-149.

Yaqin, K., Widiatilay, B., Riani, E., Masud, Z. A. and Hansen, P. D. 2008. The use of selected Biomarkers, phagocytic and cholinesterase activity to detect the effects of dimethoate on marine mussel (*Mytilus edulis*). *Hay. Jour. Bio.,* **15**(1): 32-38.

Yasmeen, S., Suryawanshi, G. D. and Dama, L. B. and Mane, U.H. 2012. Behavioural changes of freshwater bivalve *Lamellidens marginalis* due to acute toxicity of cadmium. *Dav.Int. Jour. Sci.,* **1**(2): 103-106.

2015, Perspectives in Animal Ecology and Reproduction, Vol. 10   Pages *137–143*
Editors: V.K. Gupta, Anil K. Verma and G.D. Singh
Published by: DAYA PUBLISHING HOUSE, NEW DELHI

Chapter 9

Feeding Acceptability of Non-crop Plants by the Invasive Giant African Snail, *Achatina fulica* (Bowdich) (Stylommatophora: Achatinidae)

M. Jayashankar*[1], V. Sridhar[1] and Abraham Verghese[2]

Division of Entomology and Nematology,
Indian Institute of Horticultural Research,
Hesaraghatta, Bengaluru – 560 089, Karnataka, India
[2]National Bureau of Agriculturally Important Insect,
Bellary Road, Bengaluru – 560024, Karnataka, India

ABSTRACT

The feeding behavior of *Achatina fulica* (Bowdich) on 36 different species of non-crop plants was undertaken *ad-libitum*. Among the plants provided to the snails, 10 were highly accepted, 17 moderately accepted and 9 were rejected by the snails. Of the highly accepted plants, *Argyreia cuneta, Solanum nigrum* and *Sida acuta* were fed 100 per cent by snails. *Mimosa pudica, Grevillea robusta, Plumbago zeylanica, Stylosanthes hamata, Blumea* sp, *Achyranthes aspera, Kirganelia reticulate, Passiflora foetida* and *Atylosia* sp. were rejected. Planting non-crop plants in guard rows or around the agri-horticultural crops might be effective in reducing damage inflicted by *A. fulica*.

Keywords: *Feeding, Achatina fulica, Non-crop plants.*

* *Corresponding author.* E-mail: jay_zoology@rediffmail.com

Introduction

Invasive giant African snail, *Achatina fulica* (Bowdich) (Stylommatophora: Achatinidae) is one of the most damaging agricultural pests worldwide representing a potentially serious threat to natural ecosystems and human health (Cardoso, 2012). It is a serious agri-horticulture pest in India inflicting serious crop damage to a wide spectrum of crops (Veeresh *et al.,* 1979; Thangavelu and Bijoy, 1983; Javaregowda, 2004; Thakur, 2004; Naik *et al.,* 2008; Vanitha *et al.,* 2011) although the afore-quoted references quantify the economic loss of crop plants, damage inflicted by *A. fulica* to non-crop plants are minimal and restricted to *ad libitum* studies (Raut and Ghose, 1983). Trap cropping is considered as most worthwhile for pests that are abundant and destructive in most years. Hence the present study to assess the feeding acceptability of native flora by *A.fulica* was undertaken in laboratory conditions. Subsequently to aid in devising future pest control strategies, using non-crop plants as barriers/trap crops.

Materials and Methods

Maintaining Snails

Snails were maintained in cubical glass terraria (58 x 35 x 30 cm) with native soil collected along with the snails and filled up to 3-4 cm in each terrarium provided with several ventilation holes in the roof cover.

Laboratory Bioassays

Different plant parts of flora in the natural habitat were collected and given a mild wash under tap water. The plant parts were then air dried and fed *ad-libitum* and strict hygienic conditions were maintained within the terrarium. The snails were active in the period of acclimatization. Laboratory experiments were conducted in glass terraria regularly sprinkled with water to maintain humidity. No-choice preference assays were used to assess relative preference of potential trap crops. Five snails were allowed to feed on different plant parts *viz.,* leaf, pod/fruit, stem, flower/ bud and root, placed at equidistant for equal access, the experiment was replicated five times.

The extent of feeding of different plant parts was estimated on a 3 points scale (Table 9.1) based on standard devised by Raut and Ghose (1983).

Table 9.1: Scores and Sign Notations Based on Acceptability Indices of Plant Parts Offered to the Snails

Per cent Point Scale and Sign	Observation
0 per cent (-)	None of the food was eaten
1-50 per cent (±)	Food was moderately accepted
51-100 per cent (+)	Food was highly accepted

Results and Discussion

Feeding preference by *A.fulica* on 36 different species of non-crop plants quantified as acceptability index is represented in Table 9.2. Among the plants provided to the snails, 10 were highly accepted, 17 were moderately accepted and 9 were rejected by the snails. Of the highly accepted plants, *Argyreia cuneta, Solanum nigrum* and *Sida acuta* were fed 100 per cent by snails. *Mimosa pudica, Grevillea robusta, Plumbago zeylanica, Stylosanthes hamata, Blumea* sp., *Achyranthes aspera, Kirganelia reticulate, Passiflora foetida* and *Atylosia* sp. were not preferred at all (Table 9.1). Rank abundance of different families from the data (Figure 9.1) indicate Fabaceae with highest number of food plants (8) offered and one each belonging to nine families Meliaceae, Passifloraceae, Plumbaginaceae, Proteaceae, Ranunculaceae, Rhamnaceae and Santalaceae.

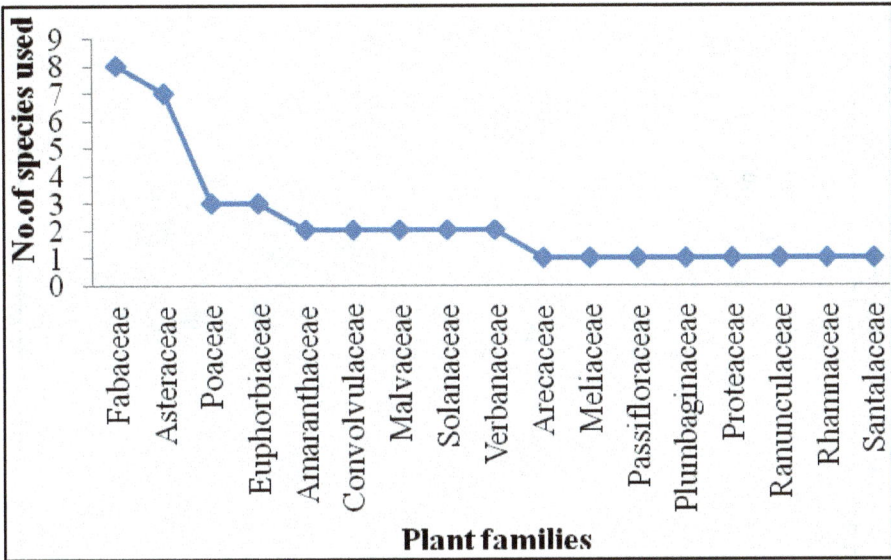

Figure 9.1: Rank of Abundance of Food Plant Families.

The expanding host range of *A.fulica* and its potential to cause economic damage to agri-horticultural crops deserves attention. The present observations contribute to such alerts indicating that *A.fulica* could already have had a negative impact on the flora in the region and it is recommended that the situation be closely monitored. Based on the food preference studies on *A. fulica,* Raut and Ghose (1983) observed that availability of food plants in the feeding zone influences feeding in different age. As suggested by Raut and Ghose (1983) the damage to economic crop plants by *A. fulica* can be reduced through the introduction of non-crop plants as barrier or trap crops.

Identifying trap/barrier crops is a necessary step in pest management as they are an environmentally compatible approach used to protect main crop from a pest or complex of pests (Jayanthi *et al.,* 2009). Protection is rendered either at the cost of phytophagy or as roosting hosts (McQuate, 2011) by the barrier/trap crops. Barwal and Dhiman (2002) found *Tagetes* spp. as excellent trap crops against the snail, *Machrochlamys glauca.* Feeding preference by *A.fulica* by Thakur (2004) recorded the heaviest damage incurred by different life stages of *A.fulica* on okra from seedling

Table 9.2: Acceptability Indices of Food Plants Provided to *A. fulica*

Sl.No.	Plant Species	Family	Leaf	Leaf (A.I) per cent	Pod/Fruit	Stem	Flower/Bud	Root
1.	*Argyreia cuneta*	Convolvulaceae	+	100	No	±	No	No
2.	*Sida acuta*	Malvaceae	+	100	No	No	No	No
3.	*Synedrella nodiflora*	Asteraceae	+	95.60	No	±	No	–
4.	*Lantana camara*	Verbenaceae	+	92	No	No	No	No
5.	*Sauropus androgynus*	Phyllanthaceae	+	91.40	No	No	No	No
6.	*Azadirachta indica*	Meliaceae	+	87.80	No	No	No	No
7.	*Cocculus hirsutus*	Ranancualales	+	85.70	No	–	No	No
8.	*Desmodium heterocarpon*	Fabaceae	+	84.70	±	–	+	No
9.	*Alternantheria sessile*	Amaranthaceae	+	82	No	±	±	–
10.	*Santalum album*	Santalaceae	+	50.10	No	–	–	–
11.	*Eupatorium odaratum*	Asteraceae	±	49.80	No	–	–	No
12.	*Vinca rosea*	Malvaceae	±	45.90	–	+	+	–
13.	*Apulda mutica*	Poaceae	±	37.80	No	No	No	No
14.	*Vernonia cinerea*	Asteraceae	±	37.50	No	±	–	–
15.	*Stachystarpheta indica*	Verbenaceae	±	31.80	No	–	No	No
16.	*Crotolaria retusa*	Fabaceae	±	30	–	–	–	No
17.	*Parthenium hysterophorus*	Asteraceae	±	30	No	No	No	No
18.	*Zizyphus sp.*	Rhamnaceae	±	26.90	–	No	No	No
19.	*Crotalaria pallida*	Fabaceae	±	26	No	–	–	No
20.	*Bidens pilosa*	Asteraceae	±	23.10	No	–	–	–
21.	*Tephrosia purpurea*	Fabaceae	±	21.40	No	+	No	No
22.	*Cassia tomentosa*	Fabaceae	±	20	–	–	–	No

Contd...

Table 9.2–*Contd...*

Sl.No.	Plant Species	Family	Leaf	Leaf (A.I) per cent	Pod/Fruit	Stem	Flower/Bud	Root
23.	*Setaria pumila*	Poaceae	±	20	No	No	No	No
24.	*Cenchrus ciliaris*	Poaceae	±	19	No	No	No	No
25..	*Ageratum conyzoides*	Asteraceae	±	15.70	No	–	No	No
26.	*Acalypha* sp.	Euphorbiaceae	±	11.80	No	No	No	No
27.	*Ipomoea obscura*	Convolvulaceae	±	2.50	No	–	+	No
28.	*Mimosa pudica*	Fabaceae	–	0	–	–	No	No
29.	*Grevillea robusta*	Proteaceae	–	0	No	–	No	No
30.	*Plumbago zeylanica*	Plumbaginaceae	–	0	–	–	–	No
31.	*Stylosanthes hamata*	Fabaceae	–	0	No	–	No	No
32.	*Blumea* sp.	Asteraceae	–	0	No	No	No	–
33.	*Achyranthes aspera*	Amaranthaceae	–	0	No	–	–	No
34.	*Kirganelia reticulate*	Euphorbiaceae	–	0	No	0 per cent	No	No
35.	*Passiflora foetida*	Passifloraceae	–	0	No	±	No	No
36.	*Atylosia* sp.	Fabaceae	–	0	–	–	–	No

+: Most preferred; ±: Preferred; –: Rejected; No: Not offered.

stage compared to other crops *viz., Cucumis sativus, Solanum melongena, S. lycopersicum, Carica papaya, Tagetes patula, Catharanthus roseus, Luffa aegyptiaca, Vigna unguiculata, Cucurbita pepo, Amaranthus tricolor* and *Tricosanthes dioica*. Naik *et al.* (2008) recorded highest feeding rate in pepper and banana by different size groups of *A.fulica*. Similar studies conducted using non-crop plants under laboratory conditions have revealed host preference and differential feeding rate by *Achatina fulica* (Raut and Ghose, 1983), *A. achatina* (Otchoumou and Kouassi, 2005) and *M. nuda* (Oli and Pandey, 1999).

In the present observations 27 species of test plants offered have moderate to high acceptability, indicating the usage of these plants as barrier/trap crops in agri-horticultural ecosystems prone to *A.fulica* menace. However, further research is necessary in order to establish which of the chemical compounds contained in those plants influence the feeding behaviour of the snails and by what mechanism they act. Further research also needs to clarify if the differences represent real variation between species or simply reflect differences between species selected for the present study. While the possibility of using the preferred plants as barrier/trap crops needs to be explored, equal importance needs to be focused on the possibility of the non-attractiveness of unpreferred plants. Prasad *et al.* (2004) recommend the use of *Annona glabra* softwood cutting as a repellent fence to protect nursery beds from *A. fulica* due to their annonacin content. After thorough investigations attempts to use non-crop plants in guard rows or around the agricultural fields, may perhaps serve in effective reduction of damage to crop plants by *A. fulica* in the areas experiencing the pestiferous snail menace.

Acknowledgements

The first author is grateful to Department of Zoology and Dr.M.S.Reddy, Bangalore University for the facilities and support provided.

References

Barwal. R.N., and Dhiman.M.R., 2002.Evaluation of Marigold, *Tagetes* spp. Cultivars as trap crop for the temperate snail, *Macrochlamys glauca* Pfr. *Pest Management in Horticultural Ecosystems*, **8 (2):** 133-134.

Javaregowda., 2004. Incidence of the giant African snail, *Achatina fulica* (Bowdich), on horticultural crops. *Pest Management and Economic Zoology*, **12(2):** 221-222.

Jayanthi, P.D.K., Verghese, A and Nagaraju, D.K.2009.Studies on feeding preference of adult fruit sucking moth, *Eudocima (Otheris) materna* (L.): A clue for devising trap cropping strategies.*Pest Mangement in Horticultural Crops*, **15(2):** 107-113.

Mead, A.R., 1979. Economic malacology with particular reference to *Achatina fulica*. *In: Fretter, V. and Peake, J. (eds.) Pulmonates*, **2B.** *Academic Press, London,* 150 pp.

McQuate, G. T., 2011. Assessment of attractiveness of cassava as a roosting plant for the melon fly, *Bactrocera cucurbitae*, and the Oriental fruit fly, *B. dorsalis. Journal of Insect Science* (Madison). **11:** 30, available online: insectscience.org/11.30.

Naik, M. I., Ravi Kumara., Manjunatha, M., and Pradeep, S., 2008. Studies on host preference and feeding rate of giant African snail, *Achatina fulica* (Gastropoda:

Achatinidae) in areca and intercrops of areca. *Mysore Journal of Agricultural Sciences*, **42 (3)**: 468-471.

Oli, B.P., and Pandey, I., 1999. Food and feeding behaviour of the land snail *Macrochlamys nuda* (Pfeiffer) (Stylommatophora: Ariophantidae) from Kumaon Himalyan Forests. *Zoos Print Journal*, **1418**: 82-86.

Otchoumou, A., N'Da, K., and Kouassi, K. D., 2005. The edible African snails farming: inventory of wild vegetables consumed by *Achatina achatina* (Linne 1758) and dietary preferences. *Livestock Research for Rural Development*, **17(3):**

Prasad, G.S., Singh, D.R., Senanai, S., and Medhi, R.P., 2004. Eco-friendly way to keep away pestiferous Giant African snail, *Achatina fulica* Bowdich from nursery beds. *Current Science*, **87 (12):** 1657-1659.

Raut.S.K., and Ghose, K.C., 1983. The role of non-crop plants in the protection of crop plants against the pestiferous snail *Achatina fulica*. *Malacological Review*, **16(1-2):** 95-96.

Srivastava, P. D., 1992. Problem of land snail pests in Agriculture: A study of the giant African snail. Concept publishing company, New Delhi, 234 pp.

Thakur, S., 2004. Food Consumption and growth potential of giant African snail, *Achatina fulica.Journal of Ecobiology*, **16(6):** 455-461.

Thangavelu, K., and Bijoy, K. S., 1983.Giant African snail, *Achatina fulica* Bowdich (Pulmonata: Gastropoda) as a serious pest of mulberry from the north-eastern region. *Indian Journal of Agricultural Sciences*, **53(9):** 871-872.

Vanitha, K., Karuppuchamy, P., and Sivasubramanian, P., 2011. Feeding preference of *Achatina fulica* attacking vanilla and its management through barrier substances. *Pest Management in Horticultural Ecosystems*, **17(1):** 38-41.

Veeresh, G. K., Rajagopal, D., and Puttarudraiah, M., 1979. First record of African giant snail, *Achatina fulica* (Bowdich) (Mollusca: Gastropoda) as a serious pest of ornamental crops in Bangalore. *Current Research*, **8**: 202-204.

2015, Perspectives in Animal Ecology and Reproduction, Vol. 10   *Pages 145–158*
Editors: **V.K. Gupta, Anil K. Verma and G.D. Singh**
Published by: **DAYA PUBLISHING HOUSE, NEW DELHI**

Chapter 10

Sublethal Waterborne Nickel Toxicity in an Indian Major Carp *Cirrhinus mrigala*: Hematological and Biochemical Responses

M. Ramesh*, B. Sathyavathy,
R.K.Poopal and V. Maruthappan
Unit of Toxicology, Department of Zoology, School of Life Sciences,
Bharathiar University, Coimbatore – 641 046, Tamil Nadu, India

ABSTRACT

Metals from anthropogenic sources have been recognized as important contaminants in aquatic ecosystems. Nickel (Ni) is a grey-listed metal found naturally in the planet's core. A low level of Ni is found to be essential for the normal growth of many organisms. However, elevated level of Ni may affect the behaviour, survival, growth, and reproduction of many organisms. Fish have been found to be good indicators of heavy metal contamination in aquatic systems because they occupy different tropic levels and are of different sizes and ages. In the present study fingerlings of *Cirrhinus mrigala* were exposed to sublethal concentration of nickel (nickel sulphate) for 15 days to assess the response pattern of selected haematological and biochemical parameters. The median lethal concentration of nickel sulphate to the fish *C. mrigala* was found to be 85.90 mg/L. During sublethal treatment (8.5 mg/L) erythrocyte (RBC), haemoglobin (Hb) and plasma protein levels were decreased in nickel exposed fish throughout the study period and on the other hand, the leucocyte count (WBC) and plasma

---

* *Corresponding author.* E-mail: mathanramesh@yahoo.com

glucose level were increased against the toxicity. From our result, we conclude that haematological and biochemical alterations could be used to understand an overall physiology and health condition of fish under metal toxicity.

Keywords: *Biochemical, Cirrhinus mrigala, Ecological indicators, Haematology, Nickel, Sublethal.*

Introduction

Metals are natural components found everywhere in the biosphere. Their concentration in the aquatic ecosystem increases through natural and human activities. Metals are the major contaminant and culprit in the aquatic ecosystem that causes serious health risk to aquatic biota and human (Sathya *et al.,* 2012; Poopal *et al.,* 2013). Metals such as nickel, zinc, cobalt, copper etc., are essential for normal biological function (Muyssen *et al.,* 2004; Ojo and Wood, 2007). Nickel (Ni) is a grey-listed or transition metal and a microelement which is ubiquitous in the earths crust (Mason, 1996; Barceloux, 1999; Jiang *et al.,* 2013). Ni species occurs about 0.01 per cent in the environment and are classified based on their chemical nature. Monosulfide, subsulfide, carbonates and oxides are water insoluble whereas chloride, sulphate and nitrate are water soluble nickel compounds. Carbonyl is one of the nickel compounds which are colourless and volatile in nature (Schaumlöffel, 2012). A low level of Ni is detected in tissues of many animals and it is found to be essential for the normal growth of many organisms (WHO 1991; Pane *et al.,* 2003a; Woo *et al.,* 2009). Natural sources of Ni in aquatic ecosystem include leaching, weathering of rocks and volcanic process (Pane *et al.,* 2003a). Anthropogenic by-products like electroplating, steel alloys, batteries, discharge of mining, waste incinerators, power plants, paint factories and aircraft industry play a major role in entry of elevated level of Ni in aquatic ecosystem (WHO 1991; Pane *et al.,* 2003a; Jiang *et al.,* 2013).

Among the Ni species Ni^{2+} is predominant in aquatic ecosystem (WHO 1991; Chau and Kulikovsky-Cordeiro, 1995). Waterborne Ni species have affinity towards inorganic and organic compounds; some of their combination may produce synergistic or antagonistic effects (Sveceviius, 2010). Ni is genotoxic, immunotoxic and carcinogenic and it could affect behaviour, survival, growth, and reproduction in living organisms at elevated levels (Wong *et al.,* 1993; Andrew *et al.,* 2001; Kasprzak *et al.,* 2003; Vandenbrouck *et al.,* 2009). Ni species is listed as a dangerous substance, priority chemical, human carcinogen, and possible human carcinogen by European Commission List II (ECL-II), Canadian Environmental Protection Act, WHO and U.S. Public Health Service respectively (USPHS, 1993; Hughes *et al.,* 1994; Bubb and Lester, 1996). The toxicity report of Ni on aquatic ecosystems is scanty (Pane *et al.,* 2003a).

Fish is considered as an ideal indicator in the field of aquatic toxicology research because fish is the top most organism of the aquatic food web and highly sensitive to the slight environmental change. Fish is well known for its nutrient content especially essential proteins, polyunsaturated fatty acids and liposoluble vitamins. Unfortunately, fish could accumulate aquatic contaminants such as metals, pesticides, pharmaceuticals etc., This may be one of the carriers of toxicants into human through diet. These unique characters in fish made to use the fish as a biological model by the toxicology researchers (Sathya *et al.,* 2012; Poopal *et al.,* 2013). Hence fish is suitable organism to know the health of the aquatic environment.

Biomarker is the measured response of biological system caused by toxic substances. Hematological and biochemical are commonly used parameters in the field of toxicology. The pathophysiological reflection of the whole body can be studied by measuring the changes in haematological parameters (Lavanya *et al.,* 2011; Sathya *et al.,* 2012). The alteration in hematological parameters such as RBC, WBC count, Hb, Hct, MCV, MCH and MCHC are widely used to assess the toxic stress, integrity of the immune system and tissue damage (Kavitha *et al.,* 2010; Saravanan *et al.,* 2012). Blood biochemical parameters are often used to analyse of metabolic disorder and diagnosis of disease process. Biochemical parameters especially protein and glucose are frequently used to analyse the stress at unfavourable condition (Vutukuru, 2003; Abhijith *et al.,* 2012).

Hence, the present study is aimed to investigate the sublethal toxicity of nickel in an Indian major carp *Cirrhinus mrigala* using certain haematological and biochemical parameters to observe the toxic effects. The fish *C. mrigala* is an edible and endemic species to Indo-Gangetic riverine systems. It is commercially important and cultivated in many of Southeast Asian countries. The endpoints obtained from this study will be useful to monitor the sublethal effects of metals in aquatic ecosystem.

Materials and Methods

Procurement of Fish and Maintenance in the Laboratory

Healthy fingerlings of *C. mrigala* with an average weight of 8.0 ± 0.3 g and length of 6.0 ± 0.2 cm were purchased from Tamilnadu Fisheries Development Corporation Limited, Aliyar Fish Farm, Aliyar, Tamilnadu, India and used as an experimental animal model. They were shifted to laboratory in polythene bags which is filled with aerated water. Then they were stocked/acclimatized in a 1000 L capacity cement tank for 30 days and once in every day they fed *ad libitum* with rice bran and ground nut oil cake in dough form. Three forth of the water was replaced every day (after 1h from feeding) to remove untaken feed and deposition of faecal materials. Dechlorinated tap water with temperature $30.0 \pm 1.0\,^{\circ}C$, pH 7.4 ± 1 units, salinity 0.26 ± 0.1 ppt, total alkalinity 36.0 ± 0.4 mg/L, dissolved oxygen 7.2 ± 0.04 mg/L, total hardness 17.1 ± 0.8 mg/L, calcium 4.02 ± 0.50 mg/L and magnesium 2.40 ± 0.2 mg/L were used throughout the experimental period. Healthy fingerlings from the acclimatization tank were shifted to 200 L capacity clean glass aquarium tanks; these fingerlings are used as a stock for experimental schedule.

Median Lethal Concentration of Nickel Sulphate

Median lethal tolerance limit (24 h) of fingerlings, *C. mrigala* against the toxicity of nickel sulphate was studied. For the commencement of the study, 60 healthy fingerlings from the stock were kept starved for 48 h. The study was performed in the 50 L water capacity of circular plastic tubs. To each tub, different concentration of nickel sulphate and 10 starved fingerlings were introduced. Simultaneously, a control group (toxicant free) was also maintained under identical conditions. The 24 h median lethal concentration (50 per cent mortality) (Finney, 1978) of nickel sulphate to *C. mrigala* was found to be 85.90 mg/L.

Sublethal Study and Sampling

For the sublethal study, six groups (5 experimental and 1 control) with 100 fingerlings in each group were maintained. Calculated concentration of 8.59 mg/L (1/10th of 24 h LC50 value) of nickel sulphate was added to five experimental groups. Simultaneously one control group was maintained without adding of nickel sulphate. During this study period fingerlings were fed *ad libitum* once in a day (one hour prior to replacement of water). This experimental setup was performed for 15 days, at the end of every 5 days fingerlings from control and nickel sulphate treated tanks were collected and sacrificed for haematological and biochemical analysis. Cardiac blood sampling was done by using the prechilled and heparin moisture 26 gauge needles fitted plastic disposable syringe. The blood sample was immediately transferred into the heparin rinsed vials to prevent blood coagulation. The RBCs, WBCs and Hb were calculated by using whole blood of the fingerlings. Then the remaining pooled blood samples were centrifuged at 93.9 g for 20 min which was used for estimation of glucose and protein.

Haematological Studies

Haemoglobin level was estimated by Cyanmethaemoglobin method and expressed as g/dl (Drabkin, 1946). RBC and WBC counts were calculated by the method of Rusia and Sood (1992)

Biochemical Studies

Plasma glucose was estimated following *O*-Toluidine method (Cooper and Mc Daniel, 1970) and the value was expressed as mg/100 mL. Plasma protein estimation was done by the method established by Lowery *et al.* (1951), and expressed as mg/mL.

Statistical Analysis

The statistical significances between the control and nickel sulphate treated fingerlings were analyzed by using Student's t-test, their significance are calculated at $p < 0.05$ level.

Results

Nickel Sulphate Toxicity on *C. mrigala*

There was no mortality in the control group. In the nickel sulphate treated groups a concentration dependent mortality was observed. During the treatment period the fish shows various behavioural changes like convulsions, fast swimming, mucus secretion, jerky movement and death. At the end of 24 h, 50 per cent mortality was noted at 85.90 mg/L concentration of nickel sulphate treated group. For sublethal study, 8.59 mg/L (1/10th of 24 h LC_{50} value) of nickel sulphate was taken. Based on the Chi-square test the fingerlings used in the present study was known to be homogenous.

Haematological Alterations

The Figures 10.1–10.3 reveals the haematological alterations of fingerlings exposed to 8.59 mg/L concentration of nickel sulphate. The RBC level of nickel sulphate treated fingerlings was significantly decreased at all the exposure days (15 days) when compared to control group (Figure 10.1). A maximum percent decrease of 68.36 per cent was noted at the end of 15th day of exposure. The WBC level of fingerlings exposed to 8.59 mg/L of nickel sulphate was significantly increased throughout the study period when compared to control group (Figure 10.2). A minimum increase (14.39 per cent) was found at the end of 5th day. Hb concentration of the fingerlings was decreased significantly at all the sampled day when compare to control groups (Figure 10.3). At the end of 15th day a maximum percentage change of 32.83 per cent was recorded.

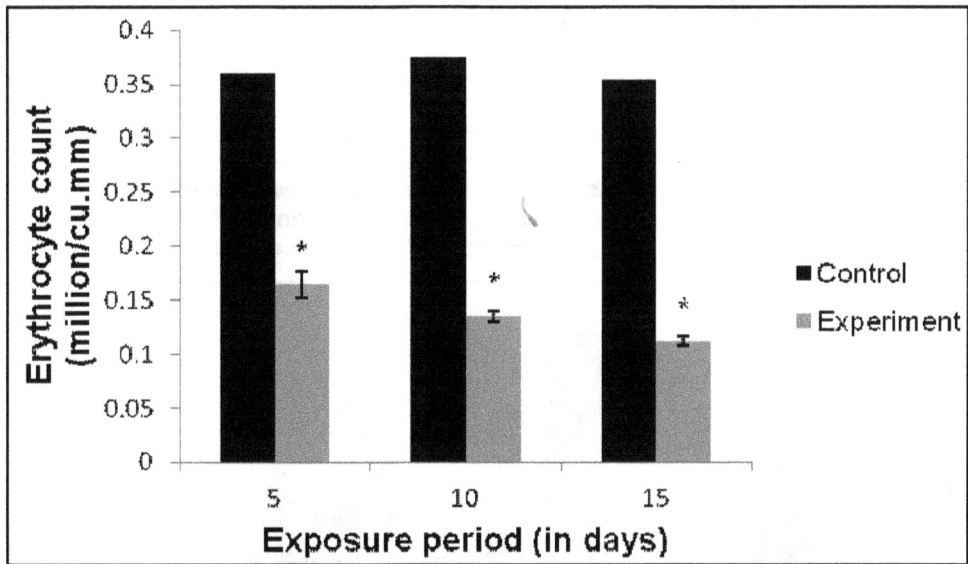

Figure 10.1: Changes in the Erythrocyte Count of *Cirrhinus mrigala* Exposed to 8.59 mg/L Concentration of Nickel Sulphate. Values are mean ± S.E. of five individual observations. Significant at p < 0.05 (based on t-test).

Biochemical Changes

The biochemical response in nickel sulphate exposed and control fingerlings were presented in Figs. 4-5. The plasma glucose level of nickel sulphate exposed fish was significantly increased when compare to control group (Figure 10.4). A maximum percent increase of 54.10 was noted at the end of 15th day of exposure. However, plasma protein level was found to be decreased in nickel sulphate treated fingerlings when compare to control group (Figure 10.5). A maximum percent decrease of 20.86 was noted at the end of 15th day of exposure.

Discussion

Metals causes adverse effects on behaviour, metabolic and physiology of fish

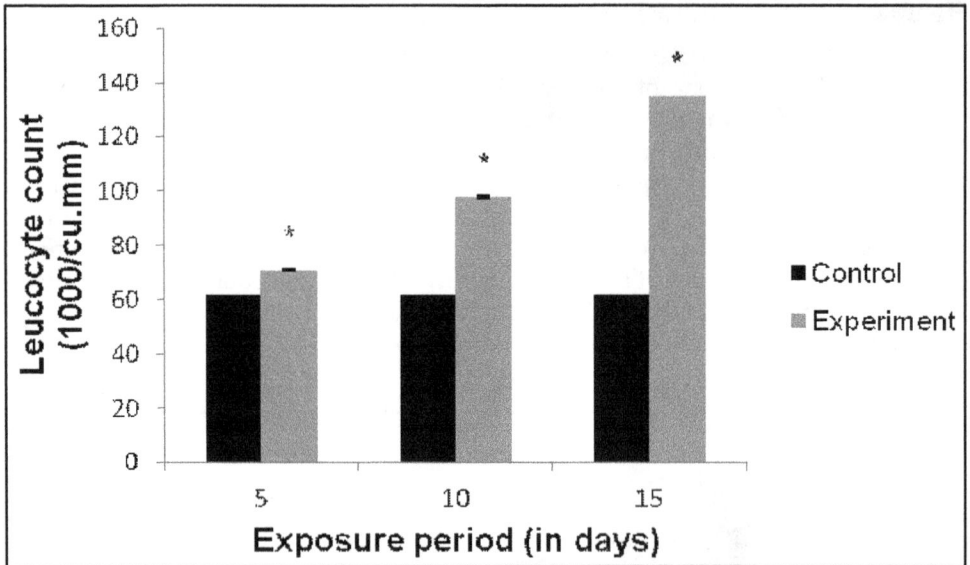

Figure 10.2: Changes in the Leucocyte Count of *Cirrhinus mrigala* Exposed to 8.59 mg/L Concentration of Nickel Sulphate. Values are mean ± S.E. of five individual observations. Significant at p < 0.05 (based on t-test).

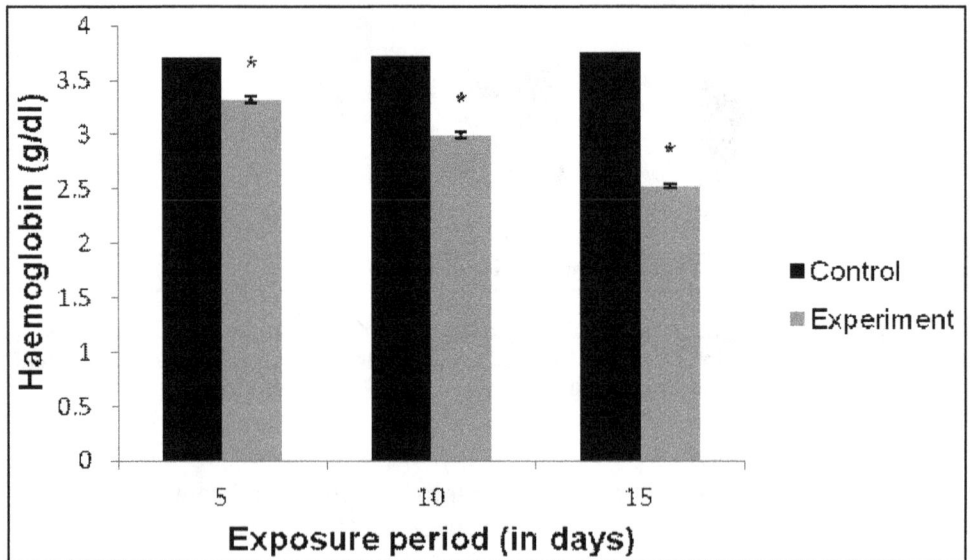

Figure 10.3: Changes in the Haemoglobin Content of *Cirrhinus mrigala* Exposed to 8.59 mg/L Concentration of Nickel Sulphate. Values are mean ± S.E. of five individual observations. Significant at p < 0.05 (based on t-test).

(Soengas *et al.,* 1996). The toxicity data in this study show that nickel sulphate is toxic to *C. mrigala.* The observed behavioural changes may be due to the toxic or abnormal environment created by nickel. Toxicity of the nickel on aquatic organisms depends

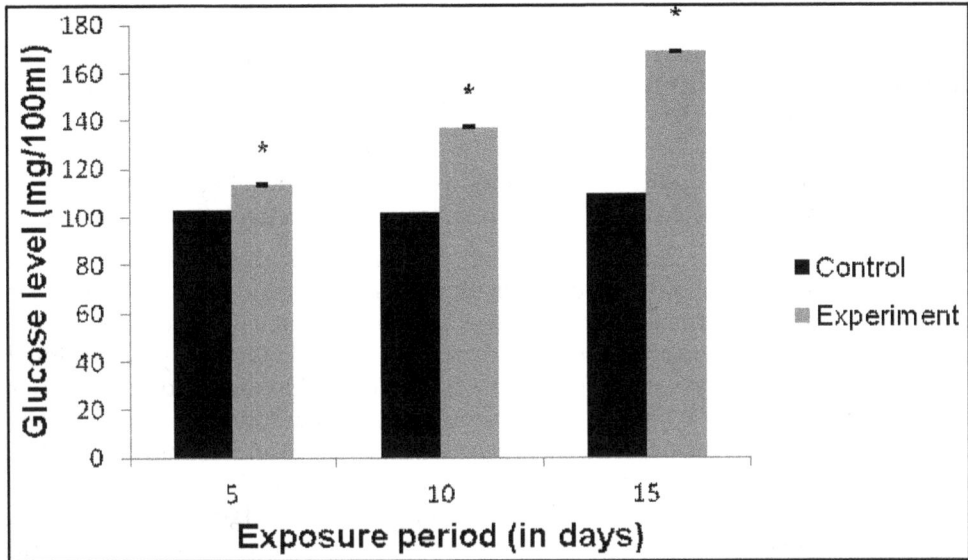

Figure 10.4: Changes in the Glucose Level of *Cirrhinus mrigala* **Exposed to 8.59 mg/L Concentration of Nickel Sulphate. Values are mean ± S.E. of five individual observations. Significant at p < 0.05 (based on t-test).**

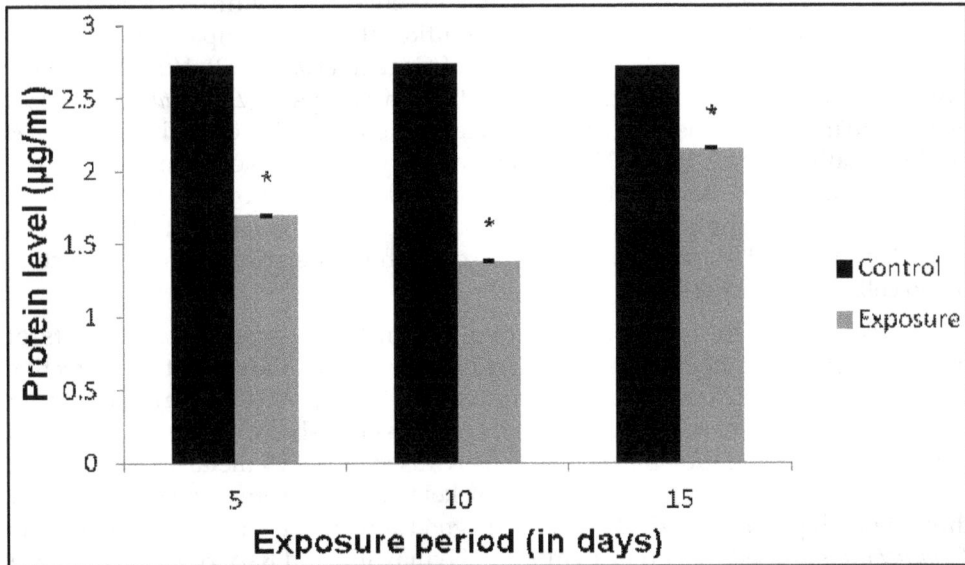

Figure 10.5: Changes in the Protein Level of *Cirrhinus mrigala* **Exposed to 8.59 mg/L Concentration of Nickel Sulphate. Values are mean ± S.E. of five individual observations. Significant at p < 0.05 (based on t-test).**

on the sex, species, water chemistry, chemical nature of nickel and other environmental factors (Blaylock and Frank, 1979; USEPA, 1986). The mechanisms of Ni toxicity were different for vertebrates and invertebrates (Pane *et al.,* 2003b). For example, in teleost rainbow trout, *Oncorhynchus mykiss,* the metal Ni acts as a respiratory toxicant (Pane *et al.,* 2003a; Pane and Wood, 2004), whereas in the cladoceran *Daphnia magna,* Ni acts as a ionoregulatory toxicant which may disrupts Mg homeostasis (Pane *et al.,* 2003b). Rothstein (1973) suggested that in nickel chloride exposed groups the death of the fish may be due to the toxic effect of nickel chloride on the biochemical processes related to cellular metabolic pathways and to other inclusions. In the present study the observed mortality of fish might have resulted from ionoregylatory disturbances.

Haematology is the index to know the physiological status of fish and proven valuable for fisheries scientists to report on the status of environmental health (Elahee and Bhagwant, 2007). Haemoglobin is a special protein which is responsible for the pigmentation of the blood and efficient in oxygen carrying. In this study the observed significant decrease in haemoglobin level of fish exposed to nickel toxicity reveals anaemic condition. Similar results were found in the *C. carpio* and *C. gariepinus* when exposed to nickel (Vinodhini and Narayanan, 2009; Ololade and Oginni, 2010). This situation is due to fragile of erythrocytes, lack of oxygen binding capacity and cell swelling, haemic metabolism (Witeska and Kosice, 2003). The significant decrease in Hb in *Labeo rohita* exposed to sodium selenite indicates release of oxygen radical brought about by sodium selenite (Ramesh *et al.,* 2014). Erythrocytes are specialised cells, typically elliptical and biconcave which exchanges respiratory gases to the body cells. Changes in the erythrocytic profile envisage a compensation of oxygen deficit in the body due to gill damage. In the present study erythrocyte count of fish exposed to nickel toxicity was decreased significantly when compare to the control group. Our result was similar to the result of Cockell *et al.* (1992), Vutkuru (2005), Kavitha *et al.* (2010), and Lavanya *et al.* (2011) in *O. mykiss, L. rohita* and *C. catla* exposed to metals. This is due to erythropoiesis and internal haemorrhages (Joshi *et al.,* 2002; Latimer *et al.,* 2003). The reduction in erythrocyte caount may be caused either by the inhibition of erythropoiesis or by the destruction of red blood cells (Ramesh, 2001). In the present investigation the decrease in Hb and RBCs counts might have resulted from resulted inhibition of erythropoiesis process and destruction of red cells due to nickel toxicity.

Leukocytes cells in the blood, involved in the principal components of immunological functions. The elevation of their values indicates the protective response under stress condition. Remyla *et al.* (2008) reported that the alterations in leucocyte number are sensitive indicators of stress in fish. WBC counts are useful methods of detecting sublethal effects in fish caused by heavy metals. In the present study the WBCs count of fish exposed to nickel toxicity was increased significantly throughout the study period. Our result is in good agreement with the report found in *C. punctatus* exposed to mercuric chloride (Hymavathi and Rao, 2000), *C. batrachus* treated with mercuric chloride (Joshi *et al.,* 2002), and *C. catla* exposed to arsenic trioxide (Lavanya *et al.,* 2011). Increases in WBCs are related to depression of leucopoiesis, immunosuppression, alteration in the cell membrane or disintegration of WBC (Abhijith *et al.,* 2012). According to El-Sayed *et al.* (2007) and Ates *et al.* (2008), elevation of WBC may resulted due to the lymphopoiesis or by the stimulatory effect of the toxic agent. The increase in WBC count in *Labeo rohita* exposed to sodium selenite indicates a generalized immune response and a protective response to the toxicant (Ramesh *et al.,* 2014). In the present study, the observed increase in WBC count indicates the protective mechanism of the fish against nickel toxicity or it may be due to increase in the population of neutrophils, acidophils and basophils. Thus the haematological responses are used as a sensitive bio indicator of nickel toxicity in freshwater fish.

Metal pollution in aquatic ecosystem causes measurable changes in cellular or molecular level. Due to its complex formation metal inhibit the normal metabolism of cell. To overcome these abnormal situations certain physiological and biochemical changes could occur (Vutukuru, 2003; Gagnon *et al.,* 2006). Plasma glucose and protein are the commonly used biochemical parameters for its well known response during abnormal conditions (Nemcsok and Boross, 1982; Kavitha *et al.,* 2010; Saravanan *etal.,* 2012; Ramesh *etal.,* 2014). In the present study plasma glucose level of the fish fingerlings exposed to nickel was found to be increased significantly when compare to control groups. Hyperglycemia condition is due to energy requirement required during metabolism or increased secretion of cortisol to cope up the nickel toxicity in fish. Generally in teleostean species, glucocorticoids and catecholamine are secreted excessively during the stress condition. In which catecholamine is responsible for the hyperglycaemic situation in fish (Wedemeyer, 1969). Imbalance of output and uptake of glucose in hepatic cells may increase blood glucose (Afaghi *et al.,* 2007). Similar results were found in *C. carpio* exposed to hexavalent Cr, *H. fossilis* exposed to nickel, *C. gariepinus* treated with copper, *C. mrigala* exposed to silver nitrate (Nath and Kumar, 1988; Vuren *et al.,* 1994; Velma *et al.,* 2009; Sathya *et al.,* 2012).

Proteins are an important biochemical parameter, utilized for major physiological events and it is also considered as an index of fluid volume disturbances (Goss and Wood 1988; Martinez *et al.,* 2004). In the present study a significant decrease of plasma protein was noted in the fish exposed to nickel toxicity. Similar result were found in *L. rohita* (Palaniappan, and Vijayasundaram, 2009) and in *C. mrigala* (Sathya *etal.,* 2012) exposed to metals. The reduction in plasma protein is resulted due to the direct utilization, impaired protein synthesis, nephrosis, formation of lipoproteins, increased lipolysis, tissue repair and alteration in protein biosynthesis enzyme (Bradbury *etal.,* 1987; Ghosh and Chatterjee, 1989; Abhijith *etal.,* 2012). Ramesh *etal.* (2014) reported that the reduction in plasma protein level in sodium selenite treated fish may be due to impaired protein synthesis or their possible utilization for metabolic demands. Hadson (1988) and Shah and Altindang (2005) reported that dissolved forms of heavy metals in the aquatic environment are easily taken up by aquatic organisms where they are strongly bound with sulfhydril groups of proteins and accumulation in their tissues. In the present study the observed decrease in plasma protein level might have resulted from reduced protein synthesis or direct utilization of protein in gluconeogenesis to produce energy during stress.

From the investigation, it is concluded that nickel sulphate at sublethal concentration has altered the haematological and biochemical parameters of the fish. These parameters can be effectively used as non specific biomarkers against anthropogenic stress.

References

Abhijith, B.D., Ramesh, M., and Poopal, R.K. 2012. Sublethal toxicological evaluation of methyl parathion on some haematological and biochemical parameters in an Indian major carp *Catla catla*. *Comp. Clin. Pathol.,* **21**: 55-61.

Afaghi, A., Zare, S., Heidari, R., Asadpoor, Y., and Malekzadeh Viayeh, R. 2007. Effects of copper sulfate on the levels of glucose and cortisol in common carp. *Cyprinus carpio. Pak. J. Biol. Sci.,* **10(10)**: 1655-1660.

Andrew, A.S., Klei, L.R., and Barchowsky, A. 2001. Nickel requires hypoxia-inducible factor-1a, not redox signaling, to induce plasminogen activator inhibitor-1. *Am. J. Physiol. Lung Cell. Mol. Physiol.,* **281**: L607-L615.

Ates, B., Orun, I., Talas, Z.S., Durmaz, G., and Yilmaz, I. 2008. Effects of sodium selenite on some biochemical and hematological parameters of rainbow trout (*Oncorhynchus mykiss Walbaum* 1792) exposed to Pb^{2+} and Cu^{2+}. *Fish Physiol. Biochem.,* **34**: 53-59.

Barceloux, D.G., 1999. Nickel. *Clin. Toxicol.,* **37**: 239-258.

Blaylock, B.G., and Frank, M.L. 1979. A comparison of the toxicity of nickel to the developing eggs and larvae of carp (*Cyprinus carpio*). *Bull. Environ. Contam. Toxicol.,* **21**: 604-611.

Bradbury, S.P., Symonik, D.M., Coats, J.R., and Atchison, G.J. 1987. Toxicity of fenvalerate and its constituent isomers to the fat-head minnow (*Pimephales promelas*), bluegill (*Lepomis macrochirus*). *Bull. Environ. Contam. Toxicol.,* **38**: 727-35.

Bubb, J.M., and Lester, J.N. 1996. Factors controlling the accumulation of metals within fluvial systems. *Environ. Monitor. Assess.,* **41**: 87-105.

Chau, Y.K., and Kulikovsky-Cordeiro, O.T.R. 1995. Occurrence of nickel in the Canadian environment. *Environ. Rev.,* **3**: 95-120.

Cockell, K.A., Hilton, J.W., and Bettger, W.J. 1992. Hepatobiliary and haematological effects of dietary di sodium arsenate heptahydrate in juvenile rainbow trout (*Oncorhynchus mykiss*). *Comp. Biochem. Physiol. C,* **103**: 453-458.

Cooper, G.R., and McDaniel, V. 1970. The determination of the glucose by the Orthotoluidine method. In: *Standard Methods of Clinical Chemistry,* Vol. 6, Academic Press, New York, p. 159-170.

Drabkin, D.L., 1946. Spectrometric studies, XIV-the crystallographic and optical properties of the hemoglobin of man in comparison with those of other species. *J. Biol. Chem.,* **164**: 703-723.

Elahee, K.B., and Bhagwant, S. 2007. Hematological and gill histopathological parameters of three tropical fish species from a polluted lagoon on the west coast of Mauritius. *Ecotoxicol. Environ. Safe.,* **68**: 361-371.

El-Sayed, Y.S., Saad, T.T., and El-Bahr, S.M. 2007. Acute intoxication of deltamethrin in monosex Nile tilapia, *Oreochromis niloticus* with special reference to the clinical, biochemical and haematological effects. *Environ. Toxicol. Pharmacol.*, **24**: 212-217.

Finney, D.J., 1978. Statistical Methods in Biological Assay (3rd edn.). Griffin Press, London, UK, p. 508.

Gagnon, A., Jumarie, C., and Hontela, A. 2006. Effects of Cu on plasma cortisol and cortisol secretion by adrenocortical cells of rainbow trout, *Oncorhynchus mykiss*. *Aquat. Toxicol.*, **78**: 59-65.

Ghosh, T.K., and Chatterjee, S.K. 1989. Influence of nuvan on the organic reserves of Indian freshwater murrel *Channa punctatus*. *J. Environ. Biol.*, **10**: 93-99.

Goss, G.G., and Wood, C.M. 1988. The effects of acid and acid/aluminium exposure on circulating plasma cortisol levels and other blood parameters in the rainbow trout, *Salmo gairdneri. J. Fish. Biol.*, **32**: 63-76.

Hadson, P.V., 1988. The effect of metal metabolism uptake, disposition and toxicity in fish. *Aquat. Toxicol.*, **11**: 3-18.

Hughes, K., Meek, M. E., Chan, P. K. L., Shedden, J., Bartlett, S., and Seed, L.J. 1994. Nickel and its compounds: evaluation of risks to health from environmental exposure in Canada. *J. Environ. Sci. Health. C*, **12**: 417-433.

Hymavathi, V., and Rao, L.M. 2000. Effect of sublethal concentration of lead on the haematology and the biochemical constituents of *Channa punctatus. Bull. Pure Appl. Sci.*, **19**: 1-5.

Jiang, J.L., Wang, G.Z., Mao, M.G., Wang, K.J., Li, S.J., and Zeng, C.S. 2013. Differential gene expression profile of the calanoid copepod, *Pseudodiaptomus annandalei*, in response to nickel exposure. *Comp. Biochem. Physiol., C*, **157**: 203-211.

Joshi, P.K., Bose, M., and Harish, D. 2002. Changes in certain hematological parameters in a siluroid catfish, *Clarias batrachus* (L.) exposed to cadmium chloride. *Pollut. Res.* **21(2)**: 119-131.

Kasprzak, K.S., Sunderman Jr., F.W., and Salnikow, K. 2003. Nickel carcinogenesis. *Mutat. Res.*, **533**: 67-97.

Kavitha, C., Malarvizhi, A., Senthil Kumaran, S., and Ramesh, M. 2010. Toxicological effects of arsenate exposure on hematological, biochemical and liver transaminases activity in an Indian major carp, *Catla catla. Food Chem. Toxicol.*, **48**: 2848-2854.

Latimer, K.S., Mahaffey, E.A., and Prase, K.W. 2003. Veterinary laboratory medicine. Clinical Pathology (4th Edn.). Iowa State Press, Iowa, p. 3-45.

Lavanya, S., Ramesh, M., Kavitha, C., and Malarvizhi, A. 2011. Hematological, biochemical and ionoregulatory responses of Indian major carp *Catla catla* during chronic sub-lethal exposure to inorganic arsenic. *Chemosphere*, **82**: 977-985.

Lowry, O.H., Rosebrough, N.J., Farr, A.L., and Randall, R.I. 1951. Protein measurement with Folin phenol reagent. *J. Biol. Chem.*, **193**: 265-275.

Martinez, C.B.R., Nagae, M.Y., Zaia, C.T.B., and Zaia, D.A.M. 2004. Morphological and physiological acute effects of lead in the neotropical fish *Prochilodus lineatus*. *Braz. J. Biol.,* **64**: 797-807.

Mason, C.F., 1996. Biology of Freshwater Pollution (3ʳᵈ edn.). Longman, London, p. 1-4.

Muyssen, B.T.A., Brix, K.V., DeForest, D.K., and Janssen C.R. 2004. Nickel essentiality and homeostasis in aquatic organisms. *Environ. Rev.,* **12**: 113-131.

Nath, K., and Kumar, N. 1988. Hyperglycemic response of *Heteropneustes fossilis* exposed to nickel. *Acta Hydrochim. Hydrobiol.,* **16(3)**: 333-6.

Nemcsok, J., and Boross, L. 1982. Comparative studies on the sensitivity of different fish species to metal pollution. *Acta Biol. Hung.,* **33**: 23-27.

Ojo, A.A., and Wood, C.M. 2007. *In vitro* analysis of the bioavailability of six metals via the gastro-intestinal tract of the rainbow trout (*Oncorhynchus mykiss*). *Aquat. Toxicol.,* **83**: 10-23.

Ololade, I.A., and Oginni, O., 2010. Toxic stress and hematological effects of nickel on African catfish, *Clarias gariepinus*, fingerlings. *J. Environ. Chem. Ecotoxicol.,* **2(2)**: 014-019.

Palaniappan, P.L.R.M., and Vijayasundaram, V. 2009. The effect of arsenic exposure and the efficacy of DMSA on the proteins and lipids of the gill tissues of *Labeo rohita*. *Food Chem. Toxicol.,* **47**: 1752-1759.

Pane, E.F., and Wood, C.M. 2004. Mechanistic analysis of acute, Ni-induced respiratory toxicity in the rainbow trout (*Oncorhynchus mykiss*): an exclusively branchial phenomenon. *Aquat. Toxicol.,* **69**: 11-24.

Pane, E.F., Richards, J.G., and Wood, C.M. 2003a. Acute waterborne nickel toxicity in rainbow trout (*Oncorhynchus mykiss*) occurs by a respiratory rather than ionoregulatory mechanism. *Aquat. Toxicol.,* **63**: 65-82.

Pane, E.F., Smith, C., McGeer, J.C., and Wood, C.M. 2003b. Mechanisms of acute and chronic waterborne nickel toxicity in the freshwater cladoceran, *Daphnia magna*. *Environ. Sci. Technol.,* **37**: 4382-4389.

Poopal, R.K., Ramesh, M., and Dinesh, K.P.B. 2013. Short-term mercury exposure on Na$^+$/K$^+$-ATPase activity and ionoregulation in gill and brain of an Indian major carp, *Cirrhinus mrigala*. *J. Trace Elem. Med. Biol.,* **27**: 70-75.

Ramesh, M., 2001. Toxicity of copper sulphate on some haematological parameters of freshwater teleost *Cyprinus carpio* var. *communis. J. Indian Fish. Assoc.,* **28**: 131-136.

Ramesh, M., Sankaran, M., Veera-Gowtham, V., and Poopal, R.K. 2014. Hematological, biochemical and enzymological responses in an Indian major carp *Labeo rohita* induced by sublethal concentration of waterborne selenite exposure. *Chem. Biol. Interact.,* **207**: 67-73.

Remyla, S.R., Ramesh, M., Sajwan, K.S., and Senthil Kumar, K. 2008. Influence of zinc on cadmium induced haematological and biochemical responses in a freshwater teleost fish *Catla catla. Fish Physiol. Biochem.,* **34(2)**: 169-174.

Rothstein, A., 1973. In: W.W. Miller, T.W. Clarkson (Eds.), Mercury, Mercurials and Mercaptans, Charles C. Thomas, Springfield, IL, p. 68-95.

Rusia, V., and Sood S.K. 1992. Routine hematological tests. In: Kanai, L., Mukerjee, I. (Eds.), *Medical Laboratory Technology*. Vol. I, Tata McGraw Hill, New Delhi, p. 252-258.

Saravanan, M., Usha Devi, K., Malarvizhi, A., and Ramesh, M. 2012. Effects of Ibuprofen on hematological, biochemical and enzymological parameters of blood in an Indian major carp, *Cirrhinus mrigala. Environ. Toxicol. Pharmacol.,* **34**: 14-22.

Sathya, V., Ramesh, M., Poopal, R.K., and Dinesh, B. 2012. Acute and sublethal effects in an Indian major carp *Cirrhinus mrigala* exposed to silver nitrate: gill Na^+/K^+-ATPase, plasma electrolytes and biochemical alterations. *Fish Shellfish Immunol.,* **32(5)**: 862-868.

Schaumlöffel, D., 2012. Nickel species: analysis and toxic effects. *J. Trace. Elem. Med. Biol.,* **26(1)**: 1-6.

Shah, S.L., and Altindag, A. 2005a. Effects of heavy metal accumulation on the 96-h LC50 values in tench *Tinca tinca* L., 1758. *Turk. J. Vet. Anim. Sci.,* **29**: 139-144.

Soengas, J.L., Agra-Lago, M.J., Carballo, B., Andres, M.D., and Veira, J.A.R. 1996. Effects of an acute exposure to sublethal concentrations of cadmium on liver carbohydrates metabolism of Atlantic salmon, *Salmo salar. Bull. Environ. Contam. Toxicol.,* **57(4)**: 625-631.

Sveceviius, G., 2010. Acute toxicity of nickel to five species of freshwater fish. *Polish J. Environ. Stud.,* **19(2)**: 453-456.

U. S. Public Health Service (USPHS), 1993. Toxicological profile for nickel. U.S. Public Health Service, Agency for Toxic Substances and Disease Registry, Atlanta, Georgia. Report TP-92/14, p. 158.

U.S. Environmental Protection Agency (USEPA), 1986. Ambient water quality criteria for nickel. EPA 440/5-86-004. Technical report. Washington, DC. http://nepis.epa.gov/exe/zyPURL.cgi

Van Vuren, J., Van der Merwe, M., and Du Preez, H. 1994. The effect of copper on the blood chemistry of *Clarias gariepinus* (clariidae). *Ecotoxicol. Environ. Safe,* **29(2)**: 187-99.

Vandenbrouck, T., Soetaert, A., van der Ven, K., Blust, R., and De Coen, W. 2009. Nickel and binary metal mixture responses in *Daphnia magna*: Molecular fingerprints and (sub) organismal effects. *Aquat. Toxicol.,* **92**: 18-29.

Velma, V., Vutukuru, S., and Tchounwou, P.B. 2009. Ecotoxicology of hexavalent chromium in freshwater fish: a critical review. *Rev. Environ. Health,* **24(2)**: 129-46.

Vinodhini, R., and Narayanan M. 2009. The impact of toxic heavy metals on the hematological parameters in common carp (*Cyprinus carpio* l.). *Iran. J. Environ. Health. Sci. Eng.,* **6(1)**: 23-28.

Vutukuru, S.S., 2003. Chromium induced alterations in some biochemical profiles of the Indian major carp, *Labeo rohita* (Hamilton). *Bull. Environ. Contam. Toxicol.,* **70**: 118-123.

Vutukuru, S.S., 2005. Acute effects of hexavalent chromium on survival, oxygen consumption, hematological parameters and some biochemical profiles of the Indian Major Carp, *Labeo rohita. Int. J. Environ. Res. Public. Health,* **2**: 456-462.

Wedemeyer, G., 1969. Stress induced ascorbate depletion and cortisol production in two salmonid fishes. *Comp. Biochem. Physiol.,* **29**: 1247-1251.

Witeska, M., and Kosciuk, B. 2003. Changes in common carp blood after short-term zinc exposure. *Environ. Sci. Pollut. Res.,* **3**: 15-24.

Wong, C.K., Chu, K.H., Tang, K.W., Tam, T.W., and Wong, L.J. 1993. Effects of chromium, copper and nickel on survival and feeding behaviour of *Metapenaeus ensis* larvae and postlarvae (Decapoda: Penaeidae). *Mar. Environ. Res.,* **36(2)**: 63-78.

Woo, S., Yum, S., Park, H.S., Lee, T.K., and Ryu, J.C. 2009. Effects of heavy metals on antioxidants and stress-responsive gene expression in Javanese medaka (*Oryzias javanicus*). *Comp. Biochem. Physiol. C,* **149**: 289-299.

World Health Organization (WHO), 1991. Nickel. Environmental Health Criteria, 108: p. 383.

2015, Perspectives in Animal Ecology and Reproduction, Vol. 10   *Pages 159–170*
Editors: **V.K. Gupta, Anil K. Verma and G.D. Singh**
Published by: **DAYA PUBLISHING HOUSE, NEW DELHI**

Chapter 11

An Overview on the Caspian White Fish Biology, Ecology and Reproduction

Peiman Zandi[*1], *Saikat Kumar Basu*[2]
and Shahram Khademi Chalaras[1]

*[1]Department of Agronomy, Faculty of Agriculture, Takestan Branch,
Islamic Azad University, P.O. Box 466, Takestan, Iran
[2]Department of Biological Sciences, University of Lethbridge,
Lethbridge, AB Canada T1K 3M4*

ABSTRACT

The Caspian Sea, the world's largest landlocked lake, is encircled by five Caspian states; Russia, Azerbaijan, Iran, Turkmenistan and Kazakhstan. This unique aquatic environment is inhabited by the sturgeons and a number of other bony fish species. The Caspian Kutum (*Rutilus frisii kutum* Kamensky, 1901), also known as white fish is only found in the Iranian waters. It is the one of the most important migratory anaderomous bony fish which is placed at the top of North-Iranian food basket and is especially consumed by the people of Caspian coastal zone for its high nutritional quality. There are about 12,000 fishermen directly involved in catching the kutum which represents almost 78 per cent of the captured bony fishes in the Iranian coastal division. However disruption of the natural reproduction of the species due to rapid habitat destruction in major parts of rivers, estuaries and in the Anzali Lagoon, over fishing, illegal fishing practices, out of season fishing, fluctuations in the sea level, environmental pollution and dramatic increase of natural predators and other exotic species have threatened this species with the potential risk of extinction, if proper steps are not adopted. The Iranian Fisheries Research Organization (IFRO) therefore needs to play an

* *Corresponding author.* E-mail: z_rice_b@yahoo.com

important and effective role for conserving the popular kutum species through implementation of restocking/breeding programs.

Keywords: *Caspian sea, Distribution, Migration type, Reproduction, Rutilus frisii kutum, Spawning.*

Abbreviations

IFRO: Iranian Fisheries Research Organization

SAV: Submerged Aquatic Vegetation

Introduction

The Caspian (Khazar) term refers to an ancient tribe that once ruled in the north of the Caspian Sea in the northern direction of Caucasus Mountains (Khanipour and Valipour, 2009). The Caspian Sea, which comprises a surface area of 3733000 km², is the greatest landlocked lake in the world shared by five Caspian coastal states (Figure 11.1). There are approximately about 115 fish species and their sub-species including five unique sturgeon species with great commercial importance (Valipour and Khanipour, 2006). These species include the beluga sturgeon (*Huso huso*), the Persian sturgeon (*Acipenser persicus*), the Russian sturgeon (*Acipenser gueldenstaedtii*), the ship sturgeon (*Acipenser nudiventris*) and the stellate sturgeon (*Acipenser stellatus*) as well as the bony fishes like the Kutum (*Rutilus frisii kutum*, Kamensky,1901), breams (*Abramis*), mullets, barbus, carps, salmons, and kilka fish (Khanipour and Valipour, 2009). The starting point of the bony fish catching in the Iranian coastal waters goes

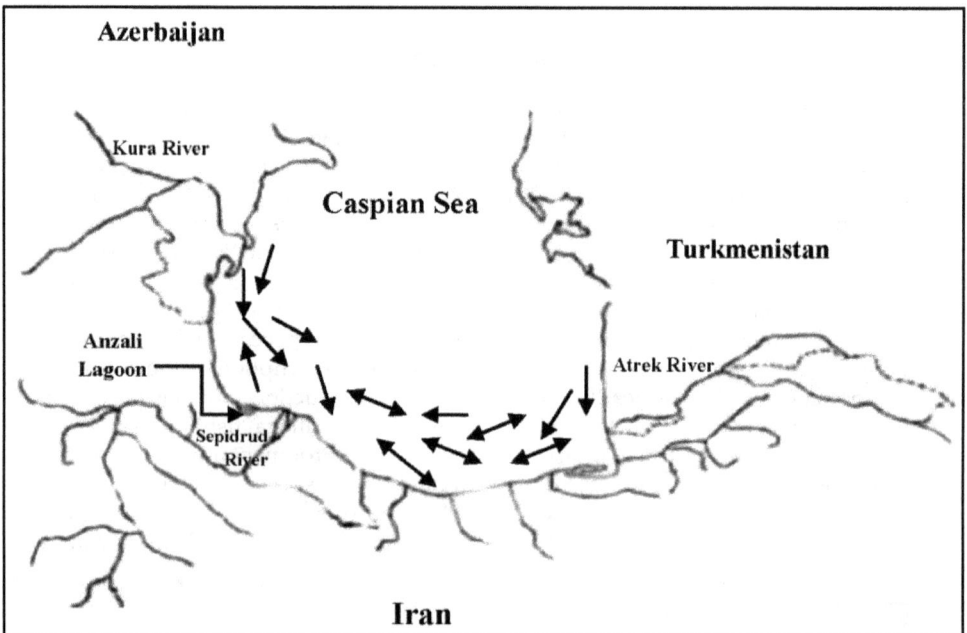

Figure 11.1: Map Showing the Distribution of *Rutilus frisii kutum* (White fish) in the Iranian Waters of Caspian Sea (Map not to scale).

back to 1927 (Arabi *et al.,* 2012). During the past three decades several factors including rapid habitat destruction in major parts of rivers, estuaries and in the Anzali Lagoon, over fishing, illegal fishing practices, out of season fishing, accidental introduction of exotic species like invasive jellyfish (*Mnemiopsis leidyi* first recorded in 1999), fluctuations in the sea level, pollution significantly changed the environmental conditions of the Caspian Sea and finally could possibly impact different ecological components of this unique aquatic environment negatively (Rohi *et al.*,2010; Fazli *et al.,* 2013).

The Caspian white fish (*Rutilus frisii kutum*) (Figure 11.2) is an important endemic bony fish species with great commercial value that mainly inhabits the southern part of the Caspian Sea (900 km coastline) rather than its North and Volga river (Ourjani *et al.,* 2011; Fazli *et al.,* 2013). This species is considered popular among fishermen, coastal dwellers and the individuals of the Caspian basin states for its superior nutritional quality and taste (Kuliev, 1997; Afraei Bandpei *et al.,* 2010). This is a fast growing species with short growing cycle, and medium size (length 45-70 cm, weight d"5 kg) and belongs to the Cyprinidae family. The life span of the species varies between 9-10 years under natural conditions and their sexual maturity alters between 2-3 (males) and 3-4 (females) years (Adeli Mosabbab and Piri, 2005; Abdolmaleki and Ghaninejad, 2007; Afraei Bandpei *et al.,* 2010). The species is a true migratory or anaderomous fish which has two types of migration; spring and autumn type (Keivany *et al.,* 2012). The spring types of the species begins their migration from southern parts of the sea into the estuaries and rivers for spawning in the months of March and April, sometimes up to May (Berg,1964; Afraei Bandpei *et al.,* 2011) and their harvest starts in early October and ends in the late March every year (Khanipour and Valipour, 2009). However, the autumn type usually starts its migration to the rivers between October to February (Magomedov *et al.,* 1987; Keivany *et al.,* 2012). It is important to note that the spawning during the migratory period is dependent on the water temperature, ecological conditions and sea processes (Valipour and Khanipour, 2006; Fazli *et al.,* 2013).

The Caspian Kutum locally known as the Mahisefid (White fish) constitutes a total of 78 per cent of harvested bony fish capture and a major percentage of the local fishermen's income (~ e" 60 per cent) during the fishing seasons in the region (Afraei Bandpei, 2010; Shafiei Sabet *et al.,* 2010). According to a report by FAO (2003), the average annual Caspian Kutum catch in the Iranian coasts has been16,000 ton in 2006 compared to a continues 10-year (1991-2001, 9600 ton per year) fishing period,

Figure 11.2: The Mature Caspian White Fish (*Rutilus frisii kutum* Kamensky, 1901).

which in turn demonstrates an increase of 60 per cent in the fishing rate. Since 1982, due to a number of factors such as ecological changes in the sea, environmental pollution, decrease in the natural spawning grounds in the rivers, estuaries and in the Anzali lagoon due to habitat destruction, decrease in the sea water level since 1950s (Valipour and Khanipour, 2006) and the appearance of *Mnemiopsis leidyi* (Sea walnutan unpopular, accidentally introduced, exotic predator of zooplankton and fish eggs) in the sea as well as over fishing, illegal fishing and off season harvests there has been a gradual decline in the fish population. The Iranian Fisheries Research Organization (IFRO) responded to this alarming threat of the species decline by establishing artificial propagation and/or restocking programs releasing upto 200 million fries with mean body weight of 1 g into the estuaries and the river systems (Azari Takami *et al.,* 1990; Afraei Bandpei *et al.,* 2009; Shafiei Sabet and Imanpoor, 2011; Arabi *et al.,* 2012).

The successful rearing of larvae and fries in the nursery ponds, situated in the restocking centers, has come up at an appropriate time to help in the long term sustenance of the species in their natural habitat (Ourjani *et al.,* 2011). The most important factor for rapid growth and sound health of the fish larvae is availability of quality diet (*e.g.* zooplankton and artificial diet) in a way that can assure the required nutritional need of the growing fry populations (Fazli *et al.,* 2013). Zooplanktons (*e.g.* rotifers, cladoceranse and copepods) have been found to be suitable sources of food items for the first-feeding larvae and can be used in the aquaculture system (Szlauer and Szlauer, 1980). Briefly, the restocking program subsequently enhanced the mean annual catch of kutum to the highest level (17200 t) in 2007-2008 (Fazli *et al.,* 2013). In other word, the artificial propagation of Kutum by IFRO, initiated in 1982, successfully conserved the natural population stocks and preserved the genetic pool of such valuable fish species for long term survival and sustenance of the species concerned.

Feeding and Reproduction

The species has an omnivore nutritional behavior and unlike other similar omnivores has a short digestive tract; hence, only feeds on limited food items (Afraei Bandpei *et al.,* 2009). Depending of several factors including season, month, habitat and fish size the feeding behavior of kutum changes (Afraei Bandpei, 2010). During early stages of development, it is only nourished by phytoplankton, zooplankton and insect larvae; however, as it grows and starts to migrate to the sea, its nutrition is largely supplied by a group of Bivalve molluscs. A summary of the biodiversity of different prey items, consumed by kutum, is presented in Figure 11.3. There are five principal components (*i.e.* mouth, alimentary canal, anterior gut, posterior gut and anus), that constitutes the digestive tract in Kutum. The species lack a true stomach. The high fecundity rate is an important feature in such aquatic organisms (Afraei Bandpei, 2010). The rate of absolute fecundity varies between 19717-147696 eggs (in average 74774 eggs; 272 eggs/g). Egg color is usually pale/golden yellow and are rarely pistachio green. The egg's diameter varies between 0.12-0.19 cm. The weight of mature kutum, depending on gender, differs from 0.4-2.1 kg in males and 0.5-3.3 kg in females. The sexual maturity in females is attained 1-2 years later than the males (*i.e.* females 3-4 years; males 2-3 years).The male/female proportion in rivers at the time of natural reproduction varies between the ratios of 3.2:1- 6.6:1(Zarin Kamar, 1996).

Zooplankton	**Rotatoria:** *Lepadella* sp., *Pompholyx* sp., *Philodina* sp., *Keratella* sp., *Brachionus* sp., *Asplanchna* sp., *Monostyla* sp., *Synchaeta* sp., *Philodina* sp., *Rotaria* sp., *Polyarthra* sp. **Copepoda:** *Cyclops* sp., *Diaptomus* sp., *Daphnia* sp., *Moina* sp. **Prorozoa:** *Arcella* sp., *Lionotus* sp., *Marituja* sp., *Diffugia* sp. **Cladosera:** *Daphina* sp., *Moina* sp.
Phytoplankton	**Chrysophyta:** *Synura* sp., *Dinobryon* sp., *Chlorella* sp., *Pediastrum* sp., *Ankistrodesmus* sp., *Selenastrum* sp., *Volvux* sp., *Spirogyra* sp., *Ulothrix* sp., *Pandorina* sp., *Dictiosphaerium* sp., *Oedogonium* sp., *Cladophora* sp., *Schroederia* sp., *Actinastrum* sp., *Scenedesmus* sp., *Chlamydomonas* sp. **Bacillariophyta:** *Cyclotella* sp., *Melosira* sp., *Stephanodiscus* sp., *Synedra* sp., *Naviculs* sp., *Fragilaria* sp., *Tabellaria* sp., *Asterionella* sp., *Pseudo-nitzschia* sp., *Gyrosigma* sp., *Gomphonema* sp., Diatoms spp., *Dentcula* sp. **Cynophyta:** *Merismopedia* sp., *Microcystis* sp., *Lyngbya* sp., *Anabaena* sp., *Rophidiopsis* sp., *Nostoc* sp., *Phormidium* sp., *Gomphosphaerium* sp., *Anabaenopsis* sp., *Oscillatoria* sp., *Aphanizomenon* sp., *Spirulina* sp., *Chrococcus* sp., *Gloeotrichia* sp. **Euglenphyta:** *Euglena* sp., *Phacus* sp., *Lepocinclis* sp., *Trachaelomonas* sp. **Pyrrhophyta:** *Ceraimm* sp., *Glenodinium* sp., *Gymnodinum* sp., *Peridinium* sp. **Cryptophyta:** *Mallomonas* sp., *Cryptomonas* sp.

Plankton (branching to Zooplankton and Phytoplankton)

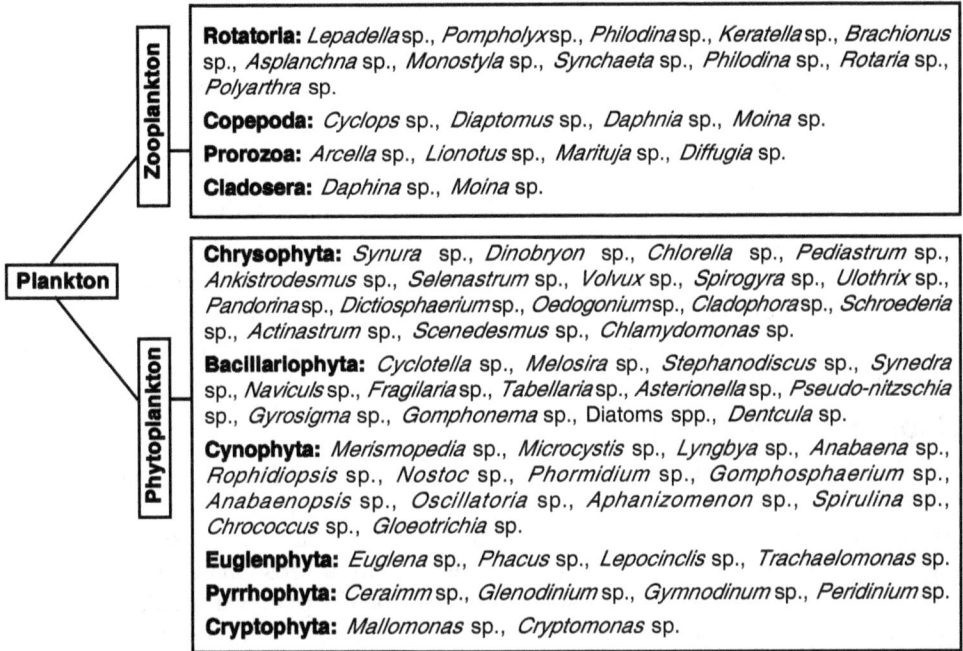

**Figure 11.3: A Selective List of Major Natural Food Sources of
Rutilus frisii kutum in the Caspian Sea and the Neighboring Rivers.**

Both types of kutum (autumn and spring) when ready to spawn (in the autumn and spring seasons), proceed to migrate first to the Anzali lagoon and to the nearest river systems leading to the southern side of the Caspian Sea for spawning on submerged aquatic vegetation (SAV) or on pebbly beds (Zarin Kamar, 1996).

The most important factors directly involved in the movement of both male and female fishes into the natural spawning sites, once they attain reproductive readiness, are water flow, water clarity, dissolved oxygen level, seasonal temperature in the freshwater ecosystem and coastal processes (Afraei Bandpei *et al.,* 2011). During this period, the mature male and female spawners gradually undergo certain morphological transitions. Males develop into a cone-shaped body, and generate epithelial colourless tubercles (nearly white) that are spread on both sides of their body on the head; and the females develop a smooth-surfaced body and bilaterally symmetrical distended abdomen (Zarin Kamar, 1996). These fishes most often spawn in the shallow waters; in such a way that, each female is flanked by at least 3-6 males. After being stimulated by the surrounding males, the females rapidly rub their lower abdominal muscles on the gravel bed, thereby putting pressure on the pectoral area, created by a severe female convulse and the eggs are shed into the water. The males simultaneously release the sperms on those recently shed eggs, causing them to be fertilized in a broadcast fertilization method (Shafiei Sabet *et al.,* 2010). From the time, the sperm enters the micropyle, the eggs begin to absorb water and get swelled; meanwhile, cell division is started. At this time, the eggs are extremely adhesive ensuring their firm establishment on the same location (Khanipour and Valipour,

2009). The eggs can simply stick to the pebbles and to the aquatic vegetation in their spawning ground. For about 14-16 days they remain as such during which the embryos continue their growth and development within the eggs. At the of this period, the egg cell membrane reputes and the tiny larvae emerge from the hatched eggs in the spawning beds (Haghighi, 2006). Normally the incubation period last for around 9-10 days. Initially, the young larvae feed on yolk pouch reserves for a very short period (4-5 days), after which they begin to prowl for various planktons in the river systems and lagoon networks (Khanipour and Valipour, 2009). Their growth stages continue in the same locations, until they develop into fingerlings weighing a few grams. Then, through development of their osmoregulation system, the fingerlings are able to return back to the open sea for attaining sexual maturity and performing subsequent reproduction.

Habitat and Distribution

Rutilus frisii kutum is primarily found in the south and southwest coast of the Caspian Sea and is an economically valuable species in the region spreading from Atrek River located in the Caucasus region (western coasts of the central Caspian region) to the southern coasts of Turkmenistan (Adeli Mosabbab and Piri, 2005). These species are barely found in other areas of the Caspian Sea such as the north Caspian and the Volga River, and there is no official report and evidence on their migration to these regions (Emadi, 1977). In the past years, these species were found in abundance in the regions between the Kura River in Azerbaijan and the Sepidrud River in Iran. However, the mass release of kutum fingerlings all over the Iranian coasts of the Caspian Sea in the past 25 years brought about a noticeable difference in the distribution of this species; and their numbers in the coasts of the Mazandaran and Golestan have enhanced dramatically (Haghighi, 2006). Variations in the kutum abundance are highly relevant to the variations in the season and temperature (Afraei Bandpei, 2010). In the early winter, due to lower temperatures in the coastal waters of the Guilan Province to that in the Mazandaran Province, this species migrates to the warmer eastern regions moving away from the coast and migrating to the deeper regions of the sea.

As spring approaches, in late February, this species gets prepared for its spawning and they are seen in larger numbers in the coastal areas reaching a peak in late March. Nowadays, the Anzali Lagoon in Iran and the Ghazel Aghaj Lagoon in Azerbaijan that used to be the main spawning places for kutum, are not capable to support kutum breeding and have hence lost their significances (Afraei Bandpei, 2010). Previously, most of the rivers in the Iranian coast of the Caspian Sea were used as the natural spawning places by the kutum. However, now, only a few rivers (such as Chalusrud, Lemir, Shalmanrud, Khoshkud, Sefidrud, Shirrud and Babolsar) are used as major spawning grounds for the spring migration and also for the artificial spawning of this species (Haghighi, 2006). In addition to the rivers in the Iranian coasts, the rivers in the Caucasus regions such as Atrek, Samur and Kura are also important spawning grounds for the kutum (Afraei Bandpei, 2010). However, spawning migrations of this species to these regions have decreased exceedingly due to the pollution, over fishing of spawners and lack of artificial breeding and rehabilitation programs of stocks via release of fingerlings.

Rutilus frisii kutum lives the remaining periods of spring and summer feeding and growing in the shallow coasts of the Caspian Sea where there is a plenty of benthic fauna (Afraei Bandpei *et al.,* 2009). In late summer, when the water temperature goes up and the epilimnion develops to the depth of 30 m, these fish abandon the shallow coastal areas and move to the deeper layers close to the thermocline in order to return to the shallow coastal regions (<20 m) during the autumn circulation for feeding (Khanipour and Valipour, 2009). This species migrates in shoals along the coastline from the west to the east or vice versa during autumn. Such migrations are completely dependent on the local atmospheric conditions and sea currents regulated by winds. During late autumn and early winter when the surface layers become cold and during the thermal stratification, the kutum species gradually abandon the shallow coastal regions and move back again to deeper areas; therefore, this species is rarely found in the shallow regions especially between mid December and late January (Khanipour and Valipour, 2009; Haghighi, 2006).

Migration and Population Diversity

After migration to the sea, the fish starts feeding and reaches to sexual maturity and for the purpose of natural spawning, enters the freshwater habitats like Anzali lagoon and the available neighboring river systems in the northern flatlands (Emadi, 1977). The species possess two distinct types of migration; spring and autumn migratory types (Kazanchev, 1981; Keivany *et al.,* 2012).

I. The Autumn Type of Kutum

As late September coincides with the favorable conditions, the autumn type of kutum (both male and females) begins to migrate to the Anzali lagoon through available navigable channels (Afraei Bandpei *et al.,* 2011). Spawners will mostly migrate towards those rivers leading into the lagoon such as Siah darvishan, Pasikhan, Masouleh, and in case of good hydrological conditions in Pirbazar as well (Khanipour and Valipour, 2009). The group usually overwinters in the wetlands (*i.e.* Anzali lagoon) especially in the Central, West and Shyjan districts in the East at deeper areas. For accessing warmer water in the late winter, they migrate to the rivers having aquatic vegetation like common reeds (*Phragmites australis*) and bulrush (*Typha latifolia*) and then spawns on the aquatic plants (Haghighi, 2006). That is why this form of kutum is known as phytophilous. At the time of spawning each female fish will be stimulated by two or three males and both eggs and sperm will be concurrently released in the water afterwards (Keivany *et al.,* 2012). This results in fertilizing the eggs.

After deposition of eggs on these aquatic vegetations the fertilized ones adhere to them and the embryonic development occurs in naturally oxygen-enriched environment. Then, the larvae emerge from the hatched eggs in about two weeks and start to swim freely. Because of existence of several predators in the lagoon, the larvae have little chance for growth and success to migrate to the sea (Haghighi, 2006; Fazli *et al.,* 2013). Moreover, certain ecological conditions of the lagoon that is sometimes affiliated with serious depletion of oxygen lead to actual mortality of a large number of larvae and fingerlings. The kutum overwintering normally take place in deeper parts of the sea bed and their feeding activities are low in the winter months (Khanipour and Valipour, 2009). When the regional weather conditions are favorable (late January to early February), the spring type begins moving to the coastal regions and resume feeding in order to store enough energy for their further spawning migratory action toward the rivers and lagoons (Afraei Bandpei *et al.*,2011).

The autumn type of kutum will migrate back to the sea after spawning in late winter and early spring. As soon as they reach the sea, they will resume active nutrition on prey items to compensate for the lost energy (Zarin Kamar, 1996). Larvae produced in the lagoon have better growth than those of spring type due to the extensive available food resources (Khanipour and Valipour, 2009). These larvae generally migrate towards the sea after about 25-35 days and reaching to the weight of 2-3 g in order to spend their growth period through feeding on various natural foods in the shallow areas of the Caspian Sea (Afraei Bandpei *et al.*, 2011).

II. The Spring Type of Kutum

Currently, the main population of kutum in the Caspian Sea belongs to the spring population, which consists of more than 98 per cent of its stocks in the Caspian Sea (Khanipour and Valipour, 2009). In spring, coupled with thermal circulation, when the water temperature in different depth levels (especially at the surface and bottom) becomes almost uniform; the spring type of this species moves from the deeper area to the shallower parts near the coastal regions (Khaval, 1999; Afraei Bandpei *et al.*, 2011). Usually in late winter and early spring the fishes that have not yet reached sexual maturity remain in the shallow coastal areas for feeding and growth and the adults migrates to the rivers for spawning (Khanipour, 1989). The migration of the kutum into the Sefidrud River for breeding is regulated by several factors such as temperature, sea processes, water flow, water transparency (Secchi disk depth), and the concentration of the dissolved oxygen (Khaval, 1999; Khanipour and Valipour, 2009). The mature fishes frequently migrate to these streams and rivers for spawning so that their early growth stages (ontogeny) are spent in it. Locating the native river network is possibly conducted by their lateral line (Khanipour, 1989; Khaval, 1999).

The spring type of kutum by crossing the riverine estuary and entering the river continue their path towards the upstream points (~500-800 m after estuary) for depositing their eggs on the gravel and stony beds; and are hence called Lithophylus fish (Afraei Bandpei, 2010). Indeed, this type of kutum on its arrival into the river is in the late IV stage of sexual maturity (Emadi, 1977; Saeedi, 2003). However, within a short period of time, by entering and reaching the spawning grounds they attain their complete sexual maturity and are then ready for spawning. The inception of such process in spring type begins with 8°C water temperature and riches the peak at 13-14°C (Berg, 1964). Usually a single female fish is flanked by 3-4 males (Khanipour and Valipour, 2009). In response to the provoking activities of the males, the females sharply rub their lower abdomen and pectoral area on the nearest gravel bed. The engaged fishes in spawning process at a specific time release their sperms and eggs in a clear and richly oxygenated water (Khanipour and Valipour, 2009; Keivany *et al.*, 2012). Once the eggs are shed in between the graves the released sperms will go after them for fertilization. The fertilized eggs due to their naturally adhesive properties adhere to stones at the waterbed.

During the spawning period sometimes their dorsal fin and backs are protruded from the water level and this is because of females feature seek shallow and clear points in the river (Khanipour and Valipour, 2009). In this period, the female spawner changes its mates several times by choosing new and vigorous ones (Afraei Bandpei *et al.,* 2011). The spawning procedure is so intensive and energy consuming that in some cases results in injury when they move to gravel bottom (Afraei Bandpei *et al.,* 2011). Due to depletion of their energy on account of natural reproduction, the fishes during this period are easily captured by the fishermen. The spawners that could finally reach safely back to the sea resume their feeding activities for compensating their energy loss hrough the spawning process (Afraei Bandpei *et al.,* 2009). Meanwhile, the adhesive fertilized eggs gradually absorb water and become turgid and after about 16 days at a temperature of ~14-16°C, the larvae emerge from the eggs (Haghighi, 2006). Once the yolk pouch is absorbed by the larvae, they start to swim freely in the river for 30-45 days before they finally move back to the open sea (Khanipour, 1989; Khaval, 1999).

Current Status

The growth of kutum in the Caspian Sea is largely dependent on several factors such as available food reserves, heredity, bioenvironmental factors, pollution, diseases etc. Recently IFRO reported that there are dramatic differences between the growth rate in kutum's population in the past (*i.e.* 1970's) and recent decades; in a way that the amount of increase in the length and weight of this species in terms of age effect, has been scaled down substantially (Abdolmaleki, 2006). This finding may be greatly attributed to the annual artificial breeding and restocking programs developed by the IFRO. Other factors which probably contributed to the situation are those of overfishing and eradication of larger species by applying improper fishing practices.

Currently, the Iranian methods of choosing male and female spawners for artificial reproduction and rearing programs are not in accordance with any selective-based approach; which may ultimately result in their gene bank to be gradually shifted. With an overview on the growth variation trend in kutum in the Caspian Sea during 1971-2005, the mean length in adult kutum (3-year classes) was decreased from 42.2-33 cm (Khanipour and Valipour, 2009). Moreover, such decreasing trends were also recorded for about all year classes in the species and for different years. The recorded observations by Khanipour (2005) shows that the mean length and weight in captured kutum in the Iranian parts of the Caspian Sea since 1991-2003 has declined due to failure in their natural reproduction and also through the artificial breeding programs for restocking the species. It is therefore important to establish natural ponds where the selection of parents is allowed in a semi-natural breeding approach and also for providing the favorable conditions for the natural spawning in the rivers and in the Anzali lagoon.

Conclusion

Rutilus frisii kutum is a fish of Cyprinidae family which distributed throughout the Caspian Sea, from Atrek River (Turkmenistan) to Kura River (Azerbaijan). This fish is of great commercial value in the southern parts of the Caspian Sea, consumed by the local people around the coastline. The species is typically is medium sized, reaching to 45-55 cm in length (sometimes 70 cm), weighting up to 4-5 kg. It is important to note that the study of age, growth behavior and reproduction of such fishes is an obligatory action for successful propagation, sustainability, preservation, consistency of their exploitation and management. Due to its economic importance, the fish biology is well documented in the form of primary literature sources. Most of studies are focused on age structure, ecology, fertility, feeding, fecundity, fishing, reproduction, and spawning of the kutum. In the present work, we studied the habitats and distribution, biology, feeding, growth, migration types, incubation procedure and fecundity during the breeding season to study the impacts of these important ecological factors on the life cycle of the kutum.

Over fishing, elimination of spawning grounds along with severe artificial reproduction and release of fingerlings have led to an outstanding fluctuation in the stocks of the species within the past decades. The population stock of kutum at present time has quantitatively enhanced in comparison to the past decades. However, the quality of growth trend in the species is still quite low. The decrease in the growth could be attributed to the artificial breeding programs by IFRO. The uncontrolled exploitation rate of this valuable species in the Caspian Sea needs to be strictly regulated and managed by all the local governments around the Sea for long term sustenance of this valuable fish species.

References

Abdolmaleki, S., 2006. Trends in stocks fluctuation of *Rutilus frisii* Kutum in the Caspian Sea. *Iranian Scientific fisheries Journal*, **25(2)**: 87-100.

Abdolmaleki, S., and Ghaninejad, D. 2007. Stock assessment of the Caspian Kutum *Rutilus frisii* Kutum in the Iranian coastal waters of Caspian Sea. *Iran. J. Fish. Sci.*, **16(1)**: 113-116.

Adeli Mosabbab, Y., and Piri, K. 2005. Monitoring of release fries of *juvenile vobla* (*Rutilus rutilus*) at the Iranian coast of the Caspian Sea with the purpose of restoring of its stocks. In: Proceeding of XIII International Conference on Fisheries Oceanology, 12-17 Sep., Svetlogorsk, Kanliningrad, Russia, pp.12-17.

Afraei Bandpei, M.A., 2010. Population dynamics of Kutum, *Rutilus frisii* kutum in the Caspian Sea. PhD thesis, University Sains, Malaysia, 243 pp.

Afraei Bandpei, M.A., Mansor, M., Abdolmalaki, S.H., Keymaram, F., Mohamad Isa, M., and Janbaz, A.A. 2010. Age and growth of kutum (*Rutilus frisii* kutum, Kamensky, 1901) in southern Caspian Sea. *International Aquatic Res.*, **2**: 25-33.

Afraei Bandpei, M.A., Mashhor, M., Abdolmalaki, S., and Abdel-fattah el-sayed, M. 2009. Food and feeding habits of the Caspian Kutum, *Rutilus frisii* kutum (Cyprinidae) in Iranian waters of the Caspian Sea. *Cybium*, **33(3)**: 193-198.

Afraei Bandpei, M.A., Mashhor, M., Abdolmaleki, S.H., Najafpour, S.H., Bani, A., Pourgholam, R., Fazli, H., Nasrolahzadeh, H., and Janbaz A.A. 2011. The Environmental Effect on Spawning Time, Length at Maturity and Fecundity of

Kutum (*Rutilus frisii* kutum, Kamensky, 1901) in Southern Part of Caspian Sea, Iran. *Iranica Journal of Energy and Environment*, **2(4)**: 374-381.

Arabi, M.H.G., Sedaghat, S., Hosseini, S.A., and Fakhri, A. 2012. Age and Growth of Kutum, *Rutilus frisii* kutum (Kamenskii 1901) in Tajan River (Southern Caspian Sea to Iran). *Global Veterinaria*, **9(2)**: 211-214.

Azari Takami, A.H., Razavi Sayad, B., and Hosseinpour, N. 1990. Review of artificial propagation and breeding fish in white *Rutilus frisii* kutum. *J. Veterinary Medicine*, Tehran University, **45**: 52-45.

Berg, L.S., 1964. Freshwater Fishes of the USSR and adjacent Countries. Vol. ii, 4th edit, Jerusalem: Israel Program for Scientific Translations, 496 pp.

Emadi, H., 1977. *Rutilus frissi* kutum: its past and present status in the northern waters of Iran. Guilan Fisheries Research Center, Bandar Anzali, Iran, 15 pp.

FAO, 2003. Induced spawning of Indian major carp sand maturation of a perch and a catfish by murrel gonadotropin releasing hormone, pimozide and calcium. In: *Fish stat Plus. Fishery Statistics*, (Eds.) S. Halder, S. Sen, S. Bhattachearaya, A.K. Ray, A. Ghosh and A.G. Jhingan 1991, pp. 373-382.

Fazli, H., Daryanabard, G.R., Abdolmaleki, S. and Bandani, G.A. 2013. Stock Management Implication of Caspian kutum (*Rutilus frisii* kutum Kamensky, 1901) in Iranian Waters of the Caspian Sea. *ECOPERSIA,* **1(2)**: 179-190.

Haghighi, D.T., 2006. Larval rearing in *Rutilus frissi* kutum with formulated diets. Inland water Aquaculture Research Center, Bandar Anzali, 185 pp.

Kazanchev, E.N. 1981. Fishes of the Caspian Sea and its drainage basin. Translated by Shariati, A. 1992. Iranian Fisheries Research Organization (I.F.R.O.) Publication, Tehran, Iran, 171 pp.

Keivany, Y., Zare, P., and Kalteh, L. 2012. Age, Growth and Reproduction of the Female Kutum *Rutilus kutum* (Kamensky, 1901) (Teleostei: Cyprinidae), in Gorgan-Rud Estuary, Northern Iran. *Research in Zoology*, **2(3)**: 7-14.

Khanipour, A.A., 1989. Limnological study (in terms of migration of *Rutilus frissi* kutum) of the Polrud and Khoshkrud Rivers Tehran University, College of Natural Resources, Karaj, Iran, 152 pp.

Khanipour, A.A., and Valipour, A. 2009. Kutum, Jewel of the Caspian Sea, Iranian Fisheries Research Organization (I.F.R.O.) Publication, Tehran, Iran, 97 pp.

Khaval, A., 1999. Migration in *Rutilus frissi* kutum to the Sefidrud River. Department of Breeding and Rearing, Iranian Fisheries Organization, Tehran, Iran, 12 pp.

Kuliev, Z.M., 1997. Carps and perches of the southern and Middle Caspian (structure of the population, ecology, distribution and measures for population restocking, PhD. dissertation, Baku, Azerbaijan, pp. 14-15.

Magomedov, G.M., Aliev, D.A., and Proskurova, V.O. 1987. On the spawning of kutum in the region of the Dagestan coast of the Caspian Sea. Rybnoe Khozyaistvo, USSR, pp.37-39.

Ourjani, H., Khalili, K.J., Ebrahimi, G., and Jafarpour, S.A. 2011. Determination of the optimum transfer time of kutum (*Rutilus frisii* kutum) larvae from live food to artificial dry feed. *Aquatic Int.*, **19(4)**: 683-691.

Roohi, A., Kideys, A.E., Sajjadi A., Hashemian, A., Pourgholam, R., Fazli, H., Ganjian Khanari A., and Eker-Develi, E. 2010. Changes in biodiversity of phytoplankton, zooplankton, fishes and macrobenthos in the Southern Caspian Sea after the invasion of the ctenophore *Mnemiopsis Leidyi. Biol. Invasions*, **12(7)**: 2343-2361.

Saeedi, A.A., 2003. Qualitative and quantitative study of *Rutilus frissi* kutum breeding in the Shirrud and Tajan Rivers. Caspian Sea Ecology Research Center, Iran, 70 pp.

Shafiei Sabet, S., and Imanpoor, M.R. 2011. Study on Ovarian Follicle Structure and Rhythm of Gonad Development (*Rutilus frisii* kutum) from Caspian Sea, *World Applied Sciences J.*, **13(3)**: 431-437.

Shafiei Sabet, S., Imanpoor, M.R., Bagher, A.F., and Gorgin, S. 2010. Histological Study of Ovarian Development and Sexual Maturity of Kutum (*Rutilus frisii* kutum Kamenskii, 1901). *World Applied Sciences J.*, **8(11)**: 1343-1350.

Szlauer, B., and Szlauer, L. 1980. The use of Lake Zooplankton as feed for carp (*Cyprinus carpio* L.) fry in pond culture. *Acta Ichthyol Piscat.*, **10(1)**: 79-102.

Valipour, A., and Khanipour, A. 2006. Artificial breeding in autumn form of *Rutilus frissi* kutum. Inland water Aquaculture Research Center, Bandar Anzali, Iran, 40 pp.

Zarin Kamar, H., 1996. Feeding physiology and feeding habits in *Rutilus frissi* kutum in the Bandar Anzali region. Islamic Azad University, North Tehran Branch, Tehran, Iran, 164 pp.

2015, Perspectives in Animal Ecology and Reproduction, Vol. 10   Pages *171–182*
Editors: V.K. Gupta, Anil K. Verma and G.D. Singh
Published by: DAYA PUBLISHING HOUSE, NEW DELHI

Chapter 12

Herb-Feed Supplements for Health and Disease of Fish

Govind Pandey[1]*

[1]Professor/Principal Scientist of Pharmacology and Toxicology,
College of Veterinary Science and Animal Husbandry, Rewa
(Nanaji Deshmukh Veterinary Science University, Jabalpur),
Madhya Pradesh, India

ABSTRACT

Herbs can act as immunostimulants, conferring the non-specific defense mechanisms of fish and elevating the specific immune response. Therefore, the herbal drugs are used not only against diseases but also as growth promoters, stress resistance boosters and preventatives of infections. It has been proved that herb-feed supplements enhance the growth of fishes and protect them from various diseases. Inclusion of herb-feed supplements/additives in diets often provides cooperative action to various physiological functions. Beneficial role of vitamins C and E have been found in fish nutrition, reproduction, growth and related indices. In addition, vitamins C and E are credited with modulating the stress response in fish. The biological role played by vitamins C and E is very vital for the sustained growth and health of fish. The dietary vitamins have antibody enhancement effects in fish. The herbal drugs, *viz.*, ginger, nettle and mistletoe have been used as an adjuvant therapy in rainbow trout fish through feed. The disease resistant of catla fish has been produced through the immersion herbal treatment (neem, garlic and turmeric) of spawn. Many fish farmers and ornamental fish hobbyists buy the bulk of their feed from commercial manufacturers. Small ornamental fish farms with an assortment of fish require small amounts of various diets with particular ingredients. Most feed mills will

* *Corresponding author.* E-mail: drgovindpandey@rediffmail.com

only produce custom formulations in quantities of more than one ton, and medicated feeds are usually sold in big bags. However, the small quantities of fish feeds can be made quite easily in the laboratory, classroom, or at home, with common herbal ingredients and with simple kitchen or laboratory equipment. Henceforth, this chapter will provide the knowledge of some herbal feed supplements, which are beneficial for fish health and may act against fish diseases. Besides, it also covers some fish feed formulations and feeding technology, including feeds for fish larvae, fish feed ingredients, common fish feed stuffs, animal and plant sources of feeds for culture fish, and fish feeding methods.

Keywords: Feed supplements, Fish, Fish feeds, Health, Herbs/Herbal drugs, Disease.

Introduction

The herbal drugs (herbs or medicinal plants) are used not only to cure various diseases but they are also utilized as growth promoters, stress resistance boosters and preventatives of infections. The phytoconstituents like tannins, alkaloids and flavonoids present in herbs can have antimicrobial activity. The herbs can also act as immunostimulants, conferring the non-specific defense mechanisms of fish and elevating the specific immune response. The non-specific immune functions like bacteriolytic activity and leukocyte function of fish have been improved by some herbs (Pandey *et al.,* 2012a and b). Treatment with medicinal plants having antibacterial activity is a potentially beneficial alternative in the aquaculture. These herbs mitigate many of the side effects which are associated with synthetic antimicrobials. Recently, research has been initiated to evaluate the feasibility of herbal drugs in fish diseases (Madhuri *et al.,* 2012; Pandey *et al.,* 2012b).

Usefulness of herbs in treatment of diseases is attaining success, because the herbal treatment is cost effective, eco-friendly and has minimal side effects. Traditional herbal drugs may have the potential immunostimulation. Therefore, the use of herbs is an alternative to antibiotics in fish health management. Many reports have shown that the herbal supplements enhance the growth of fishes and protect them from different diseases (Johnson and Banerji, 2007). Vitamins C and E exhibit the beneficial role in fish nutrition, reproduction, growth and related indices. They can also modulate the stress response in fishes. Deficiency of vitamins in fish under aquaculture can cause biochemical dysfunction, leading to tissue and cellular level clinical manifestations. Many morphological and functional abnormalities have been seen in different fish species deprived of vitamins C and E. Vitamin C is synthesized in animals from either D-glucose or D-galactose as part of the glucuronic acid pathway. Branching from L-gulonic acid, the biosynthetic pathway of vitamin C comprises three consecutive steps: (a) enzymatic lactonization of L-gulonic acid catalyzed by L-gulonolactone hydrolase; (b) oxidation of L-gulonolactone catalyzed by L-gulonolactone oxidase (GLO); and (c) spontaneous isomerization of 2-keto-L-gulonolactone leading to vitamin C. Generally, the animals lacking GLO are not able to synthesize vitamin C and so they depend on it dietary source. The fishes retaining numerous ancestral characters, *viz.,* lamprey, shark, ray, lungfish and sturgeon, have GLO in the kidney; while teleost fishes are deficient in GLO (Tatina *et al.,* 2010).

The fish can not synthesize the essential amino acids (EAAs), which usually remain inadequate but they are required for the growth and development of body. The 'fish meal' contains complete EAAs which meet the protein requirement of most fish species. Since the fish meal is costly as a feed ingredient, the use of nonconventional feed stuffs has been reported with good growth and better cost benefit values. The use of nonconventional feed stuffs of herb/plant origin had been limited as a result of the presence of alkaloids, glycosides, oxalic acids, phytates, protease inhibitors, haematoglutinin, saponegin, momosine, cyanoglycosides and linamarin to mention a few despite their nutrient values and low cost implications. The nonconventional feed resources (NCFRs) are feeds that are not usually common in the markets, and are not the traditional ingredients used for commercial fish feed production. They are very cheap, byproducts or waste products from agriculture, farm made feeds and processing industries, and are able to serve as a form of waste management in enhancing good sanitation. These include all types of feed stuffs from animals (silkworm, maggot, termite, grub, earthworm, snail, tadpoles, etc.), plant wastes (jack bean, cottonseed meal, soybean meal, cajanus, chaya, duckweed, maize bran, rice bran, palm kernel cake, groundnut cake, brewers waste, etc.), and wastes from animal sources and processing of food for human consumption like animal dung, offal, visceral, feathers, fish silage, bone, blood). All these can be recycled to improve their value if there are economically justifiable and technological means for converting them into usable products (Abowei and Ekubo, 2011; Pandey, 2013).

The commercially prepared diets for channel catfish and salmonids have been developed for the specific nutritional requirements of these species, their production systems and their life stages. Some studies have been performed for tilapia production also. For all other species, including freshwater and marine ornamentals, nutritional management is based on a combination of application of knowledge generated for the species mentioned above and the experience of successful aquarists. The successful maintenance of 'difficult' species is often influenced by the aquarist's success in obtaining or rearing specialized food items. For example, the sea horses and sea dragons (sygnathid family) have long, tubular mouth parts. These fishes are not physically capable of ingesting typical commercial fish feeds. Normally, the feeds for fry and fingerlings frequently exceed 50 per cent crude protein. As growth rate decreases with fish age, the protein levels in diets are decreased accordingly. The levels of protein on grow out diets often approach or exceed 40 per cent crude protein, while the maintenance diets may contain as little as 25 to 35 per cent. Many fishes require live food when they are hatched because their mouth parts are so small (Abowei and Ekubo, 2011).

In fish feeds, the fish meal should be a major protein source. There are essential amino and fatty acids that are present in fish meal but not present in tissue from terrestrial plants or animals. The low cost formulations in which fish meal has been eliminated and replaced by less expensive proteins from terrestrial sources (soybeans) are not recommended for fish. Fish meals and fishery byproducts have high lipid content, so the rancidity can be a problem if foods are not properly stored. In addition, fishes require long chain fatty acids (C20 and C22) which are not found in tissue from terrestrial organisms. There is a high oil content associated with carotenoid pigments, so vitamin E supplementation is recommended. Many fishes need dietary ascorbic acid (vitamin C). The ascorbic acid added to fish feed should be phosphorylated to stabilize the vitamin and increase storage time. Additionally, vitamins A, D, E and B complex should be added to fish feeds. Concentration of vitamin E is generally inadequate, especially in diets which are high in fat (Abowei and Ekubo, 2011).

Since the fish feeds normally contain relatively high amounts of fish meal and/ or fish oil, it is very susceptible to rancidity. In addition, ascorbic acid is highly volatile, but critical to normal growth and development of most species of fish. Thus, the fish feeds should be purchased frequently, ideally at least once a month and more frequently if possible. The fish feeds should be stored in a cool, dry place and should never be kept on hand for more than three months. Refrigeration of dry feeds is not recommended because of the high moisture content of that environment. The fish feeds which do not contain stabilized ascorbic acid are not recommended for fish. Pellets are typically the most complete diets. They are cooked, and, if marketed as a complete ration, the nutrition in each particle should be uniform. For larger animals, a very small pellet may be unacceptable. The semi-moist diets are soft and compact. Most of these are expensive, but they tend to be high quality diets and may be an excellent choice for some species of fish (Abowei and Ekubo, 2011; Pandey, 2013).

The technology involved in rearing of live fish feeds is having a positive impact on larval rearing, a frequent bottleneck for commercialization of 'new' species. The rotifers are the smallest live food that is regularly used for larval rearing. The newly hatched brine shrimps are larger but still quite small, and are commonly used in fish hatcheries. The cultured live foods can provide a source of high quality nutrition, but care must be taken to avoid perpetuation of infectious disease. The fish should be fed based on a percentage of body weight. For maintenance, 51.0 per cent body weight per day is sufficient. The fish should probably be fed at least 5 days per week. One feeding per day is plenty for most fishes. Rearing of young stock requires small meals fed more frequently. This is often accomplished using automatic feeders on commercial farms. The advances in the diagnosis and correction of nutritional diseases should be significant over the next few years as there seems to be a great deal of research activity in this area (Abowei and Ekubo, 2011).

Effects of Herb-Feed Supplements in Fish

The herb-supplements/additives included in the fish feeds usually maintain and improve the physiological functions. Ahilan *et al.* (2010) reported that the herbs have significant role in aquaculture. The herbal growth promoters in the carp fish feeds showed beneficial effects. There was a significant difference between different herbal additives on the effect of growth rate in goldfish. The synergistic effects of herbs have been found in *Clarias gariepinus* (Turan and Akyurt, 2005) and Japanese flounder (Ji *et al.*, 2007). The dietary vitamins elicited antibody enhancement effects in salmon fish. The disease resistance properties of vitamin C and E have been noticed in fish. The disease resistance and humoral antibody production in rainbow trout was directly and positively related to vitamin C levels in the trout diet. In a 6-month-old white sturgeon fish fed a diet devoid of vitamin C, the tissue total vitamin C concentrations were not decreased, indicating that white sturgeon GLO produced adequate amounts of vitamin C to meet the fish needs (Tatina *et al.,* 2010).

The dietary vitamin C prevented the appearance of vitamin E deficiency signs in Atlantic salmon in a dose-dependent manner. Vitamin E concentrations in liver declined with increasing the dietary C vitamin in hybrid striped bass. The liver vitamin E concentrations of normal lake sturgeon and vitamin E deficient yellow perch increased in fish fed with high dietary vitamin C concentrations. There are two interaction mechanisms between vitamins C and E: a synergistic simultaneous protection effect of lipid and aqueous phases against oxidation, and the action of vitamin C on vitamin E regeneration in the tissues. The reports elicit that vitamin C protects the fish against vitamin E deficiency (Pandey *et al.,* 2012b). It has also been reported that the supra-dietary levels of vitamins C and E can increase the antibody production and immunity in juvenile milk fish to formalin-killed *Vibrio vulnificus* bacteria (Azad *et al.,* 2007).

The vitamins C and E act as antioxidants, protecting the cellular macromolecules (DNA, protein, lipids) and other antioxidant molecules from uncontrolled oxidation by 'oxygen free radicals' during normal metabolism or under the conditions of oxidative challenge like infection, stress and pollution. So, both these vitamins are known to have protective effects against 'free radicals'. The interactions between vitamins C and E have been observed, as vitamin C spares vitamin E by regenerating it from tocopheroxyl radicals. The interaction between them also influences the beneficial effects they induce in cultured fish. Vitamin C/E sparing action in channel catfish showed the variability seen in its sensitivity to vitamin E deficiency. Thus, due to their potential for interaction, the dietary need for vitamins C and E is usually considered together. In a study, the highest number of leucocytes (WBC) was observed in starlet (a kind of fish) fed with diet containing 100 mg kg^{-1} vitamin E and 400 mg kg^{-1} vitamin C. The highest number of RBC was noticed in different diets other than the basal diet, which indicates that the diets containing different levels of vitamins C and E have significant influence on the RBCs. However, different levels of vitamins C and E did not show any significant influence on glucose and protein. The fish fed with different levels of dietary vitamins C and E did not show any significant differences in cortisol and glucose values, which might be due to the stress avoided during experiment. Different levels of vitamin C also had no effect on cholesterol amount; but the diets containing excessive amounts of vitamin E and the basal diet without any vitamin supplement had the lowest significant amount of cholesterol, indicating that only an optimum amount of vitamin E can increase the cholesterol, and the excessive amounts can lead to its decline. Thus, the dietary levels of vitamins C and E have effects on some of the haematological and biochemical values of starlet fish. Similar results were received in channel catfish fed with different vitamin C and iron concentrations, and in *S. aurata* fed with different concentrations of vitamins C and E (Tatina *et al.,* 2010). Likewise, the collaborating findings have been found in great sturgeon fish fed with different vitamin C levels (Falahatkar, 2005), and in pirarucu fish fed with different vitamin C and E concentrations (Andrade *et al.,* 2007).

By increasing the vitamin E in diet, the WBC count of fish is increased accordingly (Tatina *et al.,* 2010). There was found significant difference in WBC among the treatments of dietary vitamin C to great sturgeon (Falahatkar, 2005). The high vitamins C and E concentrations could stimulate protein production in fish, suggesting an important role of both vitamins in the modulation of plasma proteins. It was also observed that the plasma glucose concentrations in pirarucu fish elevated in 800 and 1200 mg vitamin E kg^{-1} treatments. So, it is not possible to confirm whether hyperglycemia is an advantage for these animals, since there is still a lack of standardization on vitamin E supplementation (Andrade *et al.,* 2007). Further, the high levels of dietary vitamin E may decrease the amount of triglyceride and cholesterol in humans. The cortisol and glucose could increase in teleost exposed to stress (Pandey *et al.,* 2012b). *Piaractus mesopotamicus* fish fed with diets containing 100 or 450 mg kg^{-1} vitamin E did not show a glycemic alteration compared to fish fed diets without this supplementation (Belo *et al.,* 2005).

The enhanced phagocytosis, and cellular and humoral defense mechanisms were observed against fish pathogens in rainbow trout after feeding with ginger, nettle and mistletoe adjuvant herbal therapy. The traditional Chinese medicines in yellow croaker elevated the non-specific defense mechanism and increased the disease resistance of fish against the bacterial pathogens. The disease resistant of *Catla catla* (catla) fish was produced by neem, garlic and turmeric herbal treatments. The *Aloe vera* herb was found effective as a disease suppressing and antibacterial agent in juvenile rock fish. The survival rates of challenged fishes were in the increasing trend when there was an increase in the concentrations of all herbal additives used in the experiment. A significant difference was seen between the different herbal additives at different concentrations on the survival rate of goldfish (Ahilan *et al.,* 2010; Pandey *et al.,* 2012b).

Feeding of *Labeo rohita* (rohu) fishes with aqueous root extract of *Achyranthes aspera* plant in diet significantly (P<0.05) increased the serum anti-proteases level than the fishes fed with control diet. The *Catla catla* fishes were fed a diet containing seeds of *A. aspera* (0.5 per cent) for 4 weeks prior to and after ip injection with chicken erythrocytes (RBCs). The haemagglutination antibody titers, serum globulin levels, anti-trypsin activities, and RNA/DNA ratio of spleen and kidney were significantly higher in test group than the control group. The results showed that this herb enhances the immunity of catla. Furthermore, after 4 weeks of feeding of *A. aspera* seeds (0.5 per cent) with diet, the *Cyprinus carpio* fishes were immunized with chicken RBC. The antigen-specific antibody response, total serum protein, serum albumin and globulin, lysozyme, serum a1-protease inhibitor and a2-macroglobulin and RNA/DNA ratio of spleen and kidney were significantly (P<0.05) higher, suggesting that the immune response of fish was enhanced when fed with diet containing *A. aspera*. After 2 weeks of feeding of *A. aspera* seeds (0.01 per cent, 0.1 per cent and 0.5 per cent) with diet, the *L. rohita* fingerlings were immunized with heat-killed *Aeromonas hydrophila* bacteria, and after a further 2 weeks, the fishes were experimentally infected with *A. hydrophila*. After 7 days, the superoxide anion production, serum bactericidal activity, lysozyme, alkaline phosphatase (ALP), serum protein and albumin:globulin ratio (A/G) were increased (towards normal); while the serum glutamate oxaloacetate transaminase (SGOT) and serum glutamate pyruvate transaminase (SGPT) levels were decreased (towards normal) in *L. rohita* fingerlings treated with *A. aspera*. The results showed that the herb *A. aspera* stimulates the immunity and increases resistance to infection in *L. rohita*. Further, after 4 weeks of *A. aspera* feeding to *C. catla*, the significant (P<0.05) enhanced bovine serum albumin (BSA)-specific antibody titers were found. The efficiency of antigen clearance was also enhanced (Chakrabarti and Vasudeva Rao, 2006; Vasudeva Rao *et al.*, 2005; Madhuri *et al.*, 2012).

The herb-feed supplements promote the cellular lipid and fatty acid utilization and protein accumulation resulting in good growth performance in *Pagrus major* fish (Ji *et al.*, 2007). The growth increase in *L. rohita* fish fed with herbal supplemented diet was due to improved food utilization and high protein synthesis (Johnson and Banerji, 2007). Among the two different herbal feed supplements, the *Phyllanthus niruri* fed group recorded higher specific growth rate, followed by *Aloe vera* fed group (Ahilan *et al.*, 2010). Goldfish, *Carassius auratus* fed with mixed herbal supplementation diets significantly restored the altered haematological (*viz.*, WBC, RBC, haemoglobin, haematocrit value, mean corpuscular volume and mean corpuscular haemoglobin concentration), biochemical (*viz.*, total protein, glucose and cholesterol) and immunological parameters, and triggered the innate immune system against *A. hydrophila* bacteria (Harikrishnan *et al.*, 2010).

Fish Feed Ingredients of Plants/Herbs Source

Many fish farmers and ornamental fish hobbyists buy the bulk of their feed from commercial manufacturers. Small ornamental fish farms with an assortment of fish require small amounts of various diets with particular ingredients. Most feed mills will only produce custom formulations in quantities of more than one ton, and medicated feeds are usually sold in big bags. However, the small quantities of fish feeds can be made quite easily in the laboratory, classroom, or at home, with common herbal ingredients and with simple kitchen or laboratory equipment. The nutrients essential to fish are the same as those required by most other animals (Abowei and Ekubo, 2011; Pandey, 2013).

The nutrients necessary to fish include water, proteins (amino acids), lipids (fats, oils, fatty acids), carbohydrates (sugars, starch), vitamins and minerals. In addition, pigments (carotenoids) are commonly added to the diet of salmonid and 'ornamental aquarium' fishes to enhance their flesh and skin colouration, respectively. Soybean meal, legumes and wheat gluten are excellent plant sources of protein. The vegetable oils from canola, sunflower and linseed are common sources of lipids in fish feeds. Cooked carbohydrates from flours of corn, wheat or other 'breakfast' cereals are relatively inexpensive sources of energy that may spare protein (which is more expensive) from being used as an energy source. The variety and amount of vitamins and minerals are so complex that they are usually prepared synthetically, and are available commercially as a balanced and premeasured mixture called 'vitamin or mineral premix'. This premix is added to the diet in generous amounts to ensure that adequate levels of vitamins and minerals are supplied to meet dietary requirements. A variety of natural and synthetic pigments or carotenoids are available to enhance the coloration in the flesh of salmonid fish, and the skin of freshwater and marine ornamental fish. The pigments most frequently used supply the colours red and yellow. The synthetically produced pigment, astaxanthin, is the most commonly used additive (100-400 mg/kg). Cyanobacteria (blue green algae, *e.g., Spirulina*), palm oils and extracts from marigold, red peppers and *Phaffia* yeast are excellent plant sources of pigments. Another important ingredient in fish diets is a 'binding agent', which provides stability to the pellet and reduces leaching of nutrients into the water. Carbohydrates (starch, cellulose, pectin), and various other polysaccharides, such as extracts or derivatives from plants (gum arabic, locust bean) and sea weeds (agar, carageenin and other alginates) are popular binding agents. The preservatives, such as antimicrobials and antioxidants are often added to extend the shelf life of fish diets and reduce the rancidity of the fats. Vitamin E is an effective but expensive antioxidant which can be used in laboratory prepared formulations. The amino acids glycine and alanine, and the chemical betaine are also known to stimulate strong feeding behaviour in fish. Basically, the attractants enhance feed palatability and its intake. The fiber and ash (minerals) are a group of mixed materials found in most feed stuffs. In experimental diets, fiber is used as a filler, and ash as a source of calcium and phosphorus. In practical diets, both should be no higher than 8 to 12 per cent of the formulation. The live, frozen or dried algae are common feed stuffs used in ornamental fish diets. Fresh leafy or cooked green vegetables are often used in fish feeds. Although vegetables are composed mainly of water, they contain some ash, carbohydrates and certain vitamins. Kale, dandelion greens, parsley and turnip greens are examples of relatively nutritious vegetables. Many domestic and agricultural wastes, *e.g.,* corn bran, guinea corn bran, rice bran, wheat bran, palm kernel cake, groundnut cake, cotton seed cake, soybean, vegetable oil, palm oil, etc. are the main plant ingredients needed for compounding artificial fish diet to reduce the cost of fish production (Abowei and Ekubo, 2011).

Common Conventional and Unconventional Plant Feeds for Fish

Some common conventional feedstuffs from plant sources are groundnut cake, soybean meal, palm kernel meal, brewers dried yeast, brewers dried grain, maize and wheat offal. Groundnut cake contains about 45 per cent crude protein but lacks the essential amino acid, lysine. When moldy, it becomes poisonous due to the presence of mycotoxin called 'aflatoxin'. Soybean meal is fast gaining increasing acceptability. It has balanced amino acids and can replace a substantial part of fish meal. Use of this feedstuff is, however, limited due to its high fat content and presence of trypsin inhibitor. Palm kernel meal contains a high quantity of crude fiber; crude protein is 17 per cent. Palm kernel meal is only useful when its crude fiber content is high. Brewers dried yeast is a byproduct of brewery industry. It contains sufficient quantity of crude protein but limited in amino acids methionine and cystine. Brewers dried grain is readily available and contains similar protein levels as palm kernel cake; crude fiber content is high and so this is in limited use. Maize is palatable and free from anti-nutritional factors. Its energy content is high. This limits the use in fish feed. Wheat offal has nutritional property similar to palm kernel meals. Thus, these two can be used interchangeably but scarcely together. Wheat offal is very scarce due to adverse government policy (Abowei and Ekubo, 2011).

The 'unconventional fish feeds' are potential feed ingredients. They can be of 'animal' or 'plant' source. The plant sources of fish diets include leaf protein, leaf meal, aquatic macrophytes, cultivable pulses such as mucuna bean, yam beans, bread beans, winged beans or any legume ornamental that can yield pods with seeds. The leaves contain diverse levels of protein, which can produce an inexhaustible and inexpensive source of nutrient for fish. Nutritionally valued leaves are cassava (*Manihot esculenta*), pawpaw (*Carica papaya*), pineapple (*Ananas comosus*), groundnut (*Arachis hypogea*), soyabean/soybean (*Glycine max*) and plantain (*Musa paradisica*). Aquatic macrophytes are common aquatic plants found growing on water surface. These include rooted flowering plants like grasses and sedges that are commonly seen along the rim of freshwater bodies; rotted flowering plants with submerged leaves like ceratophyllum, and with floating leaves like water lilies (nymphaea); free floating plants such as duckweed, water lettuce, water hyacinth and salvinia, a water fern. Water hyacinth are so wide spread that they constitute a menace to shipping and fishing activities but can be used as feed component for fish. A large number of legumes (unconventional pulses) are used as cover crops or ornamentals. They are not eaten by reasons of suspected content of toxic substances. For example, mucuna beans, broad beans, sword beans, winged beans, yam beans, etc. Their protein contents range from 18 to 20 per cent, fat 3 to 10 per cent and carbohydrate 50 to 60 per cent, making them easily gelatinisable. The toxic substances in them are hydrogen cyanide and trypsin inhibitors. These can be removed by applying heat during processing, which can be done by toasting (groundnut fashion), boiling, steam cooking or drying (Abowei and Ekubo, 2011).

Fish Feed Preparation and Feeding Technology

Fish feeds are prepared, may be dry with final moisture content of 6 to 10 per cent; semi-moist with 35 to 40 per cent water; or wet with 50 to 70 per cent water content. Most feeds used in intensive production systems or in home aquaria are commercially produced as dry feeds. Dry feeds may consist of simple loose mixtures of dry ingredients, *e.g.*, mash or meals, to more complex compressed pellets or granules. Pellets are often broken into smaller sizes known as 'crumbles'. The pellets or granules can be made by cooking with steam or by extrusion. Depending on the feeding requirements of fish, pellets can be made to sink or float. Flakes are another form of dry food and a popular diet for aquarium fishes. They consist of a complex mixture of ingredients, including pigments. They are made into slurry, which is cooked and rolled over drums heated by steam. Semi-moist and wet feeds are made from single or mixed ingredients, *e.g.*, trash fish or cooked legumes, and can be shaped into cakes or balls. There is no single method for preparation of fish feed; however, most methods begin with the formation of a dough like mixture of ingredients. Dough is started with blends of dry ingredients which are finely ground and mixed. The dough is then kneaded and water is added to produce desired consistency for whatever fish is going to be fed. The same dough may be used to feed several types of fish, such as eels and small aquarium fish. Pelleting or rolling converts the dough into pellets or flakes, respectively. The amount of water, pressure, friction and heat greatly affects pellet and flake quality, *e.g.*, excess water in the mixture results in a soft pellet. Too little moisture and pellet will crumble. Proteins and especially vitamins are seriously affected by high temperatures. Thus, avoid storing diet ingredients at temperatures at or above 70°C and do not prepare dry feeds with water at temperature higher than 92°C (Abowei and Ekubo, 2011; Pandey, 2013).

To make the own fish feed, the tools are used primarily for chopping, weighing, measuring ingredients, and for blending, forming and drying the feed. The multipurpose kitchen shears, hand graters, a paring knife, a 5 inches serrated knife, a 6 to 8 inches narrow blade utility knife and a 10 inches chef knife for cutting, slicing and peeling can be used. A couple of plastic cutting boards protect the counter and facilitate the handling of raw ingredients. Heat resistant rubber spatulas, wooden and slotted spoons, long handled forks and tongs are very good for handling and mixing ingredients. A basic mortar and pestle, electric blender, food processor or coffee grinders are very useful to chop or puree ingredients. A food mill and strainer such as a colander or flour sifter help discard coarse material and obtain fine food particles. For weighing and measuring ingredients, dry and liquid measuring cups and spoons, and a food or laboratory bench scale are required. Other utensils include plastic bowls (1½, 3, 5, and 8 quarts) for weighing and mixing ingredients, a thermometer and a timer. A 3-quart saucepan and 10 inches stockpot are good for heating gelatins and cooking raw foods like vegetables and starches. The ingredients and blends may be cooked in a small electric or gas burner. A few trivets to put under hot pans will protect counters and table tops. Ingredients may be mixed by hand using a rotary beater or wire whisk; however, an electric mixer or food processor is more efficient. After mixing, a dough is formed which can be fashioned into different shapes. A pasta maker, food or meat grinder will extrude the dough into noodles or 'spaghetti' of different diameters. As the noodles emerge from the outside surface of the die, they can be cut off with a knife to the desired length or crumbled by hand, thus making pellets. A potato ricer also serves to extrude the dough into noodles of the same size. For making flakes, a traditional hand cranked or electric pasta maker will press out the dough into thin sheets. The pellets or thin sheets can be placed on a cookie sheet and dried in a household oven on low heat or in a forced air oven. A small food dehydrator also performs the task quite well. To add extra oil and/or pigments to pellets, a handheld oil atomizer or sprayer can is useful. To separate pellets into different sizes, a set of sieves (*e.g.*, 0.5, 0.8, 1.0, 2.0 and 3.0 mm) is needed. The freezer bags serve to store the prepared feeds, and using a bag vacuum sealer will greatly extend the shelf life of both ingredients and the feed. The feed can be stored double bagged in the freezer but should be discarded after 6 months. One simple formulation, which is used traditionally to feed ornamental fish in ponds, consists of a mixture of 30 per cent ground and processed oats or wheat and 50 per cent of pellets from a commercial manufacturer. By weight, about 23 per cent of fish oil, and a 0.3 per cent vitamin and a 1 per cent mineral premix are added to the mixture. This mixture is blended with water and can be formed into dough balls of different sizes (Abowei and Ekubo, 2011).

With increased growth rate of stocked fish, the available natural foods in the pond become inadequate to support the fish population. So, there is need to supplement the natural food with artificial feeds for enhanced fish growth. The selection of fish for culture with supplementary diet depends on the crude protein requirement of the fish and the unit cost of the feed ingredient available in a given locality. This consideration is important because the fish feed accounts for 40 to 70 per cent of operating cost of a fish farm with an intensive management system. The feeding is one of the most important aspects of the fish culture. The range of type of food consumed by fish is greater than for other groups of vertebrates. Different technical methods are used for in administering feed to culture fish. Supplementary feeds are given to fish in addition to the natural food organisms in the water body. The supplementary feeds contain all the essential nutrients, *e.g.*, proteins, carbohydrates, lipids, vitamins and minerals required for fish growth. These feeds are introduced into the pond by broadcasting and spot feeding. In broadcasting, the feed is spread over the pond. Spreading is enhanced by drifting of the pond water. This method often contaminates the pond. Spot or spontaneous feeding involves placing the feed in a bag and tied on a spot. Alternatively, the feed is introduced at a particular time. The fish is fed (34 per cent body weight) twice daily, preferably morning and evening (Abowei and Ekubo, 2011; Pandey, 2013).

References

Abowei, J.F.N., and Ekubo, A.T., 2011. A review of conventional and unconventional feeds in fish nutrition. *Br. J. Pharmacol. Toxicol.,* **2(4)**: 179-191.

Ahilan, B., Nithiyapriyatharshini, A., and Ravaneshwaran, K., 2010. Influence of certain herbal additives on the growth, survival and disease resistance of goldfish, *Carassius auratus* (Linnaeus). *Tamilnadu J. Vet. Ani. Sci.,* **6(1)**: 5-11.

Andrade, J.I.A., Ono, E.A., Menezes, G.C., Brasil, E.M., and Roubach, R., 2007. Influence of diets supplemented with vitamins C and E on pirarucu (*Arapaima gigas*) blood parameters. *Comp. Biochem. Phys.,* **146**: 576-580.

Azad, I.S., Dayal, J.S., Poornima, M., and Ali, S.A., 2007. Supra dietary levels of vitamins C and E enhance antibody production and immune memory in juvenile milkfish *Chanos chanos* (Forsskal) to formalin-killed *Vibrio vulnificus. Fish Shellfish Immunol.,* **23**: 154-163.

Belo, M.A.A., Schalch, S.H., Moraes, F.R., Soares, V.E., Otoboni, A.M., and Moraes, J.E.R., 2005. Effect of dietary supplementation with vitamin E and stocking density on macrophage recruitment and giant cell formation in the teleost fish *Piaractus mesopotamicus. J. Comp. Pathol.,* **133**: 146-154.

Chakrabarti, R., and Vasudeva Rao, Y., 2006. *Achyranthes aspera* stimulates the immunity and enhances the antigen clearance in *Catla catla. Int. Immunopharmacol.,* **6(5)**: 782-790.

Falahatkar, B., 2005. *The effect of dietary vitamin C on some of hematologic biochemistry and growth indexes of great sturgeon (Huso huso)*. Ph.D. thesis. Tarbiat Modaarres University, Tehran, Iran (Persian).

Harikrishnan, R., Balasundaram, C., and Heo, M.S., 2010. Herbal supplementation diets on hematology and innate immunity in goldfish against *Aeromonas hydrophila. Fish Shellfish Immunol.,* **28(2)**: 354-361.

Ji, S.C., Takaoka, O., Jeong, G.S., Lee, S.W., Ishimaru, K., Seoka, M., and Takii, K., 2007. Dietary medicinal herbs improve growth and some non-specific immunity of red seabream *Pagrus major. Fish. Sci.,* **73**: 63-69.

Johnson, C., and Banerji, A., 2007. Influence of extract isolated from the plant *Sesuvium portulacastrum* on growth and metabolism in freshwater teleost, *Labeo rohita* (Rohu). *Fishery Technol.,* **44(2)**: 229-234.

Madhuri, S., Sahni, Y.P., and Pandey, Govind. 2012. Herbal feed supplements as drugs and growth promoter to fishes. *Int. Res. J. Pharm.,* **3(9)**: 30-33.

Pandey, Govind. 2013. Feed formulation and feeding technology for fishes. *Int. Res. J. Pharm.,* **4(3)**: 23-30.

Pandey, Govind, Madhuri, S., and Mandloi, A.K., 2012a. Medicinal plants useful in fish diseases. *Pl. Arch.,* **12(1)**: 1-4.

Pandey, Govind, Madhuri, S., and Sahni, Y.P. 2012b., Beneficial effects of certain herbal supplements on the health and disease resistance of fish. *Novel Sci.: Int. J. Pharmace. Sci.,* **1(7)**: 497-500.

Tatina, M., Bahmani, M., Soltani, M., Abtahi, B., and Gharibkhani, M., 2010. Effects of different levels of dietary vitamins C and E on some of hematological and biochemical parameters of starlet (*Acipenser ruthenus*). *J. Fish. Aqu. Sci.,* **5**: 1-11.

Turan, F., and Akyurt, I., 2005. Effects of red clover extract on growth performance and body composition of African catfish, *Clarias gariepinus. Fish. Sci.,* **71**: 618-620.

Vasudeva Rao, Y., Das, B.K., Jyotyrmayee, P., and Chakrabarti, R., 2005. Effect of *Achyranthes aspera* on the immunity and survival of *Labeo rohita* infected with *Aeromonas hydrophila. Fish Shellfish Immunol.,* **20**: 263-273.

2015, Perspectives in Animal Ecology and Reproduction, Vol. 10   Pages *183–193*
Editors: V.K. Gupta, Anil K. Verma and G.D. Singh
Published by: DAYA PUBLISHING HOUSE, NEW DELHI

Chapter 13

Effect of Starvation and Refeeding on Certain Biochemical and Haematological Parameters of the Climbing Perch, *Anabas testudineus*

Kuldeep Kumar[1] and A.K. Pandey[2]*

*[1]Central Institute of Freshwater Aquaculture,
Kausalyaganga, Bhubaneswar – 751 002, Odisha, India
[2] National Bureau of Fish Genetic Resources,
Canal Ring Road, Lucknow – 226 002, U.P., India*

ABSTRACT

12 weeks experiment was conducted to record the effect of starvation (9 weeks) and refeeding (3 weeks) on moisture content, muscle protein and fat content, serum protein, blood glucose, liver glycogen and differential lecucocyte counts in *Anabas testudineu*s. Fishes were starved for 9 weeks but feeding was done during 10th, 11th and 12th week using pelleted diet. The moisture content increased 9.7 per cent over initial value while tissue protein and fat decreased significantly during the 9th week of starvation. Serum protein and blood glucose and liver glycogen also declined during the starvation period and appeared to be good indicators of starvation. Among leucocytes, thrombocytes, neutrophils and lymphocytes were highly sensitive to starvation. The significance of the findings have been discussed in details.

Keywords: *Starvation, Refeeding, Biochemical and haematological parameters, Anabas testudineus.*

---

* *Corresponding author.* E-mail: akpandey.ars@gmail.com; kuldeepkumar.kk@gmail.com

Introduction

When food is not available in the ponds due to overstocking, natural breeding and unfavourable weather conditions and also during unequal availability of supplementary feed to different individuals in intensive aquaculture, fish often encounter starvation (Abdel-Tawwab *et al.,* 2006; Perez-Jimenez *et al.,* 2011; Sridee and Boonanuntansarn, 2012). Utilization of body reserves can, however, maintain the fish for some time from disposition during non-feeding phases but prolonged starvation causes metabolic disorders that would not normalize even after provision of feed. In fish, basal energy consumption is low as they derive physical support from the environment. This enables them to perform the routine metabolic activities without recourse of their white muscles which may be needed in event of pursuit or escape. Though feeding below the desired level often cause starvation, fish do not hibernate as in case of worm-blooded animals. Starvation causes pronounced changes in the fish (Phillips *et al.,* 1960; Kamra, 1966; Swallow and Fleming, 1969; Cre'ac'h and Gas, 1971; Kawatsu, 1974; Bazhenova and Shcherbina, 1975; Joshi, 1979; Mustafa and Mittal, 1982: Barton *et al.,* 1988; Rajyasree and Naidu, 1989; Kumar *et al.,* 1990; Tzeng and Yu, 1992; Yamada *et al.,* 1994; Borah and Yadav, 1996; Johansen and Overturf, 2006; Caruso *et al.,* 2010). Variations in blood glucose and liver glycogen have been found during starvation in fish (Sidiqqui, 1975; Moorthy *et al.,* 1980; Mahajan and Dheer, 1983; Gillis and Ballantyne, 1996; Perez-Jimenez *et al.,* 2011) while gradual fall in blood glucose level in Indian major carps during starvation and gradual rise during recovery period has been recorded (Datta *et al.,* 1985). Barton *et al.* (1988) observed plasma glucose to be generally higher in fed fish than in starved ones. Kumar *et al.* (1990) reported the increasing and decreasing trends in serum protein and blood glucose levels in *Clarias batrachus* during starvation. Mukhopadhyay *et al.* (1991) found depletion of endogenous protein prior to lipid reserves in fingerlings of *Labeo rohita* and *Cirrhinus mrigala* showing greater lability of protein for energy purposes during the period of starvation. An attempt has, therefore, been made to study the effect of starvation and refeeding on some biochemical and haematological parameters of the commercially important freshwater air-breathing perch, *Anabas testudineus.*

Materials and Methods

Healthy *Anabas testudineus* (Bloch) average length 12.04 cm (range 9.5-16.5 cm) and weight 31.94 g (range 20-65g) were collected from village ponds near Central Institute of Freshwater Aquaculture, Bhubaneswar (India). They were kept in plastic pools after a bath in 500 ppm $KMnO_4$ solution to avoid infection. Fishes were acclimatized for 15 days under the laboratory conditions before initiation of the experiment. During acclimatization period, they were fed with pelleted (diameter 2 mm) diet comprising groundnut oil-cake 50 per cent, rice bran 23 per cent, fish meal 25 per cent, calcium diphosphate 1.5 per cent, sodium chloride 0.3 per cent, trace mineral mixture 0.1 per cent and vitamin mixture 0.1 per cent. Initially, the fishes were starved for 9 weeks, thereafter, the feeding resumed on the same feed.

Fishes were killed on 0, 1, 3, 5, 7, 9, 10R, 11R, 12R weeks (R = recovery period) and the blood samples were collected according to the method of Steuche and

Schoettger (1967) and used immediately for various haematological tests. Thin smears of blood were also prepared on alcohol cleaned slides and fixed in methanol after air-drying. The combinations of Wright and Giemsa method was used for staining. The moisture content, protein and fat content of the fish muscles was analyzed by following the standard methods (AOAC, 1980) while the water parameters were analyzed as per APHA, AWWA, WPCF (1981). Blood glucose was analyzed by GOD-PAP method while serum protein by Biurette method and liver glycogen using anthrone reagent (Oser, 1965). The data were subjected to analyses of variance (ANOVA) with F and T tests to determine the significance of variations in different parameters between the treatments and periods (Fischer and Yates, 1963; Fischer, 1970).

Results and Discussion

Water Quality Regime

The water quality parameters in the experimental plastic pools were within the permissible limits throughout the experimental period - temperature 29.5-32.0°C, pH 7.1-7.4, total alkalinity 118-123 ppm, free CO_2 0.5-0.7 ppm, total hardness (as $CaCO_3$) 103.6-118.4 ppm, electrical conductivity 345-367 milli mhos/cm, dissolved oxygen 4.2-4.6 ppm, P_2O_5 0.03-0.05 ppm, ammonia-nitrogen 0.03-0.04 ppm and nitrate-nitrogen 0.3-0.4 ppm.

Moisture Content

The variations in moisture content of fish under experiment are summarized in Table 13.1. The initial moisture content of 67.56 per cent ranged narrowly between 66.62-69.88 per cent during 12 weeks of the experiment in control. During starvation, it increased and registered 9.7 per cent increase over initial value during 9th week of starvation. Later during 10th and 11th week, it decreased and suddenly increased during 12th week. Significant differences were found during weeks 1, 9 and 12.

Tissue Protein

In control, tissue protein contents varied narrowly between 48.41 and 50.66 per cent (dry weight) throughout the experimental period (Table 13.1). It decreased to 37.89 per cent at the end of 9 weeks of starvation registering 25.60 per cent reduction over the day 0 value. During recovery period, it again increased but did not reach the value recorded in the control. Significant differences were observed on 7, 9, 10R and 12R weeks.

Tissue Fat

The initial value of tissue fat (29.4 per cent) ranged from 28.14 to 30.58 per cent during the course of experiment in control (Table 13.1). It decreased sharply to 23.63 per cent in fish under starvation at the end of 9 week. During recovery period, however, an increase was recorded. Significant differences were registered during 7, 9, 10R, 11R and 12R weeks.

Serum Protein

The initial serum protein value was 3.62g/dl which remained within a narrow range of 3.36-3.7 g/dl in control fishes (Table 13.1) whereas it increased to 4.14 g/dl

Table 13.1: Effects of Starvation and Refeeding on Biochemical Composition of the Climbing Perch, *Anabas testudineus*

Group	0 Week	1 Week	3 Week	5 Week	7 Week	9 Week	Recovery		
							10 Week	11 Week	12 Week
Moisture content of muscles (per cent)									
Control	67.56±2.81	69.62±2.21	68.86±2.30	69.04±0.95	68.04±0.95	67.89±2.49	66.72±1.07	67.96±2.40	67.57±1.66
Experimental	67.56±2.81	71.70±1.87[a]	68.96±2.78	68.16±1.19	69.09±1.62	74.12±1.82[c]	61.56±2.49	69.74±2.27	71.17±3.93[a]
Tissue protein (per cent)									
Control	50.53±3.54	49.90±2.42	50.66±1.18	48.41±0.92	48.60±1.93	49.83±2.50	48.50±1.12	49.65±2.33	50.50±1.36
Experimental	50.53±3.54	48.43±2.13	47.05±2.83	43.07±2.50[a]	40.24±1.49[b]	37.59±1.54[c]	40.48±1.61[c]	43.39±2.40[a]	44.75±1.00[a]
Tissue fat (per cent)									
Control	29.41±1.48	29.37±1.09	28.140.80	30.58±2.57	29.10±1.37	28.02±0.86	26.57±0.99	30.24±0.62	29.33±1.12
Experimental	29.41±1.48	29.90±3.21	28.08±1.61	27.40±1.80	26.00±2.77[a]	23.63±3.06[b]	25.15±1.72[b]	25.38±1.52[a]	26.25±1.35[a]
Serum protein (g/dl)									
Control	3.62±0.14	3.44±0.16	3.56±0.19	3.63±0.19	3.36±0.15	3.43±0.18	3.50±0.17	3.74±0.21	3.65±0.21
Experimental	3.62±0.14	4.14±0.22	3.42±0.12	2.71±0.19	2.47±0.19	1.93±0.25[a]	2.23±0.18[a]	2.74±0.16	3.16±0.21
Blood glucose (mg/dl)									
Control	80.56±1.98	81.24±1.34	82.00±2.58	78.57±2.42	76.98±2.94	80.95±1.95	83.43±2.07	81.59±2.64	82.23±2.57
Experimental	80.56±1.98	76.29±2.74	70.61±2.80	64.20±2.12[a]	57.15±3.37[b]	57.72±2.94[b]	58.70±4.16[b]	67.32±3.64	76.86±2.07
Liver glycogen (mg/g)									
Control	14.62±0.59	15.32±0.87	15.90±0.59	14.56±0.58	16.10±0.80	15.50±0.68	14.36±0.70	15.08±0.87	14.26±0.72
Experimental	14.62±0.59	13.20±0.73	11.42±1.01[a]	10.14±0.94[b]	8.62±1.05[c]	7.52±0.98[c]	8.12±0.56[c]	10.42±1.32[b]	13.60±0.87

Significant response: [a] $P > 0.05$; [b] $P < 0.01$; [c] $P < 0.001$.

during week 1 registering 14.36 per cent increase above the initial but decreased sharply to 1.93 g/dl (46.68 per cent below the initial value) in week 9. The recovery in the decreasing serum protein was observed during week 12R. Significant differences in serum protein were observed in during weeks 5, 9, 10R and 11R.

Blood Glucose

Initially, the blood glucose was 80.56 mg/dl but ranged between 76.98-83.43 mg/dl in the controls during the experimental period (Table 13.1). Under starvation, it decreased sharply till week 9, registering a value of 51.72 mg/dl (35.79 per cent below the initial value). During week 12R, it increased and recovered. Significant differences were observed on weeks 5, 7, 9, 10R and 11R.

Liver Glycogen

Initially, the liver glycogen content of *Anabas testudineus* was 14.02 mg/g which ranged between 14.26-16.10 mg/g in controls and decreased gradually to 7.52 mg/g during starvatin. When feeding was restored, it again increased and reached a value near to the control. Significant differences were observed during week 5, 7, 9 and 10R (Table 13.1).

Differential Leucocytes Count

The leucocyte ranged between – large lymphocytes (9.2-12.0 per cent), small lymphocytes (17.61-21.8 per cent), neutrophils (19.0-23.2 per cent), monocytes (1.4-3.6 per cent), eosinophils (2.0-3.4 per cent), basophils (0.6-1.6 per cent) and thrombocytes (39.4-45.2 per cent) among controls but the starved *Anabas testudineus* showed marked variations (Table 13.2). The large lymphocytes declined and increased only during recovery period. The neutrophils increased till 9th week and subsequently reduced on restoring the feed. During starvation period, the monocytes ranged between 1.6-4.6 per cent, eosinophils 2.6-3.2 per cent and basophils 0.6-1.0 per cent and these three types of cells did not show any marked variation during the entire course of experiment. The thormbocyte population after initial increase declined to 26 per cent during week 9. During recovery period, these cells increased and stabilized further. The effects due to starvation were significant in case of small lymphocytes ($F_{1,7}$=6.031; P<0.05), neutrophils ($F_{1,7}$=7.808; P<0.05) and thrombocytes ($F_{1,7}$=8.748; P <0.05).

Starvation is a common feature experienced by fish every year during the winter season even in the tropics. It is also common due to overstocking, natural breeding and continuous unfavourable weather conditions. At such time, body constituents are mobilized for survival which has been well documented by Siebert *et al.* (1964). The rate of depletion of different constituents depend on their organ sources. The protein, lipid and water show striking relationships. An increase in one leads to depletion of the other as in the cases of moisture and lipid fractions (Jacquot, 1961). Starvation influences the haematology of fish (Smallwood, 1916) and variations in different parameters of blood have been studied by several workers. In the present study, the moisture content of body tissue increased to 74.12 per cent (9.70 per cent higher than the initial value) in starved fish while it remained within the limits of 66.62-69.88 per cent in the controls. Restoration of feeding caused a reduction in moisture content. Similar observations were made by Love (1958) and Suttan (1968)

Table 13.2: Effects of Starvation and Refeeding on differential Leucocyte Counts of the Climbing Perch, *Anabas testudineus*

Group	1 Week	3 Week	5 Week	7 Week	9 Week	10 Week	11 Week	12 Week
Lymphocytes (large)								
Control	10.6±1.14	9.4±1.14	9.2±1.09	10.8±2.77	12.0±2.55	10.2±1.92	9.2±2.86	11.2±3.11
Experimental	9.2±2.28	10.0±2.12	6.8±2.39	8.0±2.55	8.2±2.77	11.2±2.39	11.0±2.74	9.2±2.39
Lymphocytes (small)								
Control	20.8±1.30	19.2±1.92	18.0±1.58	18.6±2.07	17.6±2.30	206±2.07	20.4±3.05	21.8±3.27
Experimental	19.4±1.82	21.1±1.39	23.8±4.44	23.6±4.61	25.2±3.35	22.0±3.53	22.8±3.35	21.0±3.81
Neutrophils								
Control	21.2±1.09	22.2±1.79	21.6±3.05	19.8±1.48	23.2±2.28	21.6±4.45	19.0±3.67	22.6±2.07
Experimental	18.2±2.39	24.6±2.30	27.8±2.39	31.0±3.53	32.8±3.49	31.2±2.86	28.0±2.55	21.0±2.55
Monocytes								
Control	2.8±0.84	2.4±1.14	2.0±1.00	2.2±0.84	3.6±1.14	2.4±1.14	3.4±1.51	1.4±1.14
Experimental	2.2±0.84	3.0±1.87	1.6±0.89	4.6±2.07	3.6±1.14	4.0±0.71	3.0±1.58	3.4±1.14
Eosinophils								
Control	2.2±1.30	2.8±1.48	3.4±1.14	2.6±1.14	2.8±1.48	2.0±1.58	2.6±1.52	2.2±1.64
Experimental	3.2±1.30	2.6±1.52	3.0±1.22	3.2±1.48	3.2±1.48	2.4±1.67	2.4±1.14	3.0±1.00
Basophils								
Control	1.0±0.71	0.6±0.55	0.6±0.89	1.6±1.52	1.4±1.14	1.0±1.00	1.4±0.89	1.0±1.41
Experimental	1.0±1.00	0.6±0.89	0.8±1.30	1.0±1.00	1.0±0.71	0.8±0.84	1.6±1.52	1.86±1.52
Thrombocytes								
Control	41.6±2.88	43.4±4.23	45.2±3.63	44.4±3.85	39.4±2.61	42.2±2.39	44.0±5.15	39.8±3.56
Experimental	46.8±2.94	38.0±7.55	36.2±3.96	28.6±1.14	26.0±8.63	28.4±6.31	31.2±5.49	40.8±7.98

in the muscle of starved *Gadus morhua*. Sorvachev (1957) recorded an increase in the water content of *Cyprinus carpio* muscle to 80 per cent by starving them for three months but Cre'ac'h and Cournede (1965) found 91 per cent water content after eight months of starvation. In the present study, protein content of the muscle tissue on dry weight basis decreased sharply with a reduction of 25.6 per cent during week 9 indicating that *Anabas testudineus* has a tendency to mobilize protein during the period of starvation. Restoration of feed again increased the protein content while in control it remained within a range of 48.41-50.66 per cent. The crude fat content remained within the narrow range of 28.02-30.58 per cent on dry weight basis in control while it reduced to 23.63 per cent during week 9 of starvation and restoration of feeding again caused enhancement of fat content. The present studies confirm the observations of Phillips *et al.* (1960). As reported by Atwater (1888 - cited by Jacquat, 1961), a marked inverse relationship between lipid and moisture contents were discernible in *Anabas testudineus*. Love (1958) also reported that during severe starvation, the moisture content of *Gadus morhua* increased to 88 per cent with the corresponding fall in protein content. Stirling (1976) also documented that starvation caused a rapid initial decrease in carbohydrate and a progressive decline in lipid in all the tissues. He also reported an inverse relationship for both lipid and protein with respect to moisture content of the whole fish. The present work in the freshwater climbing perch support these observations.

The serum protein levels in *Anabas testudineus* varied within a narrow range of 3.36-3.74 g/dl in controls, in the starved fish (after an initial rise during week 1), it reduced to 46.68 per cent in 9 weeks of starvation while restoration of feeding improved the serum protein content. Bailey *et al.* (1942) found that protein mobilization of fast-swimming fish like *Scomber scombrus* is more rapid with higher activity of catheptic enzyme and rapid autolysis of the muscle as compared to *Cyprinus carpio* or *Gadus morhua*. Barnes *et al.* (1963), Kosmina (1966), Robertson *et al.* (1967), Kawatsu (1974), Mishra and Mishra (1979) and Mahajan and Dheer (1983) also documented the reduction in serum protein content due to starvation. The present results in *Anabas testudineus* are in agreement with these observations. Blood glucose levels decreased sharply from initial value of 80.50 mg/dl throughout the starvation period, a 35.79 per cent decrease was recorded over the initial value during week 9 and restoration of feeding resulted in the elevation of blood glucose but perhaps, three weeks time was not sufficient to restore the value to the normal level. Biochemically, the blood glucose level may be a good indicator of starvation because in control, the glucose levels did not fall below 76.98 mg/dl. Mahajan and Dheer (1983) and Datta *et al.* (1985) also reported similar trends in murrels and Indian major carps, respectively. Liver glycogen content of *Anabas testudineus* registered a 48.56 per cent decrease over the initial during 9 weeks of starvation. Restoration of feeding, however, could enhance it to 13.6 mg/g which was close to the initial value. Kamra (1966), Ahsan and Ahsan (1975), Mishra and Mishra (1979) and Mahajan and Dheer (1983) also recorded reduction in the liver glycogen content during starvation. Fish have low glycogen reserves in the liver. Starvation causes their reduction from the beginning and it is usually utilized before protein or lipid. This might have caused the sharp reduction of about 48.56 per cent in glycogen levels. The liver glycogen acts as reserve for the whole organism. Normally, the liver maintains a substantial reserves of glycogen over and above its own metabolic requirements readily convertible to free glucose by the process of glycogenolysis and maintain the level of blood glucose making the fluctuations in the blood glucose minimal. Most of the end products of protein digestion are absorbed from the intestine into the portal blood and thus conveyed to liver which may be used up rapidly in the early stages of starvation (Maheswari Devi *et al.,* 1991).

As regards differential leucocyte counts of *Anabas testudineus,* certain blood cells like thrombocytes, lymphocytes, particularly small lymphocytes, and neutrophils showed clear variations during starvation. After an initial increase in the percentage of thrombocytes, a reduction was conspicuous while on the other hand, neutrophil percentage showed clear increase till feeding was restored. A slow elevation in small lymphocytes was also observed in the climbing perch. Mahajan and Dheer (1983), Sakthivel and Sampath (1989) and Kumar *et al.* (1990) observed increase in neutrophils, decrease in thrombocytes and variations in the percentage of lymphocytes in *Channa punctatus, Cyprinus carpio* and *Clarias batrachus,* respectively. The present observations in *Anabas testudineus* are in agreement with the findings of these workers. It may be concluded that these three cells - thrombocytes, neutrophils and lymphocytes are highly sensitive to starvation. Similar observations have also been reported by John and Mahajan (1979), Joshi (1979) and Agrawal and Mahajan (1980). Delaney *et al.* (1976) found that the thrombocytes increased while neutrophils decreased during starvation in *Protopterus aethiopicus.*

Acknowledgement

We are grateful to the Director, Central Institute of Freshwater Aquaculture (ICAR), Bhubaneswar for providing the laboratory facilities to carry out the work.

References

Abdel-Tawwab, M., Khattab, Y.A.E., Ahmad, M.H. and Shalaby, A.M.E., 2006. Compensatory growth, feed utilization, whole-body composition, and haematological changes in starved juvenile Nile tilapia, *Oreochromis niloticus* (L.). *J. Appl. Aquacult.,* **18 (3):** 17-36.

Agrawal, N.K. and Mahajan, C.L., 1980. Haematological changes due to vitamin C deficiency in *Channa punctatus* (Bloch). *J. Nutr.,* **110:** 2172-2181.

Ahsan, S.N. and Ahsan, J., 1975. Changes in liver glycogen in starved and normally fed, growth hormone treated *Clarias batrachus* (Linn.). *J. Ann. Zool.,* **11:** 53-58.

AOAC. 1980. *Official Methods of Analysis. 13ᵗʰ Edn.* Association of Analytical Chemists, Washington, D.C. 1018 p.

APHA, AWWA, WPCF. 1981. *Standard Methods for the Examination of Water and Wastewater. 15ᵗʰ Edn.* American Public Health Association, Washington, D.C. 1134 p.

Bailey, B., Koram, P. and Bradley, H.C., 1942. The autolysis of muscle of highly active and less active fish. *Biol. Bull. Mar. Biol. Lab., Woods Hole,* **83:** 129-136.

Barnes, H., Barnes, M. and Finlayson, D.M., 1963. The metabolism during starvation of *Balanus balanoides. J. Mar. Biol. Ass. U.K.,* **43:** 213-223.

Barton, B.A., Schreck, C.B. and Fowler, L.G., 1988. Fasting and diet content affect stress-induced change in plasma glucose and cortisol in juvenile chinook salmon. *Prog. Fish-Cult.,* **50**: 16-22.

Bazhenova, K. Ya and Shcherbina, M.A., 1975. Chemical composition of carp fingerlings muscles during winter starvation. *J. Hydrobiol.,* **11**: 60-62.

Borah, S. and Yadav, R.N.S., 1996. Biochemical and haematological responses to starvation in an air-breathing freshwater teleost, *Heteropneustes fossilis* (Bloch). *Indian J. Fish.,* **43**: 307-311.

Caruso, G., Maricchiolo, G., Muglia, U., Genovese, L. and Denaro, M.G., 2010. Changes in some physiological parameters of red progy, *Pagrus pagrus,* during a fasting-refeeding experiment. *Biol. Mar. Mediterr.,* **17**: 266-267.

Cre'ac'h, Y. and Courne'de, C., 1965. Contribution to the study of enforced starvation in the carp, *Cyprinus carpio* (Linn.). Variation in the amount of water and nitrogen in the tissues. *Bull. Soc. Hist. Nat. Troulouse,* **10**: 361-370.

Cre'ac'h, Y. and Gas, N., 1971. Some biochemical and structural changes in the muscle of carp undergoing starvation. *J. Physiol.* (*Paris*), **63**: 33-35.

Datta, N.C., Sen, P.R. and Mohanty, A.N., 1985. Studies on the blood glucose of major Indian carp fingerlings, rohu, *Labeo rohita* (Ham.) and mrigala, *Cirrhinus mrigala* (Ham.) in starvation and its recovery with the provision of graded protein diets. In: *Proceedings of the International Conference on Warmwater Aquaculture of Finfish.* pp. 157-176. Brigham Young University, Hawaii Campus, USA.

Delaney, R.G., Shub, C. and Fishman, A.P. 1976. Haematological observations on the aquatic and aestivating African lungfish, *Protopterus aethiopicus. Copeia,* **1976**: 423-434.

Fisher, R.A., 1970. *Statistical Methods for Research Workers.* Oliver and Boyd, Edinburgh and London. 362 p.

Fisher, R.A. and Yates, F.I., 1963. *Statistical Tables for Biological, Agricultural and Medical Research. 6ᵗʰ Edn.* Hafner, New York. 146 p.

Jacquot, R., 1961. Organic constituents of fish and other aquatic animals. In: *Fish as Food* (Ed.) G. Borgstrom. pp. 145-209. Academic Press, London and New York.

Johansen, K.A. and Overturf, K., 2006. Alterations in expression of genes associated with muscle metabolism and growth during nutritional restriction and refeeding in rainbow trout. *Comp. Biochem. Physiol.,* **144B**: 119-127.

John, M.J. and Mahajan, C.L., 1979. The physiological response to fishes to a deficient diet of cyanocobalamine and folic acid. *J. Fish Biol.,* **141**: 127-133.

Joshi, B.D., 1979. Effects of starvation on some haematologic values in some freshwater fishes. *Indian. J. Exp. Biol.,* **17**: 968-969.

Kamra, S.K., 1966. Effect of starvation and refeeding on some liver and blood constituents of Atlantic cod (*Gadus morhua* L.). *J. Fish. Res. Bd. Can.,* **23**: 975-982.

Kawatsu, H., 1974. Studies on the anaemia. IV. Further note on the anaemia caused by starvation in rainbow trout. *Bull. Freshwater Res. Lab.* (Tokyo), **24**: 89-94.

Kosmina, V.V., 1966. Electrophoretic changes in protein of blood serum of fish during prolonged starvation. *J. Hydrobiol.*, **2 (4)**: 74-77.

Kumar, K., Kumar, D., Dey, R.K. and Murjani, G., 1990. Starvation induced haematological changes in an air-breathing freshwater fish, *Clarias batrachus* (Linn.). In: *The Second Indian Fisheries Forum* (May 27-31, 1990). p. 71. College of Fisheries, Mangalore.

Love, R.M., 1958. Studies on the North Sea cod. III. Effects of starvation. *J. Sci. Food. Agric.*, **9**: 617-620.

Mahajan, C.L. and Dheer, T.R., 1983. Haematological and haematopoetic responses to starvation in an air-breathing fish, *Channa punctatus* (Bloch). *J. Fish Biol.*, **22**: 111-123.

Maheswari Devi, K., Gopal, V. and Gopal, R., 1991. Liver-somatic index of *Channa striatus* as a biomonitoring tool of heavy metal and pesticide toxicity. *J. Ecotoxicol. Environ. Monit.*, **1**: 135-141.

Mishra, R.N. and Mishra, N.K., 1979. Effects of starvation on *Heteropneustes fossilis* (Bloch). A biochemical study. *Indian J. Exp. Biol.*, **17**: 1224-1229.

Moorthy, C.V.N., Murthy, V.K., Reddy, G.V., Haranath, V.B., Reddanna, P. and Govindappa, S., 1980. Effect of starvation and refeeding on tissue proximate analysis of hepatic and muscular tissues of *Tilapia mossambica* (Peters). *Indian J. Fish.*, **27**: 215-219.

Mukhopadhyay, P.K., Mohanty, S.N., Das, K.M., Sarkar, S. and Patra, B.C., 1991. Growth and changes in carcass composition in young of *Labeo rohita* and *Cirrhinus mrigala* during feeding and starvation. In: *Proceedings of Fourth Asian Fish Nutrition Workshop* (Ed). S.S. de Silva. pp. 87-91. Asian Fisheries Society, Manila, Philippines.

Mustafa, S. and Mittal, A., 1982. Protein, RNA and DNA levels in liver and brain of starved catfish, *Clarias batrachus. Jap. J. Ichthyol.*, **28**: 396-400.

Oser, B.L., 1965. *Hawk's Physiological Chemistry. 14ᵗʰ Edn.* McGraw-Hill, New York. 1472 p.

Perez-Jimenez, A., Romero, C.E.T. and Hernandez, G.C., 2011. Metabolic responses to food deprivation in fish. In: *Biology of Starvation in Humans and Other Organisms* (Ed.) T. C. Merkin. pp. 303-346. Nova Sci. Pub., New York.

Phillips, A.M. Jr., Livingston, D.L. and Dumas, R.F., 1960. Effect of starvation and feeding on the chemical composition of brook trout. *Prog. Fish-Cult.*, **22**: 147-154.

Rajyasree, M. and Naidu, K.R.P., 1989. Starvation induced changes in biochemical aspects of hepatic tissue of fish, *Labeo rohita. Indian J. Fish*, **36**: 339-341.

Robertson, I., Love, R.M. and Cowie, W.P., 1967. Studies on the North Sea cod. V. Effects of starvation. 3. Electrophoresis of the muscle myogens. *J. Sci. Food Agric.*, **18**: 563-565.

Sakthivel, M. and Sampath, K., 1989. Haematological responses of *Cyprinus carpio* in relation to starvation. *Geobios,* **16**: 61-65.

Siddiqui, N., 1975. Variations in chemical constituents of blood plasma of *Clarias batrachus* (L.) during starvation. *Curr. Sci.,* **44**: 126-127.

Siebert, G., Schitt, A. and Bottke, I. 1964. Enzymes of amino acid metabolism in cod musculature. *Arch. Fisch. Wiss.,* **15**: 233-244.

Smallwood, W.M., 1916. Twenty months of starvation in *Amia calva. Biol. Mar. Biol. Lab. Woods Hole,* **31**: 453-464.

Sorvachev, K.F., 1957. Changes in proteins of carp blood serum during hibernation. *Biokhimiya,* **22**: 872-878.

Sridee, N. and Boonanuntanasarm, S., 2012. The effects of food deprivation on haematological indices and blood indicators of liver function in *Oxyleotris marmorata. Intern. J. Med. Biol. Sci.,* **6**: 254-258.

Steucke, E.W. and Schoettger, A., 1967. Comparison of three methods of sampling trout blood for measurements of haematocrit. *Prog. Fish-Cult.,* **29**: 98-101.

Stirling, H.P., 1976. Effects of experimental feeding and starvation on the proximate composition of the European bass, *Dicentracus labrax. Mar. Biol.,* **34**: 85-91.

Sutto, A.H., 1968.The relationship between ion and water contents of cod (*Gadus morhua* L.) muscle. *Comp. Biochem. Physiol.,* **24**: 149-161.

Swallow, R.L. and Fleming, W.R., 1969. The effect of starvation, feeding, glucose and ACTH on liver glycogen levels of *Tilapia mossambica. Comp. Biochem. Physiol.,* **28**: 95-106.

Tzeng, W.N. and Yu, S.Y., 1992. Effects of starvation on the formation of daily increments in the otoliths of milkfish, *Chanos chanos* (Forsskal) larvae. *J. Fish Biol.,* **40**: 39-48.

Yamada, S., Tanaka, Y., Sameshima, M. and Furuichi, M., 1994. Effects of starvation and feeding on tissue N alpha-acetylhistidine levels in Nile tilapia, *Oreochromes niloticus. Comp. Biochem. Physiol.,* **109A**: 227-283.

2015, **Perspectives in Animal Ecology and Reproduction, Vol. 10**   *Pages 195–203*
Editors: **V.K. Gupta, Anil K. Verma and G.D. Singh**
Published by: **DAYA PUBLISHING HOUSE, NEW DELHI**

Chapter 14

Dietary 17α-Methyltestosterone Supplementation Enhances Growth in Fry of Indian Major Carps Under Hatchery Conditions

A.K. Pandey[1]*, P.K. Arvindakshan[2] and B.N. Singh[2]
[1]National Bureau of Fish Genetic Resources,
Canal Ring Road, Lucknow – 226 002, U.P., India
[2]Central Institute of Freshwater Aquaculture,
Kausalyaganga, Bhubaneswar – 751 002, Odisha, India

ABSTRACT

Effect of dietary 17α-methyltestosterone (17α-MT) supplementation on growth of the fry of Indian major carps under hatchery condition was recorded. Control feed (T-1) was prepared with base ingredients like fish meal 5 per cent, groundnut oil-cake 35 per cent, soybean oil-cake 25 per cent, rice bran 25 per cent, wheat flour 10 per cent whereas 17α-MT was added in experimental feed in the dose of 2 (T-2), 4 (T-3) and 6 ppm (T-4). Crude protein content of all the four feeds were kept at uniform level of 30.17 per cent. The pelleted feed (0.2 mm diameter) was given @ 5 per cent of the body weight once daily for 90 days. The fry given 17α-MT supplemented feed recorded higher growth rate in comparison to those maintained on the control diet (T-1). 17α-MT in the dose of 6 ppm (Feed T-4) was found to be the most effective in inducing higher growth in comparison to 2 ppm (T-2) and 4 ppm (T-3). FCR was lowest whereas PER and FCE (per cent) were highest in the fry kept on 6 ppm 17α-MT supplemented feed.

Keywords: *Dietary 17α-MT-supplementation, Growth, FCR, PER, FCE, Catla catla, Labeo rohita.*

---

* *Corresponding author.* E-mail: akpandey.ars@gmail.com

Introduction

As growth improvement of cultured organisms through high protein diet has got restricted application in commercial aquaculture due to the high cost involved in using such feed in intensive culture system (Pandey *et al.,* 2012) and its dubious role in contributing to the nitrogen load in pond ecosystem (Viola and Lahav, 1991; Chakraborty and Chakraborty, 1998; Stibranyiova and Paraova, 2000), incorporation of various steroids in the diet of cultivable fishes assumes significance (Donaldson *et a.l,* 1979; Guerrero, 1979; Higgs *et al.,* 1982; Yamazaki, 1983; Mukhopadhyay *et al.,* 1986; Felix, 1989; Guerrero and Guerrero, 1997; Poczyczyriski *et al.,* 1998; Davis *et al.,* 2010). Recent studies have demonstrated that the naturally-occurring (testosterone, 11-ketotestosterone) as well as synthetic androgens (dimethazine, norethanedrolone) and estrogens have anabolic (growth-promoting) effect in salmonids, common carp, tilapia, eel and ornamental fishes (Fagerlund *et al.,* 1979; Yu *et al.,* 1979; Higgs *et al.,* 1982; Lone and Matty, 1983; Yamazaki, 1983; Degani, 1986; Howrton *et al.,* 1992; Kuwaye *et al.,* 1993; Santandreu and Diaz, 1994; Singh and Pandey, 1995; Ahmad *et al.,* 2002; James and Sampath, 2006; Davis *et al.,* 2010). Contrary to these findings, no significant growth difference, rather negative growth trend, has been recorded in the channel catfish, eel, yellow perch and *Oreochromis andersonii* treated with synthetic steroids (stanozolol and methandrostenolone) applied even at a moderate dose (Simone, 1990; Gannam and Lovell, 1991, 1992; Malison, 1986; Degani and Dosorets, 1996; Kefi *et al.,* 2013). Further, androgenic property of the hormone may alter sex of the species at a certain dose (Nagy *et al.,* 1981; Shelton *et al.,* 1981; Goudie *et al.,* 1983; Yamazaki, 1983; Smitherson, 1993; Pandian and Sheela, 1995; Singh and Pandey, 1995; Chatain *et al.,* 1999; Gale *et al.,* 1999; James and Sampath, 2006; Marjani *et al.,* 2009; Ferdous and Ali, 2011; Khalil *et al.,* 2011; Mubarik *et al.,* 2011). Possibility of application of steroids for growth improvement cannot be rejected because of the fact that the anabolic property of steroid has been proved and accepted in species like salmonids (McBride and Fagurland, 1973; Fagerlund and McBride, 1975; Fagerlund *et al.,* 1979; Ando *et al.,* 1986), common carp (Lone and Matty, 1983; Manzoor Ali and Satyanarayana Rao, 1989), tilapia (Ridha and Lone, 1990; Lone and Ridha, 1993; Singh and Pandey, 1995; Abdelghany, 1995) and also in ornamental fishes (James and Sampath, 2006). Keeping in view of the intensification of commercial carp culture, it becomes essential to develop a cost-effective diet with high acceptability and low feed conversion ratio (FCR). An attempt has, therefore, been made to evaluate the efficacy of a naturally-occurring steroid, 17a-methyltestosterone (17a-MT), in stimulating growth of the Indian major carps under hatchery conditions.

Material and Methods

Fry of *Catla catla* (weight range 0.78±0.10-0.93±0.10 g) and *Labeo rohita* (weight range 0.70±0.11-0.81±0.12 g) were procured from the induced breeding experiments and divided into four equal groups of 100 each. One group served as control whereas the rest three groups were experimental (Table 14.1). The control fry were maintained on the diet (Feed T-1) containing the ingredients like fish meal, groundnut oil-cake, soybean oil-cake, rice bran and wheat flour (crude protein content 30.17 per cent). The feed of experimental groups were supplemented with 17α-methyltestosterone

(Sigma Chemicals, USA) in the dose of 2.0 ppm (Feed T-2), 4.0 ppm (Feed T-3) and 6.0 ppm (Feed T-4), respectively (Table 14.1). The pelleted feed was given @ 5 per cent of the body weight (once daily) for 90 days. Sampling was carried out at the regular intervals of 15 days and the feed was adjusted accordingly. Physico-chemical parameters of the water were monitored during the period as pH 7.1-7.5, temperature 28-30°C, dissolved oxygen (DO) 4.1-5.2 ppm, ammonia 0.05-0.1 ppm and total alkalinity 120-132 ppm. Weight gain percentage, feed conversion efficiency (FCE), feed conversion ratio (FCR) and protein efficiency ratio (PER) of fishes from the control as well as experimental groups were calculated as per formulae of Singh (1990). The results were evaluated for statistical significance using Students "t" test.

Table 14.1: Feed Ingredients and their Composition Used for Rearing of *Catla catla* and *Labeo rohita* Fry

Ingredients (per cent)	Feed T-1	Feed T-2	Feed T-3	Feed T-4
Fish meal	05	05	05	05
Groundnut oil-cake	35	35	35	35
Soybean oil-cake	25	25	25	25
Rice bran	25	25	25	25
Wheat flour	10	10	10	10
17α-MT (ppm)	—	02	04	06
Crude protein	30.17	30.17	30.17	30.17

Results and Discussion

The effects of various doses of 17α-methyltestosterone on growth of the fry of *Catla catla* have been summarized in Table 14.2. The fry of *Catla catla* maintained on 17α-MT supplemented diet recorded higher growth rate in comparison to those maintained on the control feed. 17α-MT in the dose of 6 ppm (T-4) was found to be the most effective in inducing higher growth rate in comparison to those given 2 ppm (T-2) and 4 ppm (T-3) supplemented diets. The feed conversion ratio (FCR) was lowest

Table 14.2: Effect of different Concentrations of 17α-Methyltestosterone on Growth of *Catla catla* Fry

Parameter	Feed T-1	Feed T-2	Feed T-3	Feed T-4
Initial weight (g)	0.90±0.10	0.92±0.12	0.93±0.10	0.78±0.11
Final weight (g)	10.20±0.10	11.00±0.10	11.40±0.10	11.60±0.10
Increase in weight (g)	9.30±0.11	10.08±0.14	10.47±0.12	10.82±0.12
Weight gain per cent	13.33±0.14	19.56 ±0.18*	22.58±0.21*	35.89±0.12*
PER	0.24±0.10	0.33±0.12*	0.42±0.11*	0.54±0.12*
FCR	2.84±0.14	2.46±0.12*	2.17±0.10*	1.84±0.11*
FCE (per cent)	35.21±0.18	40.65±0.20*	46.10±0.16*	54.34±0.22*

* Significant response: P > 0.05.

in Feed T-4 whereas protein efficiency ratio (PER) and feed conversion efficiency (FCE per cent) were higher among the fish maintained on the Feed T-4. Similar observations were also recorded for the fry of *Labeo rohita* supplemented with the three doses of 17α-MT.

There exist reports that the naturally-occurring (testosterone, 11-ketotestosterone) and synthetic androgens (dimethazine, norethanedrolone) have growth-promoting effect among the teleosts (Fagerlund *et al.,* 1979; Higgs *et al.,* 1982; Hunter and Donaldson, 1983). Anabolic effect of 17α-methyltestosterone in salmonids, grouper, common carp, tilapia, eel, catfish, murrel and ornamental fishes has been recorded (Chua and Tang, 1980; Higgs *et al.,* 1982; Nirmala and Pandian, 1983; Sindhu and Pandian, 1984; Hunter and Donaldson, 1993; Lone and Matty, 1983; Yamazaki, 1983; Degani, 1986; Mukhopadhyay *et al.,* 1986; Felix, 1989; Killian and Kohler, 1991; Meyer, 1991; Kuwaye *et al.,* 1993; Santandreu and Diaz, 1994; Ron *et al.,* 1995; Singh and Pandey, 1995; James and Sampath, 2006; Montajami, 2012; Mohamed *et al.,* 2013). However, there exist reports on its ineffectiveness or rather negative effect on growth in *Poecilia reticulata, Ictalurus punctatus, Anguilla anguilla* and *Perca favescence* treated with synthetic steroids (stanozolol and methandrostenolone) (Clemens *et al.,* 1966; Simone, 1990; Gannam and Lovell, 1991, 1992; Malison, 1986). The observed enhancement in growth in fry of *Catla catla* and *Labeo rohita* kept on the semi-balanced diet supplemented with varying doses of 17α-methyltestosterone under hatchery condition for 90 days indicates anabolic effects of the steroid hormone through inducing better utilization of feed in the Indian major carps (Table 14.2) which is having implications in larval rearing of commercially important species for aquaculture. Nirmala and Pandian (1983) and Sindhu and Pandian (1984) suggested the hormone as appetite stimulant and anabolic steroid. There are indications that methyltestosterone act synergistically with growth hormone on whole body protein metabolism (Riley *et al.,* 2002; Mauras *et al.,* 2003; Birzniece *et al.,* 2011). The negative effect of methyltestosterone in some fishes has been attributed to the enhanced catabolism at higher doses of the hormone (Fagerlund *et al.,* 1979; Yamazaki, 1983; Nirmala and Pandian, 1983; Sindhu and Pandian, 1984; James and Sampath, 2006).

Acknowledgements

We are grateful the Director, Central Institute of Freshwater Aquaculture, Bhubaneswar for extending the necessary facilities to carry out the experiments. Staff of Fish Physiology Division, CIFA helped in execution of the work. Financial assistance provided by the Department of Biotechnology (Government of India), New Delhi is thankfully acknowledged.

References

Abdelghany, A. E., 1995. Effect of feeding 17α-methyltestosterone and withdrawal on feed utilization and growth of Nile tilapia, *Oreochromis niloticus* L., fingerlings. *J. Appl. Aquacult.,* **5 (3)**: 67-75.

Ahmad, M.H., Abdel-Tawwab, Shalaby, A.M.E. and Khattab, Y.A.E., 2002. Effects of 17α-methyltestosterone on growth performance and some physiological changes of Nile tilapia, (*Oreochromis niloticus* L.) fingerlings. *Egypt. J. Aquat. Biol. Fish.*, **6 (2)**: 1-23.

Ando, S., Yamazaki, F. and Hattano, M., 1986. Effect of 17 α-methyltestosterone on muscle composition of chum salmon. *Bull. Jap. Soc. Sci. Fish.*, **52**: 565-571.

Birzniece, V., Meinhardt, U.J., Umpleby, M.A., Handelsman, D.J and Ho, K.K., 2011. Interaction between testosterone and growth hormone on whole body protein anabolism occurs in the liver. *J. Clin. Endocrinol. Metab.*, **96**: 1060-1067.

Boney, S.E., Shelton, W.L., Yang, S. and Wilken, L.O., 1984. Sex reversal and breeding of grass carp. *Trans. Am. Fish. Soc.*, **113**: 348-353.

Chakraborty, S. C. and Chakraborty, S., 1998. Effect of dietary protein level on excretion of ammonia in Indian major carp, *Labeo rohita*, fingerlings. *Aquacult. Nutr.*, **4**: 47-51.

Chatain, B., Saillant, E. and Peruzzi, S., 1999. Production of monosex male populations of European seabass, *Dicentrarchus labrax* L., by use of the synthetic androgen, 17α-methyldehydrotestosterone. *Aquaculture*, **178**: 225-234.

Chua, T.E. and Teng, S.K., 1980. Economic production of estuary grouper, *Epinephlus salmoides* Maxwell, reared in floating cages. *Aquaculture*, **20**: 187-228.

Clemens, H.P., McDermitt, C. and Inslee, T., 1966. The effect of feeding methyltestosterone to guppies for 60 days after birth. *Copeia*, **2**: 280-294.

Davis, K.B. and Ludwig, G.M., 2004. Hormonal effects on sex differentiation and growth in sunshine bass, *Morone chrysops* x *Morone sexatilis*. *Aquaculture*, **231**: 587-596.

Davis, L.K., Fox, B.K., Lim, C., Lerner, D.T., Hirano, T. and Grau, *E.G.*, 2010. Effects of 11-ketotestosterone and fishmeal in the feed on growth of juvenile tilapia (*Oreochromis mossambicus*). *Aquaculture*, **305**: 143-149.

Degani, G., 1986. Effect of dietary 17β-estradiol and 17α-methyltestosterone on growth and body composition of slow-growing elvers (*Anguilla anguilla* L.). *Comp. Biochem. Physiol.*, **85A**: 243-247.

Degani, G. and Dosoretz, C., 1986. The effect of 3, 3′, 5-triiodo-L-thyronine and 17 α-methyltestosterone on growth and body composition of the glass stage of the eel (*Anguilla anguilla* L.). *Fish Physiol. Biochem.*, **1:** 145-151.

Donaldson, E.M., Fagerlund, U.H.M., Higgs, D.A. and McBride, J.R., 1979. Hormonal enhancement of growth. In: *Fish physiology. Vol. VIII* (eds. Hoar, W.S., Randall, D.J. and Brett, J.R.). pp. 456-578. Academic Press, Orlando & New York.

Fagerlund, U.H.M. and McBride, J.R., 1975. Growth increments and some flesh and gonad characteristics of juvenile coho salmon receiving diets supplemented with 17α-methyltestosterone. *J. Fish Biol.*, **7**: 305-314.

Fagerlund, U.H.M., McBride, J.R. and Stone, E., 1979. A test of 17 α-methyltestosterone as a growth promoter in the coho salmon hatchery. *Trans. Am. Fish. Soc.*, **108**: 467-472.

Faucconeau, B., 1985. Protein synthesis and protein deposition in fish. In: *Nutrition and Feeding in Fish* (eds. Cowey, B., Mackie, A.M. and Bell, J.G.). pp. 17-45. Academic Press, New York.

Felix, S., 1989. Effect of 17α-methyltestosterone on the growth of ornamental fish, *Xiphophorus maculeatus*. *Indian J. Fish.,* **36**: 263-265.

Ferdous, Z. and Ali, M.M., 2011. Optimization of hormone dose during masculinization of tilapia (*Oreochromis niloticus*) fry. *J. Bangladesh Agric. Univ.,* **9 (2)**: 359-364.

Gale, W.L., Fitzpatrick, M.S., Lucero, M., Contreras-Sanchez, W.M. and Schreck, C.B., 1999. Masculinization of Nile tilapia (*Oreochromis niloticus*) by immersion in androgen. *Aquaculture,* **178**: 349-357.

Gannam, A.L. and Lovell, R.T., 1991. Growth and bone development in channel catfish fed 17α-methyltestosterone in production ponds. *J. World Aquacult. Soc.,* **22**: 95-100.

Gannam, A.L. and Lovell, R.T., 1992. Effects of feeding 17α-methyltestosterone, 17-ketotestosterone, 17β-estradiol and 3, 5, 3'-triiodothyronine to channel catfish, *Ictalurus punctatus*. *Aquaculture,* **92**: 377-388.

Goudie, C.A., Redner, B.D., Simco, B.A. and Davis, K.B., 1983. Feminization of channel catfish by oral administration of steroid sex hormones. *Trans. Am. Fish. Soc.,* **112**: 670-672.

Guerrero, R.D. III, 1979. Use of hormonal steroids for artificial sex reversal of *Tilapia*. *Proc. Indian Natl. Sci. Acad.,* **45B**: 512-514.

Guerrero, R.D. III and Guerrero, L.A., 1997. Effects of androstenedione and methyltestosterone on *Oreochromis niloticus* fry treated for sex reversal in outdoor net enclosures. In: *Proceedings of the Fourth International Symposium on Tilapia in Aquaculture* (November 9-12, 1997) (ed. K. Fitzsimmons). pp. 772-775. Orlando, Florida.

Higgs, D.A., Fagerlund, U.H.M., Eales, J.G. and McBride, J.R., 1982. Application of thyroid and steroid hormones as anabolic agents in fish culture. *Comp. Biochem. Physiol.,* **73**: 143-176.

Howerton, R.D., Okimoto, D.K. and Grau, E.G., 1992. The effect of orally-administered 17α-methyltestosterone and 3, 5, 3'-triiodo-L-thyroxine on growth of sea water-adapted tilapia, *Oreochromis mossambicus* (Peters). *Aquacult. Fish. Manage.,* **23**: 23-128.

Hunter, G.A. and Donalson, E.M., 1983. Hormonal sex control and its application to fish culture. In: *Fish Physiology. Vol. IXB. Reproduction, Behaviour and Fertility Control* (eds. Hoar, W.S., Randall, D.J. and Donaldson, E.M.). pp. 223-303. Academic Press, Orlando & Florida.

Kefi, A.S., Kangombe, J. and Katongo, C., 2013. Effect of 17α-methyltestosterone on haematology and histology of liver and heart of *Oreochromis andersonii* (Castlnau, 1861). *J. Marine Sci. Res. Dev.,* **3 (2)**: 1-7.

Khalil, W.K.B., Hasheesh, W.S., Marie, M.-A.S., Abbas, H.H. and Zahran, E.A., 2011. Assessment the impact of 17α-methyltestosterone hormone on growth, hormone concentration, molecular and histopathological changes in muscles and testis of Nile tilapia, *Oreochromis niloticus. Life. Sci. J.,* **8**: 329-342.

Killian, H.S. and Kohler, C.C., 1991. Influence of 17α-methyltestosterone on red tilapia under two thermal regimes. *J. World Aquacult. Soc.,* **22**: 83-94.

Kuwaye, T.T., Okimato, D.K., Shimoda, S.K., Howerton, R.D., Lin, H.R., Pang, P.K.T. and Grau, E.G., 1993. Effect of 17α-methyltestosterone on the growth of euryhaline tilapia, *Oreochromis mossambicus,* in freshwater and in sea water. *Aquaculture,* **113**: 137-152.

James, R. and Sampath, K., 2006. Effect of dietary administration of methyltestosterone on the growth and sex reversal of two ornamental fish species. *Indian J. Fish.,* **53**: 283-290.

Lone, K.P. and Matty, A.J., 1983. The effect of ethylsternol on the growth, food conversion and tissue chemistry of the carp, *Cyprinus carpio. Aquaculture,* **32**: 39-55.

Lone, K.P. and Ridha, M.T., 1993. Sex reversal and growth of *Oreochromis spirulus* (Gunther) in brackishwater and sea water by feeding 17α-methyltestosterone. *Aquuacult. Fish. Manage.,* **34**: 593-602.

Malison, J.A., 1986. Growth promotion influence of sex steroids on sexually-related dimorphic growth and differentiation in yellow perch (*Perca flavescence*). *Diss. Abstr. Intern.,* **46B**: 171.

Manzoor Ali, P.K.M. and Satyanarayana Rao, G.P., 1989 Growth improvements in carp, *Cyprinus carpio* Linnaeus), sterilized with 17α-methyltestosterone. *Aquaculture,* **76**: 157-167.

Marjani, M., Jamill, S., Mostafavi, P.G., Ramin, M. and Mashinchian, A., 2009. Influence of 17α-methyltestosterone on masculinization and growth in tilapia (*Oreochromis mossambicus*). *J. Fish. Aquat. Sci.,* **4**: 71-74.

Mauras, N., Rini, A., Welch, S., Sager, B. and Murphy, S.P., 2003. Synergistic effects of testosterone and growth hormone on protein metabolism and body composition in pubertal boys. *Metabolism,* **52**: 964-969.

McBride, J.R. and. Fagerlund, U.H.F., 1973. The use of 17α-methyltestosterone for promoting weight increases in juvenile Pacific salmon. *J. Fish. Res. Bd. Can.,* **30**: 1099-1104.

Meyer, D.C., 1991. Growth, survival and sex ratios of *Tilapia hornorum, Tilapia nilotica* and their hybrid (*T. nilotica* female x *T. hornorum* male) treated with 17α-methyltestosterone. *Diss. Abstr. Intern.,* **51B**: 75.

Mohamed, A.H., Traifalgar, R.F.M. and Serrano, A.E. Jr., 2013. Optimum dosages for growth of dietary dehydroepiandrosterone in hybrid red tilapia fry *Oreochromis niloticus* x *Oreochromis mossambicus. Eurp.J. Exp. Biol.,* **3 (1)**: 255-259.

Montajami, S., 2012. Assessment of the impact of 17α-methyltestosterone on growth and survival rate of golden barb fish, *Puntius gelius* (Hamilton, 1822). *Amer.-Eur. J. Agric. Environ. Sci.,* **12**: 1052-1055.

Mubarik, M.S., Ahmed, I., Mateen, A. and Iqbal, T., 2011. 17 α-methyltestosterone induced masculinization and its effect on growth and meat quality of *Cyprinus carpio. Int. J. Agric. Biol.,* **13**: 971-975.

Mukhopadhyay, P.K., Venkatesh, B. and Das, P. 1986. Growth and some biochemical changes in *Clarias batrachus* due to methyltestosterone. *Indian J. Fish.,* **33**: 252-264.

Nagy, A., Bersenyi, M. and Sanyi, K., 1981. Sex reversal in carp (*Cyprinus carpio*) by oral administration of methyltestosterone. *Can. J. Fish. Aquat. Sci.,* **38**: 725-728.

Nirmala, A.R.C. and Pandian, T.J., 1983. Effect of steroid injection on food utilization in *Channa striatus. Proc. Indian Acad. Sci.,* **92**: 1-10.

Pandey, A.K., Sarkar, M., Mahapatra, C.T., Kanungo, G. and Arvindakshan, P.K., 2012. Effect of dietary lysine and methionine supplementation on growth of *Catla catla* and *Labeo rohita* fingerlings. *J. Exp. Zool. India,* **15**: 259-262.

Pandian, T.J. and Sheela, S.G., 1995. Hormonal induction of sex reversal in fish. *Aquaculture,* **138**: 1-22.

Poczyczyriski, P., Kozlowski, J., Mamcarz, A. and Kujawa, R., !998. Effect of simultaneous administration of 17α-methyltestosterone and of 3, 5, 3'-triiodothyronine in starter diets on rearing of whitefish (*Coregonus lavaretus* L.) larvae. *Arch. Pol. Fish.,* **6**: 51-58.

Ridha, M.T. and Lone, K.P., 1990. Effect of oral administration of different levels of 17α-methyltestosterone on the sex reversal, growth and food conversion efficiency of the tilapia, *Oreochromis spilurus* (Guenther) in brackishwater. *Aquacult. Fish. Manage.,* **21**: 391-397.

Riley, L.G., Richman, N.H., Hirano, T. and Grau, E.G., 2002. Activation of the growth hormone/insulin-like growth factor axis by treatment with 17α-methyltestosterone and seawater rearing in the tilapia, *Oreochromis mossambicus. Gen. Comp. Endocrinol.,* **127**: 285-292.

Ron, B., Shimoda, S.K., Iwama, G. and Grau, G., 1995. Relationship among ration, salinity, 17α-methyltestosterone and growth in euryhaline tilapia, *Oreochromis mossambicus. Aquaculture,* **135**: 185-193.

Santandreu, I.A. and Diaz, N.F., 1994. Effect of 17 α-methyltestosterone on growth and nitrogen excretion in masu salmon (*Oncorhynchus masou* Brevoort). *Aquaculture,* **124**: 321-333.

Shelton, W.L., Guerrero, R.D. and Lopez-Marcias, J., 1981. Factors affecting androgen sex reversal of *Tilapia aurea. Aquaculture,* **25**: 59-65.

Simone, D.A., 1990. The effect of the synthetic steroid 17 α-methyltestosterone on the growth and organ morphology of the channel catfish (*Ictalurus punctatus*). *Aquaculture,* **84**: 81-93.

Sindhu, S. and Pandian, T.J., 1984. Effect of administering 17 α-methyltestosterone in *Heteropneustes fossilis. Proc. Indian Acad. Sci.*, **93**: 511-516,

Singh, A.K. and Pandey, A.K., 1995. Effect of long photoperiod on growth, survival and sex ratio of androgenizing tilapia, *Oreochromis mossambcus.Nat. Acad. Sci. Lett.,* **18**: 227-231.

Singh, B.N., 1990. Protein requirement of young silver carp, *Hypophthalmichthys molitrix* (Val.). *J. Freshwater Biol.,* **2:** 89-95.

Smitherson, R.O., 1993. Clinical field trial for tilapia sex reversal feed begins. *Aquacult. Mgz. (*July/August): 68-70

Stibranyiova, I. and Paraova, J., 2000. The effect of lysine, methionine and threonine supplementation in practical diets with graded protein levels on growth, body composition and nitrogen excretion in juvenile African catfish (*Clarias gariepinus*). In: *Proceedings of Fish and Crustacean Nutrition Methodology and Research for Semi-intensive Pond-based Farming System Workshop* (April 1996). Szarvas (Hungary), **23**: 122-130.

Viola, S. and Lahav, E., 1991. Effects of lysine supplementation in practical carp feeds on total protein sparing and reduction of pollution. *Israeli J. Aquacult. (Bamidgeh),* **43**: 112-118.

Yamazaki, F., 1983. Sex control and manipulation. *Aquaculture,* **33**: 329-354.

Yu, T.C., Sinnhuber, R.O. and Hendricks, J.D., 1979. Effect of steroid hormones on the growth of juvenile coho salmon (*Oncorhynchus kisutch*). *Aquaculture,* **16:** 351-359.

2015, Perspectives in Animal Ecology and Reproduction, Vol. 10   *Pages 205–236*
Editors: V.K. Gupta, Anil K. Verma and G.D. Singh
Published by: DAYA PUBLISHING HOUSE, NEW DELHI

Chapter 15

Metallothioneins of Fishes as a Biomarker of Heavy Metal Pollution

*K.M. Arunkumar[1], Lazar V. Kottickal[2] and
Balu T. Kuzhivelil[1]**

*[1]Applied Biochemistry and Biotechnology Laboratory,
Department of Zoology, Christ College, Irinjalakuda,
University of Calicut, Kerala – 680 125, India
[2]Department of Environmental Science,
Central University of Kerala, Kerala – 671 314, India*

ABSTRACT

Aquatic ecosystems are getting polluted from natural and anthropogenic sources. Among industrial pollutants, the heavy metals are of primary concern because they are highly toxic and affect aquatic life. When food from these sources is consumed, human health may get impaired. In order to protect human and wild life, surveillance and pollution monitoring programs are required. Traditionally, the monitoring of aquatic environment is carried out by chemical analysis of water and organ samples. One major drawback of this chemical analysis is that, it does not provide information about the impact of the pollutants on the organisms. An improvement of the pollution monitoring can be achieved with the introduction of "biomarkers" to get early information about the level of toxicity. A molecular biomarker that express the effects of the pollutants at the initial stage can be used as "early warning tools" for the heavy metal pollution monitoring programmes. Metallothioneins (MTs) are ubiquitously expressed genes in organisms, which are over expressed under stress including heavy metal pollution. In this review, we discuss the concept of ecotoxicogenomics, with special reference to MT genes of fishes as biomarkers of heavy metal pollution.

Keywords: Aquatic ecosystem, Biomarker, Ecotoxicogenomics, Heavy metals, Metallothioneins, Pollution monitoring.

* *Corresponding author.* E-mail: balu99@rediff.com

Introduction

Aquatic environment is most vulnerable to pollution of heavy metals, polybrominated diphenyl ethers, polychlorinated biphenyls, polycyclic aromatic hydrocarbons, endocrine disruptors, pharmaceuticals, pesticides, insecticides and other pollutants due to anthropogenic activities. Heavy metals are considered to be one of the most widespread groups of micro- contaminants because of their persistent nature and slow elimination from environmental compartments (Fargasova, 1998). Majority of the heavy metals occur naturally in the environment and are released as a result of human activities and natural processes. The large scale industrial output, the fossil fuel burning, together with the waste incineration, industrial waste discharge and mining have contributed to widespread heavy metal contamination of the aquatic environment (Ma *et al.,* 2009). Contaminants in different segment of the marine environment have become a major threat to the health of the marine ecosystem due to accumulation of their residues in tissues of various species of marine organisms (Sarkar *et al.,* 2006). The process of accumulation of heavy metals or any contaminants in the tissues of the organism within a trophic level is called "bioaccumulation". Bioaccumulation of the pollutants/contaminants may lead to the accumulation of these substances into the higher trophic level through the food chain; this process is called "biomagnification" (Esser and Moser, 1982).

An evaluation of the environmental quality of aquatic ecosystems using traditional chemical analysis fails in its objectives, as the ecosystems are under the pressure of complex mixtures of contaminants. Further, the complex mixtures are not always easy to analyze. When the environment monitoring is limited to chemical analysis, the effects of pollutants on aquatic biota is not becoming clear (Lam and Gray, 2003). The process of continuous inspection on the presence of pollutants in ecosystem through the assessment of its effects on the biota is called "biomonitoring". The importance of "biomarker" approach in ecotoxicological studies may be viewed in this context. Biomarkers are acting as early warning signals against the presence of potentially toxic xenobiotics. They are useful tools for assessing either exposure or bioavailability of the toxicants and providing information about the toxicant's effects on organisms (Picado *et al.,* 2007). With the help of rapid progress in the field of "genomics" and "aquatic toxicology", we now have greater understanding about the ecosystem health as well as the effects of pollutants on aquatic organisms at molecular, cellular, tissue/organ and organism level (Snape *et al.,* 2004). The study on how the organisms adapt or resist to the polluted ecosystem conditions with the help of the genes that express in such situations results in the development of another branch of toxicological study called "ecotoxicogenomics" (Viarengo *et al.,* 2007). If the organisms cannot convert metals into less toxic metabolite form, they tend to protect themselves against metal toxicity through the method of intracellular complexation which is mainly achieved by a group of proteins, called metallothioneins (MTs) (Knapen *et al.,* 2005). Previous studies have reported that MT proteins and their gene transcripts can be induced rapidly in fish tissues upon metal exposure. Consequently, these findings have led to the proposed use of fish MTs as potential biomarkers for metal pollution and toxicity in aquatic environments (Langston *et al.,* 2002; Cho *et al.,* 2005; Bourdineaud *et al.,* 2006).

Biomarkers

Aquatic organisms accumulate pollutants from the ecosystem based on the level of the pollutants in the environment. Biomarkers are the biological indicators, which have the ability to provide specific information about the health of the ecosystem by measuring and characterizing an ecosystem or its components based on the specific biological signals from the biota (Bickham *et al.,* 2000). Sarkar *et al.* (2006) have reviewed the advantages of biomarkers over the traditional chemical analysis.

Depledge (1994) defined the term "biomarker" as "biochemical, cellular, physiological or behavioral variations that can be measured in tissue or body fluid samples, or at the level of whole organisms, to provide evidence of exposure and/or effects from one or more contaminants". Lam and Gray (2003) defined biomarkers as "quantitative measures of changes in the biological system that respond to either (or both) exposure to, and/or doses of substances that lead to biological effects and are potential tools for detecting responses at different levels of biological organization". The monitoring of biological effects has recently become an integral component of environmental monitoring programs as a supplement to the commonly used contaminant/chemical monitoring.

According to Narbonne *et al.* (1999) and Picado *et al.* (2007) the use of multiple biomarkers instead of single biomarker can ensure more detailed assessment about the effects of different pollutants present in the aquatic environment. The use of multiple biomarkers for the purpose of pollution monitoring leads to the development of the concept of global biomarker index (BI). BI is calculated as the sum of the individual biomarkers measured. The value of BI would be in direct proportional to the level of pollutants in the ecosystem. The battery of biomarkers used for environmental biomonitoring programs usually includes biomarkers of exposure, *i.e.,* parameters whose changes can be related to the organism's exposure to a specific class of pollutants (Viarengo *et al.,* 2007).

Criteria for the Selection of Biomarkers

The following criteria have been suggested by Mayer *et al.* (1992) for the selection of the most functional and appropriate biomarkers to use: 1) easy to quantify, 2) responds in a dose or time dependent manner to the toxicant and 3) sensitive and biologically important. Viarengo *et al.* (2007), have suggested the following biomarkers in the pollution monitoring: 1) molecular as well as cellular biomarkers which are rapidly activated and have the capability to give early warning signals against the effects of pollutants on the organisms, 2) biomarkers which can assess pollutants at the tissue level and 3) biomarkers at the organism level which can indicate potential survival capacity and reproductive performance of the exposed organisms.

The studies on the temporal relationships of biomarker are also recommended in addition to the dose response studies for the proper confirmation of biomarker concept (Depledge, 1994). Biomarker responses at sub-lethal level of exposure should be quantifiable sufficiently before the initiation of signals that leads to death (Depledge, 1994; Hagger *et al.,* 2006). Clements (2000) and Chambers *et al.* (2002) pointed out that molecular biomarkers are sensitive tools against pollutants as molecular responses are quick compared to other biomarkers.

Types of Biomarkers

Biomarkers can directly contribute in detecting, quantifying and understanding the significance of exposure to chemicals in the environment. These measurements may also help to assess the potential for human exposure to environmental pollutants and for predicting the human health risks. Chambers *et al.* (2002) have classified biomarkers into 3 types, *viz.*, 1) biomarkers of exposure, 2) effect and 3) those of susceptibility.

Biomarkers of exposure have the ability to measure effects of short lived chemicals and to replace the expensive chemical analysis (Hagger *et al.*, 2006). The biomarkers of effect are coupled exclusively with the mechanism of action of pollutants and are characterized to co-relate the extent of modifications of biomarker to the level of adverse effects (Chambers *et al.*, 2002). It can also provide details regarding the basics about ecological consequences of the hazards, depending on the level of specificity of the biomarker (Hagger *et al.*, 2006). Biomarkers of susceptibility differ in its features from biomarker of effect or exposure, as they do not represent the stages of dose–effect continuum but reveal an increase in the rate of transition between steps along that continuum (Dethloff *et al.*, 1999).

Metallothioneins as Biomarkers

Metallothioneins (MTs) are a class or family of low molecular weight (6,000 to 7,000 Da), cysteine rich (20 residues per protein molecule; which is 25-30 per cent of total amino acids of MT protein) and heavy-metal-binding cytosolic protein found in most major groups of organisms. They are often associated with cadmium, zinc, mercury or copper ions (Hamer, 1986; Kagi and Kojima, 1987; Chan, 1995; Kalpaxis *et al.*, 2004; Bae *et al.*, 2005; Atif *et al.*, 2006). Their presence has been established in animals, plants, fungi, protists (Piccinni *et al.*, 1994) and prokaryotes (Kagi, 1993). MTs may be induced by a variety of stimuli, including elevated concentrations of essential (Cu, Zn) and non-essential heavy metals (Cd, Hg, Pb), glucocorticoids hormones, inflammatory agents and a variety of stress conditions (Kagi, 1993; Petering *et al.*, 2006).

MT proteins have very distinct properties such as very high cysteine content; lack of aromatic amino acids and histidine (Hamer, 1986; Kagi and Kojima, 1987; Petering *et al.*, 2006). The cysteine residues are distributed in typical motifs consisting of CC, CXC or CXXC sequences (where C is Cysteine and X is an amino acid other than cysteine) (Kagi, 1993). MT peptides from diverse species of crab, oyster, fishes and mammals shows extensive homology among amino acid sequences (Hamer, 1986; Kagi and Schaffer, 1988; Unger *et al.*, 1991; Ma *et al.*, 2009). Therefore, it is suggested that MT has a conservation of heavy metal binding function throughout the evolution. The members of the MT protein superfamily have evolved through rounds of duplication and loss events leading to the current heterogeneous scenario. Difficulties in drawing straight evolutionary relationships could be in part attributed to the lack of detailed comparative studies (Trinchella *et al.*, 2012).

History of Metallothioneins

In terms of protein metabolism induced by heavy metals, metallothioneins (MTs) are the most widely studied proteins. Margoshes and Vallee (1957) while doing fractionation of horse kidney cortex with ethanol and ammonium sulfate obtained a product containing 20 to 25 milligrams of cadmium per gram dry weight of trichloroacetic acid precipitable material. The features like low sedimentation constant and high metal content of this "special material" was identified by colorimetric and emission spectrographic analysis. They suggested that these features are the indications for a low molecular weight protein, probably containing a small number of cadmium atoms per molecule.

The primary purification of MT from equine renal cortex by Kagi and Vallee (Kagi and Vallee, 1961) yielded preparations that were electrophoretically heterogeneous and contained 2.9 per cent cadmium, 0.6 per cent zinc and 4.1 per cent sulfur. They also reported the concentrations as 5.9 per cent cadmium, 2.2 per cent zinc and 8.5 per cent sulfur by using diethylaminoethyl cellulose column chromatography. They suggested that most of the sulfur in metallothionein is accounted for by its cysteine content.

From the detailed study on the absorption spectra of MT protein, Kagi and Vallee (1961) speculated that, Zn^{2+} and Cd^{2+} ions seem to compete with each other for the metal-binding sites of the protein. Metallothionein is formed through the interaction of one atom of cadmium or zinc with three sulfhydryl groups (SH) of cysteine. Three hydrogen ions are displaced from protein for each Cd^{2+} or Zn^{2+} atom bound. The displacements of hydrogen ions are from SH group of cysteine residues of MT protein. The strict dependence of these reversible changes on the Zn^{2+} ion concentration indicates that cadmium is bound approximately 3000 times more firmly than zinc. Cadmium and zinc compete for the same binding site and may be isomorphic in metallothionein. Although the sum of their concentrations in the native protein is constant, the two metals replace one another when metallothionein is dialyzed against Cd^{2+} on Zn^{2+} ions, resulting in cadmiumthionein and zincthionein respectively.

Approximately one of every three or four amino acids in this small protein molecule is a cysteine residue. The absence of tyrosine and tryptophan accounts for the lack of absorption at 280 nm. The large number of cysteine residues in the primary structure of thionein must condition its conformation. Moreover, the arrangement of three –SH groups to give a steric relationship which permits the coordination of one cadmium or zinc atom, severely restricts and circumscribes the number of possible conformational states by limiting the number of degrees of freedom of the molecule. The data on cysteine and metal composition disallows a large number of possible structures of this protein (Kagi and Vallee, 1961).

Classification of Metallothioneins

The first workshop on the metallothionein research was held in Zurich in 1979 (Kagi and Nordberg, 1979). For a long time after the first workshop, roman numbers like MT-I and MT-II were used to identify different MTs. The nomenclature system was updated in 1987 (Fowler *et al.*, 1987) by dividing all MTs into three classes. Class I included all MTs with locations of Cys closely related to those to equine MT. Class II grouped all MTs lacking or distantly related to this characteristic feature of class I. Class III is comprised of phytochelatins and vegetal γ glutamyl peptides of enzymatic origin. So, class III is not direct gene products. MT genes from various organisms are now sequenced and a new nomenclature system was developed by Binz and Kagi in 1999, which uses Arabic numbers. The official designation in the SwissProt data base and the MCBI (Medical Center for Biotechnology Information) data base uses Arabic numbers, *e.g.*, MT-1 and MT-2. However, in both data bases, older designations like MT-I or MT-II are also given as synonyms along with Binz and Kagi nomenclature system (Nordberg and Nordberg, 2009).

The classification system of Binz and Kagi (1999) is based on sequence similarity and phylogenetic relationships. In this system, MTs are subdivided into 16 MT families (vertebrate, mollusc, crustacean, echinodermata, diptera, nematoda, ciliata, fungi-I, fungi-II, fungi-III, fungi-IV, fungi-V, fungi-VI, prokaryota, plant, phytochelatins and other non-proteinaceous MT-like polypeptides). Within each family, subfamilies may be recognized, sharing a more stringent set of phylogenetic features. The vertebrate MT family consists of 12 subfamilies (m1: mammalian MT-1, m2: mammalian MT-2, m3: mammalian MT-3, m4: mammalian MT-4, m: mammalian MT, a1: avian MT-1, a2: avian MT-2, a: avian MT, r: reptilian MT, t: teleost MT, s: sharks MT, b: batracian MT). Although this classification system allows further division of subfamilies into subgroups, this has not been done for teleost MTs in contrast to the m1 subfamily of the mammals. Isoforms or allelic forms of MT are specifiable as members of subgroups, subfamilies and families. The isoforms are named as MT-A, MT-B, MT1-A, MT1-B, MT2-A, MT2-B etc. (Binz and Kagi, 1999). But Knapen and co-workers (2005) suggested that, teleost MTs should be classified in a different way. For this purpose, a phylogenetic analysis of all currently known teleost MT sequences is needed, preferably supplemented with new sequences from a wide range of taxa.

Metallothioneins of Fishes

The study of fish MTs, in particular, is interesting in the perspective of ecology, because aquatic organisms are the primary targets of toxic environmental pollutants such as Zn, Cu, Hg and Cd from mining and industrial activities. Fishes are one among those organisms having a top position in the aquatic food chain and also showing susceptibility to heavy metal pollution. They may reflect the extent of the biological effects of metal pollution in waters. Klaverkamp *et al.* (1984) have suggested that, in fishes pre-exposure to sub-lethal levels of metals results in the induction of metallothioneins (MTs) and can result in acclimatization to a potentially toxic level of these metals in waters.

According to Chan (1995), the potential use of MTs as a biomarker for monitoring metal pollution in fish focuses on using MTs as an indicator of sub-lethal concentrations of a mixture of metal ions and the biological significance of such metal ions. In contrast, the measurement of environmental or tissue concentrations of a variety of metal ions does not reflect the biological response or significance of the accumulated metals. MT genes have been isolated from a number of teleost fishes belonging to a wide array of taxa.

Metallothionein Gene Structure, Expression and Regulation

Investigating the structure of MT genes is important for a better understanding of both the physiological role, as well as the use of MTs as a biomarker. Although MT-like proteins have been identified in several fish tissues and cell lines (Bonham and Gedamu, 1984; Hamilton and Mehrle, 1986), no information about the structure of MT genes in a major taxonomic group like fish has been available until Zafarullah and Gedamu (1988) published the structure of the rainbow trout (*Salmo gairdneri*) MT gene. In rainbow trout, 2 isoforms of the MT gene was reported by Bonham and Gedamu (1984) and Bonham *et al.* (1987) after chromatographic studies of the proteins and cloning of MT cDNA. In later years, the MT gene is sequenced from a few teleost fishes.

The studies on the MT genes of the fishes revealed that it is interrupted by the presence of 2 introns with GT-AG splicing signals and is coding for 60 amino acid protein (Breathnach and Chambon, 1981; Hamer, 1986; Bonham *et al.,* 1987; Kay *et al.,* 1991; Cho *et al.,* 2009). Zafarullah and Gedamu (1988) reported that the 3'-untranslated region of the gene, contains two possible polyadenylation signals with the AATAAA and ATTAAA sequences. Kille *et al.* (1991) have suggested that the partial regression of the first polyadenylation signal permits mRNAs to be more stable for the reliable MT production. Knapen *et al.* (2005) have supported this hypothesis even in the absence of any conformational data to substantiate it, but they suggested that to find out the possible relationship between the protective mechanism of MT and differences in the 3' UTR length, a detailed study is to be carried out in future.

Various studies on the MT gene regulation revealed that, the synthesis of MTs is induced through transcriptional activation systems, composed of metal response elements (MREs), antioxidant response elements (AREs) and glucocorticoid response elements (GREs) in the enhancer/promoter region of the MT gene and direct or indirect induction of MRE/ARE/GRE-binding proteins by a variety of heavy metals, glucocorticoid hormones, interferon, interleukin-I, bacterial endotoxin, UV radiation, cytokines and oxidative stress. MT synthesis is controlled by a zinc-activated transcription factor called metal response element binding transcription factor-1 (MTF-1). MTF-1 is a multiple Zn finger protein; the only known mediator of the metal responsiveness of MT. Both Zn and Cd are potent inducers of MT gene transcription, whereas Cu is a relatively poor inducer. Cd and Cu promote transcription of MT either by displacement of Zn from cellular binding ligands, such as ZnMT, or by other indirect mechanisms. Induction occurs through activation and binding of the MTF-1 to a unique cis-element (MREs), which plays a central role in regulating both basal and metal-induced MT gene expression. Number of MRE may vary from isoforms to isoforms or species to species (Zafarullah and Gedamu, 1988; Harford and Sarkar, 1991; Dalton *et al.,* 1996; Bittel *et al.,* 1998; Syring *et al.,* 2000; Ghoshal and Jacob, 2001; Sato and Kondoh, 2002; Laity and Andrews, 2007; Cho *et al.,* 2009).

Cho *et al.* (2009) have computationally predicted various transcription factor binding sites in the two mud loach MT promoters. These included: binding motifs for activation protein 1 (AP-1), GATA factors and hepatocyte nuclear factor 3 and 5 (HNFs). In addition, some immune modulating or stress-related transcription factors such as signal transduction and activation of transcription (STAT), nuclear factor for activated T-cells (NF-AT) and heat shock factor (HSF) were also predicted. These findings agree with the previous observations on the immunological and/or pathological involvement of MT gene expression in vertebrates by Thirumoorthy *et al.* (2007). Cho *et al.* (2009) suggested further requirements of research to determine the functional role of each of the above-listed motif or element to gain a deeper insight into the transcriptional regulation of the MTs.

Zafarullah and Gedamu (1988) studied the 5'-flanking region of rainbow trout MT gene which showed strong homologies with MREs of mammalian MT genes. They also suggested that, this homology may indicate the similar (although not necessarily identical) mechanisms evolved for the regulation of MT genes in these organisms. The structure of MT and its functions in different marine animals has been reviewed by Roesijudi (1996), Formigari *et al.* (2007) and Wang and Rainbow (2010).

Structure of Metallothioneins

Typically fish MTs are 60 amino-acid polypeptides that wrap 6–7 divalent heavy metal ions at highly conserved cysteine residues (Vallee and Auld, 1993; Chang and Huang, 1996). Kay *et al.* (1991) have described the structure of MTs and found that, the major structural constraint of protein is represented by the arrangement of the 20 cysteines and by the well-defined architecture of the metal-thiolate clusters. The residues placed between the cysteine residues may also affect the structure of the protein. The high similarity, typical of vertebrate MTs allowed the grouping of MT proteins in a single gene family. However, as a result of mutational events, a number of sequence-specific characters appeared during evolution, resulting in a moderate variability of the amino acid residues other than cysteines in MTs of different organisms. Among vertebrates, the amino-terminal region exhibits typical species-specific characteristics.

Capasso *et al.* (2003) studied the structure of MT from the Antarctic fish (*Notothenia coriiceps*) and found that MT exist in the cell as metal-free form (called apo-MT or thionein) following *de novo* synthesis, prior to post-translational metallation. The apo form is often referred to as a random coil due to the lack of secondary structural features such as α-helices or β-pleated sheets. After structural studies of metal free form of MTs, Rigby and Stillman (2004) reported that, it is highly probable that apo-MT is not completely a random coil, but instead possesses some structural characteristics that facilitate the coordination of metal ions in a sequential manner leading to metal–thiolate clusters with the same connectivities. They also reported that, formation of the secondary structure is almost exclusively dependent on metal-coordination and the three-dimensional shape is determined by the network of metal-thiolate bonds in the metal binding site. Jiang *et al.* (1994) have studied the Cd, Hg, Zn and Cu thiolate clusters of MTs using spectroscopic tools and found that, MTs binds not only Zn and Cd, but also with Cu, and they suggested the possible role for metallothioneins in essential-metal metabolism by transferring MT-bound metals to other metal free or apo-metallo-proteins. They also documented the compositional changes in metallothionein, typically in the form of Zn replacement by the higher affinity metals Cd, Cu and Hg.

The structure of MTs was first described by Winge and Miklossy (1982). Vertebrate MTs are characterized by the well-known dumbbell shape, composed of two globular domains connected by a flexible linker consisting of a Lys-Lys segment. The 20 cysteinyl sulfurs of MT function as bridging and terminal ligands for the coordination of seven divalent metal ions. The metal–thiolate clusters form into two well-defined domains, named β (N-terminal) and α (C-terminal), with stoichiometries of M_3S_9 and M_4S_{11} (M stands for metal and S stands for sulfur of cysteines) respectively, for divalent metal ions. Hamer (1986) studied the transcription of MT proteins and found that, as in mammalian MTs, the β domain is encoded by exons 1 and 2, while α domain is represented by exon 3. Gehrig *et al.* (2000) studied the metallation stages of MT and found that *in vitro* Cd binding to apo-MT has been shown to involve cooperative binding to α domain, followed by binding to β domain. However, the mechanism for *de novo* metallation of metallothionein is unknown. MT exhibits an unusual primary sequence and unique 3D arrangement. Rigby and Stillman (2004) reported that, positioning the cysteine side chains on the exterior of the MT proteins gives the hydrophilic groups access to the surrounding environment for rapid scavenging of metal ions for the achievement of the optimal metallated conformation. This protects the protein from oxidation of the thiol groups, as well as from proteolytic digestion.

Harford and Sarkar (1991); Kagi (1991) and Rigby and Stillman (2004) have studied the metallation and folding pattern of MT peptide and found interesting variations in the usual folding pattern of proteins. The usual folding events of proteins are either spontaneous by localized secondary structure formation or by the assistance of folding chaperones while the polypeptide chain is still attached to the ribosome. Metallothionein, however, differs from these type of proteins in that the folding mechanism is metal-dependent. The clustered network of metal-thiolate bonds among the sulfur atoms of the cysteine residues and the metal ions, present within α and β domain consist of both bridging and terminal sulfur ligands. These metal–sulfur connections cross-link to opposite ends of the polypeptide chain, therefore, require the translation of the entire protein sequence before initiation of the metallation and subsequent folding events. As a result, the nascent chain of metallothionein must initially exist in the cell as the highly reactive, metal-free form. It would normally be expected that the metal-free protein would only exist for a short time under these conditions and it must be concluded that the conformation of the nascent peptide is highly reactive towards metallation. Indeed, metallation with 3 or 4 divalent metals or up to 6 monovalent metals must take place before degradation of the peptide begins. Otherwise, the bridging between cysteines located at either end of the peptide chain cannot take place. Thus, long-term retention of metal ions in the cell is achieved not by a long life time of the protein, but by rebinding of the released metal ions from protein degradation by the *de novo* synthesized apo-metallothionein.

Zhou *et al.* (2000); Zangger and Armitage (2002) and Vergani *et al.* (2005) have studied the domains of MT proteins and found that, zinc is preferentially located in the β domain and cadmium in α domain. Therefore, α domain may play a central role in heavy metal detoxification, whereas β domain would regulate copper and zinc homeostasis. α Domain is involved in oxidative dimerization, whereas the loosely structured β domain is responsible for metal-bridge dimerization. Oxidative dimers are reduced by reducing compounds, whereas metal bridge dimerization is reversed by dilution or addition of chelating agents. Oxidative dimerization may also occur *in vivo* under conditions of stress, such as exposure to reactive oxygen species and toxic metals.

Capasso *et al.* (2003) have differentiated the MTs of teleosts from other organism's MTs, for two peculiar features: the shift of a C-terminal cysteine and a smaller number of CK (KC) motifs (where K is lysine). They also pointed out that, they have different thermostability, different metal affinity and different redox properties with respect to mammalian MTs. All vertebrate MTs display the same arrangement of the cysteine-containing motifs, with the exception of fish MTs, in which the last CXCC motif in α domain becomes CXXXCC (Scudiero *et al.*, 1997; Knapen *et al.*, 2005). Such a synapomorphism may have important structural and possibly, functional consequences in fish MTs (Capasso *et al.*, 2003).

General Functions of Metallothioneins

Metallothioneins have been isolated from various organisms including vertebrates, invertebrates and fungi. The conserved nature and ubiquitous synthesis of MTs in a wide range of organisms suggests that these proteins have potentially significant biological functions in these organisms. The role played by MTs has been debated for a long time, but the high thermodynamic and low kinetic stability suggest that they may bind the metal moiety with a considerable stability, providing at the same time a facile metal exchange with other proteins (Templeton and Cherian, 1991). Otvos *et al.* (1993) have suggested the non-catalytic features of MT proteins. From the various studies about the function of the MT in living organisms, it has been proposed that MTs may be involved indirectly or directly in numerous cellular processes such as scavenging of free hydroxyl ions and oxyradicals, inflammation and infection, Zn and Cu homeostasis, free radicals and ionizing radiations, detoxification/protection against stress-causing agents like toxic heavy metals (Cd and Hg), growth and development (Hamer, 1986; Kagi and Schaffer, 1988; Sato and Bremner, 1993; Masters *et al.*, 1994; Lazo *et al.*, 1995; Brouwer, 1996; Roesijadi, 1996; Viarengo *et al.*, 2000; Coyle *et al.*, 2002; Knapen *et al.*, 2004).

MTs are proposed to act as the major intracellular chaperones for essential metal ions. Each monomer cooperatively binds seven divalent zinc cations in two unique cluster motifs, involving sulfhydryl groups from twenty cysteine residues (Otvos *et al.*, 1993). The binding of toxic metals to MT represents a sequestration function that renders these metals unable to interact with other proteins, such as enzymes, and thereby offers a protection against metal toxicity at the cellular level (Chowdhury *et al.*, 2005). As discussed above, it has been pointed out that MTs play a number of functions in relation to heavy metal cation homeostasis. Moreover, they display reactive oxygen species (ROS) scavenger activity as part of the antioxidant defense system of the cells (Sato and Bremner, 1993; Viarengo and Nott, 1993; Viarengo *et al.*, 1999). They act as regulators of the activity of Zn finger proteins in modulating gene expression (Zeng *et al.*, 1991; Roesijadi *et al.*, 1998) and play an important role in intracellular trafficking of Zn (Feng *et al.*, 2005). Hamilton and Mehrle (1986); Hodson (1988) and Chan (1995) have studied the heavy metal binding activities of MTs and found that, at low doses metal ions bind to MTs, whereas at higher doses the metal ions 'spill-over' to bind to high molecular-weight proteins and cause tissue damage. Obviously, faster the MT is produced, less the chance for the trace metal to 'spill over' to other physiologically important intracellular proteins.

The role of MTs in apoptosis has been aggressively investigated in mammals and the vast majority of studies show that MT plays a protective role with respect to apoptosis (Shimoda *et al.,* 2003; Santon *et al.,* 2004). However, many possibilities are likely, since MT-I and II are indeed multipurpose proteins involved in a variety of functions, which also include protein–protein and protein–nucleotide interactions, immune defense responses, regulation of Zn fingers and Zn-containing transcription factors, thermogenesis, mitochondrial respiration, body energy metabolism, cell cycle progression, angiogenesis, cell differentiation and survival (Cherian and Apostolova, 2000; Chung and West, 2004). Some of MT actions may have therapeutic relevance in a range of acute and chronic disorders, in which inflammation and oxidative stress play a central pathophysiological role (Allan and Rothwell, 2003).

Metallothioneins as Biomarkers in Biomonitoring Assays

Numerous methods have been developed to monitor heavy metal contamination of aquatic organisms, including the analyses of sediment and water of affected ecosystem. It is difficult to monitor the pollution level, when analyzing long-standing accumulated pollutants, especially when at very low levels of exposure. According to Giguere *et al.* (2003), the use of toxicant-specific biomarkers such as metallothionein has been widely employed to indicate the presence of heavy metals. On the other hand, since several pollutants can modify directly or indirectly the balance between the concentration of pro-oxidants and antioxidants, the determination of oxidative stress (DNA damage, protein oxidation, lipid peroxidation) and/or antioxidant responses in aquatic species is commonly employed as a non-specific biomarker (Bainy *et al.,* 1996; Geracitano *et al.,* 2004).

According to Tom and Auslander (2005), contaminant-dependent patterns should include dose response to model contaminants. Contaminant-independent reference expression patterns should include natural fluctuations of the biomarker level. Exposure and toxic effects of contaminants, measured in terms of the molecular responses such as transcript level variation of the gene of the organisms is called "molecular biomarkers" and if the toxicant's effect measured in terms of the biochemical responses and functioning in the cell, it can be defined as "biochemical biomarkers". Interactions of contaminants with organisms occur initially at the molecular level, and induction of MTs is an early response to contaminant exposure. The levels of contaminant-affected gene products (transcripts and proteins) are increasingly utilized as environmental biomarkers. Quantitative reverse transcriptase polymerase chain reaction (qRT-PCR) and competitive enzyme-linked immunosorbent assay (ELISA) are suggested as preferred evaluation methods for transcript and protein, respectively. The large number of isoforms and sub isoforms within the MT family makes application of immunological techniques to biological samples difficult. Quantification of the gene expression using qRT-PCR is advantageous over ELISA, as in qRT-PCR accurate measurement of mRNA level is possible by employing isoforms specific primers. MT level has become a major molecular and biochemical biomarker for monitoring metal pollution in fish. Because of the specificity and ability of MTs to bind metal ions, MTs are an ideal biomarker of exposure to monitor metal contamination in fish (Chan, 1995; Cajaraville *et al.,* 2000; Dallinger *et al.,* 2000; Ciocan and Rotchell, 2004; Petering *et al.,* 2006; Ma *et al.,* 2009).

Tom and Auslander (2005) pointed out the significance of gene expression studies on the development of molecular biomarkers. According to them, the expression of a broad selection of genes is affected by a variety of environmental stimuli. Therefore, changes of expression levels, measured in a native sentinel species may serve as biomarkers for the effect of present and future environmental perturbations. Various studies on the MT proteins and gene expressions explained the inducibility of MT by heavy metal exposure. The synthesis of MT protein and MT mRNA in fish tissues is inducible by trace metals including the essential elements zinc and copper and non-essential elements lead, mercury and cadmium. The identification of MT genes in a broad range of species, combined with measuring MTs at the protein level, will allow gene expression studies which will lead to a better insight in the relative importance of transcriptional and translational processes in metal-induced protective mechanisms (Chan *et al.*, 1989; Olsson, 1993; Ren *et al.*, 2000; Vasconcelos *et al.*, 2002).

There is much interest in molecular biomarkers for the monitoring of environmental contamination using MTs (Das and Kaviraj, 1994; Pedersen *et al.*, 1997; Dallinger *et al.*, 2000; De Lafontaine *et al.*, 2000; Vasak and Hasler, 2000; Chowdhury *et al.*, 2005). Metallothioneins (either MT protein or MT mRNA, or both) as a biomarker of exposure is attractive because it is easily quantified by using sensitive molecular biology techniques, such as quantitative polymerase chain reaction (qPCR), reverse transcription-quantitative competitive PCR (RT-qcPCR) and immune assay, and it can provide sub-lethal diagnosis of contamination and evaluation of sub-lethal exposure to metal ions (Chan, 1995; Ackermann *et al.*, 2002; Inui *et al.*, 2003; Kamer *et al.*, 2003; Tom *et al.*, 2003; Funkenstein *et al.*, 2004; George *et al.*, 2004; Hayes *et al.*, 2004; Tom *et al.*, 2004).

Winge *et al.* (1974) hypothesized that pathological effects of heavy metals occur when the amount of metal entering the animal exceeds the ability of MTs to bind and detoxify it. Roch *et al.* (1982) suggested the use of hepatic MTs levels for monitoring the degree of pollution in water. Molecular and immunological probes of MTs is useful in testing the feasibility of using MT levels as a biomarker for monitoring metal pollution and determining the basic physiological significance of MTs in metal tolerance in fishes. From the various experimental evidences, it has been established that the hepatic levels of MTs in fish are dose dependently increased after exposure to cadmium, mercury, copper and zinc (Hogstrand *et al.*, 1989; Hogstrand and Haux, 1991; Atli and Canli, 2008; Navarro *et al.*, 2009; Ghedira *et al.*, 2010; Banni *et al.*, 2011; Monteiro *et al.*, 2011; Sampaio *et al.*, 2012). It is also well characterized that MTs are induced by metal in laboratory experiments (Hamilton and Mehrle, 1986; Hogstrand and Haux, 1991; Sinaie *et al.*, 2010; Shariati *et al.*, 2011; Hauser-Davis *et al.*, 2012; Kim *et al.*, 2012). From field studies, Roch *et al.* (1982); Roch and McCarter (1984); Olsson and Haux (1986); Montaser *et al.* (2010); Roy *et al.* (2011) and Krasnici *et al.* (2012) demonstrated that MTs in fishes was induced by exposure to a mixture of metal ions in contaminated water. The details regarding the various studies for the development of MTs as a molecular and biochemical biomarker in fish are given in Tables 15.1–15.4. These results indicate the importance of MTs for biomonitoring of heavy metal contaminated water bodies.

Table 15.1: Levels of MT Proteins in Response to Induced Heavy Metal Stress in Fishes

Sl.No	Name of the Fish (s)	Heavy Metal(s)	Dose(s) Selected	Tissue (s) Studied	Time of Exposure	Significance/Level of MT	References
1.	*Piaractus mesopotamicus*	Cu	400 µg/L	Gills	48 hr	Double	Sampaio *et al.,* 2012
2.	*Oreochromis niloticus*	Cu	2 mg/ml	Bile	96 hr	Double	Hauser-Davis *et al.,* 2012
3.	*Prochilodus lineatus*	Pb	5 mg/L	Liver	6, 24, 96 hr	Significant in 6 and 24 hr	Monteiro *et al.,* 2011
4.	*Oncorhynchus mykiss, Lepomis gibbosus*	Cu, Cd	2.5, 5.0 mg/L (Cu) 40 µg/L (Cd)	Liver	3 weeks (Cu) 2 weeks (Cd)	Significant	Gonc and Magalha, 2011
5.	*Thalassoma pavo*	Cd	10, 40, 60, 120mM (mg/L)	Gills	48, 96, 192 hr	Expression proportional to time and dose	Brunelli *et al.,* 2011
6.	*Acipenser persicus*	Cd	50, 400, 1,000 µg/L	Gills, liver, kidney	1, 2, 4, 14 days	Gills - less Liver-constant Kidney- significant	Shariati *et al.,* 2011
7.	*Sparus aurata*	Cu, Cd	500 µg/kg	Liver, gills, kidney	2 days	3.56 and 3.3 fold respectively	Ghedira *et al.,* 2010
8.	*Scatophagus argus*	Hg	10, 20, 30 µg/L	Gills, liver	24, 48, 72 hr	Gills-significant in 30 µg/L at 72 hr. Liver – proportional to time and dose	Sinaie *et al.,* 2010
9.	*Oreochromis niloticus*	Cu, Zn, Cd, Pb	5, 10 and 20 µM	Liver	14 days	Significant	Atli and Canli, 2008
10.	*Oreochromis niloticus*	Cd	0.001, 0.005, 0.01, 0.1, and 1 mg/L	Liver	7, 14, 21, 28 days	Increased 2-26 fold depending on the exposure and duration	Chandrasekera *et al.,* 2008
11.	*Dicentrarchus labrax*	Cu, Cd, Hg	50–250 µg/kg	Liver	48 hr	Significant increase	Jebali *et al.,* 2008
12	*Scophthalmus maximus*	Cd, Cu, Zn	8.9, 89 µM Cd 15.26,152.6µµ Cu 15.3,153 µM Zn	Gills Chloride cells	7 days	Significant.	Alvarado *et al.,* 2006
13.	*Seriola dumerilii*	Cd	50, 100, 250 µg/kg	Liver	2 days	Significant	Jebali *et al.,* 2006

Cu: Copper; Pb: Lead; Cd: Cadmium; Hg: Mercury; Zn: Zinc; hr: Hour.

Table 15.2: Field Studies on the Levels of MT Proteins in Fishes in Response to Heavy Metal Stress

Sl.No.	Name of Organism (s)	Studied Tissue (s)	Significance Based on Protein Level	References
1.	*Mullus barbatus*	Liver	Insignificant	Martinez-Gomez *et al.*, 2012
2.	*Oreochromis niloticus*	Liver, Gills	Significant	Carvalho *et al.*, 2012
3.	*Squalius cephalus*	Liver	Significant	Krasnici *et al.*, 2012
4.	*Dicentrarchus labrax, Solea senegalensis, Pomatoschistus microps*	Liver	Significant	Fonseca *et al.*, 2011
5.	*Dicentrarchus labrax, Liza aurata*	Gills, blood, liver, kidney, muscle, brain	Insignificant	Mieiro *et al.*, 2011
6.	*Labeo rohita, Catla catla, Cirrhinus mrigala*	Liver, gills, muscle, brain	Significant	Roy *et al.*, 2011
7.	*Deania hystricosa, Etmopterus princeps, Hydrolagus pallidus*	Liver, Gills, Muscle	Insignificant	Company *et al.*, 2010
8.	*Clarias gariepinus*	Liver	Significant	Mdegela *et al.*, 2010
9.	*Liza aurata*	Liver	Significant	Oliveira *et al.*, 2010
10.	*Cyprinus carpio*	Liver	Significant	Falfushynska and Stoliar, 2009
11.	*Squalius cephalus*	Gill	Insignificant	Dragun *et al.*, 2009

Table 15.3: Expression of MT Genes in Fishes in Response of Heavy Metal Stress

Sl.No	Name of the Fish (s)	Heavy Metal(s)	Dose(s) Selected	Tissue (s) Studied	Time of Exposure	Significance/Level of MT mRNA	References
1.	*Pelteobagrus fulvidraco*	Cd, Cu, Zn	Cd – 5 ppm Cu – 1 ppm Zn – 1 ppm	Liver, gonad, skin Kidney, brain, gills, intestine, muscle	6, 12, 24, 48, 96 hr	Significantly increase in liver tissues at 48 hr, followed by gill at 12 hr and intestine at 48 hr after Cd exposure.	Kim *et al.*, 2012
2.	*Sparus aurata*	Cd	0.1, 5 10 mg/L	Larvae	6 days	Significant	Sassi *et al.*, 2012
3.	*Oncorhynchus kisutch*	Cd	3.7, 347 ppb	Liver, olfactory, gills	8, 24, 48 hr	Significant in 347 ppb exposure	Espinoza *et al.*, 2012
4.	*Oncorhynchus mykiss*	Co, Zn	10 mg/L - Co 1 mg/L - Zn	Muscle	6, 12, 24, 48 hr	Expression increase with time	Ceyhun *et al.*, 2011
5.	*Danio rerio*	Cd	4 mg/L	Ovary, liver	21 day	Significantly increased than non-exposed	Banni *et al.*, 2011
6.	*Danio rerio*	Cd	0.4 mg/L	Ovary	21 day	Significantly increased than non-exposed	Chouchene *et al.*, 2011
7.	*Cyprinius carpio*	Hg	20 µg/kg	Kidney, liver	72 hr	Significant in kidney	Navarro *et al.*, 2009
8.	*Siniperca chuatsi*	Cd	1 mg/L	Gill, muscle, liver, trunk kidney, brain	7 days	significantly increased in the brain and gills	Gao *et al.*, 2009
9.	*Cyprinus carpio*	Cd	10 mg/L	Liver, brain, kidney, heart, muscle	24, 48, 96 hr	Significant level in the brain, and the lowest in the kidney and the liver	Ferencz and Hermesz, 2008
10.	*Sparus sarba*	Cd	50 µg/L, 200 µg/L	Liver	5 day	Significant	Man and Woo, 2008
11.	*Sparus sarba*	Cd	0.1mg/Kg, 1.0 mg/kg	Liver	5 day	Significant	Man and Woo, 2008
12.	*Takifugu obscures*	Cd	5 ppm (for spatial expression study)	Brain, gills, intestine, kidney, liver, muscle	6, 12, 24, 48, 96 hr	Significant time dependent expression	Kim *et al.*, 2008

Contd...

Table 15.3—Contd...

Sl.No	Name of the Fish (s)	Heavy Metal(s)	Dose(s) Selected	Tissue (s) Studied	Time of Exposure	Significance/Level of MT mRNA	References
13.	Takifugu obscures	Cd	50, 250, 500, 1000, 5000 ppb (To study concentration dependent effect)	Brain, gills, intestine, kidney, liver, muscle	24 hr	Expression proportional to dose	Kim *et al.,* 2008
14.	Hemibarbus mylodon	Cd, Cu, Zn	Cd- 2, 5, 10 μm Cu, Zn - 2, 5, 10μM	Liver, kidney, gills	48hr	Cd induced more MT expression in liver.	Cho *et al.,* 2008

Cu: Copper; Co: Cobalt; Cd: Cadmium; Hg: Mercury; Zn: Zinc; hr: Hour; ppm: Parts per million; ppb: Parts per billion.

Table 15.4: Expression of MT Genes in Fishes in Response of Heavy Metal Stress in the Field

Sl.No.	Name of Organism (s)	Studied Tissue (s)	Significance Based on mRNA Level	References
1.	*Coris julis*	Gills	Significant in polluted site	Fasulo *et al.,* 2010
2.	*Naso hexacanthus*	Liver, Gills	Significant in polluted site	Montaser *et al.,* 2010
3.	*Aphanius fasciatus*	Liver	Expression level vary	Kessabi *et al.,* 2010
4.	*Cyprinius carpio*	Liver, kidney, muscles	Significant expression in kidney	Navarro *et al.,* 2009
5.	*Campostoma oligolepis, Lepomis megalotis, Hypentelium nigricans*	Liver	Insignificant	Schmitt *et al.,* 2007
6.	*Salmo trutta*	Liver, kidney	Significant	Hansen *et al.,* 2007
7.	*Gobio gobio*	Liver	Significant	Knapen *et al.,* 2007
8.	*Barbus graellsii*	Liver	Insignificant	Quiros *et al.,* 2007

Biomarkers have been suggested as practical tools for management of the environment for a number of decades, but their inclusion has not been universally accepted because of a number of unanswered questions regarding practicality, sensitivity and reproducibility. There are also reports on MT gene have not shown long duration in expression (George, 1989; Chatterjee and Maiti, 1991; Bae *et al.*, 2005). The data in some studies did not show a time/dose dependent increase in MT mRNA and protein, which supports the concept that MT mRNA induction may have limitations as a biomarker of metal contamination in aquatic environments (Schmitt *et al.*, 2007; Quiros *et al.*, 2007; Dragun *et al.*, 2009; Company *et al.*, 2010; Mieiro *et al.*, 2011; Martinez-Gomez *et al.*, 2012). These failures in application of MT suggest the importance in the selection of model organism for the ecotoxicological studies and calls for the in-depth studies during the development of a biomarker. Careful validation of the selected biomarkers in the field is necessary before its application in the monitoring programmes.

Fishes as Model Organisms

Aquatic organisms have the ability to accumulate contaminants from the environment where they live at much higher concentrations and at the same time, showing much less spatial and temporal variability. The evaluation of environment risk assessment should also take into account the effects on the biota (Cajaraville *et al.*, 2000). Aquatic organisms are more sensitive to exposure to toxicants or pollutants than terrestrial animals. Fishes are the one of the main uptake routes of metals into the human body (Ruelas-Inzunza *et al.*, 2010). The intake of trace elements in the human body does increase with fish consumption. Awareness about heavy metal concentrations and its biological effects in different fish species can diminish the hazards to public health due to the fish consumption from the polluted water bodies.

The use of fish in monitoring programs is believed to be of importance because of its key position in the trophic chain and their high commercial value. Heavy metal incorporation in fish tissues takes place through the gills, skin and food (Freitas *et al.*, 2008). Historically, species for biomarker studies have been chosen primarily for their ease of culture and sampling rather than for their ecological relevance. However, after different studies carried out in United Kingdom, Galloway *et al.* (2004) hypothesized that by selecting diverse phyla exhibiting different feeding strategies (filter feeding, grazing, omnivory, predation) and measuring suites of biomarkers at the suborganismal and physiological levels, the ecological relevance of pollutant exposures may be more readily determined and integrated into environmental management strategies. The significance in the use of fish as a model organism is proven by various water pollution monitoring programs. Fish have been employed as sentinel organisms in regular biomonitoring programs at international level (Program for the Assessment and Control of Marine Pollution in the Mediterranean (Med Pol, UNEP); the Convention for the Protection of the Marine Environment of the North-East Atlantic (OSPAR Convention)., etc.) (Viarengo *et al.*, 2007).

Ecotoxicogenomics

The term "toxicogenomics" was coined by Nuwaysir *et al.* (1999) to explain the application of genomics to toxicology. Since then, different genomic approaches through a combination of advanced instrumental, biological and bioinformatic techniques lead to the explosion of data concerning the biochemical and molecular status of organisms in relation to polluted environment (Mukhopadhyay *et al.,* 2003; Ankley *et al.,* 2006; Deane and Woo 2010; Leonardi *et al.,* 2010; Shved and Kumeiko 2011). The application of toxicogenomics in the field of ecology resulted in a new branch of toxicological study called "ecotoxicogenomics". The studies carried out in this direction are provided in Tables 15.3 and 15.4. The sequencing of stress related genes and quantification of its mRNA from more fish species will further strengthen the field of ecotoxicogenomics.

Conclusions

Heavy metals are one of the major causes of pollution of the aquatic environment. Heavy metals get bioaccumulated by aquatic organisms. The biomagnification of these compounds through the food chain leads to the development of serious health issues in all organisms including human. Biomarkers can be used to quantify the effects of pollutants in the biota precisely and cost effectively. The development of "MT biomarker" will help to assess the impact of heavy metal pollution in the organisms which live in polluted environments, even in the initial stages of pollution. Various studies conducted in different countries for a few decades suggested that, MTs have a role to play as biomarkers, if used sensibly in well-designed sampling programs. Careful choices need to be made on the selection of organism, choice of organ and method of analysis. A multi-biomarker approach including MT and other specific biomarkers of exposure may provide a better estimate of pollution than using a single biomarker.

Acknowledgements

The authors are grateful to the Ministry of Environment and Forests (MoEF), Government of India, for providing the financial support and to Christ College, Irinjalakuda, Kerala for providing facilities.

References

Ackermann, G.E., Schwaiger, J., Negele, R.D., and Fent, K. 2002. Effects of long-term nonylphenol exposure on gonadal development and biomarkers of estrogenicity in juvenile rainbow trout (*Oncorhynchus mykiss*). *Aquatic Toxicology,* **60**: 203–221.

Allan, S.M., and Rothwell, N.J. 2003. Inflammation in central nervous system injury. *Philosophical Transactions of the Royal Society B: Biological Sciences,* **358**: 1669–1677.

Alvarado, N.E., Quesada, I., Hylland, K., Marigomez, I., and Soto, M. 2006. Quantitative changes in metallothionein expression in target cell-types in the gills of turbot (*Scophthalmus maximus*) exposed to Cd, Cu, Zn and after a depuration treatment. *Aquatic Toxicology,* **77(1)**: 64–77.

Ankley, G.T., Daston, G.P., Degitz, S.J., Denslow, N.D., Hoke, R.A., Kennedy, S.W., Miracle, A.L., Perkins, E.J., Snape, J., Tillitt, D.E., Tyler, C.R., and Versteeg, D. 2006. Toxicogenomics in regulatory ecotoxicology. *Environmental Science and Technology,* **40(13)**: 4055–4065.

Atif, F., Kaur, M., Yousuf, S., and Raisuddin, S. 2006. In vitro free radical scavenging activity of hepatic metallothionein induced in an Indian freshwater fish, *Channa punctatta* Bloch. *Chemico-Biological Interactions*, **162**: 172–180.

Atli, G., and Canli, M. 2008. Responses of metallothionein and reduced glutathione in a freshwater fish, *Oreochromis niloticus* following metal exposures. *Environmental Toxicology and Pharmacology*, **25(1)**: 33–38.

Bae, H., Nam, S.S., Park, H.S., and Park, K. 2005. Metallothionein mRNA sequencing and induction by cadmium in gills of the crucian carp, *Carassius auratus. Journal of Health Science*, **51(3)**: 284–290.

Bainy, A.C.D., Saito, E., Carvalho, P.S.M., and Junqueira, V.B.C. 1996. Oxidative stress in gill, erythrocytes, liver and kidney of Nile tilapia (*Oreochromis niloticus*) from a polluted site. *Aquatic Toxicology*, **34**: 151–162.

Banni, M., Chouchene, L., Said, K., Kerkeni, A., and Messaoudi, I. 2011. Mechanisms underlying the protective effect of zinc and selenium against cadmium-induced oxidative stress in zebrafish *Danio rerio. Biometals: An International Journal on the Role of Metal ions in Biology, Biochemistry and Medicine*, **24(6)**: 981–992.

Bickham, J.W., Sandhu, S., Hebert, P.D.N., Chikhi, L., and Athwal, R. 2000. Effects of chemical contaminants on genetic diversity in natural populations: Implications for biomonitoring and ecotoxicology. *Mutation Research*, **463**: 33–51.

Binz, P.A., and Kagi, J.H.R. 1999. Metallothionein: molecular evolution and classification. In: *Metallothionein IV*, (Eds.) C.D. Klaassen. Birkhauser Verlag, Basel, Switzerland, p. 7–13.

Bittel, D., Dalton, T., Samson, S.L., Gedamu, L., and Andrews, G.K. 1998. The DNA binding activity of metal response element-binding transcription factor-1 is activated in vivo and in-vitro by zinc but not by other transition metals. *Journal of Biological Chemistry*, **273**: 7127 –7133.

Bonham, K., and Gedamu, L. 1984. Induction of metallothionein and metallothionein mRNA in rainbow trout liver following cadmium treatment. *Bioscience Reports*, **4**: 633-642.

Bonham, K., Zafarullah, M., and Gedamu, L. 1987. The rainbow trout metallothioneins: molecular cloning and characterization of two distinct cDNA sequences. *DNA*, **6**: 519-528.

Bourdineaud, J.P., Baudrimont, M., Gonzalez, P., and Moreau, J.L. 2006. Challenging the model for induction of metallothionein gene expression. *Biochimie*, **88**: 1787–1792.

Breathnach, R., and Chambon, P. 1981. Organization and expression of eukaryotic split genes coding for proteins. *Annual Review of Biochemistry*, **50**: 349-383.

Brouwer, M., 1996. Role of metallothionein in intracellular metal metabolism and in activation of copper and zinc-dependent proteins. *Advances in Inorganic Biochemistry*, **11**: 235 – 260.

Brunelli, E., Mauceri, A., Maisano, M., Bernabo, I., Giannetto, A., De Domenico, E., Corapi, B., Tripepi, S., Fasulo, S., and Fasulo, S. 2011. Ultrastructural and immunohistochemical investigation on the gills of the teleost, *Thalassoma pavo* L., exposed to cadmium. *Acta Histochemica,* **113(2)**: 201–213.

Cajaraville, M.P., Bebianno, M.J., Blasco, J., Porte, C., Sarasquete, C., and Viarengo, A. 2000. The use of biomarkers to assess the impact of pollution in coastal environments of the Iberian Peninsula: a practical approach. *Science of the Total Environment,* **247**: 295–311.

Capasso, C., Carginale, V., Crescenzi, O., Di Maro, D., Parisi, E., Spadaccini, R., and Temussi, P.A. 2003. Solution structure of MT_nc, a novel metallothionein from the antarctic fish *Notothenia coriiceps*. *Structure,* **11(4)**: 435–443.

Carvalho, C.D.S., Bernusso, V.A., de Araujo, H.S.S., Espindola, E.L.G., and Fernandes, M.N. 2012. Biomarker responses as indication of contaminant effects in *Oreochromis niloticus*. *Chemosphere,* **89(1)**: 60–69.

Ceyhun, S.B., Aksakal, E., Ekinci, D., Erdogan, O., and Beydemir, S. 2011. Influence of cobalt and zinc exposure on mRNA expression profiles of metallothionein and cytocrome P450 in rainbow trout. *Biological Trace Element Research,* **144(1-3)**: 781–789.

Chambers, J.E., Boone, J.S., Carr, R.L., Chambers, H.W., and Straus, D.L. 2002. Biomarkers as predictors in health and ecological risk assessment. *Human and Ecological Risk Assessment,* **8**: 165–176.

Chan, K.M., 1995. Metallothionein: Potential biomarker for monitoring heavy metal pollution in fish around Hong Kong. *Marine Pollution Bulletin,* **31(4-12)**: 411–415.

Chan, K.M., Davidson, W.S., Hew, C.L., and Fletcher, G.L. 1989. Molecular cloning of metallothionein cDNA and analysis of metallothionein gene expression in winter flounder tissues. *Canadian Journal of Zoology,* **67**: 2520-2527.

Chandrasekera, L.W.H.U., Pathiratne, A., and Pathiratne, K.A.S. 2008. Effects of water borne cadmium on biomarker enzymes and metallothioneins in Nile tilapia, *Oreochromis niloticus*. *Journal of the National Science Foundation of Sri Lanka,* **36(4)**: 315–322.

Chang, C.C., and Huang, P.C. 1996. Semi-empirical simulation of Zn/Cd binding site preference in the metal binding domains of mammalian metallothionein, *Protein Engineering,* **9**: 1165–1172.

Chatterjee, A., and Maiti, I.B. 1991. Induction and turnover of catfish (*Heteropneustes fossilis*) metallothionein. *Molecular and Cellular Biochemistry,* **108(1)**: 29-38.

Cherian, M.G., and Apostolova, M.D. 2000. Nuclear localization of metallothionein during cell proliferation and differentiation. *Cell and Molecular Biology,* **46**: 347–356.

Cho, Y.S., Choi, B.N., Ha, E.M., Kim, K.H., Kim, S.K., Kim, D.S., and Nam, Y.K. 2005. Shark (*Scyliorhinus torazame*) metallothionein: cDNA cloning, genomic sequence, and expression analysis. *Marine Biotechnology,* **7**: 350–362.

Cho, Y.S., Lee, S.Y., Kim, K.Y., and Nam, Y.K. 2009. Two metallothionein genes from mud loach *Misgurnus mizolepis* (Teleostei; Cypriniformes): gene structure, genomic organization, and mRNA expression analysis. *Comparative Biochemistry and Physiology Part B: Biochemistry and Molecular Biology*, **153(4)**: 317–326.

Cho, Y.S., Lee, S.Y., Kim, K.Y., Bang, I.C., Kim, D.S., and Nam, Y.K. 2008. Gene structure and expression of metallothionein during metal exposures in *Hemibarbus mylodon*. *Ecotoxicology and Environmental Safety*, **71(1)**: 125–137.

Chouchene, L., Banni, M., Kerkeni, A., Said, K., and Messaoudi, I. 2011. Cadmium-induced ovarian pathophysiology is mediated by change in gene expression pattern of zinc transporters in zebrafish (*Danio rerio*). *Chemico-Biological Interactions*, **193(2)**: 172–179.

Chowdhury, M.J., Baldisserotto, B., and Wood, C.M. 2005. Tissue-specific cadmium and metallothionein levels in rainbow trout chronically acclimated to waterborne or dietary cadmium. *Archives of Environmental Contamination and Toxicology*, **48**: 381–390.

Chung, R.S., and West, A.K. 2004. A role for extracellular metallothioneins in CNS injury and repair. *Neuroscience*, **123**: 595–599.

Ciocan, C.M., and Rotchell, J.M. 2004. Cadmium induction of metallothionein isoforms in juvenile and adult (*Mytilus edulis*). *Environmental Science and Technology*, **38**: 1073–1078.

Clements, W.H., 2000. Integrating effects of contaminants across levels of biological organization: An overview. *Journal of Aquatic Ecosystem Stress and Recovery*, **7**: 113–116.

Company, R., Felicia, H., Serafim, A., Almeida, A.J., Biscoito, M., and Bebianno, M.J. 2010. Metal concentrations and metallothionein-like protein levels in deep-sea fishes captured near hydrothermal vents in the Mid-Atlantic Ridge off Azores. *Deep Sea Research Part I: Oceanographic Research Papers*, **57(7)**: 893–908.

Coyle, P., Philcox, J.C., Carey, L.C., and Rofe, A.M. 2002. Metallothionein: the multipurpose protein. *Cellular and Molecular Life Sciences*, **59**: 627–647.

Dallinger, R., Berger, B., Gruber, C., Hunziker, P., and Sturzenbaum, S. 2000. Metallothioneins in terrestrial invertebrates: structural aspects, biological significance and implications for their use as biomarkers. *Cell and Molecular Biology*, **46**: 331–346.

Dalton, T., Li, Q., Bittel, D., Liang, L., and Andrews, G.K. 1996. Oxidative stress activates metal-responsive transcription factor-1 binding activity. *The Journal of Biological Chemistry*, **271**: 26233 – 26241.

Das, B.K., and Kaviraj, A. 1994. Influence of potassium permanganate, cobalt chloride, and dietary supplement of vitamin B complex on the histopathological changes in gill epithelium of common carp exposed to cadmium. *Progressive Fish Culturist*, **56**: 265–268.

De Lafontaine, Y., Gagne, F., Blaise, C., Costan, G., Gagnon, P., and Chan, H.M. 2000. Biomarkers in zebra mussels (*Dreissena polymorpha*) for the assessment and monitoring of water quality of the St. Lawrence River (Canada). *Aquatic Toxicology,* **50**: 51–71.

Deane, E.E., and Woo, N.Y.S. 2010. Advances and perspectives on the regulation and expression of piscine heat shock proteins. *Reviews in Fish Biology and Fisheries,* **21(2)**: 153–185.

Depledge, M.H., 1994. The rational basis for the use of biomarkers as ecotoxicological tools. In: *Nondestructive Biomarkers in Vertebrates,* (Eds.) C. Leonzio. London (UK), CRC, p. 271–295.

Dethloff, G.M., Schlenk, D., Hamm, J.T., and Bailey, H.C. 1999. Alterations in physiological parameters of rainbow trout (*Oncorhynchus mykiss*) with exposure to copper and copper/zinc mixtures. *Ecotoxicology and Environmental Safety,* **42**: 253–264.

Dragun, Z., Podrug, M., and Raspor, B. 2009. The assessment of natural causes of metallothionein variability in the gills of European chub (*Squaliuscephalus* L.). *Comparative Biochemistry and Physiology Part C: Toxicology and Pharmacology,* **150(2)**: 209–217.

Espinoza, H.M., Williams, C.R., and Gallagher, E.P. 2012. Effect of cadmium on glutathione S-transferase and metallothionein gene expression in coho salmon liver, gill and olfactory tissues. *Aquatic Toxicology,* **110-111**: 37–44.

Esser, H.O., and Moser, P. 1982. An appraisal of problems related to the measurement and evaluation of bioaccumulation. *Ecotoxicology and Environmental Safety,* **6(2)**: 131–148.

Falfushynska, H.I., and Stoliar, O.B. 2009. Function of metallothioneins in carp *Cyprinus carpio* from two field sites in Western Ukraine. *Ecotoxicology and Environmental Safety,* **72(5)**: 1425–1432.

Fargasova, A., 1998. Comparative acute toxicity of Cu^{2+}, Cu^+, Mn^{2+}, Mo^{6+}, Ni^{2+} and V^{5+} to *Chironomus plumosus* larvae and *Tubifex tubifex* worms. *Biologia,* **53**: 315–319.

Fasulo, S., Mauceri, A., Maisano, M., Giannetto, A., Parrino, V., Gennuso, F., and D Agata, A. 2010. Immunohistochemical and molecular biomarkers in *Coris julis* exposed to environmental contaminants. *Ecotoxicology and Environmental Safety,* **73(5)**: 873–882.

Feng, W., Cai, J., Pierce, W.M., Franklin, R.B., Maret, W., Benz, F.W., and Kang, Y.J. 2005. Metallothionein transfers Zn to mitochondrial aconitase through a direct interaction in mouse hearts. *Biochemical and Biophysical Research Communications,* **332**: 853–858.

Ferencz, A., and Hermesz, E. 2008. Identification and characterization of two mtf-1 genes in common carp. *Comparative Biochemistry and Physiology Part: C Toxicology and Pharmacology,* **148(3)**: 238–243.

Fonseca, V.F., Franca, S., Serafim, A., Company, R., Lopes, B., Bebianno, M. J., and Cabral, H.N. 2011. Multi-biomarker responses to estuarine habitat contamination in three fish species: *Dicentrarchus labrax, Solea senegalensis* and *Pomatoschistus microps. Aquatic Toxicology,* **102(3-4)**: 216–227.

Formigari, A., Irato, P., and Santon, A. 2007. Zinc, antioxidant systems and metallothionein in metal mediated-apoptosis: biochemical and cytochemical aspects. *Comparative Biochemistry and Physiology Part C: Toxicology and Pharmacology*, **146(4)**: 443–459.

Fowler, B.A., Hildebrand, C.E., Kojima, Y., and Webb, M. 1987. Nomenclature of metallothionein. *Experientia Supplementum*, **52**: 19–22.

Freitas, T.B., Gerson, A.F., Ferreira, C.C., Tadeu, L.R., and Pires, T.T. 2008. Heavy metal in tissues of three fish species from different trophic levels in a tropical Brazilian river. *Water Air Soil Pollution*, **187**: 275–284.

Funkenstein, B., Dyman, A., Berta Levavi-Sivan, B., and Tom, M. 2004. Application of real time PCR for quantitative determination of hepatic vitellogenin transcript levels in the striped seabream, *Lithognathus mormyrus*. *Marine Environmental Research*, **58**: 659–663.

Galloway, T.S., Brown, R.J., Browne, M.A., Dissanayake, A., Lowe, D., Jones, M.B., and Depledge, M.H. 2004. Ecosystem management bioindicators: The ECOMAN project— A multi-biomarker approach to ecosystem management. *Marine Environment and Research*, **58**: 233–237.

Gao, D., Wang, G.T., Chen, X.T., and Nie, P. 2009. Metallothionein-2 gene from the mandarin fish *Siniperca chuatsi*: cDNA cloning, tissue expression, and immunohistochemical localization. *Comparative Biochemistry and Physiology Part: C Toxicology and Pharmacology*, **149(1)**: 18–25.

Gehrig, P.M., You, C., Dallinger, R., Gruber, C., Brouwer, M., Kagi, J.H.R., and Hunziker, P.E. 2000. Electrospray ionization mass spectrometry of zinc, cadmium, and copper metallothioneins: evidence for metal-binding cooperativity. *Protein Science*, **9**: 395–402.

George, S., Gubbins, M., MacIntosh, A., Reynolds, W., Sabine, V., Scott, A., and Thain, J.A. 2004. Comparison of pollutant biomarker responses with transcriptional responses in European flounders (*Platicthys flesus*) subjected to estuarine pollution. *Marine Environmental Research*, **58**: 571–575.

George, S.G., 1989. Cadmium effects on plaice liver xenobiotic and metal-detoxification systems: dose response. *Aquatic Toxicology*, **15**: 303–310.

Geracitano, L.A., Bocchetti, R., Monserrat, J.M., Regoli, F., and Bianchini, A. 2004. Oxidative stress responses in two populations of *Laeonereis acuta* (Polychaeta, Nereididae) after acute and chronic exposure to copper. *Marine Environmental Research*, **58**: 1–17.

Ghedira, J., Jebali, J., Bouraoui, Z., Banni, M., Guerbej, H., and Boussetta, H. 2010. Metallothionein and metal levels in liver, gills and kidney of *Sparus aurata* exposed to sublethal doses of cadmium and copper. *Fish Physiology and Biochemistry*, **36(1)**: 101–107.

Ghoshal, K., and Jacob, S.T. 2001. Regulation of metallothionein gene expression. *Progress in Nucleic Acid Research and Molecular Biology*, **66**: 357–384.

Giguere, A., Couillard, Y., Campbell, P.G.C., Perceval, O., Hare, L., Pinel-Alloul, B., and Pellerin, J. 2003. Steady-state distribution of metals among metallothionein and other cytosolic ligands and links to cytotoxicity in bivalves living along a polymetallic gradient. *Aquatic Toxicology*, **64**: 185- 200.

Gonc, P., and Magalha, P.J. 2011. Voltammetric analysis of metallothioneins and copper (II) in fish for water biomonitoring studies. *Environmental Chemistry Letters*, **9(3)**: 405–410.

Hagger, J.A., Jones, M.B., Leonard, D.R.P., Owen, R., and Galloway, T.S. 2006. Biomarkers and integrated environmental risk assessment: are there more questions than answers? *Integrated Environmental Assessment and Management*, **2(4)**: 312–329.

Hamer, D.H., 1986. Metallothionein. *Annual Review of Biochemistry*, **55**: 913-951.

Hamilton, S.J., and Mehrle, P.M. 1986. Metallothionein in fish: review of its importance in assessing stress from metal contaminants. *Transactions of the American Fisheries Society*, **115**: 596-609.

Hansen, B.H., Romma, S., Garmo, O.A., Pedersen, S.A., Olsvik, P.A., and Andersen, R.A. 2007. Induction and activity of oxidative stress-related proteins during waterborne Cd/Zn-exposure in brown trout (*Salmo trutta*). *Chemosphere*, **67(11)**: 2241–2249.

Harford, C., and Sarkar, B. 1991. Induction of metallothionein by simultaneous administration of cadmium (II) and zinc (II). *Biochemical and Biophysical Research Communications*, **177**: 224–228.

Hauser-Davis, R.A., Gonçalves, R.A., Ziolli, R.L., and de Campos, R.C. 2012. A novel report of metallothioneins in fish bile: SDS-PAGE analysis, spectrophotometry quantification and metal speciation characterization by liquid chromatography coupled to ICP-MS. *Aquatic Toxicology*, **116-117**: 54–60.

Hayes, R.A., Regondi, S., Winter, M.J., Butler, P.J., Agradi, E., Taylor, E.W., and Chipman, J.K. 2004. Cloning of a chub metallothionein cDNA and development of competitive RT-PCR of chub metallothionein mRNA as a potential biomarker of heavy metal exposure. *Marine Environmental Research*, **58**: 665–669.

Hodson, P.V., 1988. The effect of metal metabolism on uptake, disposition and toxicity in fish. *Aquatic Toxicology*, **11**: 3-18.

Hogstrand, C., and Haux, C. 1991. Binding and detoxification of heavy metals in lower vertebrates with special reference to metallothionein. *Comparative Biochemistry and Physiology Part C: Toxicology and Pharmacology*, **100C**: 137 -141.

Hogstrand, C., Lithner, G., and Haux, C. 1989. Relationship between metallothionein, copper and zinc in perch (*Perca fluviatilis*) environmentally exposed to heavy metals. *Marine Environmental Research*, **28**: 179-182.

Inui, M., Adachi, T., Takenaka, S., Inui, H., Nakazawa, M., Ueda, M., Watanabe, H., Mori, C., Iguchi, T., and Miyatake, K. 2003. Effect of UV screens and preservatives on vitellogenin and choriogenin production in male medaka (*Oryzias latipes*). *Toxicology*, **194**: 43–50.

Jebali, J., Banni, M., Gerbej, H., Boussetta, H., Lopez-Barea, J., and Alhama, J. 2008. Metallothionein induction by Cu, Cd and Hg in *Dicentrarchus labrax* liver: assessment by RP-HPLC with fluorescence detection and spectrophotometry. *Marine Environmental Research*, **65(4)**: 358–363.

Jebali, J., Banni, M., Guerbej, H., Almeida, E.A., Bannaoui, A., and Boussetta, H. 2006. Effects of malathion and cadmium on acetyl cholinesterase activity and metallothionein levels in the fish *Seriola dumerilli*. *Fish Physiology and Biochemistry*, **32(1)**: 93–98.

Jiang, D.T., Heald, S.M., Sham, T.K., and Stillman, M.J. 1994. Structures of the cadmium, mercury, and zinc thiolate clusters in metallothionein: XAFS study of Zn7-MT, Cd7-MT, Hg7-MT, and Hg18-MT formed from rabbit liver metallothionein 2. *Journal of the American Chemical Society*, **116**: 11004–11013.

Kagi, J. H. R., and Nordberg, M. 1979. Metallothionein. Birkhauser Verlag, Basel, Boston, Stuttgart, p. 1–378.

Kagi, J.H.R., 1991. Overview of metallothionein. *Methods in Enzymology*, **205**: 613–626.

Kagi, J.H.R., 1993. Evolution, structure and chemical activity of class I metallothioneins: an overview. In: *Metallothionein III*, (Eds.) K.T. Suzuki., N. Imura and M. Kimura. Birkhauser, Basel, p. 29–55.

Kagi, J.H.R., and Kojima, Y. 1987. Chemistry and biochemistry of metallothionein. In: *Experientia Supplementum Metallothionein II*, (Eds.) J.H.R. Kagi and Y. Kojima. Birkhauser, Verlag Basel, p. 25-61.

Kagi, J.H.R., and Schaffer, A. 1988. Biochemistry of metallothionein. *Biochemistry*, **27**: 8509-8515.

Kagi, J.H.R., and Vallee, B.L. 1961. Metallothionein: A cadmium and zinc containing protein from Equine renal cortex physico-chemical properties. *Journal of Biological Chemistry*, **236(9)**: 2435-2442.

Kalpaxis, D.L., Theos, C., Xaplanteri, M.A., Dinos, G.P., Catsiki, A.V., and Leotsinidis, M. 2004. Biomonitoring of Gulf of Patras, N. Peloponnesus, Greece. Application of a biomarker suite including evaluation of translation efficiency in *Mytilus galloprovincialiscells*. *Environmental Research*, **94**: 211–220.

Kamer, I., Douek, J., Tom, M., and Rinkevich, B. 2003. Metallothionein induction in RTH-149 cell line as an indicator for heavy metal pollution in a brackish environment: assessment by RT-competitive PCR. *Archives of Environmental Contamination and Toxicology*, **45**: 86–91.

Kay, J., Cryer, A., Darke, B.M., Kille, P., Lees, W.E., Norey, C.G., and Stark, J.M. 1991. Naturally occurring and recombinant metallothioneins: structure, immunoreactivity and metal-binding functions. *International Journal of Biochememistry*, **23**: 1–5.

Kessabi, K., Navarro, A., Casado, M., Said, K., Messaoudi, I., and Pina, B. 2010. Evaluation of environmental impact on natural populations of the Mediterranean killifish *Aphanius fasciatus* by quantitative RNA biomarkers. *Marine Environmental Research*, **70(3-4)**: 327–333.

Kille, P., Stephens, P.E., and Kay, J. 1991. Elucidation of cDNA sequences for metallothioneins from rainbow trout, stone loach and pike liver using the polymerase chain reaction. *Biochimica et Biophysica Acta*, **23**: 407–410.

Kim, J.H., Rhee, J.S., Dahms, H.U., Lee, Y.M., Han, K.N., and Lee, J.S. 2012. The yellow catfish, *Pelteobagrus fulvidraco* (Siluriformes) metallothionein cDNA: molecular cloning and transcript expression level in response to exposure to the heavy metals Cd, Cu, and Zn. *Fish Physiology and Biochemistry*, **38(5)**: 1331–1342.

Kim, J.H., Wang, S.Y., Kim, I.C., Ki, J.S., Raisuddin, S., Lee, J.S., and Han, K.N. 2008. Cloning of a river pufferfish (*Takifugu obscurus*) metallothionein cDNA and study of its induction profile in cadmium-exposed fish. *Chemosphere*, **71(7)**: 1251–1259.

Klaverkamp, J.F., MacDonald, W.A., Duncan, D.A., and Wagemann, R. 1984. Metallothionein and acclimation to heavy metals in fish: a review. In: *Contaminant Effects on Fisheries*, (Eds.) V.W. Cairns., P.V. Hodson and J.O. Nriagu. John Wiley and Sons, New York, p. 99-113.

Knapen, D., Bervoets, L., Verheyen, E., and Blust, R. 2004. Resistance to water pollution in natural gudgeon (*Gobio gobio*) populations may be due to genetic adaptation. *Aquatic Toxicology*, **67**: 155–165.

Knapen, D., Redeker, E.S., Inacio, I., De Coen, W., Verheyen, E., and Blust, R. 2005. New metallothionein mRNAs in *Gobio gobio* reveal at least three gene duplication events in cyprinid metallothionein evolution. *Comparative Biochemistry and Physiology Part C: Toxicology and Pharmacology*, **140(3-4)**: 347–355.

Knapen, D., Reynders, H., Bervoets, L., Verheyen, E., and Blust, R. 2007. Metallothionein gene and protein expression as a biomarker for metal pollution in natural gudgeon populations. *Aquatic Toxicology*, **82(3)**: 163–172.

Krasnici, N., Dragun, Z., Erk, M., and Raspor, B. 2012. Distribution of selected essential (Co, Cu, Fe, Mn, Mo, Se, and Zn) and nonessential (Cd, Pb) trace elements among protein fractions from hepatic cytosol of European chub (*Squalius cephalus* L.). *Environmental Science and Pollution Research International*, doi: 10.1007/s11356-012-1105-8.

Laity, J.H., and Andrews, G.K. 2007. Understanding the mechanisms of zinc-sensing by metal response element binding transcription factor-1 (MTF-1). *Archives of Biochemistry and Biophysics*, **463**: 201–210.

Lam, P.K.S., and Gray, J.S. 2003. The use of biomarkers in environmental monitoring programmes. *Marine Pollution Bulletin*, **46**: 182–186.

Langston, W.J., Chesman, B.S., Burt, G.R., Pope, N.D., and McEvoy, J. 2002. Metallothionein in liver of eels *Anguilla anguilla* from the Thames Estuary: an indicator of environmental quality? *Marine Environmental Research*, **53**: 263-229.

Lazo, J., Kondo, Y., Dellapiazza, D., Michalska, A., Choo, K., and Pitt, B. 1995. Enhanced sensitivity to oxidative stress in cultured embryonic cells from transgenic mice deficient in metallothionein I and II genes. *The Journal of Biological Chemistry*, **270**: 5506 – 5510.

Leonardi, M., Vera, J., Tarifeno, E., Puchi, M., and Morin, V. 2010. Vitellogenin of the Chilean flounder *Paralichthys adspersus* as a biomarker of endocrine disruption along the marine coast of the South Pacific. Part I: induction, purification, and identification. *Fish Physiology and Biochemistry*, **36(3)**: 757–765.

Ma, W.L., Yan, T., He, Y., and Wang, L. 2009. Purification and cDNA cloning of a cadmium-binding metallothionein from the freshwater crab *Sinopotamon henanense*. *Archives of Environmental Contamination and Toxicology*, **56(4)**: 747–753.

Man, A.K.Y., and Woo, N.Y.S. 2008. Upregulation of metallothionein and glucose-6-phosphate dehydrogenase expression in silver sea bream, *Sparus sarba* exposed to sublethal levels of cadmium. *Aquatic Toxicology*, **89(4)**: 214–221.

Margoshes, M., and Vallee, B.L. 1957. A cadmium protein from equine kidney cortex. *Journal of the American Chemical Society*, **79**: 4813–4814.

Martinez-Gomez, C., Fernandez, B., Benedicto, J., Valdes, J., Campillo, J.A., Leon, V.M., and Vethaak, A.D. 2012. Health status of red mullets from polluted areas of the Spanish Mediterranean coast, with special reference to Portman (SE Spain). *Marine Environmental Research*, **77**: 50–59.

Masters, B.A., Kelly, E.J., Quaife, C.J., Brinster, R.L., and Palmiter, R.D. 1994. Targeted disruption of metallothionein I and II genes increases sensitivity to cadmium. *Proceedings of the National Academy of Sciences USA*, **91**: 584 – 588.

Mayer, F.L., Versteeg, D.J., McKee, M.J., Folmar, L.C., Graney, F.L., McCume, D.C., and Rattner, B.A. 1992. Physiological and nonspecific biomarkers. In: *Biomarkers: Biochemical, Physiological, and Histological Markers of Anthropogenic Stress*, (Eds.) H.L. Bergman. Boca Raton (FL), Lewis. p. 5–85.

Mdegela, R.H., Braathen, M., Mosha, R.D., Skaare, J.U., and Sandvik, M. 2010. Assessment of pollution in sewage ponds using biomarker responses in wild African sharp tooth catfish (*Clarias gariepinus*) in Tanzania. *Ecotoxicology*, **19(4)**: 722–734.

Mieiro, C.L., Bervoets, L., Joosen, S., Blust, R., Duarte, A.C., Pereira, M.E., and Pacheco, M. 2011. Metallothioneins failed to reflect mercury external levels of exposure and bioaccumulation in marine fish-considerations on tissue and species specific responses. *Chemosphere*, **85(1)**: 114–121.

Montaser, M., Mahfouz, M.E., El-shazly, S.A.M., Abdel-rahman, G.H., and Bakry, S. 2010. Toxicity of heavy metals on fish at Jeddah coast KSA: Metallothionein expression as a biomarker and histopathological study on liver and gills. *World Journal of Fish and Marine Sciences*, **2(3)**: 174–185.

Monteiro, V., Cavalcante, D.G.S.M., Vilela, M.B.F.A., Sofia, S.H., and Martinez, C.B.R. 2011. *In vivo* and *in vitro* exposures for the evaluation of the genotoxic effects of lead on the Neotropical freshwater fish *Prochilodus lineatus*. *Aquatic Toxicology*, **104(3-4)**: 291–298.

Mukhopadhyay, I., Nazir, A., Saxena, D. K., and Chowdhuri, D. K. 2003. Heat shock response: hsp70 in environmental monitoring. *Journal of Biochemical and Molecular Toxicology*, **17(5)**: 249–254.

Narbonne, J.F., Daubeze, M., Clerandeau, C., and Garrigues, P. 1999. Scale of classification based on biochemical markers in mussels: application to pollution monitoring in European coast. *Biomarkers* **4**: 415–424.

Navarro, A., Quiros, L., Casado, M., Faria, M., Carrasco, L., Benejam, L., Benito, J., Diez, S., Raldua, D., Barata, C., Bayona, J.M., and Pina, B. 2009. Physiological responses to mercury in feral carp populations inhabiting the low Ebro River (NE Spain), a historically contaminated site. *Aquatic Toxicology*, **93(2-3)**: 150–157.

Nordberg, M., and Nordberg, G.F. 2009. Metallothioneins: Historical Development and Overview. In: *Metal Ions in Life Sciences: Metallothioneins and Related Chelators*, (Eds.) A. Sigel., H. Sigel and R.K.O. Sigel. Royal Society of Chemistry, p. 1-29.

Nuwaysir, E.F., Bittner, M., Trent, J., Barrett, J.C., and Afshari, C.A. 1999. Microarrays and toxicology: The advent of toxicogenomics. *Molecular Carcinogenesis*, **24**: 153–159.

Oliveira, M., Ahmad, I., Maria, V.L., Serafim, A., Bebianno, M.J., Pacheco, M., and Santos, M.A. 2010. Hepatic metallothionein concentrations in the golden grey mullet (*Liza aurata*) - Relationship with environmental metal concentrations in a metal-contaminated coastal system in Portugal. *Marine Environmental Research*, **69(4)**: 227–233.

Olsson, P.E., 1993. Metallothionein gene expression and regulation in fish. In: *Biochemistry and Molecular Biology of Fishes*, (Eds.) H.A. Hochachka and T.P. Mommsen. Elsevier Science, NY, p. 259-278.

Olsson, P.E., and Haux, C. 1986. Increased hepatic metallothionein content correlates to cadmium accumulation in environmentally exposed perch, *Perca fluviatilis*. *Aquatic Toxicology*, **9**: 231-242.

Otvos, J.D., Liu, X., Shen, G., and Basti, M. 1993. Dynamic aspects of metallothionein structure. In: *Metallothionein III*, (Eds.) K.T. Suzuki., N. Imura., and M. Kimura. Birkhauser, Basel, p. 57–74.

Pedersen, S.N., Lundebye, A.K., and Depledge, M.H. 1997. Field application of metallothionein and stress protein biomarkers in the shore crab (*Carcinus maenas*) exposed to trace metals. *Aquatic Toxicology*, **37**: 183–200.

Petering, D.H., Zhu, J., Krezoski, S., Meeusen, J., Kiekenbush, C., Krull, S., Specher, T., and Dughish, M. 2006. Apo-metallothionein emerging as a major player in the cellular activities of metallothionein. *Experimental Biology and Medicine*, **231**: 1528–1534.

Picado, A., Bebianno, M.J., Costa, M.H., Ferreira, A., and Vale, C. 2007. Biomarkers: a strategic tool in the assessment of environmental quality of coastal waters. *Hydrobiologia*, **587(1)**: 79–87.

Piccinni, E., Staudenmann, W., Albergoni, V., De Gabrieli, R., and James, P. 1994. Purification and primary structure of metallothioneins induced by cadmium in the protists *Tetrahymena pigmentosa* and *Tetrahymena pyriformis*. *European Journal of Biochemistry*, **226**: 853–859.

Quiros, L., Pina, B., Sole, M., Blasco, J., Lopez, M.A., Riva, M.C., Barcelo, D., and Raldua, D. 2007. Environmental monitoring by gene expression biomarkers in *Barbus graellsii*: laboratory and field studies. *Chemosphere*, **67(6)**: 1144–1154.

Ren, H.W., Norio, I., Masako, K., Shinpei, T., Junya, K., Gab-Soo, H., Tsuyoshi, N., Norio, M., and Keiichi, T. 2000. Two metallothioneins in the freshwater fish, crucian carp (*Carassius cuvieri*): cDNA cloning and assignment of their expression isoforms. *Biological and Pharmaceutical Bulletin*, **23**: 145–148.

Rigby, K.E., and Stillman, M.J. 2004. Structural studies of metal-free metallothionein. *Biochemical and Biophysical Research Communications*, **325(4)**: 1271–1278.

Roch, M., and McCarter, J.A. 1984. Hepatic metallothionein production and resistance to heavy metal by rainbow trout (*Salmo gairdneri*) II- held in a series of contaminated lakes. *Comparative Biochemistry and Physiology Part C: Toxicology and Pharmacology*, **77C**: 77-82.

Roch, M., McCarter, J.A., Matheson, A.T., Clark, M.J.R., and Olafson, R.W. 1982. Hepatic metallothionein in rainbow trout as an indicator of metal pollution in the Campbell River system. *Canadian Journal of Fisheries and Aquatic Sciences*, **39**: 1596-1601.

Roesijadi, G., 1996. Metallothionein and its role in toxic metal regulation. *Comparative Biochemistry and Physiology Part C: Toxicology and Pharmacology*, **113**: 117–123.

Roesijadi, G., Bogumil, R., Vasak, M., and Kagi, J.H. 1998. Modulation of DNA binding of a tramtrack zinc finger peptide by the metallothionein–thionein conjugate pair. *The Journal of Biological Chemistry*, **273**: 17425–17432.

Roesijudi, G. 1996. Metallothionein and Its Role in Toxic Metal Regulation, *Comparative Biochemistry and physiology Part C: Toxicology and Pharmacology*, **113(2)**: 117–123.

Roy, S.U., Chattopadhyay, B., Datta, S., and Mukhopadhyay, S.K. 2011. Metallothionein as a biomarker to assess the effects of pollution on Indian major carp species from wastewater-fed fishponds of East Calcutta wetlands (a Ramsar Site). *Environmental Research, Engineering and Management*, **58(4)**: 10–17.

Ruelas-Inzunza, J., Paez-Osuna, F., and Garcia-Flores, D. 2010. Essential (Cu) and nonessential (Cd and Pb) metals in ichthyfauna from the coasts of Sinaloa state (SE Gulf of California). *Environmental Monitoring Assessment*, **162**: 251–263.

Sampaio, F.G., Boijink, C.D.L., dos Santos, L.R.B., Oba, E.T., Kalinin, A.L., Luiz, A.J.B., and Rantin, F.T. 2012. Antioxidant defenses and biochemical changes in the neotropical fish pacu, *Piaractus mesopotamicus*: responses to single and combined copper and hypercarbia exposure. *Comparative Biochemistry and Physiology Part C: Toxicology and Pharmacology*, **156(3-4)**: 178–186.

Santon, A., Albergoni, V., Sturniolo, G.C., and Irato, P. 2004. Evaluation of MT expression and detection of apoptotic cells in LEC rat kidneys. *Biochimica et Biophysica Acta*, **1688**: 223–231.

Sarkar, A., Ray, D., Shrivastava, A.N., and Sarker, S. 2006. Molecular biomarkers: their significance and application in marine pollution monitoring. *Ecotoxicology*, **15(4)**: 333–340.

Sassi, A., Darias, M.J., Said, K., Messaoudi, I., and Gisbert, E. 2012. Cadmium exposure affects the expression of genes involved in skeletogenesis and stress response in gilthead sea bream larvae. *Fish Physiology and Biochemistry*, **39(3)**: 649–659.

Sato, M., and Bremner, I. 1993. Oxygen free radicals and metallothionein. *Free Radical Biology and Medicine*, **14**: 325–337.

Sato, M., and Kondoh, M. 2002. Recent studies on metallothionein: protection against toxicity of heavy metals and oxygen free radicals. *The Tohoku Journal of Experimental Medicine*, **196**: 9–22.

Schmitt, C.J., Whyte, J.J., Roberts, A.P., Annis, M.L., May, T.W., and Tillitt, D.E. 2007. Biomarkers of metals exposure in fish from lead-zinc mining areas of southeastern Missouri, USA. *Ecotoxicology and Environmental Safety*, **67(1)**: 31–47.

Scudiero, R., Carginale, V., Riggio, M., Capasso, C., Capasso, A., Kille, P., Di Prisco, G., and Parisi, E. 1997. Difference in hepatic metallothionein content in Antarctic red-blooded and haemoglobin-less fish is accompanied by accumulation of untranslated metallothionein mRNA. *Biochemical Journal*, **322**: 207–211.

Shariati, F., Esaili Sari, A., Mashinchian, A., and Pourkazemi, M. 2011. Metallothionein as potential biomarker of cadmium exposure in Persian sturgeon (*Acipenser persicus*). *Biological Trace Element Research*, **143(1)**: 281–291.

Shimoda, R., Achanzar, W.E., Qu, W., Nagamine, T., Takagi, H., Mori, M., and Waalkes, M.P. 2003. Metallothionein is a potential negative regulator of apoptosis. *Toxicological Sciences*, **73**: 294–300.

Shved, N., Kumeiko, V., and Syasina, I. 2011. Enzyme-linked immunosorbent assay (ELISA) measurement of vitellogenin in plasma and liver histopathology in barfin plaice *Liopsetta pinnifasciata* from Amursky Bay, Sea of Japan. *Fish Physiology and Biochemistry*, **37(4)**: 781–799.

Sinaie, M., Bastami, K.D., Ghorbanpour, M., Najafzadeh, H., Shekari, M., and Haghparast, S. 2010. Metallothionein biosynthesis as a detoxification mechanism in mercury exposure in fish, spotted scat (*Scatophagus argus*). *Fish Physiology and Biochemistry*, **36(4)**: 1235–1242.

Snape, J.R., Maund, S.J., Pickford, D.B., and Hutchinson, T.H. 2004. Ecotoxicogenomics: the challenge of integrating genomics into aquatic and terrestrial ecotoxicology. *Aquatic Toxicology*, **67(2)**: 143–154.

Syring, R.A., Brouwer, H.T., and Brouwer, M. 2000. Cloning and sequencing of cDNAs encoding for a novel copper-specific metallothionein and two cadmium-inducible metallothioneins from the blue crab *Callinectes sapidus*. *Comparative*

Biochemistry and Physiology Part C: Toxicology and Pharmacology, **125(3)**: 325–332.

Templeton, D.M., and Cherian, M.G. 1991. Toxicological significance of metallothionein. *Methods in Enzymology*, **205**: 11–24.

Thirumoorthy, N., Kumar, K.T.M., Sundar, A.S., Panayappan, L., and Chatterjee, M. 2007. Metallothionein: an overview. *World Journal of Gastroenterology*, **13**: 993–996.

Tom, M., and Auslander, M. 2005. Transcript and protein environmental biomarkers in fish - a review. *Chemosphere*, **59**: 155–162.

Tom, M., Chen, N., Segev, M., Herut, B., and Rinkevich, B. 2004. Quantifying fish metallothionein transcript by real time PCR for its utilization as an environmental biomarker. *Marine Pollution Bulletin*, **48**: 705–710.

Tom, M., Shmul, M., Shefer, E., Chen, N., Slor, H., Rinkevich, B., and Herut, B. 2003. Quantitative evaluation of hepatic cytochrome P4501A transcript, protein and catalytic activity in the striped sea bream, *Lithognathus mormyrus*. *Environmental Toxicology and Chemistry*, **22**: 2088–2092.

Trinchella, F., Esposito, M. G., and Scudiero, R. 2012. Metallothionein primary structure in amphibians: Insights from comparative evolutionary analysis in vertebrates. *Comptes Rendus Biologies*, **335(7)**: 480–487.

Unger, M.E., Chen, T.T., Fenselau, C.C., Murphy, C.M., Vestling, M.M., and Roesijadi, G. 1991. Primary structure of a molluscan metallothionein deduced from molecular cloning and tandem mass spectrometry. *Biochimica et Biophysica Acta*, **1074**: 371-377.

Vallee, B.L., and Auld, D.S. 1993. New perspective on zinc biochemistry: co-catalytic sites in multi-zinc enzymes. *Biochemistry*, **32**: 6493–6500.

Vasak, M., and Hasler, D.W. 2000. Metallothioneins: new functional and structural insights. *Current Opinion in Chemical Biology*, **4**: 177–183.

Vasconcelos, M.H., Tam, S.C., Hesketh, J.E., Reid, M., Beattie, J.H., 2002. Metal and tissue-dependent relationship between metallothionein mRNA and protein. *Toxicology and Applied Pharmacology*, **182**: 91–97.

Vergani, L., Grattarola, M., Borghi, C., Dondero, F., and Viarengo, A. 2005. Fish and molluscan metallothioneins. *The Federation of European Biochemical Societies Journal*, **272(23)**: 6014–6023.

Viarengo, A., and Nott, J.A. 1993. Mechanisms of heavy metal cation homeostasis in marine invertebrates. *Comparative Biochemistry and Physiology Part C: Comparative Pharmacology and Toxicology*, **104**: 355–372.

Viarengo, A., Burlando, B., Cavaletto, M., Marchi, B., Ponzano, E., and Blasco, J. 1999. Role of metallothionein against oxidative stress in the mussel (*Mytilus galloprovincialis*). *American Journal of Physiology*, **277**: R1612–R1619.

Viarengo, A., Burlando, B., Ceratto, N., and Panfoli, I. 2000. Antioxidant role of metallothioneins: a comparative overview. *Cell and Molecular Biology*, **46**: 407–417.

Viarengo, A., Lowe, D., Bolognesi, C., Fabbri, E., and Koehler, A. 2007. The use of biomarkers in biomonitoring: a 2-tier approach assessing the level of pollutant-induced stress syndrome in sentinel organisms. *Comparative Biochemistry and Physiology Part C: Toxicology and Pharmacology*, **146(3)**: 281–300.

Wang, W.X., and Rainbow, P. S. 2010. Significance of metallothioneins in metal accumulation kinetics in marine animals. *Comparative Biochemistry and Physiology Part C: Toxicology and Pharmacology*, **152(1)**: 1–8.

Winge, D., Krasno, J., and Colucci, P. 1974. Cadmium accumulation in rat liver: correlation between bound metal and pathology. In: *Trace Element Metabolism in Animals*, (Eds.) W.G. Floekstra., J.W. Suttie., H.E. Ganther and W. Mertz. University Park Press, Baltimore, p. 500-502.

Winge, D.R., and Miklossy, K.A. 1982. Domain nature of metallothionein. *The Journal of Biological Chemistry*, **257**: 3471–3476.

Zafarullah, M., and Gedamu, L. 1988. Structure of the rainbow trout metallothionein B gene characterization of its metal-responsive region. *Molecular and Cellular Biology*, **8(10)**: 4469–4476.

Zangger, K., and Armitage, I.M. 2002. Dynamics of inter-domain and intermolecular interactions in mammalian metallothioneins. *Journal of Inorganic Biochemistry*, **88**: 135–143.

Zeng, J., Heuchel, R., Schaffner, W., and Kagi, J.H. 1991. Thionein (apo-metallothionein) can modulate DNA binding and transcription activation by zinc finger containing factor Sp1. *Federation of European Biochemical Societies Letters*, **279**: 310–312.

Zhou, Y., Li, L., and Ru, B. 2000. Expression, purification and characterization of beta domain and beta domain dimer of metallothionein. *Biochimica et Biophysica Acta*, **1524**: 87–93.

2015, Perspectives in Animal Ecology and Reproduction, Vol. 10   *Pages 237–267*
Editors: **V.K. Gupta, Anil K. Verma and G.D. Singh**
Published by: **DAYA PUBLISHING HOUSE, NEW DELHI**

Chapter 16

Forest Game and Habitat Disturbances in Southwest Ghana: A 1993-1995 Survey

Lars H. Holbech*

Department of Animal Biology and Conservation Science,
University of Ghana, P.O. Box LG 67,
Legon, Accra, Ghana

ABSTRACT

Whilst bushmeat is an important component of the subsistence farming economy of rural people in the West African rainforest biome, the effects upon game production of a disturbance-complex of forest fragmentation, selective logging, shifting cultivation, plantation conversion and bushmeat hunting remain poorly understood. This study aimed to identify game species that could be sustainably harvested and 'farmed' through the application of land-use systems by collecting baseline information on their populations and hunting pressure in forest habitats of southwest Ghana. During 1993-1995, my survey evaluated the relative abundance and diversity of 60 forest game species in 22 forest habitats subjected to different fragmentation and logging histories, and various agro-silvicultural practises. Field methods comprised multiple-signs transect surveys and interviews with wildlife users living close to the reserves. Overall, 63 per cent of game species were found to becoming alarmingly scarce in the study region. Overall game abundance appeared higher in heavily logged forests than in moderately logged or unlogged forests. Sub-canopy monkeys, bush pigs, small antelopes and carnivore-omnivores seemed to benefit mostly from logging. Forest size appeared to be more important for maintaining populations of endangered/rare species than did selective logging *per se*. No simple patterns were established between logging intensity, forest size and hunting pressure.

---

* *Corresponding author.* E-mail: lholbech@ug.edu.gh, l.holbech@gmail.com

Exotic timber, cash crop and rubber plantations supported *c.* 50 per cent of natural forest game diversity, and small antelopes and rodents thrived here. The relatively high game diversity of anthropogenic forest landscapes was attributed to the presence of natural forest trees in plantations, and old secondary forest and forest remnants in matrix ecotones. Results are compared to those from similar studies in the Guineo-Congolian forest biome, and the implications for integrated wildlife and land-use management are discussed in local and regional contexts.

Keywords: *Bushmeat, Fragmentation, Ghana, Hunting, Land-use, Logging, Plantations, Rainforest.*

Introduction

The importance of wild game as a source of both food and income for rural forest people in Ghana is well documented (Asibey, 1977; Ntiamoa-Baidu, 1987). Since the early 1960s bushmeat exploitation in Ghanaian forests has reached unprecedented levels for the Upper Guinea forest biome and resulted in dramatic reductions of particularly primates and large-bodied forest dependants (Martin, 1991; Oates, 1999; Barnes, 2002). This alarming trend has gone alongside rural infrastructural development, mainly driven by mechanised selective logging and the development of large-scale industrial plantations of exotic trees, rubber, and high-yielding cocoa and oil palm varieties. In effect, the Ghana high forest zone has since the late 1970s been highly fragmented into a patchwork of protected areas, managed for their timber and/or wildlife resources (Hawthorne and Abu-Juam, 1995). The over 200 high forest reserves constitute < 20 per cent of the originally c. 82,000 km² forest cover, of which reserves established to specifically protect wildlife (*i.e.* resource reserves, wildlife sanctuaries and national parks) represent only some 7 per cent (Martin, 1991; Hawthorne and Abu-Juam, 1995).

Forest reserves (FRs) and resource reserves (RRs) in Ghana are in principle protected from both illegal logging and hunting by respectively the Ghana Forestry Commission (GFC) and Ghana Wildlife Division (GWD), whereas national parks are wholly GWD-protected against any human activities, except research. This means that in FRs and RRs legal logging operations may be granted to timber concessionaires by the GFC, just as hunting may be permitted by the GWD on certain species in FRs, but not in RRs. Since 2000 however, neither logging nor hunting have been permitted in any GWD-protected RRs. Recent satellite images (*e.g.* Google Earth) depict that contiguous protected forest areas do not exceed 1,500 km², and most areas are < 500 km², with isolates < 100 km² constituting roughly 10 per cent of the protected forest estate. Moreover, many reserves are slender in shape, producing a high ratio of forest edge habitat to deep-forest core areas. Therefore, the management challenges to species conservation in Ghanaian forest are immense.

In this study, Ghanaian forest game is limited to any species of reptile, bird and mammal found in the forest zone that is hunted for its meat or an economical return (Holbech, 1996). These species range in size from small squirrels and up to elephants, though the latter is normally only targeted by ivory poachers. To safeguard forest game in Ghana, and West Africa at large, extensive knowledge is required to identify and evaluate the factors that influence distribution and diversity of these animals in anthropogenic forest landscapes. As ongoing studies from the Congolian forest biome have emphasised, the complex of collateral disturbance factors needs to be understood holistically, to curb and mitigate their negative impacts on biodiversity and ecosystem functions (Bowen-Jones and Pendry, 1999). Compared with long-term studies in the Congolian forest biome, few published studies on forest game and habitat disturbances exist from West Africa (Fimbel, 1994; Fa *et al.*, 2005). One major reason for this information bias is related to difficulties in censusing large forest game, a result of decades of excessive hunting which has rendered naturally shy and cryptic animals problematic to study through audio-visual counts. Some conservationists describe Ghanaian forests as "empty" or "silent", concluding that several large-bodied game and upper-canopy primates in particular are seriously threatened by extirpation, despite the fairly good condition of the vegetation (Struhsaker and Oates, 1995; Oates, 1999). This "defaunation syndrome", which was first observed in the Neotropics (Redford, 1992) and Central Africa (Wilkie and Carpenter, 1999), may have eliminated an endemic red colobus subspecies from Ghana (Oates *et al.*, 2000), though it probably still persist in Eastern Côte d'Ivoire (McGraw, 2005).

The present study evaluated populations of 60 game species in relation to; 1) forest fragmentation; 2) selective logging; and 3) agro-silviculture. These populations were assessed by direct on-ground transect counts, as well as indirect data collection through interviews with hunters. Hunting levels were also assessed to determine causal effects on wildlife from either habitat disturbance or hunting *per se*; a thorough analysis of wildlife utilisation, including a sustainability evaluation, is presented in Holbech (2014) based on data in Holbech (1996). The study was conducted from September 1993 to November 1995, and is parallel to avifaunal studies (Holbech, 2005 and 2009). Results are compared to experiences from the Guineo-Congolian forest biome, and discussed in local and regional contexts with emphasis on off-reserve integrated wildlife and land-use management as a conservation tool for maintaining high diversity and production of African rainforest game. Although the data of the present survey are over 20 years old, they serve as important historical baseline information on forest game abundances related to habitat disturbances, which can be used for future comparative research purposes and prospective development of management tools for conservation of forest wildlife and biodiversity in Ghana and the sub-region at large.

Study Area and Sites

The overall study area of Ghana's Western Region covers almost 24,000 km², where some 7,500 km² and 12,500 km² respectively are reserved and unreserved forest, the rest being urban areas and roads (Otsyina and Asare, 1991). Sixteen 'timber protected' forest reserves (FRs), one 'wildlife protected' resource reserve (Ankasa RR), three exotic plantations (*Gmelina arborea* and *Cedrela odorata*), and two neglected cash-crop plantations (coconut and shaded cocoa) were selected (Figures 16.1 and 16.2), covering wet evergreen (WE) and moist evergreen (ME) forest zones in Ghana (Hall and Swaine, 1976). Both exotic and cash crop plantations had interspersing of a diverse secondary plant community giving them a luxuriant appearance. The two forest zones differ only slightly in precipitation and forest structure; WE-zone receives

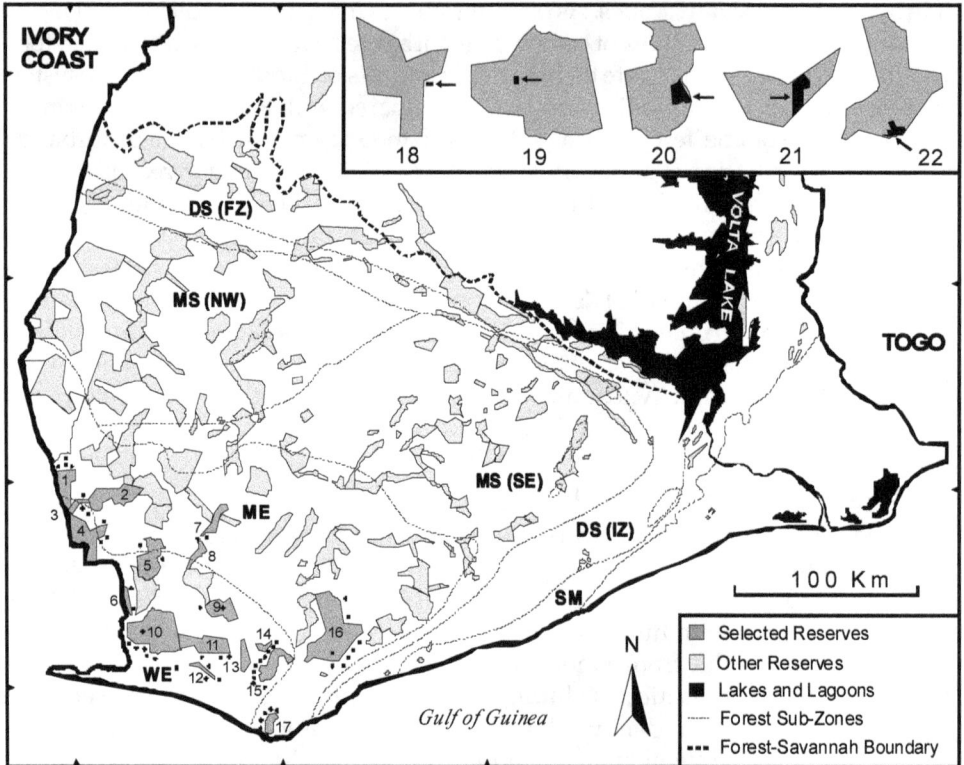

Figure 16.1: Map of Southern Ghana Showing High Forest Reserves (FR), Forest Sub-zones as well as 22 Selected Forest Habitats and 55 Interview Sites (% = towns, villages and hamlets).

The position of plantation habitats is indicated by arrows on respective enlarged reserves in the upper right corner.

Forest sub-zones after Hall and Swaine (1976): WE: Wet Evergreen; ME: Moist Evergreen; MS: Moist Semi-deciduous; DS: Dry Semi-deciduous; SE: South-east; NW: North-west; FZ: Fire Zone; IZ: Inner Zone; SM: Southern Marginal.

Forest habitats: 1: *Dadiaso FR*, 2: *Yoyo FR*, 3: *Disue FR*, 4: *Boin FR*, 5: *Tano Nimri FR*, 6: *Jema-Assemkrom FR*, 7: *Bura FR*; 8: *Mamiri FR*, 9: *Fure FR*, 10: *Ankasa Resource Reserve*, 11: *Draw FR*, 12: *Ebi FR*, 13: *Ndumfri FR*, 14: *Neung North FR*, 15: *Neung South FR*, 16: *Subri FR*, 17: *Cape Three Point FR*, 18: *Boin Cocoa*, 19: *Ankasa Coconut*, 20: *Tano Cedrela*, 21: *Neung Cedrela*, 22: *Subri Gmelina*.

>2,000 mm annually and supports closed-canopy forest reaching an average height of 35-40 m, ME-zone gets some 1,750-2,000 mm per year and has a slightly discontinuous canopy of 35-45 m (Hall and Swaine, 1976). Both faunal and floral diversity is highest in the WE-zone, although the majority of large vertebrates occur in both zones (Martin, 1991; Hawthorne and Abu-Juam, 1995). All selected forest sites lie in a lowland area with altitudes between 25-300 m.a.s.l. Table 16.1 describes the 22 selected areas, detailing forest fragmentation, logging disturbance, hunting

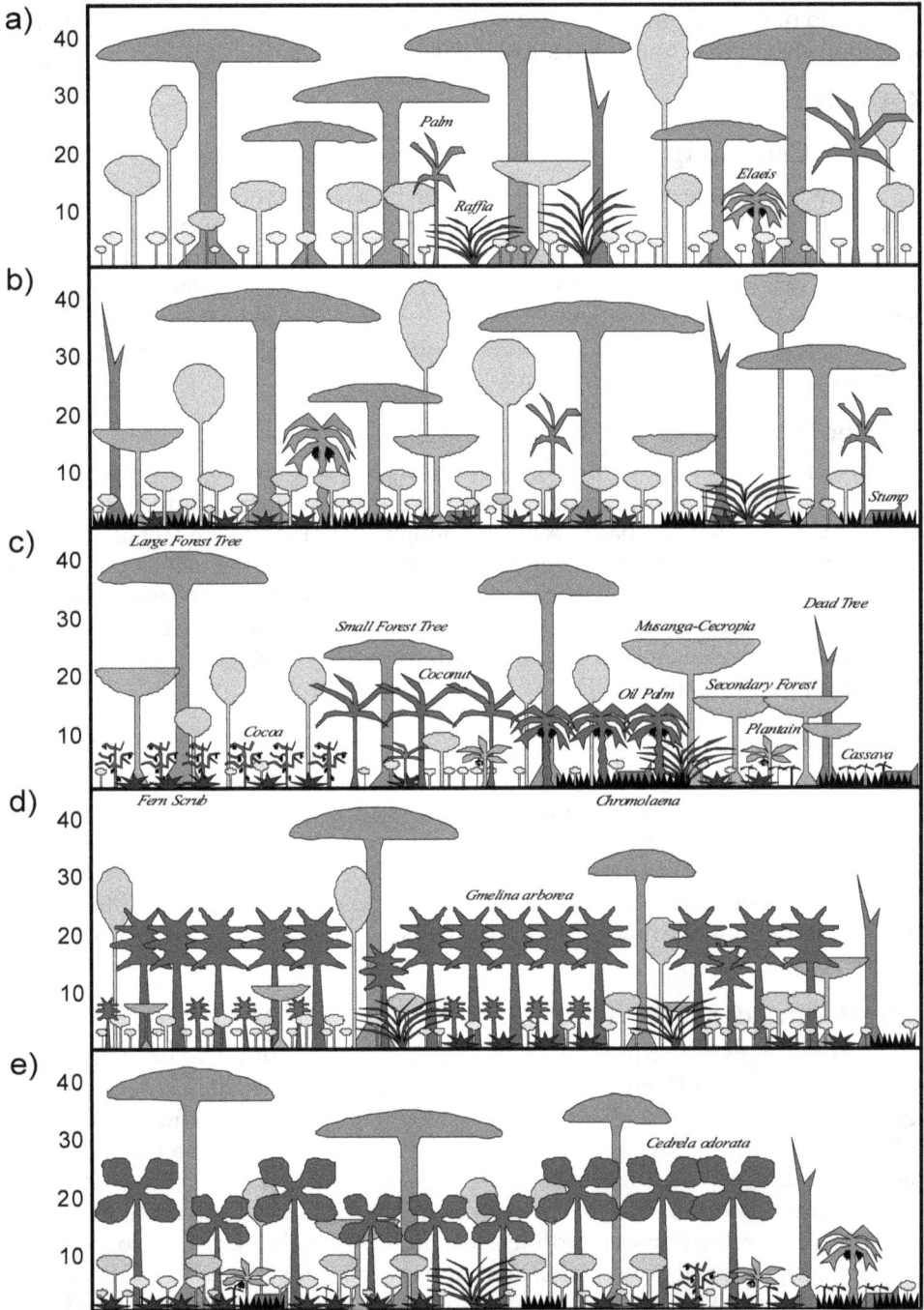

Figure 16.2: Profile Diagram of (a) Mature (Primary) Forest, (b) Selectively Logged Forest, (c) Cocoa and Coconut Plantations Interspersed with Mixed-Crop Farms and Shaded by Secondary Forest and Remnant Forest Trees, (d) *Gmelina* Plantation, and (e) *Cedrela* Plantation of Southwest Ghana, 1993-1995.

activity and human population density. Detailed information on logging history and plantation structure is found in Holbech (2005 and 2009).

Methods

Transect Censusing

All transects were cut prior to the censuses by a 6-8 person team comprising the author, a Ghana Wildlife Division (GWD) senior officer and local hunters and farmers with good game-tracking skills. Each transect was 1-3 km long and 1 m wide, and prepared so as to facilitate easy and silent movement. Three key kinds of data fields were recorded on transects: signs of logging, hunting, and animal presence. Transect censusing was not conducted in Neung South and Ndumfri forest reserves, only informant data were acquired from here.

Logging Disturbance and History

Any visible signs of ongoing or historic logging were recorded systematically within 5 m from the transect midline; these signs included logs, stumps, skid tracks, hauling roads, and loading bases. A logging index (LI) was calculated for all transects in a particular forest, based on a weighted score index according to the quantitative disturbance impacts on the vegetation structure (Holbech, 2005). The index does not account for the difference in regeneration time since last logging. Logging history for each reserve was based on Hawthorne and Abu-Juam (1995) supplemented by Forest Management Plans from the GFC.

Hunting Activity Index

Five hunting activity descriptors were recorded along transects within 5 m from the transect midline: 1) hunting trails, 2) cable snares, 3) spent shotgun-cartridges, carbide powder or batteries, 4) gunshots, and 5) encounters with hunters, non-timber-forest-product (NTFP) collectors and their dogs. Records were transformed to encounter frequencies (scores) per km walked transect, and the descriptors were pooled, representing the hunting index (HI) for each of the 20 study sites (Appendix 3; Table 16.1).

Animal Abundance Indices

Multiple-signs censusing was applied, including both audio-visual detection of animals, and the counting of dung piles, tracks, trails and diggings (Appendix 1). Sightings and vocalisations were primarily gathered from primates, diurnal antelopes and rodents. Dung piles, track and trail counting produced data on elephants, antelopes and carnivore-omnivores, and diggings assessed bush pig abundance. Diggings were detectable up to 5 m from transects, but tracks and dung piles only within 2.5 m from the transect midline. Nest building by chimps was also included. Cumulated signs for all transects in each habitat were transformed to an abundance index per km of walked transect (*i.e.* an animal signs encounter frequency), making a direct comparison on the habitat level possible.

Table 16.1: Characteristics of the 22 Sampling Sites in Southwest Ghana (1993-1995), Showing Four Forest Categories (1-4) and Two Plantations Types: Cash Crop (CCP) and Exotic Tree (ETP)

Forest Site	Habitat Category[a]	Forest Zone[b]	Size (km²)	Distance (km)[c]	Reg. Time (Years)[d]	Logging Index[e]	Hunting Index[f]	HPD (km^{-2})[g]
Jema	1	WE	66.0	3	no logging	0.0	3.5	35
Disue	1	WE-ME	23.6	0	15	0.0	0.8	35
Cape	1	WE-ME	51.0	15	19	5.0	4.9	185
Ebi	1	WE	25.9	2	6	10.0	2.5	60
Mamiri	1	ME-WE	45.3	0	21	16.8	1.6	45
Ndumfri	2	WE	72.5	4	4	–	–	85
Neung South	2	WE	112.7	3	6	–	–	85
Neung North	2	WE	45.0	3	8	35.6	3.0	85
Fure	2	WE	158.2	0	5	59.1	1.4	45
Bura	2	ME	103.1	0	2	131.3	2.3	45
Ankasa	3	WE	348.7	0	20	18.5	0.8	60
Yoyo	3	ME	235.7	0	14	19.9	1.3	35
Dadiaso	3	ME-WE	171.2	0	13	6.0	0.8	35
Subri	4	*ME*	587.9	0	0	59.0	3.6	40
Draw	4	WE	235.0	0	0	60.6	2.4	60
Tano Nimri	4	WE-ME	205.9	0	2	45.8	2.0	35
Boin	4	WE-ME	277.7	0	21	43.2	1.5	35
Ankasa Coconut	CCP	WE	1.5	Within	–	–	2.8	60
Boin Cocoa	CCP	WE-ME	0.5	0.5	–	–	10.0	35

Contd...

Table 16.1–*Contd...*

Forest Site	Habitat Category[a]	Forest Zone[b]	Size (km²)	Distance (km)[c]	Reg. Time (Years)[d]	Logging Index[e]	Hunting Index[f]	HPD (km⁻²)[g]
Subri Gmelina	ETP	ME	48.0	Within	–	–	2.5	40
Tano Cedrela	ETP	WE-ME	9.8	Within	–	–	2.5	35
Neung Cedrela	ETP	WE	6.4	Within	–	–	2.2	85

a: 1 = area (a) < 160 km², tree extraction (te) < 0.9 ha⁻¹; 2 = a < 160 km², te > 0.9 ha⁻¹; 3 = a > 160 km², te < 0.9 ha⁻¹; 4 = a > 160 km², te > 0.9 ha⁻¹; b: From Hall and Swaine (1976): WE = Wet Evergreen (> 2,000 mm); ME = Moist Evergreen (1,750-2,000 mm); c: To nearest protected forest area (0 = linked to other reserve); d: Time since last logging occurred in the reserve; e: Based on transect records of logs, stumps, skid trails, hauling roads, loading stations (Holbech, 2005); f: Based on transect records of gunshots, cartridges, snares, trails and hunters (Holbech, 1996); g: Human population density, based on extrapolation from 1984-census (Ghana Statistical Service Department).

Informant Interviews

Information was gathered from hunters living close to the reserves included in the study. A detailed questionnaire on animal encounters was developed prior to the survey, and accompanied by coloured species drawings, based on Haltenorth and Diller (1988). The main questions for each species were: i) last encounter (days, weeks, months, years), and ii) habitat preference (reserved forest; unreserved forest = remnants and old secondary forest; farms and farm-bush = neglected farms and young re-growth; rubber plantations). Respondents were selected randomly to get a full representation across age, literacy and ethnicity. The author interviewed in English, or through the GWD-officer as a translator of local dialects. I did not use any reward system, but performed traditional customary rights. To assess whether respondents were able to distinguish between closely related species, indigenous names were always used and a brief or (for rare species) detailed description was required. Interviews with hunters lasted between 45 minutes and two hours (mean = 75 minutes). One major disadvantage of questionnaires was the distinct suspicion of interviewers expressed by some respondents living close to the Ankasa Resource Reserve (GWD-protected wildlife reserve). Interviews were terminated if respondents seemed unreliable. To avoid immediate suspicion, we never wore uniforms, nor used stickers on my vehicle. I emphasised to respondents that the purpose of our work was purely academic.

Comparative Data Analyses

To assess the relative impact of logging and fragmentation on game populations, the 17 forest sites were divided into four categories; lightly logged small forests (< 160 km²; LI < 35 ~ 0.90 trees ha^{-1} = the mean extraction level among the 14 surveyed logged forests): Cape Three Points, Disue, Ebi, Jema and Mamiri FRs; lightly logged large forests (> 160 km²; LI < 35): Ankasa RR, Dadiaso and Yoyo FRs; moderate-heavily logged small forests (< 160 km²; LI > 35): Bura, Fure, Neung North, Neung South and Ndumfri FRs); moderate-heavily logged large forests (> 160 km²; LI > 35): Boin, Draw, Tano and Subri FRs. Site-specific faunal data from transect surveys and interviews were pooled according to these four forest categories, to enable comparison of animal abundances and diversity. Transect data from the three exotic plantations (Subri Gmelina, Tano Cedrela, Neung Cedrela) and the two cash-crop plantations (Ankasa Coconut, Boin Cocoa), were also pooled. Informant data on animal preferences for non-forest habitats (see Appendix 2) do not refer to these five plantations, but to rubber estates, and 'unreserved forest' refers to old secondary forest or forest remnants.

Rarity and Conservation Status of Forest Game

To assess the conservation status of each species, I used the mean annual encounter frequency (MAEF), derived from informant data. For each respondent the most recent encounter of a species was transformed into annual encounters per each respondent, and the mean for all respondents equal the MAEF for a particular species. For example, if a respondent's last reported encounter with an animal was two years ago, their annual encounter frequency is 0.5; if the encounter was four weeks ago, the figure is 13.0 encounters per year. If the MAEF for an animal is 0.02 encounters per year, it means that out of 100 respondents only two have encountered it a year ago, or alternatively only 10 out of 100 respondents have encountered it five years ago. The highest possible MAEF for a species is 365.0, which means that all respondents had encountered it one day ago. MAEF for particular species are assumed to be directly linked to how commonly animals are encountered by an average hunter, and hence reflecting how common/rare animals genuinely are. Species with a MAEF < 5.0 were termed endangered in the Western Region and Ghana at large. Four other categories were considered: *rare* ~ MAEF = 5.1-20, *uncommon* ~ 20.1-40, *common* ~ 40.1-100, and *widespread* ~ > 100. The categorisation of each species is presented in Appendix 2.

Transformations of Informant Statistics

A high MAEF should reflect high animal abundance, assuming that the detection ability of an average informant was similar across the four forest categories. Four factors *inter alia* may influence MAEF-calculation based on informants; their age, experience, number of traps, and particularly frequency of hunting. When comparing mean values of these four factors across the four forest categories none of them were statistically different (χ^2 range = 0.21-5.19, $p > 0.05$; Table 2). However, species specific MAEF for the four forest categories were adjusted to an activity of 3.0 hunting trips per week, as hunting frequency is probably the main factor that influence how often hunters may encounter animals.

Results

Logging Disturbances

The correlation between calculated logging index (LI) and the number of counted logs and stumps per km transect (= tree extraction per ha) was highly significant (GLM: $r = 0.970$, $p < 0.001$; $df = 14$), and from the regression predicts that number of trees extracted per ha is: TE = (LI – 6)/33 (Figure 16.3a). Highest tree extraction was recorded for the recently logged Bura FR (3.8 ha^{-1}), but the average was 0.90 ha^{-1} across the 14 logged forests (Jema FR was virgin). Lightly logged small forests had mean extraction levels of 0.1 ± 0.1, moderate-heavily logged small forests 2.3 ± 1.3, lightly logged large forests 0.1 ± 0.1, and moderate-heavily logged large forests 1.2 ± 0.5. There was a significant negative correlation ($r = -0.620$, $p < 0.05$; $df = 13$) between LI and regeneration time since last logging (Figure 16.3b), and the mean regeneration time was three times higher in lightly logged forests compared to heavily logged forests (Table 16.2). Hence, lightly logged forests have had a longer time to recover than heavily logged forest, indicating that logging has intensified in recent years. However, the maximum regeneration time recorded (*c.* 20 years) was only half the current minimum cycle of 40 years in Ghana.

Faunal Records from Transect Surveys

A total of 287 records distributed across 25 mammalian and two reptilian game species were encountered from walking 342.4 km on a total of 100.6 km of cut transects in 15 reserves and five adjacent cash-crop and exotic tree plantations (Table 16.3; Appendix 1). This produced an overall animal sign encounter frequency of 0.84 km^{-1} (Table 16.5). In forest areas 243 records were obtained during 318.9 km transect

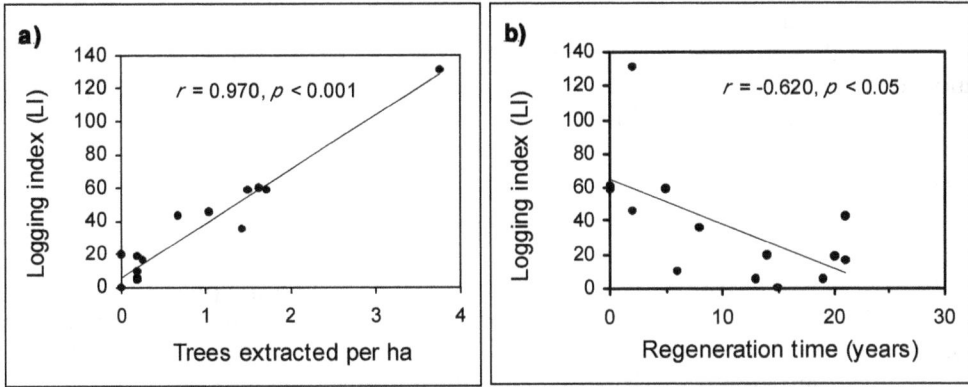

Figure 16.3: Correlations (*n* = 15 Ghanaian forests) between Logging Disturbance Index (LI) and, (a) Trees Extracted per ha; and (b) Regeneration Time since Last Logging.

walk producing an overall encounter frequency of 0.76 animal signs km^{-1} (Table 16.5). For plantations 44 records were encountered on 23.5 km – equalling 1.87 animal signs km^{-1} (Cash crop: 24/8 km = 3.00 km^{-1}; Exotic tree: 20/15.5 km = 1.29 km^{-1}). Only 21 per cent (*n* = 61) of total records were sightings, which came from a total of 19 species, primarily ungulates (39 per cent), small carnivores (29 per cent) and rodents

Table 16.2: Sampling Efforts and Exploitation Characteristics (* = Means for each habitat category) of Four Forest and Two Non-Forest Habitats, Southwest Ghana (1993-1995), Based on Combining Individual Sites within Categories of Forest. LI = Logging Index

Area Description and Sampling Efforts	Small Forest (< 160 km²)		Large Forest (> 160 km²)		Exotic Tree Plantations	Cash-Crop Plantations
	LI < 35	LI > 35	LI < 35	LI > 35		
Transect length (km)	21.5	14.4	24.2	31.2	7.3	2.0
Transect walked (km)	57.2	45.3	110.0	106.4	15.5	8.0
Area size* (km²)	42.4	98.3	251.9	326.6	21.5	1.2
Regeneration time* (yrs)	15.3	5.0	15.7	5.8	–	–
Tree extraction (ha⁻¹)	0.1	2.3	0.1	1.2	–	–
Logging index*	6.4	75.4	14.8	52.2	–	–
Hunting index*	2.7	2.2	0.9	2.4	2.4	6.4
Human population* (km⁻²)	72.0	72.0	43.3	42.5	53.3	47.5
Number of respondents	27	27	31	27	–	–
Age* (yrs)	43	43	42	46	–	–
Experience* (yrs)	17	19	11	13	–	–
Traps per respondent*	66	91	68	78	–	–
Weekly hunting trips*	3.1	3.4	2.7	3.2	–	–

(21 per cent). Only one monkey species was sighted, the Lowe's monkey (*Cercopithecus mona lowei*), though it was heard frequently in most reserves visited, unlike the Roloway monkey (*Cercopithecus diana roloway*) and the Lesser spot-nosed monkey (*Cercopithecus petaurista petaurista*), which were rarely heard. By far the most frequently heard game species was the African giant squirrel (*Protoxerus stangeri*) making up 76 per cent and 38 per cent of all vocalisations (*n* = 144) and overall records respectively.

Dung piles were identified as coming from the Bongo (*Tragelaphus euryceros*), Bay duiker (*Cephalophus dorsalis*), Black duiker (*Cephalophus niger*) and Maxwell's duiker (*Cephalophus maxwelli*). The mean density of duiker dung piles alone was 2.1 piles per ha. A faecal deposition site ('latrine') of the African civet (*Civettictis civetta*) was also detected three times. Tracks were mainly recorded from large mammals such as elephants and tragelaphines, while the Red river hog (*Potamochoerus porcus*) was detected from the signs of its 'ploughing' foraging behaviour (diggings). In summary, the quantity of transect data was very low relative to the high sampling effort expended, both regarding time spent and areas covered. Due to the scarcity of data for each of the 20 sites surveyed, a direct comparative analysis between these is statistically inappropriate, and comparative analyses are restricted to the six overall habitat categories (Table 16.1).

Faunal Records from Informants

A total of 112 hunter-farmer interviews in 55 towns, villages and hamlets provided information on 42 mammal, nine reptile and at least nine bird species commonly used as bushmeat (Appendix 2). These species had been observed, killed or traded by the informants. All species recorded during transect surveys were also reported by informants. The mammal list included virtually all forest species above 0.5-1 kg and up to the Forest elephant *Loxodonta (africana) cyclotis*. The only non-forest mammals reported were the Cane rat (*Thryonomys swinderianus*) and Crested porcupine (*Hystrix cristata*) that are savannah-woodland and coastal scrub species, now found deep within the Ghanaian forest zone. Generally, respondents did not distinguish between Emin's (*Cricetomys emini*) and Gambian (*Cricetomys gambianus*) giant pouched rats, or between the Tree pangolin (*Phataginus tricuspis*) and the Long-tailed pangolin (*Uromanis tetradactyla*).

The reptile list included the African python (*Python sebae*), Broad-fronted crocodile (*Osteolaemus tetraspis*), Nile crocodile (*Crocodilus niloticus*), Nile monitor (*Varanus niloticus*) and five chelonians; the African helmeted turtle (*Pelomedusa subrufa*), the West African mud turtle (*Pelusios castaneus*), and three *Kinixys* spp. hinged tortoises, which were not always distinguished by hunters. Among these reptiles, only the Broad-fronted crocodile and hinged tortoises are truly forest dependent. The bird list included Yellow-necked picathartes (*Picathartes gymnocephalus*), White-breasted guineafowl (*Agelastes meleagrides*), Crested guineafowl (*Guttera pucherani*), Great blue turaco (*Corythaeola cristata*), four large undistinguished casqued hornbills (*Ceratogymna elata/atrata* and *Bycanistes cylindricus/subcylindricus*), and the Crowned eagle (*Stephanoaetus coronatus*). The Palm-nut vulture (*Gyphohierax angolensis*), francolins, and several other larger birds (*e.g.* hawks, buzzards, egrets, pigeons, ducks, rails) were also hunted occasionally, predominantly in areas depleted of large game.

Rarity and Conservation Status of Forest Game

Out of 60 commonly hunted game species, 16 (26.5 per cent) were considered endangered, *e.g.* Forest elephant, Forest buffalo (*Syncerus caffer nanus*), Yellow-backed duiker (*Cephalophus sylvicultor*), Water chevrotain (*Hymenoschus aquaticus*), Giant hog (*Hylochoerus meinertzhageni*), Leopard (*Panthera pardus*), Golden cat (*Felis aurata*), Ratel (*Mellivora capensis*), Spot-necked otter (*Lutra maculicollis*), Giant pangolin (*Smutsia gigantea*), Western chimpanzee (*Pan troglodytes verus*) and notably Miss Waldron's red colobus (*Piliocolobus badius waldronae*) (Appendix). Twelve (20 per cent) and 10 (16.5 per cent) species were respectively rare and uncommon, with the Bongo as a candidate to become endangered. None of the reptiles were considered abundant, and most species were uncommon or rare. Hence, 63 per cent of forest game species can be considered as becoming alarmingly scarce in the Western Region and Ghana at large. Common and widespread forest game comprised thus 37 per cent, and typical species were Bushbuck (*Tragelaphus scriptus*), Black and Maxwell's duiker, rodents and small game (weight < 3 kg). The picathartes was highly endangered, with only two respondents having recorded it some 2-3 years ago in the hilly areas of Ndumfri FR. This species however, has its main distribution outside of the Western Region, and several isolated populations exist elsewhere in Ghana (Birdlife International, 2009). The White-breasted guineafowl was known from most reserves, though it was much rarer than the Crested guineafowl. Large casqued hornbills as well as the Great blue turaco were abundant and widespread, though hornbills increasingly were becoming restricted to the forest interior, and tended to avoid anthropogenic forest landscapes.

Forest Game in Fragmented and Selectively Logged Forest

Transect data showed no significant differences with respect to total relative abundance ($\chi^2 = 1.41$, $p > 0.70$, $df = 3$), species recorded ($\chi^2 = 2.43$, $p > 0.30$, $df = 3$), and ecological diversity ($\chi^2 = 0.526$, $p > 0.90$, $df = 3$) when comparing the mean encounter frequencies from the four forest categories (Table 16.3). Similarly, there were no significant differences (χ^2 range = 0.24-2.55, $p > 0.05$) between the four categories when comparing single species or guilds (Table 16.3). Hence, there were no obvious correlations between selective logging and forest area versus game abundance and diversity in general. However, the data suggest that small lightly logged or largely undisturbed forests contained the lowest abundance and diversity of game. Both moderate-heavily logged small forests and lightly logged or unlogged large forest had similar abundance and diversity of game.

In contrast to the statistically scarce transect data, MAEF derived from informants varied greatly among the four forest categories for the majority of species and guilds. Overall abundance (MAEF-based) was significantly higher ($\chi^2 = 45.03$ and 134.1, $p < 0.001$) for both large and small moderate-heavily logged forests compared to lightly logged large and small forests, and this trend was apparent for four guilds; sub-canopy monkeys ($\chi^2 = 19.54$ and 167.02, $p < 0.001$), small carnivores/omnivores ($\chi^2 = 31.85$ and 58.10, $p < 0.001$), other mammals ($\chi^2 = 8.08$ and 7.05, $p < 0.01$), and birds ($\chi^2 = 64.47$ and 22.15, $p < 0.001$). Note that endangered/rare game attained their highest MAEF in moderate-heavily logged large forests, whereas lightly logged large forests had lowest MAEF for these animals.

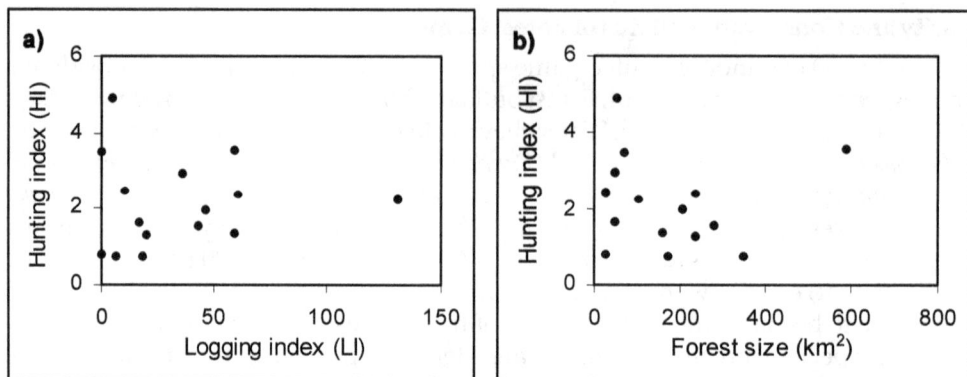

**Figure 16.4: Correlations (*n* = 15 Ghanaian forests) between
Hunting Activity Index (HI) and, (a) Logging Disturbance Index (LI); and,
(b) Forest Fragmentation (Size of forest in km²).**

Summarising transect and informant data, trends suggest that logging had no negative effect on overall game abundance and diversity, and that some game guilds may benefit from logging. As indicated by the high levels of hunting activities measured throughout the study area (Table 16.2), human population density and consequently hunting pressures likely have more adverse impacts on game populations than both logging and fragmentation (Holbech, 2014). However, there was no simple relationship between the three quantified disturbance factors, as is shown in Figures 16.4a and 16.4b.

Forest Game in Exotic Tree and Cash-Crop Plantations

A total of 13 forest game species were recorded during transect surveys in luxuriant exotic plantations (gmelina and cedrela) and cash-crop plantations ('rustic cocoa' and neglected coconut), including four antelope species, Red river hog, Tree hyrax (*Dendrohyrax dorsalis*), two sub-canopy monkeys, two carnivores and three rodents, whereas no reptilian game was detected (Table 16.3). Both habitat types had 11 mammal species each, and species overlap was 82 per cent. Compared to the relatively small sampling effort (23.5 km walked on total 9.3 km transect), the species richness was considerable and contained several species that constitute the bulk of bushmeat emanating from the Ghanaian forest zone, particularly Bushbuck, Maxwell's and Black duiker, Giant rats and Cane rat (Crookes *et al.,* 2005). Moreover, overall relative abundance was significantly higher ($\chi^2 = 13.47, p < 0.001$) in neglected cash-crop plantations (mean 3.00 records km⁻¹) compared to average abundance levels in forest (mean = 0.76 records km⁻¹), and this pattern was mainly attributed to three rodent species. Overall relative abundance in exotic plantations was 72 per cent higher than the forest average, although the difference was insignificant ($\chi^2 = 1.42, p > 0.20$).

The relative abundance of Bushbuck and small antelopes was comparatively high in both plantation types; notably, Black duiker abundance was higher in plantations than in any of the four forest categories. Overall diversity was similar in

Table 16.3: Relative Abundance (Records per 10 km walked transect), Number of Species (in parentheses) and Ecological Diversity (D_{Mg} = (S − 1) · (lnN)$^{-1}$) of Forest Game Recorded in Six Habitat Categories in Southwest Ghana, 1993-1995. MTE = mean tree extraction (ha^{-1}).

	Small Forest (< 160 km²)		Large Forest (> 160 km²)		Exotic Tree Plantations	Cash-Crop Plantations
	MTE=0.1	MTE=2.3	MTE=0.1	MTE=1.2		
Forest elephant	0	0	0.27	0.19	0	0
Large antelopes[1]	0	0.66 (2)	0.55 (3)	0	0.64 (1)	1.25 (1)
Small antelopes[2]	1.22 (3)	1.99 (4)	2.09 (4)	1.32 (3)	3.87 (3)	2.50 (2)
Red river hog	0.17	0.66	0	0.85	1.29	1.25
Monkeys[3]	0.87 (1)	0.66 (2)	0.63 (2)	1.51 (2)	1.94 (2)	2.50 (2)
Carnivores[4]	0.70 (3)	1.32 (1)	0.27 (3)	0.56 (2)	0.65 (1)	1.25 (1)
Terrestrial rodents[5]	0	0.22 (1)	0.18 (1)	0.09 (1)	1.94 (2)	10.00 (2)
Arboreal rodents[6]	2.45 (1)	2.21 (2)	5.18 (1)	1.69 (2)	2.58 (1)	10.00 (1)
Other mammals[7]	0	0.66 (1)	0	0.09 (1)	0	1.25 (1)
Reptiles[8]	0	0.44 (2)	0.18 (1)	0.19 (2)	0	0
Endangered or rare[9]	0	0.66 (3)	0.73 (3)	0.66 (3)	0	0
Total records (N)	31	40	103	69	20	24
Total abundance	5.42	8.83	9.36	6.48	12.91	30.00
Total species (S)	9	16	16	15	11	11
Diversity (D_{Mg})	2.33	4.07	3.24	3.31	3.34	3.15

1: Bongo, Bushbuck, Yellow-backed duiker; 2: Bay duiker, Black duiker, Maxwell's duiker, Royal antelope; 3: Roloway, Lowe's, Lesser spot-nosed monkey; 4: African civet, African palm civet, Blotched genet, Marsh mongoose, Cusimanse; 5: Brush-tailed porcupine, Emin's giant rat, Cane rat; 6: Giant forest squirrel, Pel's anomalure, Lord Derby's anomalure; 7: Tree pangolin, Tree hyrax; 8: Broad-fronted crocodile, Nile monitor lizard; 9: Elephant, Bongo, Yellow-backed duiker, Roloway monkey, Broad-fronted crocodile.

both plantations (mean = 3.25 ± 0.13), and comparable to forests (mean = 3.24 ± 0.71). However, elephants, bongos and upper-canopy monkeys were not recorded in these anthropogenic habitats. Informant data on habitat preferences (Appendix 2) were comparable with transect data, showing that Bushbuck, small antelopes and terrestrial rodents thrive in unreserved forest, plantations and farmland. The mean hunting activity index (HI) in cash crop plantations (6.4 ± 5.1) was almost threefold the mean of forests (2.2 ± 1.2) or exotic plantations (2.4 ± 0.2) (Table 16.2).

Discussion

The Importance and Reliability of Informant Data on Forest Game Abundances

This extensive study (345 days, walking of 342 km on > 100 km transect) clearly demonstrated the limited usefulness of both direct and indirect transect censusing of

Table 16.4: Mean Annual Encounter Frequency (MAEF), Species Number (in parentheses) and Ecological Diversity ($D_{Mg} = (S - 1) \cdot (lnN)^{-1}$) of Forest Game Based on Interviews with 112 Hunter-Farmers, in Relation to Four Forest Categories in Southwest Ghana, 1993-1995. MTE: Mean tree extraction (ha^{-1}). MAEF-values are adjusted to 3.0 hunting trips per each forest category.

Forest Game	Small Forest (< 160 km²)		Large Forest (> 160 km²)		All Four Categories Compared (χ^2-test)
	MTE=0.1	MTE=2.3	MTE=0.1	MTE=1.2	
Forest elephant	0.2	0	1.4	0.2	2.73, $P > 0.30$
Forest buffalo	0.3	0.05	0.1	0.05	0.34, $P > 0.95$
Large antelopes[1]	57.0 (3)	57.5 (3)	35.1 (3)	71.1 (3)	12.06, $P < 0.01$
Small antelopes[2]	134.2 (5)	168.0 (5)	97.8 (5)	155.2 (5)	20.34, $P < 0.001$
Bush pigs[3]	34.3 (2)	39.0 (2)	8.0 (2)	16.4 (2)	26.37, $P < 0.001$
Water chevrotain	0.2	0.2	0.1	0.4	0.21, $P > 0.95$
Chimpanzee	3.5	0.1	0.7	6.2	9.00, $P < 0.05$
Top-canopy monkeys[4]	29.2 (3)	27.3 (3)	11.0 (3)	19.5 (3)	9.51, $P < 0.05$
Sub-canopy monkeys[5]	148.6 (4)	235.2 (4)	81.3 (4)	349.6 (4)	197.85, $P < 0.001$
Large carnivores/omnivores[6]	15.8 (3)	46.8 (4)	23.3 (4)	20.5 (4)	21.53, $P < 0.001$
Small carnivores/omnivores[7]	160.9 (4)	279.3 (4)	112.2 (4)	259.1 (4)	93.58, $P < 0.001$
Otters[8]	5.2 (2)	0.3 (2)	1.1 (2)	10.2 (2)	14.72, $P < 0.01$
Terrestrial rodents[9]	292.0 (3)	282.7 (3)	287.8 (3)	230.5 (3)	9.08, $P < 0.05$
Arboreal rodents[10]	227.3 (3)	185.2 (3)	163.3 (3)	184.5 (3)	11.35, $P < 0.01$
Giant pangolin	0.3	0.05	0.1	0.1	0.27, $P > 0.95$
Other mammals[11]	326.5 (4)	403.3 (4)	299.6 (4)	368.2 (4)	17.93, $P < 0.001$
Reptiles[12]	90.4 (5)	77.2 (5)	66.3 (5)	102.5 (5)	8.83, $P < 0.05$
Birds[13]	133.8 (6)	228.6 (7)	111.7 (6)	194.0 (6)	51.99, $P < 0.001$
Non-forest game[14]	75.9 (3)	123.3 (2)	72.1 (2)	62.3 (3)	26.63, $P < 0.001$
Endangered/rare[15]	106.4 (19)	80.9 (20)	48.6 (20)	123.7 (20)	35.61, $P < 0.001$
Total abundance (MAEF = N)	1,735.6	2,154.1	1,373.0	2,050.6	203.17, $P < 0.001$
Total species number (S)	55	55	55	56	0.01, $P > 0.99$
Diversity (D_{Mg})	7.24	7.04	7.47	7.21	0.01, $P > 0.99$

1: Bongo, Bushbuck, Yellow-backed duiker; 2: Bay duiker, Ogilby's duiker, Black duiker, Maxwell's duiker, Royal antelope; 3: Giant hog, Red river hog; 4: Red colobus, Pied colobus, Diana monkey; 5: Olive colobus, Mona monkey, Lesser spot-nosed monkey, White-naped mangabey; 6: Leopard, Golden cat, African civet, Ratel; 7: African palm civet, Blotched genet, Marsh mongoose, Cusimanse; 8: Spot-necked otter; 9: Brush-tailed porcupine, Emin's/Gambian giant rat; 10: Giant forest squirrel, Pel's anomalure; 11: Potto, Tree/Long-tailed pangolin, Tree hyrax; 12: African python, Broad-fronted crocodile, Nile monitor lizard, Mud turtles (2 species); 13: White-breasted guineafowl, Crested guineafowl, White-necked picathartes, Great blue turaco, Casqued hornbills (3 species), Crowned eagle; 14: Nile crocodile; Crested porcupine, Cane rat; 15: See Appendix 2 (21 species).

Ghanaian rainforest game. The scarcity of encounters with animals may have reflected low game densities, but most likely also an increased sensitivity of wildlife to human presence, caused by decades of excessive hunting. Unprecedented hunting levels may have enforced natural secretive anti-predator behaviour, resulting in extreme elusiveness and wariness towards humans. It is known from other heavily hunted areas that duikers may change their anti-predator strategy towards hunters, *e.g.* their response patterns of freezing and whistling, thereby avoiding hunters at longer distances (Croes *et al.,* 2006). We normally detected duikers when flushed from their daybeds in an explosion of retreat but often without whistling. The best strategy for a moving duiker may be to quietly elude hunters without whistling, in order to disguise their presence, or if already stationary to remain motionless in a concealed place until extreme proximity, before a rapid escape.

The low encounter frequency of monkeys in particular (mean = 0.11 groups km^{-1}), being 10-25 per cent of late 1970s-levels in Ghanaian moist forests (Oates, 1996), indicates that these have been detrimentally hunted, and it is possible that upper-canopy species (Pied/Red colobus, Roloway monkey) may have called less frequently in order to remain undetected by hunters, a phenomenon that has been observed in West and Central Africa (Bshary, 2001; Croes *et al.,* 2006). Several informants claimed that certain species shift the bulk of their day activities towards more nocturnal habits, so that night transect surveys could have given more accurate information on the relative abundances of these animals, particularly antelopes and carnivores. Nocturnally estimated Blue duiker (*Cephalophus monticola*) densities in Cameroon are five times higher than diurnal counts, indicating that this close relative to the Maxwell's duiker is best monitored at night as it does not exhibit fugacious behaviour when spotlighted, and is easily detected at distance by eye shine (Waltert *et al.,* 2006). We did a few night walks, providing some information on facultative nocturnal mammals (*e.g.* duikers) and true nocturnals such as Tree hyrax and Potto (*Perodicticus potto*). However, it was beyond the scope my team to systematically conduct night work in addition to extensive daytime work on birds (Holbech, 2005 and 2009).

The present study attempts systematically to integrate the vast knowledge of the local people on forest game, providing an indirect measure of animal abundance through records that have been translated into mean encounter frequencies. Population ecologists may dispute the validity of this procedure and argue that respondents are unable to adequately distinguish related species. However, my GWD-counterpart and I were surprised by the preciseness and details of the descriptions many respondents demonstrated, when asked to portray rare and elusive game. They often corrected the colour drawings we presented, which were thoroughly reproduced from the somewhat imprecise illustrations in Haltenorth and Diller (1988). Hunters and farmers may actually possess detailed knowledge on feeding ecology and habitat requirements of several species little known to science; 10-20 years of experience, gathered from 3-4 weekly hunting trips, is much more than that accumulated by most ecologists working in West Africa (Abedi-Lartey, 2004). Using local people's experience of forest game has been demonstrated successfully in direct censusing of rainforest animals (*e.g.* Marks, 1994 and 1996; Noss, 1998 and 1999; Muchaal and Ngandjui, 1999) and their habitats (Abedi-Lartey, 2004).

I cannot rule out that some respondents may have been biased by external factors to manipulate information, *e.g.* by under-reporting or exaggerating yields. However, it is unlikely that all randomly selected respondents in each of the four forest categories should have biased information in the same direction. I did not reward any of my respondents and always carried out interviews in the presence of other community members, who intervened if obviously incorrect information on game encounters were given. Due to a very tense relationship between wildlife protection officers and community members in the Ankasa conservation area, some respondents here may have under-reported game encounters due to fear of subsequent unpleasant GWD-repercussions. This may partly explain the low overall abundance index of game in the category of lightly logged large forests, in which Ankasa was represented with *c.* 10 respondents. In particular, information from hamlets situated inside Ankasa may have been biased due to an adjacent GWD-camp.

Forest Game, Forest Fragmentation and Selective Logging

My survey results suggest that selective logging with average extraction levels of 1.7 ± 1.0 trees per ha (range = 0.5-3.8) for heavily logged forests had no negative effect on forest game in general. Diversity of large game and rare forest interior specialists were similar or superior in forests subject to such extraction levels, compared to lightly or virtually unlogged forests of similar size. Only elephants and buffaloes appear absent from small reserves irrespective of logging disturbances. Very large animals obviously require large areas to maintain minimum viable populations. These findings are similar to the conclusions on the impact of selective logging on avian diversity drawn from a parallel study by the author (Holbech, 2005). It is evident that hunting related to human population density has a far greater impact on the abundance and diversity of forest game than do logging and fragmentation *per se* (Holbech, 1996; Holbech, 2014).

The apparently higher abundances of small carnivores, sub-canopy monkeys, and perhaps small antelopes in logged forest may be related to two primary factors that tend to promote the survival and production of game: 1) increased primary production at ground and understory levels, to the benefit of both folivores and frugivores here; 2) enhanced avoidance by game animals of humans because of the higher availability of hideouts in the form of dense vine tangles and the debris of felled trees. In particular, Bushbuck, duikers, Royal antelope (*Neotragus pygmaeus*), sub-canopy guenons, White-naped mangabey (*Cercocebus atys lunulatus*) and Olive colobus (*Procolobus verus*) may benefit from both increased food abundance as well as better concealment opportunities. Nummelin (1990) also found that two duikers, Red river hog and elephants were more abundant in lightly or heavily logged forest in Uganda, where extraction levels ranged between 1 and *c.* 3 trees per ha. Other West African studies have found that elephants prefer logged forest (Short, 1983; Merz, 1986) or secondary regrowth (Barnes *et al.*, 1991). Elephants are attracted to pioneer trees, particularly *Musanga cecropioides* and *Cecropia peltata* (Dudley *et al.*, 1992), which dominate along hauling roads and loading bases.

Physical conditions are probably more favourable for hunters in largely undisturbed forest compared to logged forest, both in terms of unhindered and silent mobility as well as greater visibility, due to sparse ground vegetation and more 'transparent' under- and mid-stories. Vertical and horizontal visibility is often over 50 m in undisturbed forest, but rarely over 25 m in heavily logged forest, particularly where the Siam weed (*Chromolaena odorata*) and other shrubs are abundant. Many hunters stated that they preferred hunting in unlogged forest due to better mobility and visibility, but not because of higher animal abundances. The inhospitable appearance of logged forests to hunters may indicate how suitable they are for wary game.

Selective logging is generally perceived to have a substantial negative effect on both diurnal and nocturnal primates (Johns and Skorupa, 1987; Weisenseel *et al.,* 1993). Most vulnerable are large-bodied and frugivorous monkeys, but dietary diversity among 38 New and Old World primates shows no correlation with disturbance levels (Johns and Skorupa, 1987). The present study found no significant difference in overall monkey abundance between logged and unlogged forests, small or large forests. On the contrary, hunters more commonly encountered three lower strata species (Olive colobus, Lowe's and spot-nosed monkeys) in logged compared to unlogged forests. The two guenons are basically frugivorous, but their dietary flexibility and preference for secondary growth and forest edges may be an important factor for their persistence in logged habitats. The rare upper-canopy Roloway monkey was only detected in two moderately and recently logged forests (Fure and Draw), but never in Ankasa and Jema, which had minimal logging disturbance. Again, this indicates that unsustainable hunting has been a far more detrimental factor than logging and forest fragmentation *per se*, a conclusion likewise made from a study in Cameroon, though logging also appears to have a negative effect on monkeys there (Waltert *et al.,* 2002).

Unfortunately, no colobines were encountered anywhere in this study, but informant data indicate that logging has a negative impact on particularly red colobus. This species is extremely vulnerable to logging as it is an upper-canopy folivore that depends on several common timber species for food (Roth and Merz, 1986; Oates, 1996). The Pied colobus is more resilient to logging and fragmentation (Martin, 1991; Wong and Sicotte, 2006), and was commonly encountered by hunters in my study. This species and Roloway monkeys have a reported dietary diversity 2-3 times higher than red colobus, and therefore may be more resilient to logging disturbances (Martin, 1991). Moreover, both of these species have a better predator avoidance strategy, as they move down to lower strata when disturbed, whereas red colobus typically ascend to the upper canopy, thus increasing their visibility to hunters. In Ghana, hunting pressures are definitely the main cause of the assumed extinction of red colobus (Oates *et al.,* 2000), rather than selective logging and forest fragmentation.

My data did not show a positive correlation between hunting activity and logging disturbances, but rather a tendency that hunting was positively correlated with forest fragmentation and human population density (Holbech, 1996; Holbech, 2014). In this context, as the forest interior is relatively close to the reserve perimeter in small and slender-shaped reserves, easy access to these areas is relatively high. Logging roads seem to play a minor role in the increase of access within Ghanaian forests, as the majority of these roads are not maintained regularly after logging operations have ceased. This contrasts to the scenario in the Congolian forest biome, where logging roads are a major force behind large-scale commercial hunting (Wilkie *et al.,* 2000; Blom *et al.,* 2004 and 2005; Laurance *et al.,* 2006). In my study, commercial hunting, facilitated by hauling roads was only evident in areas with ongoing logging (Draw, Fure and Subri). However, roads inevitably increase access *between* forests in Ghana, thereby dispersing indiscriminate slash-and-burn deforestation. Hence, the low road density in the areas of medium and low population density in the present study certainly explains the high density of forest remnants in the anthropogenic forest landscape, and why these areas support more forest game than heavily populated areas with a relatively well-developed road network (Holbech, 1996).

Forest Game in Plantations and Anthropogenic Landscapes

Almost 50 per cent of forest game species recorded in closed forest (S = 27, 318.9 km transect), were also recorded in exotic timber and cash-crop plantations (S = 13, 23.5 km), with good representation of primates, antelopes and rodents. These complex agro-ecosystems had up to 20-25 large forest trees per ha (*c.* 50 per cent of density in natural forest), and the undergrowth was luxuriantly dense, as weeding intensity was moderate and periods between slashing rather long (Holbech, 2009). According to many hunters, several game species particularly rodents, duikers, Bushbuck and Red river hog were encountered in off-reserve habitats. Many informants pointed at high encounter rates of ungulates and guenons in neglected rubber (*Hevea*) plantations, often feeding on rubber seeds or seedlings. I have experienced bushy rubber plantations in Ghana, in which the sub-canopy vegetation structure and secondary plant community resembles that of heavily logged or senescent secondary forest. However, the majority of rubber plantations in Ghana are industrial estates most likely poor in forest wildlife.

Bushmeat studies from Ghana have indirectly shown that up to 76 per cent of market-traded bushmeat derives from anthropogenic matrix landscapes, including the meat from rodents, duikers, Bushbuck and Red river hog, some of which are considered crop pests (*e.g.* Falconer, 1992; Holbech, 1998, Cowlishaw *et al.,* 2005). My study also suggests that neglected plantations in forest ecotone landscapes may provide better food and shelter opportunities than does closed forest for several prolific forest game species with high habitat and food versatility. These species can utilise food resources in farms, secondary regrowth and bush fallow, where the dense vegetation forms good shelter from hunters. Studies from the DRC have shown that large tracts of *Musanga*-dominated post-agricultural secondary forest supported relatively high abundances of duikers and smaller game (Wilkie and Finn, 1990). Forest species like the Blue duiker (Central African pendant to Maxwell's duiker), Water chevrotain, Red river hog, mongooses, civets and Sitatunga *Tragelaphus spekei* (Central African relative of Bushbuck) displayed higher abundance in this habitat type compared to closed forest. Similarly, Fimbel (1994) found that 5-12 years old abandoned farmland was commonly used by Maxwell's duikers, guenons, chimpanzees, mangabeys, olive colobuses and bushbucks in Sierra Leone, the same species found to favour disturbed habitats of the present survey. However, the total biomass of game was higher in climax forest, primarily due to the absence of canopy monkeys in farm bush (as found in my study), whose height rarely exceeded 5-10m.

Forest Game Abundance and Hunting Activities

This extensive survey produced statistics on relative abundances (encounters per km walked transect) based on both animal and human (hunting) signs, the latter calculated as a hunting index (HI). It is therefore possible to analyse the ratio of animal to human encounters in order to assess the probability of encountering a sign from an animal or human activities (hunting in particular). The mean encounter frequency (MEF) of respectively animal and human signs based on 15 forest reserves (FR), including the Ankasa RR which was a wildlife reserve (WR) receiving virtually no protection in 1993-1995 (hence functionally a FR), and five plantation habitats, are shown in Table 16.5. For all forest habitats combined, the animal encounter frequency was 0.84 signs per km, which means that one had to walk *c.* 1.2 km in order to detect *one* animal from either audio-visual records or indirectly from tracks, feeding activities or dung. Across the 15 FRs the MEF was 0.76 sign per km, which equals *c.* 1.3 km walk to record one animal sign. In cash crop plantations the MEF was considerably higher (3.00 km⁻¹) due to a higher abundance of both arboreal and terrestrial rodents (squirrels, cane rats and giant rats) (Table 16.3). In exotic tree plantations MEF was 1.29 km⁻¹ partially due to high abundances of small antelopes and rodents.

Table 16.5: Mean Encounter Frequencies (MEF = per km walked transect) of animal (audio-visual, dung, foraging etc.) and Human Signs (Hunters, gunshots, cartridges/ carbide/batteries, snares, trails); Comparing Present Survey of Western Region (1993-1995) with Recent Data from Three Wildlife Reserves (Bia, Ankasa, Kakum) and Seven Forest Reserves (Boin, Boi-Tano, Cape three Points, Krokosua Hills, Mamiri, Tano-Nimri, Yoyo) in Ghana (Based on data from Sylvain, 2010). FR: Forest Reserve; WR = Wildlife Reserve; CCP = cash crop plantation; ETP = exotic tree plantation; ALL = all forest habitats combined. The ratio of animal to human signs is also shown.

Mean Encounter Frequency (MEF)	Western Region: 1993-1995			S. Gatti (2010a, b): 2007-2009			
	FR	CCP	ETP	ALL	WR	FR	ALL
Animal signs	0.76	3.00	1.29	0.84	0.48	0.28	0.34
Human signs	2.15	6.38	2.38	2.61	0.17	2.18	1.58
Animal/Human ratio	0.35	0.47	0.54	0.32	2.82	0.13	0.22

Signs of human presence was 2.61 per km across all forest habitats in the present survey, which means that human signs was three times higher on the average than animal signs (animal-human ratio = 0.32). For the 15 FRs the MEF of human signs was 2.15, producing an animal-human ratio of 0.35. These data witness about the exceptionally high hunting pressures in Ghanaian forests during 1993-1995, and likewise clearly reflecting the extremely low animal abundances.

Table 16.5 also lists similar data on human and animal signs obtained from over 5,800 km of transect walk in seven FRs (Boin, Boi-Tano, Cape Three Points, Krokosua Hills, Mamiri, Tano Nimri, Yoyo) and three WRs (Ankasa, Bia, Kakum) during 2007-2009 by Sylvain Gatti (Gatti, 2010a, b). Although he centred his work in WRs (5142 km ~ 89 per cent) he also collected data on several FRs, of which five were also surveyed by the author in 1993-1995. During his intensive work with GPS-tracking and mapping of animal and human signs (audio-visual and indirect sign from dung, tracks, foraging etc.), he recorded 27 mammal species and two reptile species that are used as bushmeat in Ghana, which was very similar (93 per cent) to the species recorded by the author in 1993-1995. The data on human signs from Gatti (2010a, b) were adjusted to match the criteria as defined in the present survey, such that only signs of logging, farming and mining were excluded from his data, and this transformation enabled a direct comparison with data on human (hunting) activities (HI) of the present survey for 15 reserved forests (FR/RRs) during 1993-1995.

Across the seven FRs surveyed by Gatti (2010a, b) he found a MEF of animal signs of only 0.28 km^{-1}, which is only 37 per cent of MEF found for similar FRs in 1993-1995 (Table 16.5). In contrast, MEF of human signs of 2.18 km^{-1} was very similar to that recorded by the author in the present survey of 2.15 km^{-1}, indicating that whilst animal densities have declined, hunting pressures have remained constantly high. Gatti (2010a, b) when visiting FRs such as Boin, Tano-Nimri, Mamiri and Yoyo found extremely low levels of animal signs (MEF range: 0.00-0.18 km^{-1}) but very high human signs (MEF range: 0.81-5.26 km^{-1}), particularly in areas with ongoing logging (Boin and Yoyo). Compared to the heavily logged Boin and Yoyo FRs, the unlogged Cape Three Points FR recorded a relatively high MEF of human (2.57 km^{-1}) as well as animal signs (0.42 km^{-1}), indicating that logging and hunting can have synergistic negative effects on game animal abundances in FRs. Gatti (2010a, b) also surveyed three WRs and found that although MEF of animal signs (0.48 km^{-1}) was considerably higher compared to FRs (0.28 km^{-1}), the levels were only *c.* 60 per cent of those recorded for FRs by the author in 1993-1995, indicating a genuine decline in game animal abundances even in WR areas that are better protected against hunting and logging. However, recent intensification of game protection in WRs seems to have had a significant impact on hunting activities here as MEF signs of humans was as low as 0.17 km^{-1} (range: 0.05-0.34). The highest recorded MEF of human signs in WRs by Gatti (2010a, b) was 0.34 km^{-1} in Bia, a WR that was heavily logged from 1995-1999, and for which the density of logging roads and surrounding human population are considerably higher compared to Kakum, and Ankasa in particular, with MEF of human signs of respectively 0.13 and 0.05 km^{-1}. Summarising animal-human ratio MEF of the present survey and Gatti (2010a, b), it seems evident that game abundances are still declining, and that hunting activities are only partially controlled in WRs, but continues unabated in FRs.

Forest Game and Off-Reserve Integrated Wildlife and Land-Use Management

My study suggests that modified forest habitats play a major role in maintaining bushmeat production in the Ghanaian high forest zone, as indicated by other recent studies in Ghana (Mendelson *et al.,* 2003; *e.g.* Cowlishaw *et al.,* 2005). In effect, off-reserve land-use management is therefore equally important to largely isolated 'protected' forests that are heavily exploited for timber and bushmeat, in the management of game production. In particular, corridor areas linking-up otherwise isolated reserves could be managed to improve forest habitat connectivity; specifically, marginal agricultural areas that require intensive labour and fertiliser treatment (*e.g.* swamps, marshes, hilly ridges and steep slopes) could be managed in a more 'extensive' fashion with 'wildlife friendly' plantation types for non-timber-forest-product (NTFP) harvesting, including bushmeat. Such a stratified land-use management system would be able to maintain high bushmeat availability in cultivated areas, and thus minimise the pressure on game populations in protected areas, where the incentives for hunting then would be reduced. Future studies should therefore focus on estimating densities of commonly utilised forest game in extensively managed agro-silvicultural systems, and evaluate the productivity of combined crop and game harvesting through cost-benefit analyses (*e.g.* Holbech, 2001). This however, requires that conservationists and wildlife mangers recognise that high off-reserve habitat diversity, driven by appropriate wildlife productive land-use practises, should be maintained to allow the utilisation of a diverse range of NTFPs, including bushmeat. In such a system, local communities need not be deprived of utilising off-reserve bushmeat sources that are often crop pests; these sources should be much more productive than forest-alien livestock or poorly domesticated bushmeat species, *e.g.* cane rats.

One of the current major drawbacks to the planning of integrated wildlife and land-use management in the Western Region of Ghana is the uncontrollable influx of immigrant cocoa farmers from other regions. These immigrants often engage in indiscriminate forest clearance, charcoal-production, and cultivate non-shaded 'sun cocoa'. Due to high rainfall, poor soil fertility and fungal attacks ('black pot disease'), sun cocoa typically dies off as early as 9 years old, particularly when planted on agricultural marginal lands (Holbech, 2001). Prior to the late 1980s, indigents of the Western Region grew mostly 'rustic cocoa', shaded by a variably intact native forest canopy throughout the plantation period. In that system, cocoa trees were often mixed with oil palms (*Elaeis guineensis*) and raffia palms, and many arboreal birds and larger mammals could be harvested as bushmeat, including anomalures and palm civets. In contrast, sun cocoa offers few food or shelter opportunities for other fauna than crop damaging bush squirrels and murids. Today, natives of the southern Western Region of Ghana maintain 'wildlife friendly' coconut and rubber cultivation with rather long weeding interims, in which a diverse secondary plant community seems to coexist with these acceptably yielding crops (Holbech, 2001). According to hunters, such luxuriant plantations are rich in secondary forest game, including bushbucks, Maxwell's and black duikers. Many of these disturbance-resilient opportunists are also rated as the most popular wildlife by the people, both in terms of palatability and with regard to their cash value in relation to the perceived crop damage that these species cause. According to many farmers only cane rats and bush squirrels constitute a serious problem to their crops, and are rather difficult to control by means of existing legal hunting methods. Ironically, it is exactly these problem animals that are particularly favoured by indiscriminate tree felling and sun cocoa cultivation.

Acknowledgements

This study was financed by a Ph.D.-grant provided by Danida (No. 104.Dan.8/606), in collaboration with University of Copenhagen, Denmark and University of

Ghana (UGL). I thank the Ghana Wildlife Division (GWD) and Ghana Forestry Commission for granting research permits, and special appreciation for invaluable field assistance from the late Mr. Joseph Amponsah (GWD). Professor John F. Oates provided constructive criticism and editing of an earlier manuscript, and Sylvain Gatti (WAPCA) kindly provided me with valuable comparative data sets. Finally, I am sincerely indebted to the numerous local communities in the Western Region of Ghana whose members were helpful in many ways.

References

Abedi-Lartey, M., 2004. *Bushmeat Hunters Do Better – Indigenous vs Scientific Habitat Evaluation*. Unpublished MSc. Thesis. International Institute for Geo-Information Science and Earth Observation, Enschede, The Netherlands. 98 pp.

Asibey, E.O.A., 1977. Expected effects of land-use patterns on future supplies of bushmeat in Africa South of the Sahara. *Environmental Conservation*, **4(1)**: 43-49.

Barnes, R.F.W., 2002. The bushmeat boom and bust in West and Central Africa. *Oryx*, **36 (3)**: 236-242.

Barnes, R.F.W., Barnes, K.L., Alers, M.P.T. and Blom, A., 1991. Man determines the distribution of elephants in the rain forests of northeastern Gabon. *African Journal of Ecology*, **29**: 54-63.

BirdLife International, 2009. *Species Factsheet: Picathartes gymnocephalus*. URL: http://www.birdlife.org.

Blom, A., van Zalinge, R., Mbea, E., Heitkönig, I.M.A. and Prins, H.H.T., 2004. Human impact on wildlife populations within a protected Central African forest. *African Journal of Ecology*, **42**: 23-31.

Blom, A., van Zalinge, R., Heitkönig, I.M.A. and Prins, H.H.T., 2005. Factors influencing the distribution of large mammals within a protected central African forest. *Oryx*, **39 (4)**: 1-8.

Bowen-Jones, E. and Pendry, S., 1999. The threat to primates and other mammals from the bushmeat trade in Africa, and how this threat could be diminished. *Oryx*, **33 (3)**: 233-246.

Bshary, R., 2001. Diana monkeys, *Cercopithecus diana*, adjust their anti-predator response behaviour to human hunting strategies. *Behavioural Ecology and Sociobiology*, **50**: 251-256.

Cowlishaw, G.C., Mendelson, S. and Rowcliffe, J.M., 2005. Evidence for post-depletion sustainability in a mature bushmeat market. *Journal of Applied Ecology*, **42**: 460–468.

Croes, B.M., Laurance, W.F., Lahm, S.A., Tchignoumba, L., Alonso, A., Lee, M.E., Campbell, P. and Buij, R., 2006. The influence of hunting on antipredator behavior in Central African monkeys and duikers. *Biotropica*, **39 (2)**: 257-263.

Crookes, D.J., Ankudey, N. and Milner-Gulland, E.J., 2005. The value of a long-term bushmeat market dataset as an indicator of system dynamics. *Environmental Conservation*, **32 (4)**: 333–339.

Dudley, J.P., Mensah-Ntiamoah, A.Y. and Kpelle, D.G., 1992. Forest elephants in a rainforest fragment: preliminary findings from a wildlife conservation project in southern Ghana. *African Journal of Ecology*, **30**: 116-126.

Fa, J.E., Ryan, S.F., and Bell, D.J., 2005. Hunting vulnerability, ecological characteristics and harvest rates of bushmeat species in afrotropical forests. *Biological Conservation*, **121**: 167–176.

Falconer, J., 1992. *Non-timber Forest Products in Southern Ghana: A Summary Report.* ODA Forestry Series 2, Overseas Development Agency, U.K. 23 pp.

Fimbel, C., 1994. The relative use of abandoned farm clearings and old forest habitats by primates and a forest antelope at Tiwai, Sierra Leone, West Africa. *Biological Conservation*, **70**: 277-286.

Gatti, S., 2010a. *Status of primate populations in Protected Areas targeted under the Community Forest Biodiversity Project.* Unpublished CFBP Report, West African Primate Conservation Action and Ghana Wildlife Division/Forestry Commission, Accra, Ghana. 42 pp.

Gatti, S., 2010b. *Mammal Surveys and Capacity Building for the Wildlife Wood Project – Ghana. Surveys in Mamiri Forest Reserve and Participation to Surveys in other Protected Areas with the WWP-Ghana Team.* Unpublished Report, Wildlife Wood Project, Zoological Society of London, London, UK. 22 pp.

Hall, J.B. and Swaine, M.D., 1976. Classification and ecology of closed-canopy forest in Ghana. *Journal of Ecology*, **64**: 913-951.

Haltenorth, T. and Diller, H., 1988. *A Field Guide to the Mammals of Africa Including Madagascar.* Third reprint. Collins, London. 400 pp.

Hawthorne, W.D. and Abu-Juam, M., 1995. *Forest Protection in Ghana.* IUCN/ODA/Ghana Forestry Department. 203 pp.

Holbech, L.H., 1996. *Faunistic Diversity and Game Production Contra Human Activities in the Ghana High Forest Zone.* Unpublished Ph.D. Thesis, University of Copenhagen, Denmark. 237 pp.

Holbech, L.H., 1998. *Bushmeat Survey - Literature Review, Field Work and Recommendations for a Sustainable Community Based Wildlife Resource Management System.* Protected Areas Development Programme – Ghana. Consultancy report. Ghana Wildlife Division/European union/ULG Northumbrian Ltd., Takoradi, Ghana. 150 pp.

Holbech, L.H., 2001. *Integrated Wildlife and Land-use Management, Off-reserve the Ankasa Resource Reserve, Western Region, Ghana.* Protected Areas Development Programme – Ghana. Consultancy report. Ghana Wildlife Division/European union/ULG Northumbrian Ltd., Takoradi, Ghana. 110 pp.

Holbech, L.H., 2005. The implications of selective logging and forest fragmentation for the conservation of avian diversity in evergreen forests of Southwest Ghana. *Bird Conservation International*, **15**: 27-52.

Holbech, L.H., 2009. The conservation importance of luxuriant tree plantations for lower storey forest birds in Southwest Ghana. *Bird Conservation International*, **19**: 287-308.

Holbech, L.H., 2014b. Bushmeat production, hunting and utilisation of forest game in southwest Ghana: A 1993-1995 survey. In either *Animal Diversity, Natural History and Conservation*, Vol 4 or 5, or *Perspectives in Animal Ecology and Reproduction*, Vol. 10.

Johns, A.D. and Skorupa, J.P., 1987. Responses of rain-forest primates to habitat disturbance: A review. *International Journal of Primatology*, **8 (2)**: 157-191.

Kingdon, J., 1997. *The Kingdon Field Guide to African Mammals*. Academic Press. London, U.K. 465 pp.

Laurance, W.F., Croes, B.M., Tchignoumba, L., Lahn, S.A., Alonso, A., Lee, M.E., Campbell, P. and Ondzeanu, C., 2006. Impacts of roads and hunting on Central African rainforest mammals. *Conservation Biology*, **20 (4)**: 1251-1261.

Marks, S.A., 1994. Local hunters and wildlife surveys: A design to enhance participation. *African Journal of Ecology*, **32**: 233-254.

Marks, S.A., 1996. Local hunters and wildlife surveys: An assessment and comparison of counts for 1989, 1990 and 1993. *African Journal of Ecology*, **34**: 237-257.

Martin, C., 1991. *The Rainforests of West Africa*. Birkhaüser Verlag, Berlin, Germany. 235 pp.

McGraw, W.S., 2005. Update on the search for Miss Waldron's Red Colobus Monkey. *International Journal of Primatology*, **26 (3)**: 605-619.

Mendelson, S., Cowlishaw, G. and Rowcliffe, J.M., 2003. Anatomy of a bushmeat commodity chain in Takoradi, Ghana. *The Journal of Peasant Studies*, **31**: 73–100.

Merz, G., 1986. Counting elephants (*Loxodonta africana cyclotis*) in tropical rain forests with particular reference to the Taï National Park, Ivory Coast. *African Journal of Ecology*, **24**: 61-68.

Muchaal, P.K. and Ngandjui, G., 1999. Impact of village hunting on wildlife populations in the Western Dja Reserve, Cameroon. *Conservation Biology*, **13 (2)**: 385-396.

Noss, A.J., 1998. The impacts of cable snare hunting on wildlife populations in the forests of the Central African Republic. *Conservation Biology*, **12 (2)**: 390-398.

Noss, A.J., 1999. Censusing rainforest game species with communal net hunts. *African Journal of Ecology*, **37**: 1-11.

Ntiamoa-Baidu, Y., 1987. West African wildlife: a resource in jeopardy. *Unasylva*, **39 (2)**: 27-35.

Nummelin, M., 1990. Relative habitat use of duikers, bush pigs and elephants in virgin and selectively logged areas of the Kibale Forest, Uganda. *Tropical Zoology*, **3**: 111-120.

Oates, J.F., 1996. Habitat alteration, hunting and the conservation of folivorous primates in African forests. *Australian Journal of Ecology*, **21 (1)**: 1-9.

Oates, J.F., 1999. *Myth and Reality in the Rain Forest: How Conservation Strategies are Failing in West Africa*. University of California Press. U.S. 310 pp.

Oates, J.F., Abedi-Lartey, M., McGraw, W.S., Struhsaker, T.T. and Whitesides, G.H., 2000. Extinction of a West African red colobus monkey. *Conservation Biology*, **14 (5)**: 1526-1532.

Otsyina, R.M. and Asare, E.O., 1991. *Western Region Agroforestry Project*. Agro-tech Consultancy Services, The Ghana Government, The Commission of European Communities, Accra, Ghana. 126 pp.

Redford, K. H., 1992. The empty forest. *Bioscience*, **42**: 412-422.

Roth, H.H. and Merz, G., 1986. Vorkommen und relative häufigkeit von säugetieren im Taï-Regenwald der Elfenbeinküste. *Säugetierkundliche Mitteilungen*, **33**: 171-193.

Short, J., 1983. Density and seasonal movements of forest elephant (*Loxodonta africana cyclotis*, Matschie) in Bia National Park, Ghana. *African Journal of Ecology*, **21**: 175-184.

Struhsaker, T.T. and Oates, J.F., 1995. The biodiversity crisis in Southwestern Ghana. *African Primates*, **1 (1)**: 5-6.

Waltert, M., Lien, J.L., Faber, K. and Mühlenberg, M., 2002. Further declines of threatened primates in the Korup Project Area, Southwest Cameroon. *Oryx*, **36 (3)**: 257-265.

Waltert, M., Heber, S., Riedelbauch, S., Lien, J.L. and Mühlenberg, M., 2006. Estimates of blue duiker (*Cephalophus monticola*) densities from diurnal and nocturnal line transects in the Korup region, Southwestern Cameroon. *African Journal of Ecology*, **44 (2)**: 290-292.

Weisenseel, K., Chapman, A.C. and Chapman, L.J., 1993. Nocturnal primates of Kibale Forest: effects of selective logging on prosimian densities. *Primates*, **34 (4)**: 445-450.

Wilkie, D.S. and Carpenter, J.F., 1999. Bushmeat hunting in the Congo Basin: an assessment of impacts and options for mitigation. *Biodiversity and Conservation*, **8:** 927-955.

Wilkie, D.S. and Finn, J.T., 1990. Slash-burn cultivation and mammal abundance in the Ituri Forest, Zaire. *Biotropica*, **22 (1)**: 90-99.

Wilkie, D.S., Shaw, E., Rotberg, F., Morelli, G. and Auzel, P., 2000. Roads, development, and conservation in the Congo Basin. *Conservation Biology*, **14 (6):** 1614-1622.

Wong, S.N.P. and Sicotte, P., 2006. Population size and density of *Colobus vellerosus* at the Boabeng-Fiema Monkey Sanctuary and surrounding forest fragments in Ghana. *American Journal of Primatology*, **68**: 465–476.

**Appendix 1: Records of Game Animals (25 mammals; 2 reptiles) Distributed on
Direct (Audio-visual) and Indirect (Dung, tracks, diggings)
Signs during 342 km Transect Walk on 100.6 km Cut Transect in
15 Forest and 5 Plantation Habitats of Southwest Ghana, 1993-1995**

Species	Visual	Audio	Dung	Tracks	Diggings	Total
Loxodonta africana cyclotis				5		5
Tragelaphus euryceros			4	1		5
T. scriptus	3			2		5
Cephalophus sylvicultor	1					1
C. dorsalis	3		5			8
C. niger	6		7			13
C. maxwelli	6		30	1		37
Neotragus pygmaeus	3					3
Potamochoerus porcus					16	16
Cercopithecus diana roloway		5				5
C. mona lowei	2	25				27
C. petaurista petaurista		5				5
Civettictis civetta			3			3
Nandinia binotata	1					1
Genetta tigrina pardina	2					2
Atilax paludinosus	3					3
Crossarchus obscurus	12					12
Dendrohyrax dorsalis		4				4
Uromanis tetradactyla	1					1
Atherurus africanus	1					1
Thryonomys swinderianus			7			7
Cricetomys emini	6		1			7
Anumalurus peli	1					1
A. derbianus	1					1
Protoxerus stangeri	4	105				109
Osteolaemus tetraspis	2					2
Varanus niloticus	3					3
Total species	19	5	7	4	1	27
Total records	61	144	57	9	16	287

Appendix 2: Conservation Status, Encounter Frequencies and Habitat Preferences of 42 Mammal, Nine Reptile and Nine Bird Species, Based on Interviews with 112 Farmer-Hunters in Southwest Ghana, 1993-1995. See text for details of how most recent animal encounter, encounter percentage and mean annual encounter frequency (MAEF) were calculated. Endangered forest dependent species in bold. E: Endangered, R: Rare, U: Uncommon, C: Common, W: Widespread; *: Non-forest species. RF: Reserved forest; UF: Unreserved forest (remnants and old secondary forest), FB: Farms and farm-bush (neglected farms and young regrowth), RP: Rubber plantations. Nomenclature after Kingdon (1997).

Species	Status	Animal Encounters		MAEF	Habitat Preference (per cent)			
		Most Recent	Per cent		RF	UF	FB	RP
Forest elephant	E	1 week	22.3	0.4	73.0	14.5	12.5	0
Forest buffalo	E	6 months	17.9	0.1	81.4	18.6	0	0
Bongo	R	1 day	53.6	5.2	89.7	7.5	1.3	1.5
Bushbuck	C	1 day	97.3	56.7	24.7	13.1	54.6	7.6
Yellow-backed duiker	E	1 week	38.4	1.5	82.7	13.3	3.3	0.7
Bay duiker	U	1 day	87.5	22.9	64.6	17.7	15.2	2.5
Ogilby's duiker	R	1 day	42.0	11.4	78.3	19.1	2.6	0
Black duiker	C	1 day	94.6	41.9	28.1	10.3	56.6	5.0
Maxwell's duiker	C	1 day	97.3	65.7	39.5	20.4	31.8	8.3
Royal antelope	U	1 day	94.6	39.9	24.7	11.6	54.8	8.9
Water chevrotain	E	1 month	20.5	0.3	83.3	16.7	0	0
Giant hog	E	3 days	25.9	1.4	92.1	5.8	2.1	0
Red river hog	U	1 day	86.6	32.4	54.8	14.0	23.1	8.1
Western chimpanzee	E	3 days	40.2	2.7	79.6	11.2	9.2	0
Red colobus	E	6 months	24.1	0.1	84.5	15.5	0	0
Pied colobus	R	3 days	83.9	13.9	83.8	11.6	4.6	0
Olive colobus	U	1 day	86.6	39.1	57.1	15.8	24.0	3.1
White-naped mangabey	U	1 day	83.9	22.9	68.5	11.2	19.5	0.8
Lowe's monkey	C	1 day	100.0	87.2	41.2	18.5	33.2	7.1
Lesser spot-nosed monkey	C	1 day	93.8	73.9	45.2	23.3	22.9	8.7
Roloway monkey	R	2 days	83.9	15.4	87.9	10.2	1.9	0
Potto	C	1 day	100.0	95.5	22.2	23.5	45.6	8.7
Leopard	E	3 days	26.8	1.6	96.7	3.3	0	0
Golden cat	E	0.5 year	8.0	< 0.1	66.7	25.0	8.3	0
African civet	U	1 day	93.8	33.1	15.8	11.7	63.2	9.3
African palm civet	C	1 day	93.8	67.1	39.6	16.4	36.3	7.7
Blotched genet	U	1 day	83.9	39.8	40.7	17.0	33.9	8.4

Contd...

Appendix 2– *Contd...*

Species	Status	Animal Encounters MAEF			Habitat Preference (per cent)			
		Most Recent	Per cent		RF	UF	FB	RP
Marsh mongoose	U	1 day	57.1	23.0	29.5	25.7	31.4	13.4
Cusimanse	W	1 day	95.5	129.2	41.6	20.4	28.9	9.1
Spot-necked otter	E	3 days	44.6	4.7	54.9	13.9	21.4	9.8
Ratel	E	2 weeks	21.4	0.7	93.3	4.5	2.2	0
Tree hyrax	W	1 day	100.0	247.3	21.1	20.1	50.0	8.8
Giant pangolin	E	1 year	25.0	0.1	93.3	6.7	0	0
Long-tailed/Tree pangolin	C	1 day	100.0	46.3	41.1	14.2	37.0	7.7
Crested porcupine*	E	1 month	7.1	0.1	91.7	8.3	0	0
Brush-tailed porcupine	C	1 day	97.3	76.9	27.4	15.4	48.1	9.1
Cane rat*	W	1 day	99.1	110.5	0	5.4	85.6	9.0
Giant rats (2 spp.)	W	1 day	100.0	236.2	25.8	20.3	43.9	10.0
Pel's anomalure	C	1 day	89.3	49.1	30.9	15.0	48.7	5.5
Giant forest squirrel	W	1 day	98.2	174.0	23.4	21.3	46.8	8.5
African python	U	1 day	81.3	23.9	25.1	24.9	40.9	9.1
Nile crocodile*	E	3 months	19.6	1.9	–	–	–	–
Broad-fronted crocodile	R	2 days	80.4	17.5	69.3	10.8	15.2	4.7
Nile monitor lizard	C	1 day	86.6	47.2	33.9	16.5	42.7	6.9
Chelonians (5 spp.)	R	1 day	69.6	14.1	40.0	18.0	36.7	5.3
White-breasted guineafowl	R	1 day	33.0	7.8	87.8	12.3	0	0
Crested guineafowl	U	1 day	61.6	21.2	76.0	16.3	6.4	1.3
Casqued hornbills (4 spp.)	C	1 day	76.8	86.7	71.8	15.3	8.2	4.7
Great blue turaco	C	1 day	84.8	93.8	51.8	19.1	24.1	5.0
Crowned eagle	R	1 day	52.7	11.8	83.1	6.1	10.0	0.8
White-necked picathartes	E	2 years	1.8	< 0.1	100.0	0	0	0
Total	60	–	–	2,196.3	–	–	–	–

Appendix 3: Signs of Hunting Recorded for 20 Forest Habitats in Southwest Ghana (1993-1995), with Statistics of Survey Efforts; Days of Work, Total Cut Transect Length (km), and Total Distance (km) Covered on Transect Walks. Hunting index (HI) calculated as the sum of five hunting sign descriptors.
CCP: Cash crop plantations; ETP: Exotic tree plantations.

Forest Habitat	Days	km Tran- sect	km Walked	Human Signs per km Walked Transect					HI
				Snares	Trails	Cartridge	Gun	Hunter	
Ankasa	62	14.8	62.5	0.37	0.08	0.13	0.13	0.05	0.75
Dadiaso	24	4.9	25.9	0.35	0.00	0.12	0.31	0.00	0.77
Disue	10	3	7.5	0.40	0.00	0.13	0.27	0.00	0.80
Yoyo	17	4.5	21.6	0.83	0.19	0.05	0.19	0.05	1.30
Fure	25	7.6	24.1	0.41	0.62	0.08	0.17	0.08	1.37
Boin	42	10.5	46.7	0.51	0.51	0.24	0.24	0.04	1.54
Mamiri	13	4	12.8	0.63	0.47	0.08	0.23	0.23	1.64
Tano Nimri	19	4.8	17.6	0.68	0.80	0.11	0.40	0.00	1.99
Bura	13	4	12.8	1.09	0.86	0.16	0.16	0.00	2.27
Draw	32	10.5	30.1	0.86	0.83	0.17	0.53	0.00	2.39
Ebi	15	5	13.8	1.45	0.72	0.00	0.29	0.00	2.46
Neung North	8	2.8	8.4	0.12	2.50	0.12	0.00	0.24	2.98
Jema	18	4.5	11.7	1.54	1.54	0.26	0.17	0.00	3.50
Subri	13	5.4	12	1.42	1.75	0.00	0.25	0.17	3.58
Cape	14	5	11.4	2.81	1.84	0.00	0.26	0.00	4.91
All forests (n = 15)	325	91.3	318.9					Mean ± STD:	2.15 ± 1.20
Ankasa Coconut	4	1	4	2.00	0.50	0.25	0.00	0.00	2.75
Boin Cocoa	4	1	4	2.75	0.50	0.50	3.75	2.50	10.00
All CCP (n = 2)	8	2	8					Mean ± STD:	6.38 ± 5.13
Subri Gmelina	4	2.8	5.2	0.00	0.38	0.38	1.35	0.38	2.50
Neung Cedrela	4	2.6	5	0.00	2.00	0.00	0.20	0.00	2.20
Tano Cedrela	4	1.9	5.3	1.13	1.13	0.19	0.00	0.00	2.45
All ETP (n = 3)	12	7.3	15.5					Mean ± STD:	2.38 ± 0.16
All plantations (n = 5)	20	9.3	23.5					Mean ± STD:	3.98 ± 3.37
All habitats (n = 20)	345	100.6	342.4					Mean ± STD:	2.61 ± 2.03

2015, **Perspectives in Animal Ecology and Reproduction, Vol. 10**   *Pages **269–299***
Editors: **V.K. Gupta, Anil K. Verma and G.D. Singh**
Published by: **DAYA PUBLISHING HOUSE, NEW DELHI**

Chapter 17

Bushmeat Production, Hunting and Utilisation of Forest Game in Southwest Ghana: A 1993-1995 Survey

Lars H. Holbech*

Department of Animal Biology and Conservation Science,
University of Ghana, P.O. Box LG 67,
Legon, Accra, Ghana

ABSTRACT

Ghana is the quintessence of how human population growth has reduced forest cover to a patchwork of variably isolated remnants within an ocean of anthropogenic landscapes. Three major anthropogenic disturbance factors possibly influence tropical forest game populations: fragmentation (patch size and isolation), habitat disturbance (agro-silviculture), and hunting (bushmeat). These factors are mutually linked, yet the mode and magnitude of interrelations remains insufficiently understood. In the case of Ghana, independent studies suggest that the immediate threat to forest game conservation is not forest fragmentation and habitat disturbance, but bushmeat hunting. During 1993-1995, my study evaluated abundance, diversity and hunting pressures of forest game, applying combined transect and informant data from respectively 22 moist forest habitats and 130 bushmeat users in southwest Ghana. Out of 30 analysed species only 23 per cent appeared to be sustainably harvested, including four rodents, and three other habitat-versatile prolific opportunists with a high preference for forest edge, secondary growth and farmland. Lack of large-bodied forest game, particularly primates and ungulates, has shifted pressures to smaller animals, including birds. It is evident that several large forest game are locally extirpated and in danger of facing national extinction. In effect, constructive and sustainable conservation policies need to apply community participatory off-reserve

* *Corresponding author.* E-mail: lholbech@ug.edu.gh and l.holbech@gmail.com

integrated wildlife and land-use management. The attitudinal basis and level of conservation awareness among the local people for successful implementation of such interventions were also evaluated. Although the data presented in this study are 20 years old they serve as important baseline information on game abundances and hunting levels, which can be used for future comparative research purposes, and the development of management tools for sustainable utilisation and long-term conservation of forest game in Ghana and the sub-region at large.

Keywords: *Bushmeat, Community, Conservation, Rainforest, Game, Ghana, Hunting, Sustainable.*

Introduction

The socio-economic importance of bushmeat in the Guineo-Congolian forest biome rates among the highest worldwide, and Ghana is a prime sub-regional example (Ntiamoa-Baidu, 1998; Fa *et al.,* 2002). The past two decades of mechanised logging, indiscriminate slash-and-burn farming, and bushmeat hunting in Central African wilderness areas, introducing indiscriminate slash-and-burn farming and bushmeat hunting (Wilkie *et al.,* 1992), is a chapter in Ghana's forest history that commenced over 50 years ago (Martin, 1991). The extensive fragmentation and intensive utilisation of the Ghanaian high forest zone pose tremendous challenges for biodiversity conservation. The over 200 reserved high forest areas constitute *c.* 20 per cent of the originally *c.* 82,000 km² forest cover, of which wildlife protected areas (resource reserves, wildlife sanctuaries and national parks) represent only some 7 per cent (Martin 1991; Hawthorne and Abu-Juam, 1995). Recent satellite images (*e.g.* Google Earth) demonstrate that contiguous forest areas do not exceed 1,500 km², and most areas are < 500 km². Whereas botanical diversity is largely safeguarded in the geographically widespread forest network of Ghana (Hawthorne and Abu-Juam, 1995), the situation for larger vertebrates and mammals in particular, appears gloomy (Struhsaker and Oates 1995; Oates, 1999).

Three major anthropogenic disturbance factors contribute to the current biodiversity crisis in Ghana: 1) forest fragmentation due to habitat loss by swidden agriculture deforestation, 2) forest habitat disturbance by logging and tree plantation development, and 3) 'defaunation' by subsistence and commercial bushmeat hunting. In Africa, commercial logging is a major drive for early phases of subsistence farming and hunting (Wilkie *et al.,* 2000), yet the combined long-term effects of the three key disturbance factors remain insufficiently understood (Newmark, 2008). This is particularly true for the Upper Guinea forest biome, of which Ghana constitutes the easternmost part. Previous Ghanaian studies suggest that low-extraction (1-3 trees felled per ha) selective logging has limited negative effects on avifauna (Holbech, 2005 and 2014a), small mammals (Holbech, 2013) and forest game in general, whereas patch size and degree of isolation appear to be more important for persistence of large-bodied forest dependent game (Holbech, 1996 and 2014b). Neglected or extensively managed plantations of exotic timber, rubber, coconut and cocoa supported *c.* 50 per cent of mature forest game diversity in southwest Ghana (Holbech 1996; Holbech, 2014b). These findings support those of Struhsaker and Oates (1995), and suggest that the immediate threat to forest wildlife in Ghana is bushmeat hunting, rather than land-use related habitat losses. This effective 'defaunation' has introduced a new concept, 'the empty forest syndrome' – floristically diverse forests almost devoid of large-bodied mammals, reptiles and birds (Redford, 1992; Oates, 1999). An extreme example is the probable extinction of an Upper Guinea endemic primate (Oates *et al.,* 1997 and 2000).

The present study evaluated bushmeat hunting and trade in relation to human population densities in southwest Ghana from September 1993 to November 1995, and is parallel and complementary to a survey that focused on the impacts of forest fragmentation and habitat disturbances (logging and agro-silviculture) on forest game (Holbech 1996; Holbech, 2014b). Data collection included transect records of forest game and hunting activities, as well as interviews with wildlife users on hunting methods, bushmeat harvest rates and trade, including a sustainability analysis of estimated yields for 30 forest game species. The general attitude and awareness of the local people towards animal conservation and community participatory integrated wildlife and land-use management, as a means of sustainable bushmeat utilisation, was also assessed.

Given the rapidly escalating destruction of Ghanaian forest habitats and biodiversity, as well as the factual scarcity of published studies related to hunting and bushmeat from the region, justify the importance of making available 20-years old data for future comparative research purposes, and the development of appropriate management tools for conservation of forest wildlife and important game species in Ghana. The present study should therefore be regarded as providing baseline data for contribution to the overall understanding of how forest fragmentation, habitat modification, and hunting relate to human population growth, migration and community development, and how the individual and combined effects of these anthropogenic parameters impact on the production and utilisation of wild game and bushmeat in the most forested region of Ghana: the Western Region.

Study Area and Sites

The study area, Ghana's Western Region, covers almost 24,000 km², of which some 7,500 km², 12,500 km² and 4,000 km² respectively are reserved forest, rural off-reserve, and urban areas (Otsyina and Asare, 1991). In this context urban areas are regarded as cities, towns, villages, hamlets and other densely populated areas with scarce vegetation cover unsuitable for forest game animals (Table 17.1). Based on LANDSAT-images (1989-91) I arbitrarily estimated that forest remnants and intensively farmed areas (industrial plantations and mixed food crops) constituted *c.* 2,500 km² (20 per cent) and 7,500 km² (60 per cent) respectively of rural off-reserve areas, with the remaining 2,500 km² (20 per cent) as a complex mosaic ecotone of secondary forest, bush fallow and abandoned tree plantations (Table 17.1). Sixteen forest reserves (FRs), one resource reserve (Ankasa RR) and five plantations associated with FRs or RRs were selected (Figure 17.1). Forest reserves and resource reserves in Ghana are in principle protected from both illegal logging and hunting by respectively the Ghana Forestry Commission (GFC) and Ghana Wildlife Division (GWD), whereas national parks are wholly GWD-protected against any such activities. This means that in FRs and RRs legal logging operations may be granted to timber concessionaires

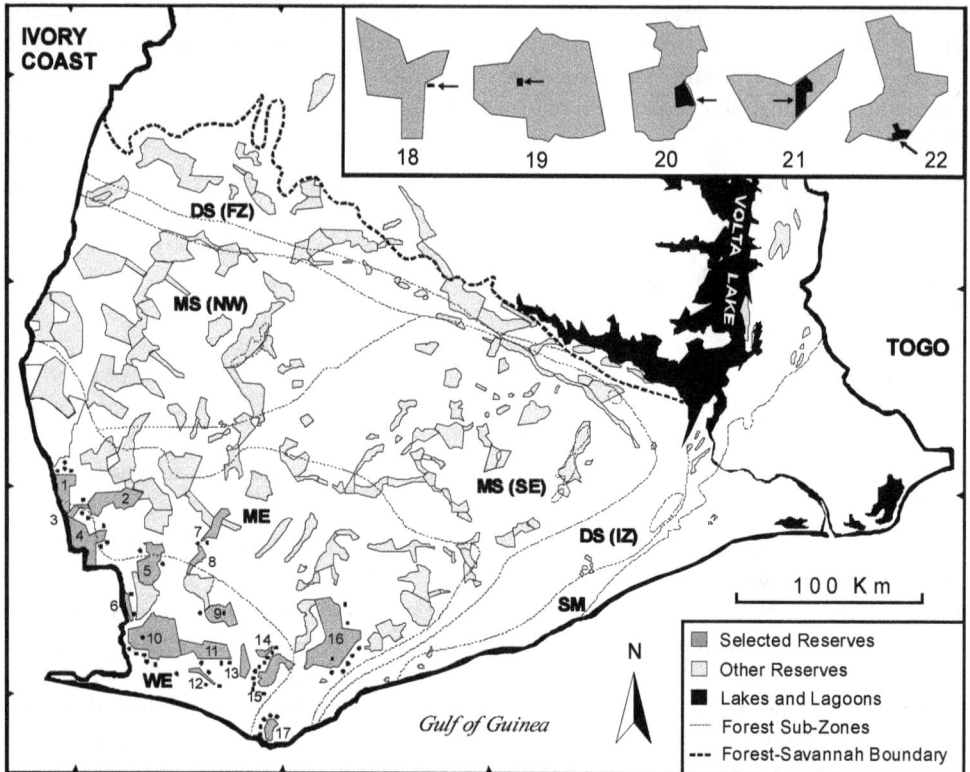

Figure 17.1: Map of Southern Ghana Showing High Forest Reserves (FR), Forest Sub-zones as well as 22 Selected Forest Habitats and 55 Interview Sites (% = Towns, villages and hamlets): The position of plantation habitats is indicated by arrows on respective enlarged reserves in the upper right corner.

Forest sub-zones after Hall and Swaine (1976): WE: Wet Evergreen; ME: Moist Evergreen; MS: Moist Semi-deciduous; DS: Dry Semi-deciduous; SE: South-east; NW: North-west; FZ: Fire Zone; IZ: Inner Zone; SM: Southern Marginal.

Forest habitats: 1: *Dadiaso FR*; 2: *Yoyo FR*; 3 *Disue FR*; 4: *Boin FR*; 5: *Tano Nimri FR*; 6: *Jema-Assemkrom FR*; 7: *Bura FR*; 8: *Mamiri FR*; 9: *Fure FR*; 10: *Ankasa Resource Reserve*; 11: *Draw FR*; 12: *Ebi FR*; 13: *Ndumfri FR*; 14: *Neung North FR*; 15: *Neung South FR*; 16: *Subri FR*; 17: *Cape Three Point FR*; 18: *Boin Cocoa*; 19: *Ankasa Coconut*; 20: *Tano Cedrela*; 21: *Neung Cedrela*; 22: *Subri Gmelina*.

by the GFC, just as hunting may be permitted by the GWD on certain species in FRs, but not in RRs. Since 2000 however, neither logging nor hunting have been permitted in any GWD-protected RRs.

The 22 study sites covered wet evergreen (WE) and moist evergreen (ME) sub-types of Ghanaian forests (Hall and Swaine, 1976). The two zones differ only slightly in precipitation and forest structure; WE-zone receives > 2,000 mm of rain annually and the closed canopy of its mature forest reaches an average height of 35-40 m, ME-zone gets some 1,750-2,000 mm per year and has a slightly discontinuous canopy of

Table 17.1: Quantitative Outline (Size of area) of Three Major Land-Use Categories in the Western Region According to Otsyina and Asare (1991), with Crude Size Estimates of Sub-categories Based on LANDSAT-Images (1989-91). Suitable area for game animals subject to sustainability analysis of harvest rates ($n = 30$) are also shown according to four major habitat preference categories (see Table 17.7 and text for further details).

Land-use category	Reserved Forest Areas		Off-Reserve Areas (Total = 12,500 km²)		Urban Areas	
Sub-category	*– Forest reserves* *– Resource reserves* *– National parks*	*Forest remnants*	*Intensive Farming:* *– Industrial plantations* *– Mixed food crops*	*Mosaic Ecotone:* *– Secondary forest* *– Bush fallow* *– Abandoned plantations*	*Open areas:* *– Cities* *– Towns* *– Villages* *– Hamlets*	
Size (km²)	7,500	2,500	7,500	2,500	4,000	
Habitat category	Game animal species ($n = 30$; see Table 17.7)					Suitable area (km²)
Forest obligate	x	x				10,000
Forest generalist	x	x		x		12,500
Non-closed forest			x	x		10,000
Habitat generalist	x	x	x	x		20,000

35-45 m. Both faunal and floral diversity is highest in the WE-zone, although the majority of large vertebrates occur in both zones. All selected forest sites lie in a lowland area with altitudes between 25-300 m.a.s.l. Table 17.2 describes the 22 selected study areas, detailing forest fragmentation, logging disturbance and human population density.

Table 17.2: Description of the 22 Selected Forest Habitats in Southwest Ghana. Only informant data were obtained at Ndumfri and Neung South FRs (no transect surveys conducted).

Forest Site	Forest Zone[1]	Size (km²)	Reg. Time (Years)[2]	Logging Index[3]	Hunting Index[4]	Population (km⁻²)[5]
Boin	WE-ME	277.7	21	43.2	1.54	35
Dadiaso	ME-WE	171.2	13	6.0	0.77	35
Disue	WE-ME	23.6	15	0.0	0.80	35
Jema	WE	66.0	no logging	0.0	3.50	35
Subri	ME	587.9	0	59.0	3.58	40
Tano Nimri	WE-ME	205.9	2	45.8	1.99	35
Yoyo	ME	235.7	14	19.9	1.30	35
Ankasa	WE	348.7	20	18.5	0.75	60
Bura	ME	103.1	2	131.3	2.27	45
Draw	WE	235.0	0	60.6	2.39	60
Ebi	WE	25.9	6	10.0	2.46	60
Fure	WE	158.2	5	59.1	1.37	45
Mamiri	ME-WE	45.3	21	16.8	1.64	45
Cape	WE-ME	51.0	19	5.0	4.91	185
Ndumfri	WE	72.5	4	–	–	85
Neung South	WE	112.7	6	–	–	85
Neung North	WE	45.0	8	35.6	2.98	85
Ankasa Coconut	WE	1.5	–	–	2.75	60
Boin Cocoa	WE-ME	0.5	–	–	10.00	35
Subri Gmelina	ME	48.0	–	–	2.50	40
Tano Cedrela	WE-ME	9.8	–	–	2.45	35
Neung Cedrela	WE	6.4	–	–	2.20	85

1. From Hall and Swaine (1976): WE: Wet Evergreen (> 2,000 mm); ME: Moist Evergreen (1,750-2,000 mm).

2. Time since last logging occurred in the reserve.

3. Based on Holbech (2005).

4. Based on Holbech (1996); see Table 4 for details.

5. Extrapolation from 1984-figures (Ghana Statistical Service Department).

Methods

Transect Censusing

All transects were cut prior to censuses by a 6-8 person team comprising the author, a GWD senior officer and locally recruited hunters and farmers with good game tracking skills. Each transect was 1-3 km long and 1 m wide, and prepared to facilitate easy and silent movement. Three key data fields were recorded on transects: signs of logging, hunting, and animal presence. Transect censusing was not conducted in Neung South and Ndumfri FRs, only informant data were acquired from these reserves.

Logging Disturbance and History

Any visible signs of ongoing or historic logging were recorded systematically within 5 m from the transect midline; these signs included logs, stumps, skid tracks, hauling roads, and loading bases. A logging index (LI) was calculated for all transects in a particular forest, based on a weighted score index according to the quantitative disturbance impacts on the vegetation structure (Holbech, 2005). The index does not account for the difference in regeneration time since last logging. Logging history for each reserve was based on Hawthorne and Abu-Juam (1995) supplemented by Forest Management Plans from the GFC.

Hunting Activity Index

Five hunting activity descriptors were recorded along transects within 5 m from the transect midline: 1) hunting trails, 2) cable snares, 3) spent shotgun-cartridges, carbide powder or batteries, 4) gunshots, and 5) encounters with hunters (also noise from humans and dogs etc.). Records were transformed to scores per km walked transect, and the descriptors were pooled, representing the hunting index (HI) for each of the 20 study sites (Table 17.4).

Animal Abundance Indices

Multiple-signs censusing included audio-visual detection of animals, and the counting of dung piles, tracks, trails and diggings. Sightings and vocalisations were primarily gathered from primates, diurnal antelopes and rodents. Dung piles, track and trail counting produced data on elephants, antelopes and carnivore-omnivores, and diggings were included to assess bush pig abundance. Diggings were detectable up to 5 m from the transect midline, but tracks and dung piles only within 2.5 m from it. Nest building by chimps was also included. Cumulated signs for all transects in each habitat were transformed to an abundance index per km of walked transect, making a direct comparison on the habitat level possible.

Informant Interviews

Information was gathered from both hunters and operators of bushmeat restaurants ('chopbars'). A detailed questionnaire on animal encounters was developed prior to the survey, and accompanied by coloured species drawings. The main questions for each species were: i) last encounter (days, weeks, months, years); approximate (to nearest 5 or 50 individuals) harvest within the past ii) two years, or

iii) since bushmeat operations began (mean for hunters = c. 15 years; mean for chopbars = c. 10 years); iv) habitat preference (reserved forest; unreserved forest = remnants and old secondary forest; farms and farm-bush = neglected farms and young regrowth; rubber plantations). The distinguishing of recent (0-2 years) versus past harvest aimed at providing trends in time of animal abundance as well as hunting pressures. Other questions were on hunting methods and expenses, and the knowledge and attitude towards hunting regulations and wildlife conservation.

Respondents were selected randomly to get a full representation across age, literacy and ethnicity. The author interviewed in English, or through the GWD-officer as a translator of local dialects. I did not use any reward system, but prior to interviews I always consulted chiefs or community elders and frequently presented customary gifts to these local authorities. To assess whether respondents were able to distinguish between closely related species, indigenous names were always used and a brief or (for rare species) detailed description was required. Interviews lasted between 45 minutes and two hours (mean = 75 minutes) for hunters, and 45-75 minutes (mean = 1 hour) for bushmeat traders. One major disadvantage of questionnaires was the distinct suspicion of interviewers expressed by some respondents living close to the Ankasa Resource Reserve (GWD-protected wildlife reserve). Interviews were terminated if respondents seemed nervous and unreliable. To avoid immediate suspicion, we never wore uniforms, nor used stickers on our vehicle. We emphasised to each respondent that the purpose of my team was purely academic.

Informant Data on Animal Abundance and Annual Harvest Rates

Through interviews, hunters provided basic information on animal abundances and associated harvests from their hunting trips. This information led to two important measures for abundance and diversity of forest game: mean annual encounter frequency (MAEF) and mean annual yield (MAY). For each respondent the most recent encounter of a species was transformed into annual encounters per each respondent, and the mean for all respondents equals the MAEF for a particular species (Holbech, 1996 and 2014b). Hence, the more recently hunters had detected (audio-visual record) particular animal species, the more commonly these were assumingly recorded and therefore most likely also more genuinely common. MAEF will inter alia depend on the: 1) detectability of animals, 2) skills of hunters, and 3) frequency of hunting. Whereas detectability is species-specific and related to habitat visibility, it is also dependent on hunters' skills and experiences. I therefore mostly selected experienced and active hunters, and these were assumed to have similar game detection skills across analysed forest categories, though inevitably there is a potential bias related to variability in hunters' detection skills. The frequency of hunting (weekly trips) was a major factor that was accounted for when comparing MAEF across forest categories, by virtue of correcting MAEF-values accordingly (Holbech, 1996 and 2014b).

The mean annual yield (MAY) per each hunter was also estimated from interviews, as each respondent was asked to report on harvest rates, and crude estimates based on weekly or monthly rates were transformed into annual yields per each hunter, and the mean based on (n) hunters hereafter calculated for respective forest categories. MAY depends on both hunters' skills, hunting frequency and number of snares and traps operated. Whereas hunters' skills were assumed to be similar across forest categories, hunting frequency and snare (trap) numbers were major factors that were accounted for through correction of MAY-values (same procedure as for MAEF-values).

Forest Categorisation and Comparative Data Analyses: Human Population Density

I hypothesised that hunting pressure was positively correlated with human population density. District level population statistics were provided by the Ghana Statistical Service Department and used to estimate population densities for three variably populated districts, to which each of the 17 forest habitats (FRs/RR) were attributed (Table 17.2). The breakdown of the forests according to these categories were: sparsely populated district (35-40 km^{-2}): Boin, Dadiaso, Disue, Jema, Subri, Tano and Yoyo FRs; moderately populated district (45-60 km^{-2}): Ankasa RR, Bura, Draw, Ebi, Fure and Mamiri FRs; densely populated district (85-185 km^{-2}): Cape, Neung North, Neung South and Ndumfri FRs (Table 17.2). Site-specific faunal data sets from respectively transect surveys and interviews were pooled according to these categories, to enable comparison of abundances, diversity and harvest rates of forest game.

Transformations of Informant Statistics

The mean annual yield (MAY) per hunter for the last two years (present and past year) was calculated for each species and represented current species-specific hunting pressures. Across the three human-population-density forest categories, mean number of traps and weekly hunting trips were substantially different, and MAY-figures for forests situated in the densely populated district were corrected by a factor 0.70 ($66+68/2 \times 94^{-1}$) according to the mean trap number for the two other districts that were similar in that respect. Similarly, MAY-figures for forests in the sparsely populated zone were corrected by a factor 1.2 (3.5/3.0), according to the mean weekly trips for the two other districts which was similar in that regard.

Biological and Ecological Data of Forest Game Applied in Hunting Sustainability Analysis

To estimate total biomass of MAY-figures I used weight data from Fa and Purvis (1997) and Kingdon (1997) for mammals, The Birds of Africa Vol. 1-3 for birds, and the experience of hunters for average sizes of reptiles. To estimate species-specific populations in the study area (Western Region) for application in the sustainability analysis of harvest rates, I used the mean of published population density figures for African forests (see references in Table 17.7). To determine the suitable area available for each species in the Western Region, 'forest obligates' were those limited to reserved (*i.e.* forest reserves, resource reserves, wildlife sanctuaries and national parks) and remnant forest (10,000 km^2), 'forest generalists' to reserved, remnant forest and non-intensively farmed areas (12,500 km^2), 'non-closed forest species' to 10,000 km^2 (only Cane rat), whereas 'habitat generalists' were unlimited to 20,000 km^2 (Table 17.1). Published fecundity data were used to estimate species-specific game production for the Western Region. Species-specific game production estimates were based on the assumptions of a 1:1 sex ratio, and that 50 per cent of females were reproductive throughout the year. Maximum sustainable harvest levels were taken (following Robinson and Redford, 1994) as 60 per cent for very short-lived animals (< 5 years), 40 per cent for short-lived (5-10 years), and 20 per cent for long-lived animals (> 10 years).

Results

Hunters and Hunting Methods

All respondents (n = 112) were males, and comprised full-time 'professional hunters' (9 per cent) and part-time 'farmer-hunters' (91 per cent). The former focus on large game, spend several days on each hunting trip, operate in teams from interior forest camps, and market large quantities of smoked meat to urban retailers. These people live secretively, and are very difficult to trace. Professional hunters were only encountered in the Ankasa-Draw-Fure, Subri and Boin areas. Farmer-hunters typically use 2-4 days weekly for hunting and trapping, make only daylong trips, and their hunting activities are largely correlated with the slack farming periods that are correlated with seasonal rainfall peaks. The mean age of respondents was 43 years (range 18-70), experience 15 years (range 0.5 - 53 years), and weekly hunting trips 3.4 (range 0.5-5), the latter equalling *c.* 177 trips per year (Table 17.3).

Table 17.3: Sampling Efforts and Characteristics of Three Forest Categories in Southwest Ghana (1993-1995), According to Human Population Density (* = the mean value for all forests in a category)

Area Description and Sampling Efforts	Sparsely Populated	Moderately Populated	Densely Populated
Population density (km^{-2})	35-40	45-60	85-185
Transect length (km)	37.6	45.9	7.8
Transect walked (km)	143.0	156.1	19.8
Tree extraction* (ha^{-1})	0.6	1.6	0.5
Hunting index*	1.9	1.8	3.9
No. of respondents	39	40	33
Age* (years)	42	39	47
Experience* (years)	13	11	21
Traps per respondent*	68	66	94
Weekly hunting trips*	3.0	3.5	3.5

Four major hunting methods were identified: (a) shotgun, (b) cable snaring, (c) dogs, and (d) smoke driving. Details of these are as follows:

(a) The basic equipment of shotgun hunters consisted of shotguns purchased locally for the equivalent of 100-150 USD, 70mm cartridges (*c.* 0.3 USD each), and a torchlight (or carbide lamp). Foot paths, animal trails or hauling roads were used for distances up to 15-20 km on a single trip, and typically began late afternoon to end late the following morning. Monkeys and birds dominated daytime bags, whereas antelopes and rodents were mostly killed at night.

(b) Snaring was widespread throughout the study area. A 3mm steel cable (*c.* 0.3 USD per yard) was either used whole or split into 1mm strands, depending on the targeted animal. Baited neck-snares were set within fences at ground level or on tree branches and logs, targeting respectively terrestrial and arboreal rodents. Leg-snares were set on animal trails targeting birds, reptiles and mammals. Neck-snares were also set in rivers to trap crocodiles at their dens. The mean trap number per hunter was 76 (range 14-450), but did not specify snare types (leg or neck).

(c) Group hunting with 4-8 dogs was mostly conducted by young men and boys, and targeted terrestrial animals from giant rats up to bushbucks. The dogs, along with clubs or cutlasses, were used to kill the prey. Antelopes appeared vulnerable to this hunting method.

(d) Smoke driving of den dwellers such as giant rats, porcupines and monitor lizards was another group-hunting method. When dens were located, all openings but one were sealed, fire was started, and smoke fanned into the den. Escaping animals were killed with dogs, cutlass or club.

Hunting Activity Index

Signs of hunting were detected throughout the 20 transect-surveyed forest and non-forest habitats (Table 17.4). The hunting index (HI) was best correlated with detections per km walked transect of snares ($r = 0.750$, $p < 0.001$, $df = 19$), gunshots ($r = 0.810$, $p < 0.001$) and hunters ($r = 0.850$, $p < 0.001$); these are therefore regarded as good indicators of hunting pressure in general (Figure 17.2). The calculated hunting (activity) indices for each of the 15 reserved forest (Table 17.4) were significantly ($r = 0.666$, $p < 0.05$, $df = 14$) positively correlated with human population densities at the district level (Figure 17.3). Hence, the assumption that hunting pressures was directly linked to human population density appeared to be confirmed as a true hypothesis, unlike the relationship between hunting pressures and logging intensity which could not be established (Holbech 1996 and 2014b). These findings indicate that whilst human population density drives hunting related game animal declines, logging is not directly linked to such declines, though hunting activities may be linked positively to ongoing logging activities.

A total of 241 snares was detected on 98.6 km transect in 15 forest and three exotic plantation habitats, producing a mean density of 2.4 snares ha^{-1}. In comparison, the dung pile density of duikers on the same transects was 2.1 ha^{-1} (Holbech, 2014b), indicating the extremely high snare density recorded. The mean snare density in the two neglected cash-crop plantations (9.5 ha^{-1}) was four times higher than recorded in exotics and forests. The mean HI for forests was 2.2 ± 1.2, exotic plantations 2.4 ± 0.2, and neglected cash-crop plantations 6.4 ± 5.1 (Table 17.4), of which only the latter was significantly higher than the two former (ANOVA with Tukey post hocs: $F_{2,17} = 5.907$, $p < 0.02$). Although mean HI for densely populated district forests (3.9 ± 1.4) was twofold higher than means for moderately (1.8 ± 0.7) and sparsely (1.9 ± 1.2)

populated district forests (Table 17.3), the difference across the three forest categories was only near-to significant (ANOVA: $F_{2,12} = 3.565$; $F_{Critical} = 3.885$, $p > 0.05$).

Table 17.4: Sampling Statistics and Hunting Activities Measured in 20 Forest Habitats of Southwest Ghana, 1993-1995. The hunting index (HI) of a habitat reflects the summed frequency of the five disturbance factors measured. TD: Transect distance; WD: Walked distance.

Forest	TD (km)	WD (km)	Observations per km Walked Transect					HI
			Snares	Trails	Cartridges	Gun-shots	Hunters	
Ankasa	14.8	62.5	0.37	0.08	0.13	0.13	0.05	0.8
Dadiaso	4.9	25.9	0.35	0.00	0.12	0.31	0.00	0.8
Disue	3	7.5	0.40	0.00	0.13	0.27	0.00	0.8
Yoyo	4.5	21.6	0.83	0.19	0.05	0.19	0.05	1.3
Fure	7.6	24.1	0.41	0.62	0.08	0.17	0.08	1.4
Boin	10.5	46.7	0.51	0.51	0.24	0.24	0.04	1.5
Mamiri	4	12.8	0.63	0.47	0.08	0.23	0.23	1.6
Tano Nimri	4.8	17.6	0.68	0.80	0.11	0.40	0.00	2.0
Bura	4	12.8	1.09	0.86	0.16	0.16	0.00	2.3
Draw	10.5	30.1	0.86	0.83	0.17	0.53	0.00	2.4
Ebi	5	13.8	1.45	0.72	0.00	0.29	0.00	2.5
Neung North	2.8	8.4	0.12	2.50	0.12	0.00	0.24	3.0
Jema	4.5	11.7	1.54	1.54	0.26	0.17	0.00	3.5
Subri	5.4	12	1.42	1.75	0.00	0.25	0.17	3.6
Cape	5	11.4	2.81	1.84	0.00	0.26	0.00	4.9
All forests	91.3	318.9	0.90	0.85	0.11	0.24	0.06	2.2 ± 1.2
Ankasa coconut	1	4	2.00	0.50	0.25	0.00	0.00	2.8
Boin cocoa	1	4	2.75	0.50	0.50	3.75	2.50	10.0
Neung cedrela	2.6	5	0.00	2.00	0.00	0.20	0.00	2.2
Subri gmelina	2.8	5.2	0.00	0.38	0.38	1.35	0.38	2.5
Tano cedrela	1.9	5.3	1.13	1.13	0.19	0.00	0.00	2.5
All plantations	9.3	23.5	1.18	0.90	0.26	1.06	0.58	4.0 ± 3.4
All habitats	100.6	342.4	0.97	0.86	0.15	0.44	0.19	2.6 ± 2.0

Hunting Statistics

Mean annual encounter frequency (MAEF, Holbech 1996 and 2014b) of animals per hunter was overall 2,196.2, equivalent to 12.4 (2196.2/176.8) animals encountered per trip (Appendix). Within the past two years, mean annual yield (MAY) was overall 222.4 animals per respondent, equivalent to 1.3 animals per trip (Appendix). Hence the overall hunting rate was *c.* 10 per cent of encountered animals. Overall MAY-figure within the past >15 years was only 164.2 animals per year, suggesting a 35 per

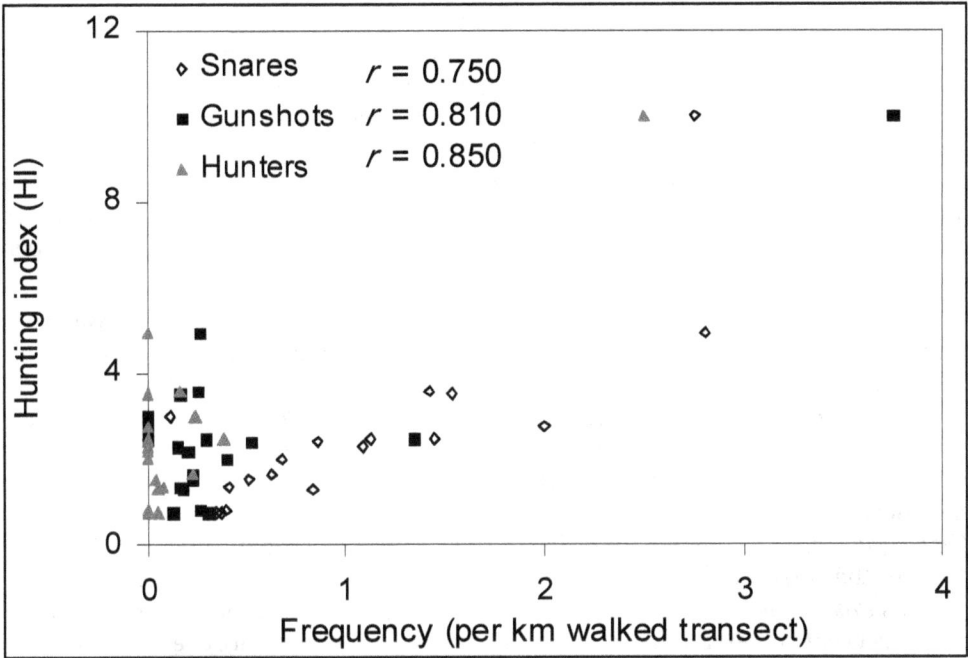

Figure 17.2: Correlations between Hunting Activity Index (HI) and Numbers of Snares, Gunshots, and Hunters Detected per km Walked Transect in 15 Forest and 5 Non-forest Habitats of Southwest Ghana, 1993-1995.

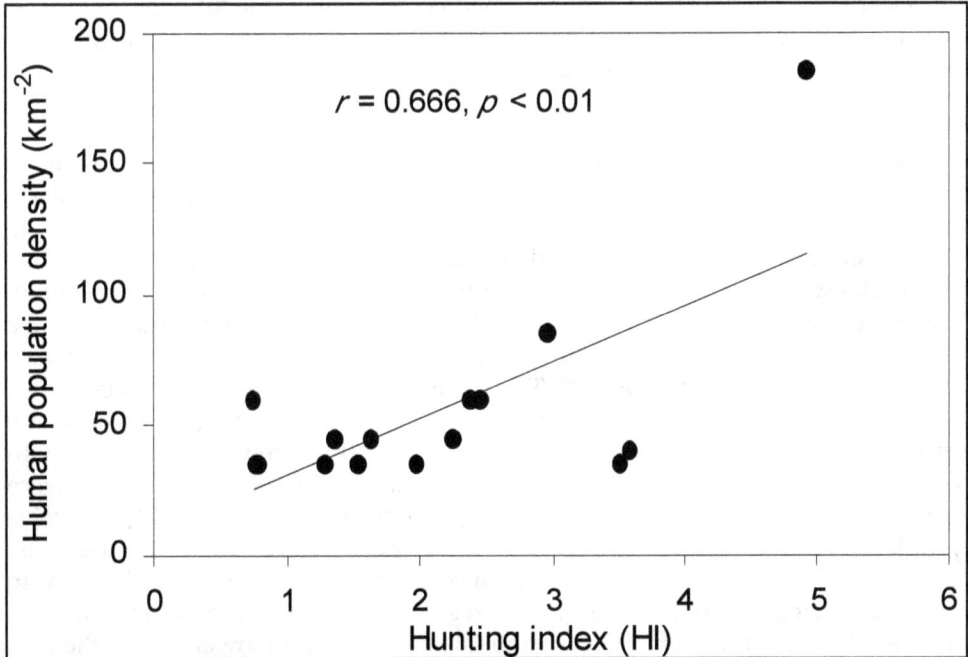

Figure 17.3: Correlation between Hunting Index and Human Population Density for 15 Reserved Forests in Southwest Ghana, 1993-1995.

cent numerical increase of total yield among hunters in the Western Region (Appendix). Overall, mammals comprised 90.9 per cent of harvests, birds 5.6 per cent, and reptiles 3.5 per cent. Commonly hunted were giant rats *Cricetomys emini/gambianus* (34.3 per cent), Cusimanse *Crossarchus obscurus* (8.1 per cent), Cane rat *Thryonomys swinderianus* (6.6 per cent), African giant squirrel *Protoxerus stangeri* (5.6 per cent), Brush-tailed Porcupine *Atherurus africanus* (4.5 per cent), and Maxwell's duiker *Cephalophus maxwelli* (3.9 per cent). Numerically, these six species together represented 63 per cent of total harvests, with rodents alone accounting for 51 per cent.

A comparison of species-specific MAY within the past two years to those of the past *c.*15 years suggests that hunting pressure has increased by >100 per cent on several species, *e.g.* Marsh mongoose *Atilax paludinosus* (195 per cent), Dwarf crocodile *Osteolaemus tetraspis* (151 per cent), Potto *Perodicticus potto* (129 per cent), Lowe's monkey *Cercopithecus campbelli lowei* (116 per cent), Tree hyrax *Dendrohyrax dorsalis* (112 per cent), casqued hornbills *Ceratogymna/Bycanistes* spp. (108 per cent), Lesser spot-nosed monkey *Cercopithecus petaurista* (107 per cent), and Ogilby's duiker *Cephalophus ogilbyi* (106 per cent). In contrast, nine species have declined in MAY between the two periods: Western Chimpanzee *Pan troglodytes* (> -1200 per cent), Red colobus *Procolobus badius waldronae* (-1200 per cent), Giant pangolin *Smutsia gigantea* (> -600 per cent), Forest buffalo *Syncerus caffer nanus* (> -100 per cent), Crested porcupine *Hystrix cristata* (> -100 per cent), Yellow-backed duiker *Cephalophus sylvicultor* (-93 per cent), Giant forest hog *Hylochoerus meinertzhageni* (-50 per cent), Honey badger *Mellivora capensis* (-50 per cent), and Bongo *Tragelaphus eurycerus* (-20 per cent). These data clearly reflected a shift from large-bodied and rare game towards small and common game, and the shift was largely attributed to dwindling abundances of the former, rather than change in hunter's behaviour.

Bushmeat Operators and Trade

The vast majority of chopbar operators (*n* = 18) were females, and the mean operating experience was *c.* 10 years (range 1-26 years). A mean of 6.5 hunters regularly supplied chopbars with bushmeat. Purchase price depended on species, size, state (fresh or smoked), and season. A total of 36 mammal, three reptile and at least four bird species were reportedly sold in 18 chopbars within the past 10 years (Appendix). The total biomass of bushmeat represented *c.* 1,600 kg per chopbar annually, with a value of *c.* 2,200 USD. The most commonly traded species were *C. emini/gambianus* (44.3 per cent), *Cephalophus maxwelli* (9.7 per cent), *Atherurus africanus* (6.5 per cent), *Thryonomys swinderianus* (6.0 per cent), and Royal antelope *Neotragus pygmaeus* (3.9 per cent). Numerically these species comprised *c.* 70 per cent of total sales, with rodents accounting for 57 per cent. Hence, several species commonly traded were also the most commonly hunted. Within the past two years, operators did not receive eight large or rare mammal species, and these species were similar to those least harvested. Hence, the tendency is that smaller game was increasingly brought to chopbars, whereas the supply of large-bodied game was steadily decreasing. However, the overall annual number of traded animals was 5 per cent lower within the past two years compared to the last *c.* 10 years. Since hunting pressure in general has increased (MAY$_{2years}$ was 35 per cent higher than MAY$_{15years}$), the downwards trend in traded animals suggests that the relative consumption by hunters has increased, and many chopbars complained about an increasing number of unstable suppliers, so that they had found it necessary to increase the number of suppliers accordingly.

Forest Game Abundance in Relation to Human Population Density

Since transect sampling efforts in forests of the densely populated district were < 20 per cent of the two other districts (Table 17.3), both total records and species detected on transects could not be readily compared between the three forest categories (Table 17.5). However, the moderately populated district had 50-90 per cent higher overall relative abundance of forest game compared to the two other district forests, although the difference was insignificant ($\chi^2 = 1.67$, $p > 0.30$, $df = 2$) (Table 17.5). Forests in the densely populated district were distinctively least diverse ($D_{Mg} = 1.56$) compared to both other zones, which were relatively equal ($D_{Mg} = 3.49$-4.16). Despite few statistical data, monkeys and antelopes appeared more abundant in forests of the medium populated district. In contrast, forests in the densely populated district displayed low abundance and diversity of large mammals, reptiles and carnivores, whereas rodents and small antelopes were relatively abundant.

Table 17.5: Relative Abundance (Records per 10 km walked transect), Species Number ( ) and Ecological Diversity ($D_{Mg} = (S - 1) \cdot (\ln N)^{-1}$) of Forest Game Animals Recorded in Forests of Districts with different Human Population Densities in Southwest Ghana, 1993-1995

Forest Game	Sparsely Populated	Moderately Populated	Densely Populated
Elephant	–	0.3 (1)	–
Antelopes	1.4 (5)	2.5 (7)	1.5 (2)
Red river hog	0.6 (1)	0.3 (1)	–
Monkeys	0.6 (1)	1.4 (3)	1.0 (1)
Carnivores	0.4 (2)	0.9 (5)	–
Other mammals	< 0.1 (1)	0.2 (1)	–
Rodents	2.1 (4)	4.2 (2)	4.1 (2)
Reptiles	0.1 (2)	0.2 (2)	–
Total records	74	156	13
Total abundance	5.2	10.0	6.6
Total species	16	22	5
Diversity (D_{Mg})	3.49	4.16	1.56

In line with the transect data, overall mean annual yield (MAY) of hunters in moderately populated district forests was significantly the highest ($c^2 = 64.61$, $p < 0.001$, $df = 2$) (Table 17.6). Primates, carnivores and rodents also had significantly higher MAY for hunters in moderately populated district forests (Table 17.6). Diversity (D_{Mg}) of MAY was fairly equal among the forests. Across the three forest categories, MAY for rare and endangered species was 2-3 times higher in moderately populated district forests compared to the other two districts, although the difference was insignificant ($\chi^2 = 4.64$, $p > 0.05$, $df = 2$) (Table 17.6). Summarising transect and informant data, it appeared that moderately populated district forests were superior habitats for forest game, whereas densely populated district forests supported depauperate guilds of game.

Table 17.6: Mean Annual Yield (MAY) per Respondent (N = 112) and Ecological Diversity ($D_{Mg} = (S - 1) \cdot (\ln N)^{-1}$) of Game Animals Recorded in Forests of Districts with different Human Population Density in Southwest Ghana, 1993-1995

Forest Game	Sparsely Populated	Moderately Populated	Densely Populated	χ^2-test df = 2
Antelopes	19.2	27.9	17.5	2.89, $p > 0.20$
Bush pigs	0.4	1.0	1.0	0.30, $p > 0.80$
Primates	21.2	41.4	22.3	9.12, $p < 0.02$
Carnivores	20.4	46.6	16.2	19.57, $p < 0.001$
Rodents	106.8	143.4	61.2	32.68, $p < 0.001$
Reptiles	6.5	13.8	5.0	5.26, p > 0.05
Birds	12.8	17.3	9.7	2.20, $p > 0.30$
Rare and endangered*	6.0	12.8	4.9	4.64, p > 0.05
Total abundance (MAY)	197.3	303.3	139.6	64.61, $p < 0.001$
Total species	43	39	40	–
Diversity (D_{Mg})	7.95	6.65	7.90	–

* See Holbech (2014b).

Sustainability Analysis of Hunting Pressures

Estimates of the hunter population in the Western Region, together with my MAY estimates of an average hunter in the Western Region, can be used to estimate species-specific hunting pressures for the Western Region as a whole. The mean snare density in forests and exotic plantations was 2.4 ha^{-1}. The mean snare density in neglected cash-crop plantations was 9.5 ha^{-1}, but limited sampling efforts (2 km transect ~ 2 ha) may not have reflected snare density in non-forest habitats as a whole. Snare density in anthropogenic forest landscapes around Ankasa RR in southwest Ghana has been estimated at *c.* 2.7 ha^{-1} (Holbech, 2001), indicating that snare density is similar in forest and non-forest areas. In my study the mean snare number per Western Region-hunter was 76, suggesting that an average hunter used *c.* 32 ha (76/2.4) of non-intensively farmed forest areas, which equalled a density of approximately 3 operating hunters per km² in the Western Region. As non-intensively farmed forest areas totalled 12,500 km² (Table 17.1), the hunter population of the Western Region could be estimated at *c.* 37,500 (~ 3 hunters per km²).

Reliable and comparable population density estimates of 30 species were gathered from 22 publications (Table 17.7). Based on those data, only seven species (23 per cent) were estimated to be harvested on a sustained basis in the Western Region, *e.g.* Red river hog (*Potamochoerus porcus*), *Dendrohyrax dorsalis*, *Atherurus africanus*, *Thryonomys swinderianus*, *Cricetomys* spp., *Protoxerus stangeri* and Great blue turaco (*Corythaeola cristata*). These species are all prolific and opportunistic forest

Table 17.7: Assessment of Sustainability of Estimated Total Harvest in Relation to Estimated Area of Suitable Habitat (see Table 17.1), Population Density, Reproduction and Maximum Sustained Yield (MSY) of 30 Forest Game in the Western Region (WR), Ghana, 1993-1995. * indicates a probably sustainable harvest; ** indicates that estimated harvest exceeds probable maximum population of that species in the Western Region.

Species	Suitable Habitat Area (km²)	Density (km⁻²)	Maximum WR Population	Fecundity (year⁻¹)	Maximum WR Production	MSY (per cent)	WR MSY	WR Harvest
Tragelaphus euryceros	10,000	0.25[9]	2,500	1.0	625	20	125	3,750**
T. scriptus	20,000	30.5[4,8,16]	610,000	1.5	228,750	20	45,750	75,000
Cephalophus sylvicultor	10,000	1.5[3,4,5,18]	15,000	1.0	3,750	20	750	3,750
C. niger	20,000	12.0[8]	240,000	1.6	96,000	40	38,400	75,000
C. dorsalis	10,000	4.3[3,4,5,14]	43,000	1.0	10,750	40	4,300	75,000**
C. ogilbyi	10,000	13.03[3,4]	130,000	1.0	32,500	40	13,000	67,500
C. maxwelli	20,000	31.4[2,8,12,13]	628,000	1.8	282,600	40	113,040	378,750
Hyemoschus aquaticus	10,000	4.8[4]	48,000	1.0	12,000	40	4,800	7,500
Potamochoerus porcus	20,000	8.2[3,4,18]	164,000	10.0	410,000	20	82,000	30,000*
Piliocolobus b. waldronae	10,000	0.1[7]	1,000	1.5	375	20	75	< 375
Colobus vellerosus	10,000	17.9[7,10,15]	179,000	1.5	67,125	20	13,425	52,500**
Procolobus verus	12,500	5.7[15]	71,250	1.5	26,720	20	5,345	123,750**
Cercopithecus diana roloway	10,000	25.2[10,15]	252,000	1.5	94,500	20	18,900	63,750
C. mona lowei	12,500	24.3[4,15]	303,750	1.5	113,905	20	22,780	255,000
C. p. petaurista	12,500	38.0[4,15]	475,000	1.5	178,125	20	35,625	307,500
Cercocebus atys lunulatus	12,500	28.7[4,15]	358,750	1.5	134,530	20	26,905	120,000
Perodicticus potto potto	12,500	8.2[4,22]	102,500	1.5	38,440	20	7,690	210,000**
Panthera pardus	10,000	0.3[4,18]	3,000	1.5	1,125	20	225	< 375
Civettictis civetta	20,000	4.1[3,4,8]	82,000	5.0	102,500	20	20,500	86,250**

Contd...

Table 17.7—*Contd...*

Species	Suitable Habitat Area (km²)	Density (km⁻²)	Maximum WR Population	Fecundity (year⁻¹)	Maximum WR Production	MSY (per cent)	WR MSY	WR Harvest
Nandinia binotata	20,000	3.6[3,4]	72,000	6.0	108,000	20	21,600	168,750**
Atilax paludinosus	20,000	2.2[4,17]	44,000	3.0	33,000	20	6,600	105,000**
Dendrohyrax dorsalis	20,000	615.1[11,18]	12,302,000	1.0	3,075,500	20	615,100	116,250*
Tree pangolins	20,000	7.6[1,4,18]	152,000	2.5	95,000	20	19,000	247,500**
Atherurus africanus	20,000	78.5[3,4,5,8,14]	1,570,000	6.0	2,355,000	20	471,000	435,000*
Thryonomys swinderianus	10,000	70.0[8]	700,000	8.0	1,400,000	60	840,000	600,000*
Protoxerus stangeri	20,000	71.4[19]	1,428,000	6.0	2,142,000	60	1,285,200	506,250*
Cricetomys emini/gambianus	20,000	181.4[3,4,5,8]	3,628,000	16.0	14,512,000	60	8,707,200	2,392,500*
Corythaeola cristata	20,000	48.0[18]	960,000	4.0	960,000	20	192,000	176,250*
Casqued hornbills (3 spp.)	12,500	9.5[20]	118,750	3.0	89,065	20	17,815	172,500**
Agelastes meleagrides	10,000	11.9[6,21]	119,000	6.0	178,500	20	35,700	37,500

1: Akpona *et al.* (2008); 2: Davies (1991); 3: Fa *et al.* (1995); 4: Fa and Purvis (1997); 5: Feer (1993); 6: Francis *et al.* (1992); 7: Harding (1984); 8: Holbech (2001); 9: Klaus-Hügi *et al.* (2000); 10: Martin (1991); 11: Milner and Harris (1999); 12: Nett (1999); 13: Newing (2001); 14: Noss (1998 and 1999); 15: Oates *et al.* (1990); 16: Plumptre and Harris (1995); 17: Ray (1997); 18: S. Sommer, unpublished data 2006; 19: Tutin *et al.* (1997); 20: Uehara (2003); 21: Waltert *et al.* (2010); 22: Weisenseel *et al.* (1993).

habitat generalists, with the four rodents numerically making up 47 per cent of killed animals, and 24 per cent and 26 per cent of the total biomass and economic value respectively (Appendix). Particularly overexploited were large-bodied game, primates, carnivores and hornbills, for some of which estimated harvest greatly exceeded estimated population size for the Western Region.

Bushmeat Socio-Economics

The MAY for an average hunter ($n = 112$) amounted to 939 kg, with a value of 912 USD (Appendix); across districts, the quantities were 694 kg (562 USD) in sparsely populated, 1,199 kg (1,228 USD) in moderately populated, and 725 kg (762 USD) in densely populated districts. The estimated kg-price of raw bushmeat varied between 0.21 and 3.00 USD, depending on demand and availability ($n = 44$ species). Guineafowl, *Neotragus pygmaeus*, Blotched genet (*Genetta tigrina pardina*) and *Atherurus africanus* were very expensive, whereas *Tragelaphus eurycerus*, Olive colobus (*Procolobus verus*), African Civet (*Civettictis civetta*), reptiles and bush pigs were low cost meat. Mean kg-price of fresh bushmeat ranged from 0.93 ± 0.53 USD kg^{-1} in sparsely populated districts to 1.08 ± 0.52 USD kg^{-1} and 1.15 ± 0.58 USD kg^{-1} in moderately and densely populated districts respectively. Hence, bushmeat prices were generally positively correlated with human population density. The overall average bushmeat kg-price was 1.05 ± 0.53 USD, which was equal to or higher than domestic meat (Appendix). There was a significant negative correlation (GLM: $r = -0.362$, $p < 0.05$; $df = 42$; Non-linear (logarithmic): $r = -0.554$, $p < 0.001$,) between animal weight and the kg-price of bushmeat (Figure 17.4).

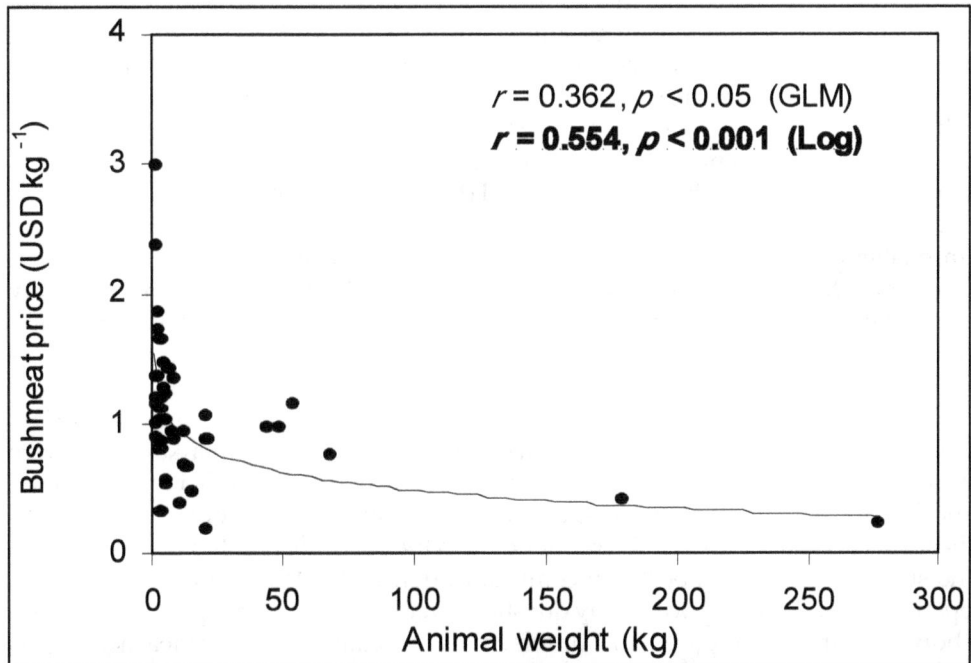

Figure 17.4: Relationship between Estimated Mean Animal Weight and Mean kg-price of 44 Bushmeat Species Sold among 112 Hunters in Southwest Ghana, 1993-1995.

Hunters claimed to sell *c.* 25 per cent of their harvests on average, representing 228 USD per family from bushmeat annually. If the dressed weight amounts to 75 per cent of gross weight (head, skin, various intestines and soft bones are eaten), the daily bushmeat available to each respondent family was *c.* 2.0 kg. With a mean family size of 7, this equalled *c.* 0.29 kg daily per person, off which 75 per cent or 0.22 kg was eaten. This consumption did not reflect the average household consumption in a community, but was only representative of those families that hunted regularly. The proportion of such families was unknown, but if it is assumed to be 20 per cent (see below) a conservative estimate is that the daily consumption per rural inhabitant of the Western Region was *c.* 50 g. The estimated 37,500 Western Region-hunters hunted an estimated total annual biomass of 35,200 tonnes valued at 34.2 million USD, representing 7-9 per cent of the annual national bushmeat trade value of 400-500 million USD (Ntiamoa-Baidu 1998). With an average family size of 7, these hunters represented > 260,000 people or *c.* 20 per cent of the rural population of the Western Region. Of the 35,200 tonnes of bushmeat hunted annually, 26,400 tonnes were reportedly consumed, and the rest (8,800 tonnes) sold to local chopbars or bushmeat traders linked to district capitals. If the bushmeat sold to local chopbars amounts to 5,000 tonnes, it means that over 3,100 (5,000/1.6) of these operate in the Western Region, *i.e. c.* one per 400 inhabitants.

Discussion

Reliability of Informant Data

Using informant data in bushmeat surveys to detect changes in wildlife abundances has recently gained more attention among wildlife ecologists, recognising the detailed knowledge wildlife users possess (Mendelson *et al.,* 2003; de Merode and Cowlishaw, 2006), and the reliability of such information is improved when applied with a proper design (Jones *et al.,* 2008). Even though some informants may have underrated yields (Schulte-Herbrugen *et al.,* 2013) in the Ankasa area, it is unlikely that all respondents in the three compared district forests categories were biased in the same direction, *i.e.* underrating or exaggerating. It is however possible, that unreliable or imprecise informants were randomly scattered across these forest categories. MAY-figures for commonly hunted species were presumably slightly exaggerated, due to the prevalence of large catches on single trips, and the perception that these species were not restricted by any wildlife regulations. In contrast, figures for rare and large game were possibly underrated, as most respondents were aware of the hunting restrictions on such species, and perhaps wished to appear more law abiding. On the other hand some primates (colobus and roloway monkeys) that were certainly rare at the time of the study may have been confused with other more common species, or exaggerated to impress upon us with their rich "folklore" knowledge of these species that were most likely abundant in the past only (likely > 5 years prior to the survey; John Oates, personal communication 2010). The appreciation for and open-minded attitude towards my questions were generally more prevalent among chopbar owners compared to hunters. This indicates that bushmeat trading is perceived as a less illegal business compared to hunting. In many communities professional hunting is seen as an inferior and unpleasant profession, yet hunters generally ranked high in the hierarchy of particularly small communities.

The Socio-Economic Importance of Bushmeat

The significance of bushmeat income for rural forest areas of Ghana was highly accentuated by my data from the Western Region. With an average bushmeat value of over 900 USD annually per each hunting family, of 7 members on the average, the annual per capita value amounts to *c.* 130 USD, which is almost 75 per cent of the annual per capita income from farm products of 175 USD around the Ankasa Resource Reserve (Appiah-Kubi, 1999). Compared to the mid-1990s annual national gross income in Ghana of 350 USD per capita, and a minimum annual wage of workers of 275 USD, the importance of bushmeat income was certainly also highlighted by the present study. The average bushmeat consumption data of roughly 0.20 kg per day per inhabitant of hunting families was comparable to estimates for the conservation areas of Bia (0.14 kg) and Ankasa (0.16 kg) conservation areas in southwest Ghana (Holbech, 1998). The estimated daily gross consumption per average rural Western Region inhabitant of 0.05 kg was close to the lower end of sub-Saharan African ranges (0.043-0.104 kg), yet less than rural figures from other parts of Central and West Africa (Wilkie and Carpenter, 1999; de Merode *et al.,* 2004).

Hunting Sustainability

The confidence of the sustainability analysis across species in the Western Region of Ghana is limited by the accuracy of estimated MAY, number of hunters, and literature-based data on animal densities and fecundity, in addition to the arbitrary 50-50 per cent assumptions on male-female ratio and proportion of reproducing females in Ghanaian populations. MAY and estimated number of snares per Western Region hunter were based on a large (n = 112) random sample of hunters across > 50 communities. The mean snare number per hunter (76) was very close to that reported by Noss (1998) (70-75, n = 17), but smaller than in the study by Fa and Yuste (2001) (113, n = 42). Mean cable snare density (2.4/km²) was also calculated from a wide range of habitats (n = 18) covering almost 100 km of transects, equivalent to a sampled area of 100 1-ha plots, though it was less than the 4.2/km² in Noss (1998). Although I specifically asked only about the number of snares set in forests and exotic-tree plantations, some respondents may have overlooked the fact that some of their snares were set in heavily farmed areas where snare density, as suggested by my few data, were comparatively high. If this is so, my numerical estimate of hunters in the Western Region is probably slightly underestimated, and off-takes likewise.

The rural population of the Western Region anno 1995 was estimated at 1.25 million (Ghana Statistical Service), and with an average family number of 7, this figure equalled *c.* 180,000 households, which means that *c.* 20 per cent of these had a hunting member. This percentage is 6 times higher than national estimates of 3-4 per cent (Crooks *et al.,* 2005), which means that rural communities situated in the Western Region, the most forested region of Ghana in which people therefore live close to wildlife, have distinctively more hunters than average rural communities of Ghana. Species-specific population densities used in calculations of maximum Western Region game animal populations were based on both disturbed (habitat and hunting) and pristine sites across tropical African moist forest. As most areas of the Western Region were generally hunted and habitat disturbed, population estimates for forest obligates are most likely overestimates, but this is far from certain for prolific habitat generalists such as terrestrial rodents, bush pigs and small antelopes. Four of the seven species that suggestively were found to be sustainably harvested in the Western Region are all species that persists in Ghanaian bushmeat markets, and this 'post depletion sustainability' of disturbance resilient species is reported elsewhere in Ghana (Cowlishaw *et al.,* 2005; Crookes *et al.,* 2005). My data however, suggest that among antelopes only Bushbuck and Black duiker were possible candidates for sustainable utilisation, whereas Maxwell's duiker more likely was heavily over-hunted. For the latter species, however, another Ghanaian study strongly indicates that this species withstands current hunting levels (Hofmann *et al.,* 1999), and new evidence (van Vliet and Nasi 2008a) has shown that the hunting sustainability criteria of the closely-related Central African Blue duiker (*Cephalophus monticola*) may be wrongly set in terms of both population density, fecundity and maximum sustained yield (MSY) compared to carrying capacity (K). Assuming that mean Western Region density of Maxwell's duikers was 70 km^{-2} with a MSY-K value of 0.6, the annual MSY for the Western Region would be 378,000 duikers which is very close to the estimated off-take of 378,750 by Western Region hunters (Table 17.7).

Integrated Wildlife and Land-Use Management and Sustainable Bushmeat Utilisation

Despite the existence of appropriate legislation, Ghanaian wildlife authorities have failed to protect large-bodied forest game sufficiently (Brashares *et al.,* 2001). Increased protection by improving human and logistic resources within protected areas (FRs in particular) appears unrealistic in the short-term (Oates, 1999), which is problematic given the immediate threat of local extinction of rare forest game and primates (Struhsaker and Oates, 1995). An alternative conservation strategy to the segregation of humans and protected wildlife could be the reduction of incentives for hunting within reserves, through the provision of alternative means of obtaining protein and income. Livestock, and in particular fowl, sheep and goats are common in the Ghanaian forest zone, but are mainly kept on small scale basis for ceremonial consumption, and as investments, rather than daily sources of protein and income. Increasing livestock will require large pastures, which are alien to the forest environment and promote increased deforestation. In this context, problematic and capital intensive domestication of game species may not be worthwhile, compared to a system of free range on-farm production that are readily available through effective and selective hunting methods.

Against this background, and with a decreasing fish supply resulting in increased fish prices (Brashares *et al.,* 2004) bushmeat has the advantage of being relatively cheap in rural areas, although prices vary according to species and season (Mendelson *et al.,* 2003). Wildlife friendly farming is therefore an alternative option for regular meat supply that would lower the incentives of on-reserve hunting. Why walk or travel long distances to conduct risky poaching in reserves, if bushmeat can be harvested right on your own land? In effect, farming practises must be compatible with a diverse and productive wildlife estate, in order to optimise the combined income from wildlife and traditional agricultural crops, *i.e.* by integrating wildlife and land-use management (Holbech, 2001; Vandermeer and Perfecto, 2007).

My data as well as many others clearly show that several species are resilient to extreme hunting pressures in Ghana, and continue to reach even urban markets (*e.g.* Cowlishaw *et al.,* 2005; Crookes *et al.,* 2005). This post-depletion sustainability of opportunistic ungulates and rodents could be used as the backbone of sustained bushmeat harvest in 'wildlife friendly' farming systems, yet to be defined and developed thoroughly in Ghana and the tropics at large. Several studies have shown that industrial plantations of cocoa, oil palm and other cash crops hold depauperate wildlife (Fitzherbert *et al.,* 2008; Donald, 2004). In contrast shaded plantations or mosaics of food crop farms and regrowth forest on abandoned farms (= wildlife friendly farming) have been documented to hold significant wildlife populations throughout the tropics (Perfecto *et al.,* 1996; Donald, 2004; Holbech, 2009), and in particular for small ungulates and scansorial rodents (van Vliet and Nasi, 2008b; Laurance *et al.,* 2008). Such productive wildlife and farming systems may counteract future collapse of bushmeat deriving from reserves as predicted by Barnes (2002).

For integrated wildlife and land-use management to be successful, community participation is imperative, from the farmer and hunter to traditional authorities. In effect, the general awareness and attitude towards wildlife in these rural communities need to be positive towards animal conservation. During my interviews with hunters and farmers (n = 66 out of 112) I assessed the awareness and observance of wildlife regulations, as well as the attitude towards integrated wildlife and land-use management (*i.e.* the willingness to adopt wildlife friendly farming by preserving forest remnants and secondary forest). The results showed that awareness of regulations was generally low (10-50 per cent), though observance of these was relatively high (80 per cent). In contrast, the willingness to adopt wildlife friendly farming unconditionally was extremely low (3 per cent), though as much as 77 per cent would do so if given some kind of monetary or logistic compensation. Only 20 per cent declined to participate in any kind of wildlife friendly farming. These results indicate that overall there is an attitudinal basis for wildlife friendly farming, but also the perceptions of unrealistic compensations. Current Ghanaian wildlife legislation, which is primarily based on colonial wildlife policies, has virtually completely transferred to central government the traditionally-founded rights towards wildlife resources, and created mistrust between local users and governmental managers of wildlife. As such the greatest legislative challenge for future conservation of wildlife resources in Ghana is to revive the former community based user rights and responsibilities for sustainable wildlife utilisation, and consequently devolve the centralised inadequate governmental control. When given the management authority of wildlife on your own land the foundation is laid for local communities to integrate bushmeat production in land-use management as a competitive alternative or complement to monoculture practises.

Acknowledgements

This study was financed by a Ph.D.-grant provided by Danida (No. 104.Dan.8/606), in collaboration with University of Copenhagen, Denmark and University of Ghana (UGL). I thank the Ghana Wildlife Division (GWD) and Ghana Forestry Commission for granting research permits, and special appreciation for invaluable field assistance from the late Mr. Joseph Amponsah (GWD), and supervision by Prof.

Y. Ntiamoa-Baidu (UGL). Professor John F. Oates provided constructive criticism and editing of an earlier manuscript. Finally, I am sincerely indebted to the local communities who were helpful in many ways.

References

Akpona, H.A., Djagoun, C.A.M.S. and Sinsin, B., 2008. Ecology and ethnozoology of the three-cusped pangolin *Manis tricuspis* (Mammalia, Pholidota) in the Lama forest reserve, Benin. *Mammalia*, **72 (3)**: 198-202.

Appiah-Kubi, K., 1999. *Socio-economic Survey of the Bia and Ankasa Protected Areas, Southwest Ghana.* Protected Areas Development Programme – Ghana. Consultancy report. Ghana Wildlife Division/European Union/ULG Northumbrian Ltd., Ghana. 87 pp.

Barnes, R.F.W., 2002. The bushmeat boom and bust in West and Central Africa. *Oryx*, **36 (3)**: 236-242.

Brashares, J.S., Arcese, P. and Sam, M.K., 2001. Human demography and reserve size predict wildlife extinction in West Africa. *Proceedings of the Royal Society of London Series B: Biological Sciences*, **268**: 2473-2478.

Brashares, J.S., Arcese, P., Sam, M.K., Coppolillo, P.B., Sinclair, A.R.E. and Balmford, A., 2004. Bushmeat hunting, wildlife declines, and fish supply in West Africa. *Science*, **306**: 1180-1183.

Cowlishaw, G.C., Mendelson, S. and Rowcliffe, J.M., 2005. Evidence for post-depletion sustainability in a mature bushmeat market. *Journal of Applied Ecology*, **42**: 460-468.

Crookes, D.J., Ankudey, N. and Milner-Gulland, E.J., 2005. The value of a long-term bushmeat market dataset as an indicator of system dynamics. *Environmental Conservation*, **32 (4)**: 333-339.

Davies, A.G., 1991. Survey methods employed for tracking duikers in Gola. *IUCN ASG Gnusletter*, **10 (1)**: 9-12.

de Merode, E. and Cowlishaw, G., 2006. Species protection, the changing informal economy, and the politics of access to the bushmeat trade in the Democratic Republic of Congo. *Conservation Biology*, **20 (4)**: 1262-1271.

de Merode, E., Homewood, K. and Cowlishaw, G., 2004. The value of bushmeat and other wild foods to rural households living in extreme poverty in the eastern Democratic Republic of Congo. *Biological Conservation*, **118**: 573-581.

Donald, P.F., 2004. Biodiversity impacts of some agricultural commodity production systems. *Conservation Biology*, **18 (1)**: 17-37.

Fa, J.E. and Purvis, A., 1997. Body size, diet and population density in afrotropical forest mammals: A comparison with neotropical species. *Journal of Animal Ecology*, **66 (1)**: 98-112.

Fa, J. E. and Yuste, J.*E.G.*, 2001. Commercial bushmeat hunting in the Monte Mitra forests, Equatorial Guinea: extent and impact. *Animal Biodiversity and Conservation*, **24 (1)**: 31-52.

Fa, J.E, Juste, J., del Val J.P. and Castroviejo, J., 1995. Impact of market hunting on mammal species in Equatorial Guinea. *Conservation Biology*, **9 (5)**: 1107-1115.

Fa, J.E., Peres C.A. and Meeuwig, J., 2002. Bushmeat exploitation in tropical forests: an intercontinental comparison. *Conservation Biology*, **16 (1)**: 232-237.

Feer, F., 1993. The potential for sustainable hunting and rearing of game in tropical forest. In: *Tropical Forests, People and Food: Biocultural Interactions and Applications to Development*, (Eds. C. M. Hladik, A. Hladik, O. F. Linares, H. Pagezy, A. Semple and M. Hadley). Man and the Biosphere Series, Parthenon Publishing, Paris, p. 691-708.

Fitzherbert, E., Struebig, M. J., Morel, A., Danielsen, F., Brühl, C. A., Donald, P. F. and Phalan, B., 2008. How will oil palm expansion affect biodiversity? *Trends in Ecology and Evolution*, **23 (10)**: 538-545.

Francis, I.S., Penford, N., Gartshore, M.E. and Jaramillo, A., 1992. The White-breasted Guineafowl *Agelastes meleagrides* in Taï National Park, Côte d'Ivoire. *Bird Conservation International*, **2**: 25-60.

Hall, J.B. and Swaine, M.D., 1976. Classification and ecology of closed-canopy forest in Ghana. *Journal of Ecology*, **64**: 913-951.

Harding, R.S.O., 1984. Primates of thee Kilimi area, Northwest Sierra Leone. *Folia Primatologica*, **42**: 96-114.

Hawthorne, W.D. and Abu-Juam, M., 1995. *Forest Protection in Ghana*. IUCN/ODA/Ghana Forestry Department, Accra, Ghana. 203 pp.

Hofmann, T., Ellenberg, H. and Roth, H.H., 1999. *Bushmeat: A Natural Resource of the Moist Forest Regions of West Africa. With Particular Consideration of two Duiker Species in Côte d'Ivoire and Ghana*. Tropical Ecology Support Program/GTZ. Eschborn, Germany. Publication number TöB F-V/7e.

Holbech, L.H., 1996. *Faunistic Diversity and Game Production Contra Human Activities in the Ghana High Forest Zone*. Unpublished Ph.D. Thesis, University of Copenhagen, Denmark. 237 pp.

Holbech, L.H., 1998. *Bushmeat survey of the Bia and Ankasa protected areas, Southwest Ghana*. Protected Areas Development Programme – Ghana. Consultancy report. Ghana Wildlife Division/European Union/ULG Consultants Ltd., Takoradi, Ghana. 149 pp.

Holbech, L.H., 2001. *Integrated wildlife and land-use management, off-reserve the Ankasa Resource Reserve, Western Region, Ghana*. Protected Areas Development Programme – Ghana. Consultancy report. Ghana Wildlife Division/European Union/ULG Northumbrian Ltd., Takoradi, Ghana. 110 pp.

Holbech, L.H., 2005. The implications of selective logging and forest fragmentation for the conservation of avian diversity in evergreen forests of Southwest Ghana. *Bird Conservation International*, **15**: 27-52.

Holbech, L.H., 2009. The conservation importance of luxuriant tree plantations for lower storey forest birds in Southwest Ghana. *Bird Conservation International*, **19**: 287-308.

Holbech, L.H., 2013. Differential responses of bats and non-volant small mammals to habitat disturbances in two tropical forest types of southwest Ghana. In: *Animal Diversity, Natural History and Conservation Vol. 2*, (Eds) V.K. Gupta and Anil K. Verma. Daya Publishing House, New Delhi, India, Chapter 15: p. 273-297.

Holbech, L.H., 2014a. Low-extraction selective logging and avifaunal recovery in the Bia moist forest of southwest Ghana: A 1990-1991 census. In: *Animal Diversity, Natural History and Conservation*, Vol. 4.

Holbech, L.H., 2014b. Forest game and habitat disturbances in southwest Ghana: A 1993-1995 survey. Chapter 16, In: *Perspectives in Animal Ecology and Reproduction*, Vol. 10.

Jones, J.P.G., Andriamarovololona, M.M., Hockley, N., Gibbons, J.M. and Milner-Gulland, E.J., 2008. Testing the use of interviews as a tool for monitoring trends in the harvesting of wild species. *Journal of Applied Ecology*, **45**: 1205-1212.

Kingdon, J., 1997. *The Kingdon Field Guide to African Mammals*. Academic Press, London, U.K. 465 pp.

Klaus-Hügi, C., Klaus, G. and Schmid, B., 2000. Movement patterns and home range of the bongo (*Tragelaphus eurycerus*) in the rain forest of the Dzanga National Park, Central African Republic. *African Journal of Ecology*, **38:** 53-61.

Laurance, W.F., Croes, B.M., Guissouegou, N., Buij, R., Dethier, M. and Alonso, A., 2008. Impacts of roads, hunting, and habitat alteration on nocturnal mammals in African rainforests. *Conservation Biology*, **22 (3):** 721-732.

Martin, C., 1991. *The Rainforests of West Africa*. Birkhaüser Verlag, Berlin, Germany. 235 pp.

Mendelson, S., Cowlishaw, G. and Rowcliffe, J.M., 2003. Anatomy of a Bushmeat Commodity Chain in Takoradi, Ghana. *The Journal of Peasant Studies,* **31 (1):** 73-100.

Milner, J.M. and Harris, S., 1999. Activity patterns and feeding behaviour of the tree hyrax, *Dendrohyrax arboreus*, in the Parc National des Volcans, Rwanda. *African Journal of Ecology*, **37**: 267-280.

Nett, D., 1999. Ansätze zur schätzung der siedlungsdichte von Maxwellduckern (Cephalophus maxwelli) in degradierten sekundärwäldern Westafrikas - territoriumgröben, habitatpräferenzen und aktivitätsrhytmen. *Unpublished M.Sc. Thesis*, University of Hamburg.

Newing, H., 2001. Bushmeat hunting and management: implications of duiker ecology and interspecific competition. *Biodiversity and Conservation*, **10 (1):** 99-108.

Newmark, W.D., 2008. Isolation of African protected areas. *Frontiers in Ecology and the Environment*, **6(6):** 321-328.

Noss, A.J., 1998. The impacts of cable snare hunting on wildlife populations in the forests of the Central African Republic. *Conservation Biology*, **12 (2)**: 390-398.

Noss, A.J., 1999. Censusing rainforest game species with communal net hunts. *African Journal of Ecology*, **37**: 1-11.

Ntiamoa-Baidu, Y., 1998. *Wildlife Development Plan Volume 6 – Sustainable Harvesting, Production and Use of Bushmeat*. Ghana Wildlife Department, Accra, Ghana. 78 pp.

Oates, J.F., Whitesides, G.H., Davies, A.G., Waterman, P.G., Green, S.M., Dasilva, G.L. and Mole, S., 1990. Determinants of variation in tropical forest primate biomass: new evidence from West Africa. *Ecology*, **71 (1)**: 328-343.

Oates, J.F., 1999. *Myth and Reality in the Rain Forest: How Conservation Strategies are Failing in West Africa*. University of California Press, Berkeley, U.S. 310 pp.

Oates, J.F., Struhsaker, T.T. and Whitesides, G.H., 1997. Extinction faces Ghana's Red Colobus Monkey and other locally endemic subspecies. *Primate Conservation*, **17:** 138-144.

Oates, J.F., Abedi-Lartey, M., McGraw, W.S., Struhsaker, T.T. and Whitesides, G.H., 2000. Extinction of a West African red colobus monkey. *Conservation Biology*, **14 (5)**: 1526-1532.

Otsyina, R.M. and Asare, E.O., 1991. *Western Region Agroforestry Project*. Agro-tech Consultancy Services. Unpublished report. Government of Ghana/The Commission of European Communities, Accra, Ghana. 126 pp.

Perfecto, I., Rice, R.A., Greenberg, R. and van der Voort, M., 1996. Shade Coffee: A disappearing refuge for biodiversity. *Bioscience,* **46 (8)**: 598-608.

Plumtre, A.J. and Harris, S., 1995. Estimating the biomass of large mammalian herbivores in a tropical montane forest: a method of faecal counting that avoids assuming a 'steady state' system. *Journal of Applied Ecology*, **32**: 111-120.

Redford, K.H., 1992. The empty forest. *Bioscience*, **42**: 412-422.

Robinson, J.G. and Redford, K.H., 1994. Measuring the sustainability of hunting in tropical forests. *Oryx*, **26 (4)**: 249-256.

Schulte-Herbruggen, B., Cowlishaw, G., Homewood, K. and Rowcliffe, M., 2013. The importance of bushmeat in the livelihoods of West African cash-crop farmers living in a faunally-depleted landscape. *PLOS ONE*, **8 (8)**: e72807.

Struhsaker, T.T. and Oates, J.F., 1995. The biodiversity crisis in Southwestern Ghana. *African Primates*, **1 (1)**: 5-6.

Tutin, C.E.G, White, L. J. T. and Mackanga-Missandzou, A., 1997. The use by rain forest mammals of natural forest fragments in an equatorial African savanna. *Conservation Biology*, **11 (5)**: 1190-1203.

Vandermeer, J. and Pefecto, I., 2007. The agricultural matrix and a future paradigm for conservation. *Conservation Biology*, **21 (1)**: 274-277.

van Vliet, N. and Nasi, R., 2008a. Why do models fail to assess properly the sustainability of duiker (*Cephalophus* spp.) hunting in Central Africa? *Oryx*, **42 (3)**: 392-399.

van Vliet, N. and Nasi, R., 2008b. Hunting for livelihood in northeast Gabon: patterns, evolution, and sustainability. *Ecology and Society*, **13 (2)**: Art. 33.

Waltert, M., Seifert, C., Radl, G., and Hoppe-Dominik, B., 2010. Population size and habitat of the White-breasted Guineafowl *Agelastes meleagrides* in the Taï region, Côte d'Ivoire. *Bird Conservation International*, **20**: 74-83.

Weisenseel, K., Chapman, A.C. and Chapman, L.J., 1993. Nocturnal primates of Kibale Forest: effects of selective logging on prosimian densities. *Primates*, **34 (4)**: 445-450.

Wilkie, D.S. and Carpenter, J.F., 1999. Bushmeat hunting in the Congo Basin: an assessment of impacts and options for mitigation. *Biodiversity and Conservation*, **8**: 927-955.

Wilkie, D.S., Sidle, J.G. and Boundzanga, G.C., 1992. Mechanized logging, market hunting, and a bank loan in Congo. *Conservation Biology*, **6 (4)**: 571-580.

Wilkie, D.S., Shaw, E., Rotberg, F., Morelli, G. and Auzel, P., 2000. Roads, development, and conservation in the Congo Basin. *Conservation Biology*, **14 (6)**: 1614-1622.

Appendix: Utilisation of Game Species Based on Interviews with Farmer-Hunters (Columns 1-6) and 'Chopbars' (Columns 7-11) in the Western Region, Ghana. Hunter data: mean annual encounter frequency (MAEF); mean annual yield (MAY) for the whole hunting period of each respondent (mean = c. 15 years) or the past two years; total biomass of yield, total value of yield, and mean kg-price of fresh uncut bushmeat (based on 0-2 years MAY). Chopbar data: percentage of respondents recording a species within the past c. 10 years; mean annual number of traded animals for the past two years; mean number of traded animals for the whole period of operation (c. 10 years on average); total yearly biomass and value of bushmeat traded within past 2 years. Nomenclature follows Kingdon (1997).

Species	Data from Hunters (n = 112)						Data from 'Chopbars' (n = 18)				
	MAEF Present	MAY c. 15 yrs	MAY 0-2 yrs	Biomass (kg)	Value (USD)	Kg-Price (USD)	Recorded Per cent	Traded 0-2 years	Traded c. 10 years	Biomass (kg)	Value (USD)
Loxodonta africana cyclotis	0.4	0.0	0.0	0.0	?	?	–	–	–	–	–
Syncerus caffer nanus	0.1	<0.1	0.0	0.0	?	?	11.1	0.0	<0.1	0.0	0.0
Tragelaphus eurycerus	5.2	0.2	0.1	27.7	6.4	0.23	22.2	0.0	0.1	0.0	0.0
T. scriptus	56.7	1.3	2.0	86.0	83.6	0.97	88.9	5.7	5.9	213.8	238.3
Cephalophus sylvicultor	1.5	0.3	0.1	5.3	6.0	1.15	11.1	0.0	<0.1	0.0	0.0
C. dorsalis	22.9	1.1	2.0	40.8	36.2	0.89	72.2	10.2	10.6	137.7	184.6
C. ogilbyi	11.4	0.9	1.8	35.1	31.1	0.89	33.3	6.3	6.2	75.6	109.0
C. niger	41.9	1.3	2.0	40.0	42.2	1.06	88.9	9.4	10.1	141.0	198.3
C. maxwelli	65.7	5.1	10.1	80.8	109.1	1.35	100.0	43.8	43.2	262.8	473.0
Neotragus pygmaeus	39.9	3.1	4.8	11.0	20.6	1.87	72.2	17.3	17.7	32.4	74.4
Hyemoschus aquaticus	0.3	0.2	0.2	3.0	1.4	0.48	11.1	0.0	0.1	0.0	0.0
Hylochoerus meinertzhageni	1.4	0.1	0.1	17.9	7.4	0.42	5.6	0.0	<0.1	0.0	0.0
Potamochoerus porcus	32.4	0.7	0.8	54.0	41.8	0.77	55.6	1.5	1.8	90.0	78.3
Pan troglodytes verus	2.7	0.1	0.0	0.0	?	?	5.6	0.0	<0.1	0.0	0.0
Piliocolobus b. waldronae	0.1	0.1	<0.01	0.4	0.4	0.88	5.6	0.0	0.1	0.0	0.0
Colobus vellerosus	13.9	1.3	1.4	16.1	10.9	0.68	38.9	0.6	1.1	5.9	4.7

Contd...

Appendix—*Contd...*

Species	Data from Hunters (n = 112)						Data from 'Chopbars' (n = 18)				
	MAEF Present	MAY c. 15 yrs	MAY 0-2 yrs	Biomass (kg)	Value (USD)	Kg-Price (USD)	Recorded Per cent	Traded 0-2 years	Traded c. 10 years	Biomass (kg)	Value (USD)
Procolobus verus	39.1	1.8	3.3	13.2	16.8	1.28	50.0	4.5	4.5	13.5	23.0
Cercocebus atys lunulatus	22.9	1.7	3.2	18.6	26.6	1.43	44.4	4.9	4.8	29.4	40.7
Cercopithecus mona lowei	87.2	3.2	6.8	29.9	31.3	1.05	88.9	10.6	13.1	35.8	48.8
C. petaurista petaurista	73.9	4.0	8.2	27.9	31.2	1.12	38.9	11.5	9.8	25.9	43.7
C. diana roloway	15.4	1.4	1.7	7.1	10.5	1.48	22.2	1.7	2.0	5.1	10.5
Perodicticus potto potto	95.5	2.4	5.6	6.7	6.7	1.00	22.2	12.8	12.6	11.5	15.4
Panthera pardus	1.6	0.0	<0.01	2.4	2.3	0.97	–	–	–	–	–
Felis aurata	<0.1	0.0	0.0	0.0	?	?	–	–	–	–	–
Civettictis civetta	33.1	1.3	2.3	28.5	19.1	0.67	77.8	4.5	4.5	47.3	37.4
Nandinia binotata	67.1	2.5	4.5	13.5	16.2	1.20	77.8	7.3	9.9	14.8	26.3
Genetta tigrina pardina	39.8	1.3	2.5	5.5	9.5	1.73	27.8	2.0	1.9	3.3	7.6
Atilax paludinosus	23.0	0.9	2.8	8.7	7.6	0.87	5.6	0.1	0.1	0.2	0.3
Crossarchus obscurus	129.2	15.1	16.1	20.9	24.2	1.15	16.7	7.7	7.7	7.5	11.6
Lutra maculicollis	4.7	0.3	0.5	2.7	3.3	1.23	–	–	–	–	–
Mellivora capensis	0.7	0.1	0.1	1.2	1.1	0.94	–	–	–	–	–
Dendrohyrax dorsalis	247.3	1.5	3.1	9.3	7.4	0.80	22.2	6.1	5.9	13.7	14.6
Smutsia gigantea	0.1	0.1	0.0	0.0	?	?	5.6	0.0	<0.1	0.0	0.0
P. tricuspis/U. tetradactyla	46.3	3.9	6.6	16.5	17.2	1.04	55.6	6.1	6.1	12.4	15.9
Hystrix cristata	0.1	<0.1	0.0	0.0	?	?	–	–	–	–	–
Atherurus africanus	76.9	6.0	11.6	32.5	53.4	1.64	94.4	30.1	28.5	63.2	138.5

Contd...

Appendix—Contd...

Species	Data from Hunters (n = 112)						Data from 'Chopbars' (n = 18)				
	MAEF Present	MAY c. 15 yrs	MAY 0-2 yrs	Biomass (kg)	Value (USD)	Kg-Price (USD)	Recorded Per cent	Traded 0-2 years	Traded c. 10 years	Biomass (kg)	Value (USD)
Thryonomys swinderianus	110.5	9.7	16.0	107.2	100.8	0.94	100.0	27.4	26.3	133.6	172.6
Cricetomys emini/gambianus	236.2	68.9	63.8	70.2	63.8	0.91	77.8	189.1	210.2	170.2	189.1
Anomalurus peli	49.1	1.8	2.0	3.2	4.4	1.38	11.1	1.8	1.8	2.2	4.0
Protoxerus stangeri	174.0	8.0	13.5	10.8	14.9	1.38	16.7	6.3	6.3	3.8	6.9
Python sebae	23.9	0.5	0.7	14.0	2.9	0.21	5.6	0.1	4.3	1.5	0.4
Crocodilus niloticus	1.9	0.0	0.0	?	?	?	–	–	–	–	–
Osteolaemus tetraspis	17.5	1.1	2.7	27.0	10.5	0.39	16.7	4.5	4.6	33.8	17.6
Varanus niloticus	47.2	2.0	3.2	16.0	9.0	0.56	16.7	4.9	0.1	18.4	13.7
Chelonians (5 spp.)	14.1	1.2	2.2	11.0	5.7	0.52	–	–	–	–	–
Agelastes meleagrides	7.8	0.6	1.0	0.6	1.8	3.00	5.6	0.1	> 0.1	0.0	0.2
Guttera pucherani	21.2	1.6	2.6	2.1	4.9	2.38	5.6	0.1	> 0.1	0.1	0.2
Casqued hornbills (3 spp.)	86.7	2.2	4.6	6.9	5.5	0.80	5.6	0.1	> 0.1	0.1	0.1
Corythaeola cristata	93.8	2.4	4.7	4.7	5.6	1.20	5.6	0.1	> 0.1	0.1	0.1
Stephanoaetus coronatus	11.8	0.4	0.6	1.8	0.6	0.33	–	–	–	–	–
Picathartes gymnocephalus	< 0.1	0.0	0.0	0.0	?	?	–	–	–	–	–
Other birds	–	0.4	0.4	0.2	0.2	1.20	–	–	–	–	–
Total (mean*)	2196.2	164.2	222.4	938.8	912.1	$1.05 \pm 0.53^{*}$	–	439.2	462.4	1,606.6	2,199.8

2015, Perspectives in Animal Ecology and Reproduction, Vol. 10   Pages *301–314*
Editors: V.K. Gupta, Anil K. Verma and G.D. Singh
Published by: DAYA PUBLISHING HOUSE, NEW DELHI

Chapter 18

The Effects of Vitamin E on Genistein-Induced Oxidative Damage on TM3 Leydig Cells

Simge Kara-Ertekin, YaseminTunalý*
and MelikeErkan
Istanbul University, Science Faculty, Department of Biology,
34134, Vezneciler, Istanbul Turkey

ABSTRACT

Soy and soy bean product is used instead of meat products because of low cost price, especially in fast food and other processed foods. In the past few years, phytoestrogens are being important with several epidemiological studies. In presence of endogenous estrogen, phytoestrogens that take as a medicine or diet may cause infertility depend on dosage. Genistein especially bind to estrogen receptor β (ER-β), acting as estrogen agonist or antagonist phytoestrogens which act as an endogenous estrogen are phytochemical. Genistein has also been demonstrated to induce apoptosis in cells as thymocytes, leukemia cells and testis cells, and has a toxic effect in consequence of cause oxidative damage. Antioxidant are spent up enough portion for being obviate oxidative stress which create free radicals and being decrease their effects to minimum. Vitamin E a dietary factor is essential for reproduction in humans and animals. It is an antioxidant in all mammalian cells. Vitamin E is a chain breaking antioxidant that prevents the propagation of free radical reaction and thus protects cells from oxidative damage. For this purpose, TM3 Leydig cells was evaluated after exposure of 10, 50 and 100 µg/ml genistein and in addition to genistein concentrations 50 µM vitamin E at 24 h and 48 h. Total cell number and cell viability was reduced at 24h and 48 h for those exposed to genistein and genistein + vitamin E. Genistein induced apoptotic cells in dose and time dependent manner, while vitamin E exposure in addition to

* *Corresponding author.* E-mail: yastun@istanbul.edu.tr

genistein decreased this detrimental effect. Protein oxidation level increased in genistein exposure. These findings suggest that genistein in higher doses cause apoptosis and protein oxidation. In addition, it was evidenced that vitamin E has a protective effect on toxicity caused by genistein. This study is thought to be a base for molecular studies which are concern with the multifunctional effect of genistein on Leydig cells.

Keywords: *Apoptosis, Cell viability, Genistein, Protein oxidation, Vitamin E.*

Introduction

In recent years, several studies have pointed out the fact that endocrine-disruption chemicals may be responsible for the impairments of the male reproductive system. Some reports suggested that environmental chemicals, introduced and spread by human activities, may affect male fertility (Oliva *et al.*, 2001; Younglai *et al.*, 2005; Xia *et al.*, 2009). Researchers also hypothesized that the male infertility has been increased because of environmental and nutritional estrogens. Soy and soy products have known as a high protein resource and a protector against the diseases; however, soy has phytoestrogens which effects on male genital organ negatively.Phytoestrogens are a diverse group of naturally occurring non-steroidal plant compounds. Because of their structural similarity with estradiol (17β-estradiol), phytoestrogens have the ability to cause estrogenic or/and anti-estrogenic effects.They consist of four main classes: isoflavones, lignans, flavonoids, and coumestants. The best-research group is the isoflavones, which are commonly found in soybean and soy products (Jian, 2009).

Genistein (4',5,7-trihydroxyisoflavone) which have the most effective estrogenic activity, is a natural isoflavone phytoestrogen found in variety of human food(Kumi-Diaka and Butler, 2000). Genistein, after entering the body, may have estrogen like action (Setchell, 2001). It possesses an ability to bind to estrogen receptors (ERs) with higher binding affinity for ER-β than for ER-α (Rimbach *et al.*, 2008). However, the cellular mechanisms of the action of genistein are not completely elucidated (Kumi-Diaka and Butler, 2000). Initially, genistein was considered to have estrogen agonist/antagonist activity. In addition to directly binding to estrogen receptor (ER), genistein may indirectly affect estrogenicity through inhibition of the cytochrome P450, which is responsible for the metabolic degradation of 17β-estradiol(Collins *et al.*, 1997; Bouker and Hilakivi-Clarke, 2000; Okazaki *et al.*, 2002). Genistein can activate a number of estrogen-responsive genes *in vitro*, including pS2 and c-fos(Bouker and Hilakivi-Clarke 2000). Genistein may act also anti-estrogenic, by use of its inhibition of estrogen-metabolizing enzymes. Genistein inhibit the enzyme 17β-hydroxysteroid oxidoreductase type I (HSOR-1) that is necessary for estradiol secretion(Collins *et al.*, 1997; Bouker and Hilakivi-Clarke, 2000; Okazaki *et al.*, 2002). Moreover, genistein have been shown to block the conversion of androgens to estrogens by inhibiting the enzyme aromatase(Collins *et al.*, 1997).

Genistein displace bound estrogen and testosterone from human sex hormone binding globulin (SHBG), and can affect the cellular levels of SHBG. (Dixon, 2004). Genistein may also inhibit enzymes involved in testosterone metabolism, including

17β-hyroxysteroid dehydrogenase and 5α-reductase. Reduced activity of these enzymes could hypothetically lead to lower levels of plasma testosterone and lower levels of DHT in androgen target cells(Allen *et al.*, 2001).

On the other hand, genistein is a specific inhibitor of protein tyrosine kinase (PTK). Genistein induces a reversible G2/M cell cycle arrest, which may be related to ability to inhibit PTK (Bouker and Hilakivi-Clarke, 2000). Furthermore, genistein inhibit DNA topoisomerase II which is involved in the processes of DNA replication, transcription and recombination which makes it one of the most important keys for DNA function as well as cell survival(Nedeljkoviæ *et al.*, 2001, Sarkar and Li, 2003). Moreover, it has been shown that genistein inhibits the activation of NF-κB and Akt signaling pathways, both of which are known to maintain a homeostatic balance between cell survival and apoptosis. Meanwhile, up regulation of Bax and down regulation of Bcl-2 may be one of the molecular mechanisms by which genistein induces apoptosis (Sarkar and Li, 2003). On the other hand, the ability of genistein to protect cells against reactive oxygen species by scavenging free radicals and reducing the expression of stress-response related genes (Sarkar and Li, 2002). The antioxidant effects were predominantly directed against oxidative damage to membrane lipids and lipoprotein particles and also against oxidative DNA damage (Wiseman *et al.*, 2004).

Vitamin E is general terms of used refer to a group of eight chemically different compounds produced by plants known as tocopherols (α, β, γ and δ) and tocotrienols (α, β, γ and δ) (Yu *et al.*, 2008). Vitamin E (VE), a dietary factor is essential for reproduction in humans and animals. It is an antioxidant present in all mammalian cells. *In vivo* and *in vitro* studies confirmed that VE had significant protective effects on oxidant-induced lipid peroxidation in cultured Leydig cells and on the ability of the cells to produce testosterone(Murugesan *et al.*, 2008).

In this study, genistein-induced cytotoxicity and protective effect of vitamin E which is an antioxidant supplementation from this damage on TM3 Leydig cells were purpose to research by determination of total cell number, cell viability, proportion of apoptosis and oxidative damage of protein.

Materials and Methods

Chemicals

Ham's F12 and Dulbecco's modified Eagle's medium (DMEM) (1:1) mixture (DMEM/F12, including 1.2 g/l sodium bicarbonate, 15 mM HEPES, 2.5 mM L-Glutamine, 0.5 mM sodium pyruvate), Horse serum (HS) and fetal bovine serum(FBS) were purchased from Wisent Bioproducts (St-Bruno, Quebec, Canada). Genistein was offered by Mikro-Gen drug co. and stock solution constituted in DMSO as the vehicle. Vitamin E (α-tocopherol) purchased from Sigma-Scientific and prepare in per cent 95 ethanol.

Cells

TM3 Leydig cells were purchased from American type culture collection (ATCC, Rockville, MD, USA). The cells were cultured in DMEM/F12 supplemented with 5 per cent HS, 2.5 per cent FBS and 1 per cent penicillin/streptomycin/amphotericin (PSA) (complete culture medium). The cells were incubated at 37°C in a humidified 5 per cent CO_2 air atmosphere. Cells were grown to confluence and trypsinized with trypsin-EDTA for use in the following experiments, using the same culture medium.

Treatment

Three separate experiments were performed to determine the effects of genistein (G) and vitamin E (VE) on the growth of TM3 cells. In each experiment, 5×10^5 cells/ well in 2.0 mL of culture medium were seeded in six-well plate at 0 h. Equal volumes of predetermined concentrations of genistein (10, 50 and 100, µg/mL) (G10, G50 and G100) and genistein + vitamin E (10, 50 and 100µg/mL+ 50µM VE) (G10+VE, G50+VE and G100+VE) or medium + DMSO and medium+DMSO-ethanol only (G0, control) were added to each well at time 0 h. The final DMSO and DMSO-ethanol concentration were the same in all wells. There were two wells per G-dose and G-dose+VE. The cells were grown in a humidified atmosphere of 5 per cent CO_2, in air at 37°C for 24 or 48h. The medium was changed every 24 h and the test genistein was replenished with each medium change. The viability of each group of cells was determined and quantitated by trypan blue exclusion assay.

Trypan Blue Cell Proliferation Assay

After treatment, wells (three each of G0-G100 and G0-G100+VE) were trypsinized and analyzed. Briefly, 0.4 mL aliquot per well of the trypsinized cells were quantitated, by trypan blue staining, in a Neubauer hemocytometer chamber. Cells were examined and counted in duplicates under light microscope at 200x (Olympus IX71). Percentage cell viability was calculated by the formula:

$$\text{Cell viability (per cent)} \frac{\text{Number of viable cells (unstained cells)}}{\text{Total number of cells (stained and unstained)}} \times 100$$

Thus the percentage of viable and non-viable cells (spontaneous and treatment-induce death) was determined, and the treatment results compared with the non-treatment control.

TUNEL Assay

TM3 cell line was cultured with predetermined concentrations of genistein and genistein+ vitamin E, as described previously. Cells were seeded 5×10^5 cells/ welldensity in 48-well plate. At pre-determined times (24 and 48h) of incubation, the adherent cells were trypsinized and re-suspended of culture medium. The cells seeded on poly L-lysine slide. Apoptotic cells were determined by a terminal deoxynucleotidyltransferase-mediated dUTP nick end-labeling method. The commercially available ApopTag® Peroxidase In Situ Apoptosis Detection Kit (Chemicon, Millipore) was used to detect DNA fragmentation. Briefly, air-dried cells were fixed in 1 per cent paraformaldehyde at room temperature for 10 min and were post-fixed in ethanol:acetic acid at -20°C for 5 min. After washing with PBS cells are holding in 3 per cent H_2O_2 solution, which is prepared with PBS, for blocking endogen peroxides. After drying carefully cells' circle on slides, cell waited on Equilibration

Buffer for 10 minutes in room temperature. After breaking 3'OH ends of DNA with this procedure, enzyme Tdt which is going to connect with broken 3'OH ends and Working Strength TdT enzyme with dUTP mixture has made freshly and in 37°C in humid place this solution were given to the cells for an hour and slides covered with coverslips. After an hour coverslips destroyed and cells were washed with PBS to stop Tdt enzyme activation. After drying cell circle on slides Anti-Digoxigenin Peroxidase were dropped on to the cells in room temperature and humid place for 30 minutes. Cells were washed with PBS, after that DAB (diaminobenzidine) was dropped on the cells for 4-6 minutes to be able to see marked cells on the microscope, after seeing that they were visible on the microscope, reaction has ended with pure water. To see unmarked cells, the opposite color methyl green has been dropped on the cells for 10 minutes. After washing pure water and n-butanol the preparation prepared with glycerin/gelatin solution and closed by coverslips. The preparation has been prepared for control and experimental groups have been examined on Olympus IX71 microscope to see normal and apoptotic cells. Apoptotic index has been established by counting 1000 cells for each preparation.

Determination of SH-Groups

After treatment with Genistein and Genistein+VE groups, cells were counted and collected by centrifugation at 3000 x g for 15 min and re-suspended in 1 ml of 0.5 M phosphate buffer pH 7.8, containing 0.1 per cent SDS. Next, 25 ml Ellman's reagent (5 mM) was added and thiol groups were measured spectrophotometrically (Zavodnik *et al.,* 2004).

Statistical Analysis

Data are presented as the mean ±S.E.M. Cell viability, formed TBARS amount, the amount of SH-groups and the number of apoptotic cells were observed in the test groups versus time and the administered dose. Then, statistical analysis was performed using SPSS 20 (IBM, SPSS software) and analysis of variance was used for comparisons among all groups. Mann Withney U and Wilcoxon test was used to determine significant differences. $P<0.001$, $P<0.01$ and $P<0.05$ significance levels were referenced to interpret the results.

Results

Effect of Genistein and Vitamin E on Cell Growth

In previous studies, showed that genistein induced a dose-dependent dual effect with regard to growth promotion and inhibition in TM3 Leydig cells. The determine effects of genistein and a combination of genistein with vitamin E on the growth and proliferation of TM3 cells. Genistein concentration 10, 50and 100µg/ml alone or in combination with 50µM vitamin E. Figures 18.1 and 18.2 demonstrates the response of TM3 cells to the treatment. According to total cell number, genistein showed a dose dependent inhibition of growth and in TM3 cell compared to the untreated cells, at 24 and 48 h incubation. The viable cells were significantly reduced in G50 and G100 doses. Furthermore treatment of cells with genistein and vitamin E the outcomes are the similar (Figure 18.2).

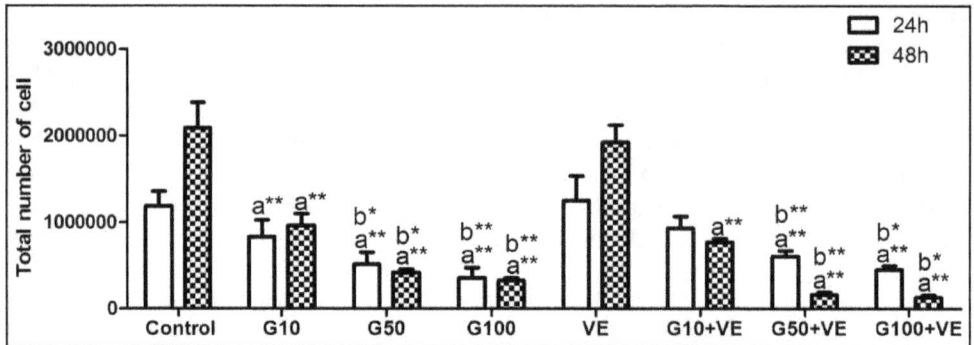

Figure 18.1: Dose- and Time-Dependent Effects of Genistein and Genistein + Vitamin E on Leydig Cell Number *In vitro.* Values are given as means ± SEM. (a) compared with control; (b) compared with G10; *: P<0.05; **: P<0.01.

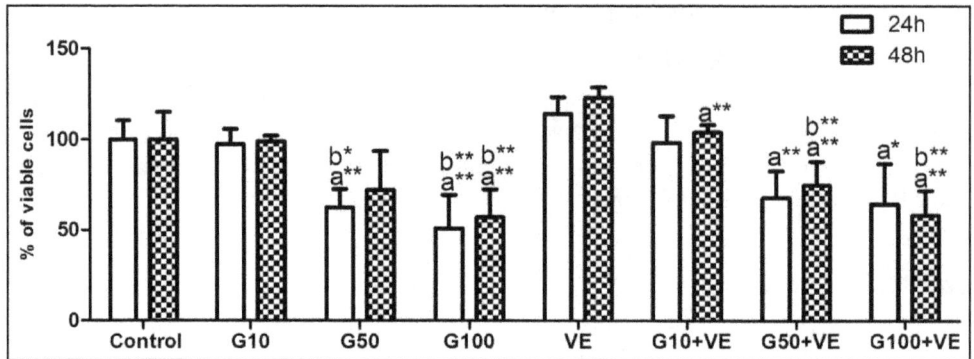

Figure 18.2: Cell Viability after 24 h and 48 h Exposure of TM3 Leydig Cells to Genistein and Genistein + Vitamin E. The data are expressed as a percentage of the control value (value obtained for untreated cells). Values are given as means ± SEM. (a) compared with control; (b) compared with G10; *: P<0.05; **: P<0.01.

Induction of Apoptosis in TM3 Cells

We investigated whether genistein induced apoptosis and a combination of genistein with vitamin E decrease of apoptosis compared treatment of alone genistein on TM3 Leydig cells. As indicated by TUNEL assay, exposure off cells to genistein and genistein+vitamin E resulted in amount of apoptosis at 24 and 48h in TM3 cell line. Treatment of TM3 cells with 10, 50 and 100µg/ml genistein for 24 and 48h led to a dose and time-dependent increase apoptotic cell number. Addition of vitamin E in this concentration, increasing apoptotic cell number significantly decreases at 24 and 48h. On the other hand, a statistically was not significant difference in apoptotic cell number was observed between the control and G10 treated cells (Figure 18.3).

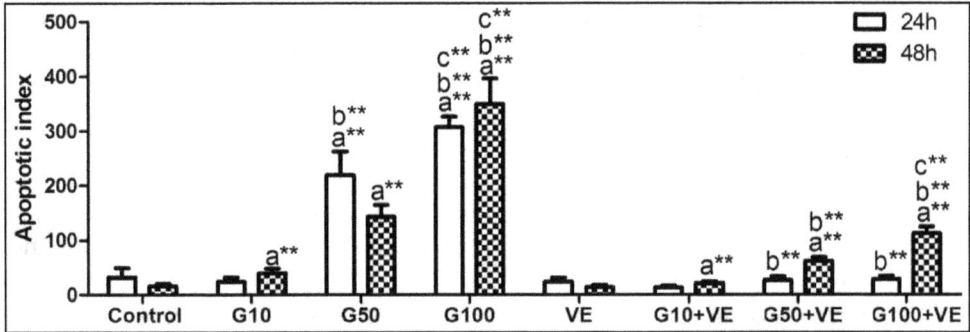

Figure 18.3: Induction of Apoptotic and Necrotic Cell Death by Genistein and Genistein + Vitamin E in TM3 Leydig Cells after 24 h and 48 h incubation. The number of cells in each of three individual experimentswas 1000. Values are given as means ± SEM. (a) compared with control; (b) compared with G10; (c) compared to G50; *: P<0.05; **: P<0.01.

Effects of Genistein on SH-groups Oxidation in TM3 Cells

Figure 18.4 provide data on the level of thiol groups (as a marker of protein oxidation) in genistein and genistein + Vitamin E-exposed Leydig cells. The thiol groups were significantly decreased in a dose-dependent manner when compared with respective control. Comparison of the Vitamin E-exposed groups and the groups treated with only Genistein showed that thiol groupsdecreased significantly in the 48-hour tests for the G100+Vitamin E-exposed group.

Discussion

Soy isoflavones have been a component of the diet of certain populations for centuries(Munro *et al.*, 2003). Genistein is a naturally occurring isoflavone, also known as phytoestrogen that has been identified predominantly in soybean (Sarkar and Li,

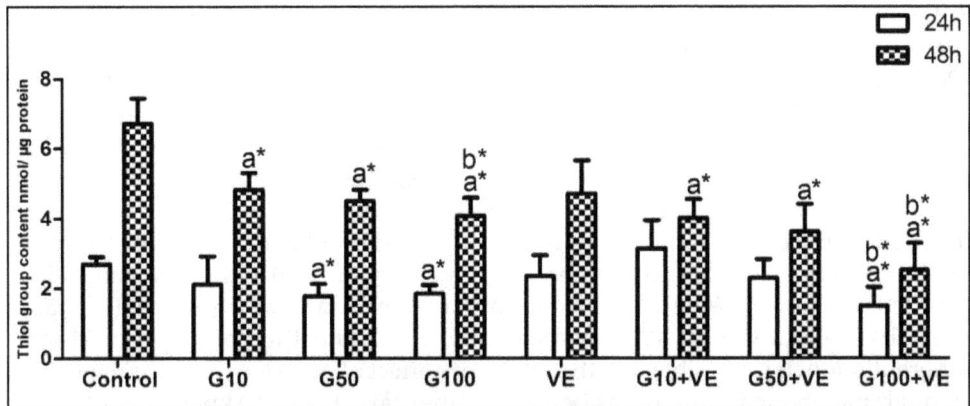

Figure 18.4: Thiol Group Content in TM3 Leydig Cells Exposed to Genistein and Genistein + Vitamin E for 24 h and 48 h. Values are given as means ± SEM. (a) compared with control; (b) compared with G10; *: P<0.05; **: P<0.01.

2002, Rucinska *et al.,* 2008). Genistein binds to estrogen receptors (ER) due to its biphenolic structure and may modulate the actions of endogenous estrogens, acting as ER agonist or antagonists (Rucinska *et al.,* 2008). In recent years, epidemiologic and experimental data suggest that consumption of diets that are rich in isoflavones may decrease cancer risk in the breast, prostate and other tissues (Polkowski and Mazurek, 2000, Faqi *et al.,* 2004). It has been known for many years that estrogen has essential physiological role on male fertility(O'Donnell *et al.,* 2001). Estrogen is synthesized and released in the male reproductive system by Sertoli, Leydig and germ cells. Moreover, estrogen can effect on development and function of these cells. Besides xenoestrogens or phytoestrogens that intake with nutrients, can affect male reproductive system adversely in presence of endogenous estrogens. In addition, consumption of phytoestrogen-rich diets can cause impaired sperm quality over recent decades may be related to increase exposure to environmental endocrine disruptors (Mitchell *et al.,* 2001).

Several studies showed that genistein exposure may be decrease Leydig cell number influencing serum testosterone and other hormone levels (Constantinou *et al.,* 1998; Shao *et al.,* 1998; Polkowski and Mazurek, 2000; Xu and Loo, 2001; Sarkar and Li, 2003; Park and Shin, 2004; Yu *et al.,* 2004; Touny and Banerjee, 2006; Chodon *et al.,* 2007a; Chodon *et al.,* 2007b; Kong *et al.,* 2013; Cui *et al.,* 2014). In the present study we first investigated the effects of genistein and genistein+vitamine E exposure on TM3 Leydig cell line.The trypan blue assays revealed a significant decrease in the total number of cells with and without vitamin E. This growth inhibitory activity may be either arrested cell cycle on G2/M or inhibition apoptosis (Constantinou *et al.,* 1998; Shao *et al.,* 1998; Polkowski and Mazurek, 2000; Xu and Loo, 2001; Sarkar and Li, 2003; Park and Shin, 2004; Yu *et al.,* 2004; Touny and Banerjee, 2006; Chodon *et al.,* 2007a; Chodon *et al.,* 2007b; Liu *et al.,* 2013; Cui *et al.,* 2014). The effects of genistein on different cell lines show inhibition of cell viability (Kumi-Diaka *et al.,* 1998; Polkowski and Mazurek, 2000; Iwase *et al.,* 2006; Lee and Park, 2013). In current study, genistein at 50μg/ml and 100μg/ml decreased TM3 cell viability in a dose-dependent manner. But no influence on cell viability was observed with lower genistein concentration (10μg/ml). Similarly, Kumi-Diaka *et al.* (1998) indicated that at a concentration of 20-100μg/ml, genistein significantly, inhibited the growth and proliferation of testicular cells at 48-72h. This growth inhibitory activity in testicular cells is an accord with the activity of genistein against a variety of hormonal and non-hormonal dependent cancer cell lines (Okura *et al.,* 1988; Naik *et al.,* 1994; Murrill *et al.,* 1996). The mechanisms of action of genistein vary within target cells; because of this multi-function nature of genistein, it is difficult to definitively elucidate its mechanisms of inhibition of cell growth and proliferation (Okura *et al.,* 1988). However, genistein is reported to inhibit tyrosine kinase and topoisomerase I and II activity in cells (Akiyama *et al.,* 1987; Okura *et al.,* 1988; Nynca *et al.,* 2013) and thus blocks growth and differentiation. It is possible that the inhibitory effect of genistein in testicular cells is through metabolic inhibition as occurs in other target cells (Akiyama *et al.,* 1987; Okura *et al.,* 1988; Naik *et al.,* 1994; Murrill *et al.,* 1996). Further research is in progress to define the mechanism of genistein inhibition of cell growth(Choi *et al.,* 2009; Fang *et al.,* 2009).

Apoptosis is the physiologically programmed cell death in which intrinsic mechanisms participate in the death of the cell(Kumi-Diaka *et al.,* 1999; Salti *et al.,* 2000; Michael McClain *et al.,* 2006). Previous studies demonstrated that genistein caused DNA strand breakage in tumor cells (Constantinou *et al.,* 1990; Kiguchi *et al.,* 1990; Salti *et al.,* 2000; Michael McClain *et al.,* 2006), arrested cell cycle progression, and apoptotic cell death in several tumor and non-tumor cells and testicular cells (Bowen *et al.,* 1993; Bowen, 1993; Spinozzi *et al.,* 1994; Kumi-Diaka *et al.,* 1998; Choi and Lee, 2004; Rucinska *et al.,* 2008; Lee and Park, 2013; Cui *et al.,* 2014). The observations in the present study, at high concentration of genistein induce apoptosis whereas lower concentration of genistein is not affected in TM3 cells. Although, addition to vitamin E low concentration of genistein, the observation is not significantly; addition to vitamin E high concentration of genistein, apoptotic cells are definitely decrease. Similarly, Kumi-Diaka *et al.* (1998) demonstrated that at 10μg/ml genistein induce apoptosis in TM3 cells at 24 h of incubation and at 50-100μg/ml genistein induce apoptosis in TM3 cells at 48-72h of incubation. Furthermore, some researchers found that genistein induced apoptosis in higher concentrations (60-120μM) on HT-29 colon cancer cells and NIH 3T3 embryonic fibroblast (Yu *et al.,* 2004; Rucinska *et al.,* 2008).These effects may be attributed, in part, to the tyrosine kinase inhibitory activity of genistein (Akiyama *et al.,* 1987). The alternative explanation suggests that the interaction between genistein and DNA topoisomerase-II may be related to the occurrence of apoptosis(Constantinou *et al.,* 1990; Constantinou and Huberman, 1995; Constantinou *et al.,* 1995; Yoon *et al.,* 2000).

Oxidative stress can damage to various biological molecules, especially DNA, proteins and lipids. This damage may lead to the formation of reactive oxygen species (ROS) within the cell. ROS can cause oxidation in the thiol groups in proteins (Shacter, 2000; Dalle-Donne *et al.,* 2003). Oxidative protein damage is characterized with a decrease of protein-thiol groups (Hu, 1994). Rucinska *et al.* (2008) reported that genistein can cause protein oxidation even lower concentrations (5 μM); however, genistein in higher concentrations (90 μM) increased approximately 50 per cent protein oxidation when compared to control. Our results indicate that genistein increased protein oxidation only in 50 and 100 μg/ml concentrations at 24h; however the level of protein oxidation was significantly higher in 10, 50 and 100 μg/mlconcentrations at 48h, which suggests that genisteineven if lower concentrations have a greater effect over the long-term. In addition, vitamin E can protect the cell from genistein mediated protein oxidation in higher concentrations.

In conclusion, the present study indicates thatgenistein has a toxic effect on TM3Leydig cells, and this toxic effect reduces cell number and cell viability, and vitamin E treatment (in addition to genistein) may reduce this toxic effect. In addition, we found that genistein caused apoptosis and protein oxidation while vitamin E might have a protective role against this damage.

Acknowledgements

This work was supported by Istanbul University Scientific Research Projects (T-2859).

References

Akiyama, T., Ishida, J., Nakagawa, S., Ogawara, H., Watanabe, S., Itoh, N., Shibuya, M., and Fukami, Y. 1987. Genistein, a specific inhibitor of tyrosine-specific protein kinases. *J Biol Chem.*, **262**: 5592-5595.

Allen, S. W., Mueller, L., Williams, S. N., Quattrochi, L. C., and Raucy, J. 2001. The use of a high-volume screening procedure to assess the effects of dietary flavonoids on human cyp1a1 expression. *Drug Metab Dispos.*, **29**: 1074-1079.

Bouker, K. B. and Hilakivi-Clarke, L. 2000. Genistein: does it prevent or promote breast cancer? *Environmental Health Perspectives,* **108**: 701.

Bowen, D. J., Urban, N., Carrell, D., and Kinne, S. 1993. Comparisons of strategies to prevent breast cancer mortality. *J Soc Issues,* **49**: 35-60.

Bowen, I. D. 1993. Apoptosis or programmed cell death? *Cell Biol Int.,* **17**: 365-380.

Chodon, D., Banu, S. M., Padmavathi, R., and Sakthisekaran, D. 2007a. Inhibition of cell proliferation and induction of apoptosis by genistein in experimental hepatocellular carcinoma. *Mol Cell Biochem.,* **297**: 73-80.

Chodon, D., Ramamurty, N., and Sakthisekaran, D. 2007b. Preliminary studies on induction of apoptosis by genistein on HepG2 cell line. *Toxicol In Vitro,* **21**: 887-891.

Choi, E. J. and Lee, B. H. 2004. Evidence for genistein mediated cytotoxicity and apoptosis in rat brain. *Life Sci.,* **75**: 499-509.

Choi, J. N., Kim, D., Choi, H. K., Yoo, K. M., Kim, J., and Lee, C. H. 2009. 2'-hydroxylation of genistein enhanced antioxidant and antiproliferative activities in mcf-7 human breast cancer cells. *J Microbiol Biotechnol.,* **19**: 1348-1354.

Collins, B. M., McLachlan, J. A., and Arnold, S. F. 1997. The estrogenic and antiestrogenic activities of phytochemicals with the human estrogen receptor expressed in yeast. *Steroids,* **62**: 365-372.

Constantinou, A. and Huberman, E. 1995. Genistein as an inducer of tumor cell differentiation: possible mechanisms of action. *Proc Soc Exp Biol Med.,* **208**: 109-115.

Constantinou, A., Kiguchi, K., and Huberman, E. 1990. Induction of differentiation and DNA strand breakage in human HL-60 and K-562 leukemia cells by genistein. *Cancer Res.,* **50**: 2618-2624.

Constantinou, A., Mehta, R., Runyan, C., Rao, K., Vaughan, A., and Moon, R. 1995. Flavonoids as DNA topoisomerase antagonists and poisons: structure-activity relationships. *J Nat Prod.,* **58**: 217-225.

Constantinou, A. I., Kamath, N., and Murley, J. S. 1998. Genistein inactivates bcl-2, delays the G2/M phase of the cell cycle, and induces apoptosis of human breast adenocarcinoma MCF-7 cells. *Eur J Cancer,* **34**: 1927-1934.

Cui, S., Wienhoefer, N., and Bilitewski, U. 2014. Genistein induces morphology change and G2/M cell cycle arrest by inducing p38 MAPK activation in macrophages. *Int Immunopharmacol.,* **18**: 142-150.

Dalle-Donne, I., Rossi, R., Giustarini, D., Milzani, A., and Colombo, R. 2003. Protein carbonyl groups as biomarkers of oxidative stress. *Clin Chim Acta.,* **329**: 23-38.

Dixon, R. A. 2004. Phytoestrogens. *Annu Rev Plant Biol.,* **55**: 225-261.

Fang, S.-C., Hsu, C.-L., Lin, H.-T., and Yen, G.-C. 2009. Anticancer effects of flavonoid derivatives isolated from Millettia reticulata Benth in SK-Hep-1 human hepatocellular carcinoma cells. *Journal of Agricultural and Food Chemistry,* **58**: 814-820.

Faqi, A. S., Johnson, W. D., Morrissey, R. L., and McCormick, D. L. 2004. Reproductive toxicity assessment of chronic dietary exposure to soy isoflavones in male rats. *Reprod Toxicol.,* **18**: 605-611.

Hu, M. L. 1994. Measurement of protein thiol groups and glutathione in plasma. *Methods Enzymol.,* **233**: 380-385.

Iwase, Y., Fukata, H., and Mori, C. 2006. Estrogenic compounds inhibit gap junctional intercellular communication in mouse Leydig TM3 cells. *Toxicol Appl Pharmacol.,* **212**: 237-246.

Jian, L. 2009. Soy, isoflavones, and prostate cancer. *Mol Nutr Food Res.,* **53**: 217-226.

Kiguchi, K., Constantinou, A. I., and Huberman, E. 1990. Genistein-induced cell differentiation and protein-linked DNA strand breakage in human melanoma cells. *Cancer Commun.,* **2**: 271-277.

Kong, D., Xing, L., Liu, R., Jiang, J., Wang, W., Shang, L., Wei, X., and Hao, W. 2013. Individual and combined developmental toxicity assessment of bisphenol A and genistein using the embryonic stem cell test in vitro. *Food Chem Toxicol.,* **60**: 497-505.

Kumi-Diaka, J. and Butler, A. 2000. Caspase-3 protease activation during the process of genistein-induced apoptosis in TM4 testicular cells. *Biol Cell.,* **92**: 115-124.

Kumi-Diaka, J., Nguyen, V., and Butler, A. 1999. Cytotoxic potential of the phytochemical genistein isoflavone (4', 5', 7-trihydroxyisoflavone) and certain environmental chemical compounds on testicular cells. *Biol Cell.,* **91**: 515-523.

Kumi-Diaka, J., Rodriguez, R., and Goudaze, G. 1998. Influence of genistein (4', 5, 7-trihydroxyisoflavone) on the growth and proliferation of testicular cell lines. *Biol Cell.,* **90**: 349-354.

Lee, Y. K. and Park, O. J. 2013. Soybean isoflavone genistein regulates apoptosis through NF-kappaB dependent and independent pathways. *Exp Toxicol Pathol.,* **65**: 1-6.

Liu, X., Sun, C., Jin, X., Li, P., Ye, F., Zhao, T., Gong, L., and Li, Q. 2013. Genistein Enhances the Radiosensitivity of Breast Cancer Cells via G2/M Cell Cycle Arrest and Apoptosis. *Molecules,* **18**: 13200-13217.

Michael McClain, R., Wolz, E., Davidovich, A., and Bausch, J. 2006. Genetic toxicity studies with genistein. *Food Chem Toxicol.,* **44**: 42-55.

Mitchell, J. H., Cawood, E., Kinniburgh, D., Provan, A., Collins, A. R., and Irvine, D. S. 2001. Effect of a phytoestrogen food supplement on reproductive health in normal males. *Clin Sci (Lond).,* **100**: 613-618.

Munro, I. C., Harwood, M., Hlywka, J. J., Stephen, A. M., Doull, J., Flamm, W. G., and Adlercreutz, H. 2003. Soy isoflavones: a safety review. *Nutr Rev.,* **61**: 1-33.

Murrill, W. B., Brown, N. M., Zhang, J. X., Manzolillo, P. A., Barnes, S., and Lamartiniere, C. A. 1996. Prepubertal genistein exposure suppresses mammary cancer and enhances gland differentiation in rats. *Carcinogenesis,* **17**: 1451-1457.

Murugesan, P., Muthusamy, T., Balasubramanian, K., and Arunakaran, J. 2008. Polychlorinated biphenyl (Aroclor 1254) inhibits testosterone biosynthesis and antioxidant enzymes in cultured rat Leydig cells. *Reprod Toxicol.,* **25**: 447-454.

Naik, H. R., Lehr, J. E., and Pienta, K. J. 1994. An in vitro and i *n vivo* study of antitumor effects of genistein on hormone refractory prostate cancer. *Anticancer Res.,* **14**: 2617-2619.

Nedeljkoviæ, A., Raduloviæ, S., and Bjelogrliæ, S. 2001. Pleiotropic effect of genistein makes it a promising cancer protective compound. *Archive of Oncology,* **9**: 171-174.

Nynca, A., Nynca, J., Wasowska, B., Kolesarova, A., Kolomycka, A., and Ciereszko, R. E. 2013. Effects of the phytoestrogen, genistein, and protein tyrosine kinase inhibitor-dependent mechanisms on steroidogenesis and estrogen receptor expression in porcine granulosa cells of medium follicles. *Domest Anim Endocrinol.,* **44**: 10-18.

O'Donnell, L., Robertson, K. M., Jones, M. E., and Simpson, E. R. 2001. Estrogen and spermatogenesis. *Endocr Rev.,* **22**: 289-318.

Okazaki, K., Okazaki, S., Nakamura, H., Kitamura, Y., Hatayama, K., Wakabayashi, S., Tsuda, T., Katsumata, T., Nishikawa, A., and Hirose, M. 2002. A repeated 28-day oral dose toxicity study of genistein in rats, based on the 'Enhanced OECD Test Guideline 407' for screening endocrine-disrupting chemicals. *Arch Toxicol.,* **76**: 553-559.

Okura, A., Arakawa, H., Oka, H., Yoshinari, T., and Monden, Y. 1988. Effect of genistein on topoisomerase activity and on the growth of [Val 12]Ha-ras-transformed NIH 3T3 cells. *Biochem Biophys Res Commun.,* **157**: 183-189.

Oliva, A., Spira, A., and Multigner, L. 2001. Contribution of environmental factors to the risk of male infertility. *Hum Reprod.,* **16**: 1768-1776.

Park, O. J. and Shin, J. I. 2004. Proapoptotic potentials of genistein under growth stimulation by estrogen. *Ann N Y Acad Sci.,* **1030**: 410-418.

Polkowski, K. and Mazurek, A. P. 2000. Biological properties of genistein. A review of in vitro and in vivo data. *Acta Pol Pharm.,* **57**: 135-155.

Rimbach, G., Boesch-Saadatmandi, C., Frank, J., Fuchs, D., Wenzel, U., Daniel, H., Hall, W. L., and Weinberg, P. D. 2008. Dietary isoflavones in the prevention of cardiovascular disease–a molecular perspective. *Food Chem Toxicol.,* **46**: 1308-1319.

Rucinska, A., Roszczyk, M., and Gabryelak, T. 2008. Cytotoxicity of the isoflavone genistein in NIH 3T3 cells. *Cell Biol Int., 32*: 1019-1023.

Salti, G. I., Grewal, S., Mehta, R. R., Das Gupta, T. K., Boddie, A. W., Jr., and Constantinou, A. I. 2000. Genistein induces apoptosis and topoisomerase II-mediated DNA breakage in colon cancer cells. *Eur J Cancer, 36*: 796-802.

Sarkar, F. H. and Li, Y. 2002. Mechanisms of cancer chemoprevention by soy isoflavone genistein. *Cancer Metastasis Rev., 21*: 265-280.

Sarkar, F. H. and Li, Y. 2003. Soy isoflavones and cancer prevention. *Cancer Invest., 21*: 744-757.

Setchell, K. D. 2001. Soy isoflavones—benefits and risks from nature's selective estrogen receptor modulators (SERMs). *J Am Coll Nutr., 20*: 354S-362S; discussion 381S-383S.

Shacter, E. 2000. Quantification and significance of protein oxidation in biological samples. *Drug Metab Rev., 32*: 307-326.

Shao, Z. M., Wu, J., Shen, Z. Z., and Barsky, S. H. 1998. Genistein exerts multiple suppressive effects on human breast carcinoma cells. *Cancer Res., 58*: 4851-4857.

Spinozzi, F., Pagliacci, M. C., Migliorati, G., Moraca, R., Grignani, F., Riccardi, C., and Nicoletti, I. 1994. The natural tyrosine kinase inhibitor genistein produces cell cycle arrest and apoptosis in Jurkat T-leukemia cells. *Leuk Res., 18*: 431-439.

Touny, L. H. and Banerjee, P. P. 2006. Identification of both Myt-1 and Wee-1 as necessary mediators of the p21-independent inactivation of the cdc-2/cyclin B1 complex and growth inhibition of TRAMP cancer cells by genistein. *Prostate, 66*: 1542-1555.

Wiseman, H., Casey, K., Bowey, E. A., Duffy, R., Davies, M., Rowland, I. R., Lloyd, A. S., Murray, A., Thompson, R., and Clarke, D. B. 2004. Influence of 10 wk of soy consumption on plasma concentrations and excretion of isoflavonoids and on gut microflora metabolism in healthy adults. *Am J Clin Nutr., 80*: 692-699.

Xia, Y., Zhu, P., Han, Y., Lu, C., Wang, S., Gu, A., Fu, G., Zhao, R., Song, L., and Wang, X. 2009. Urinary metabolites of polycyclic aromatic hydrocarbons in relation to idiopathic male infertility. *Hum Reprod., 24*: 1067-1074.

Xu, J. and Loo, G. 2001. Different effects of genistein on molecular markers related to apoptosis in two phenotypically dissimilar breast cancer cell lines. *J Cell Biochem., 82*: 78-88.

Yoon, H. S., Moon, S. C., Kim, N. D., Park, B. S., Jeong, M. H., and Yoo, Y. H. 2000. Genistein induces apoptosis of RPE-J cells by opening mitochondrial PTP. *Biochem Biophys Res Commun., 276*: 151-156.

Younglai, E. V., Holloway, A. C., and Foster, W. G. 2005. Environmental and occupational factors affecting fertility and IVF success. *Hum Reprod Update, 11*: 43-57.

Yu, W., Jia, L., Wang, P., Lawson, K. A., Simmons-Menchaca, M., Park, S. K., Sun, L., Sanders, B. G., and Kline, K. 2008. *In vitro* and *in vivo* evaluation of anticancer actions of natural and synthetic vitamin E forms. *Mol Nutr Food Res.,* **52**: 447-456.

Yu, Z., Li, W., and Liu, F. 2004. Inhibition of proliferation and induction of apoptosis by genistein in colon cancer HT-29 cells. *Cancer Lett.,* **215**: 159-166.

Zavodnik, I. B., Lapshina, E. A., Zavodnik, L. B., Labieniec, M., Bryszewska, M., and Reiter, R. J. 2004. Hypochlorous acid-induced oxidative stress in Chinese hamster B14 cells: viability, DNA and protein damage and the protective action of melatonin. *Mutat Res.,* **559**: 39-48.

2015, Perspectives in Animal Ecology and Reproduction, Vol. 10   *Pages 315–329*
Editors: V.K. Gupta, Anil K. Verma and G.D. Singh
Published by: DAYA PUBLISHING HOUSE, NEW DELHI

Chapter 19

Testicular Development of Male Offsprings Exposed to Acrylamide and Ascorbic Acid During Gestation and Lactation

*Esra Yazicioglu and Melike Erkan**
*Istanbul University, Science Faculty, Department of Biology,
34134, Vezneciler, Istanbul Turkey*

ABSTRACT

Acrylamide, which has widespread use in industry, occurs spontaneously by application high termal process to carbohydrate-containing foods when they are prepared for consumption. In addition to carcinogenic and neurotoxic effects,acrylamide has harmfull effects on the male reproductive system. In this study,we aimed to reveal the effects of acrylamide and ascorbic acid individually and combined on the development of testes in male *Mus musculus Balb/c* mice by using histological, histometricalandbiochemical methods. Dams were daily exposed to 14mg/kg acrylamide (Ac) and/or 1.5 g/kg ascorbic acid (AA) by gavagebetween gestation day 6 (GD 6) and the postnatal day 21 (PND 21). The gonadosomatic index of all acrylamide exposed groups was reduced except for lactation group (L-Ac).There were multinuclear giant cells, degenerative cells, atrophic and also maturation arrested tubules in the acrylamide groups. Leydig, Sertoli and spermatogenic cell numbers reduced in the same groups. While catalase activity decreased in the acrylamide treated groups during gestation and lactation, lipid peroxidation level increased in the same period. Moreover, we did not

---

* *Corresponding author.* E-mail: merkan@istanbul.edu.tr

observed any significant differences between the acrylamide exposed groups and both acrylamide and ascorbic acid exposed groups. These findings suggest that acrylamide has toxic effect on male reproductive system on the other hand ascorbic acid has no protective effect on toxicity caused by acrylamide.

Keywords: *Acrylamide, Ascorbic acid, Catalase, Gestation, Lactation, Lipid peroxidation, Testicular toxicity.*

Introduction

The susceptibility of the male reproductive system to toxic insults leading to reduced fertility and disfunction of male reproductive system has recently become a major concern. Nutritional, socio-economic, lifestyle and environmental factors have been attributed to comprimising male reproductive health. In recent years, several studies have suggested that the trend of decreasing male fertility might be due to exposure to nutritional toxicants.

Ac is a compound resulting from the reaction between carbohydrates and amino acids in nutrients treated with high temperatures (120°C) (Stadler *et al.,* 2002).Ac is commonly used in dye, paper, textiles and the cosmetic industry (Parzefal, 2008). It has been shown that acrylamide may cause infertility due to its toxic effect on the male reproduction system. Acrylamide cause a decrease in sperm count and DNA damage on germ cells in different phases and, induced histopathological lesions, such as formation of multinuclear giant cells. Vacuolation in late spermatids were also observed in seminiferous tubules (Sakamoto and Hashimoto, 1986; Sublet *et al.,* 1989; Sega *et al.,* 1990; Tyl *et al.,* 2000; Ma *et al.,* 2011). In addition, acrylamide adversely affects reproductive success of male rats, by reducing mating-fertilizing ability and the rate of maintenance of pregnancy. It has been determined that acrylamide cause an increase in birth anomalies, a decrease in organ weight, and morphological changes in the offsprings of dams that were administered acrylamide during gestation and lactation as it has low molecular weight and high solubility in water(Dearfielt *et al.,* 1995; Friedman *et al.,* 1999; Exon, 2006). In previous studies about developmental toxicity of acrylamide on nervous and reproductive systems, it has been shown that the toxicity in the testis and epididymis might be due to the oxidative damage(Friedman *et al.,* 1999; Wise *et al.,* 1995; Tyl *et al.,* 2000; Shipp *et al.,* 2006; Takahashi *et al.,* 2008; 2009; 2011).

Ascorbic acid as a strong reductive agent is functional in the biosynthesis of collagen I and collagen IV which are basic collagens of seminiferous tubules basal lamina structure, and also hydroxylation step of steroidogenesis (Luck *et al.,* 1995). It has also an important role in the prevention or reduction of oxidation of biological macromolecules such as DNA, proteins, lipids and carbohydrates by free oxygen radicals (Mukhopadhyay *et al.,* 1994; May, 1999; Young and Woodside, 2001). In the absense or deficiency of ascorbic acid an increase in abnormal sperm count and a decrease in sperm count, sperm motility and agglutination has been observed (Dawson *et al.,* 1990).

Furthermore, ascorbic acid is effective on the reorganization of ovaries, collagen synthesis throughout the follicular-luteal cycle and with all that synthesis of progesterone and oxytocin (Luck *et al.*, 1995). Ascorbic acid is a water-soluble vitamin that can pass to offsprings via placenta and also breast milk (Norkus *et al.*, 1981).

This study was conducted to reveal the individual and combined effects of acrylamide and ascorbic acid on the development of testes in *Mus musculus Balb/c* by using histological, histometrical and biochemical methods.

Materials and Methods

Chemicals

Acrylamide (CAS 79-06-1, >99 per cent purity) and ascorbic acid (CAS 50-81-7) other chemicals were purchased from Sigma Chemical Company (St Louis, Missouri, USA).

Animal Husbandry

Thirty adult female *Mus musculus Balb/c* mice (25-30 g) were obtained from Istanbul University, Institute of Experimental Medicine. Animals were given food and water ad libitum. The animal room was maintained at 20-25°C and 50-60 per cent relative humidity with 12 h light-dark cycles.

Study Design and Treatment

On GD 6, dams were randomly divided into ten groups of three dams each and gavaged with saline, acrylamide (Ac) and/or ascorbic acid (AA). Control group (C) was gavaged with saline (0,9 NaCl) on GD 6 though PND 21. Experiment groups were administered 14 mg/kg/day acrylamide and/or 1,5g/kg/day ascorbic acid. Gestation groups (G-Ac, G-AA, G-Ac+AA) were exposed to acrylamide and/or ascorbic acid from GD 6 to birth. Lactation groups (L-Ac, L-AA, L-Ac+AA) were given acrylamide and/or ascorbic acid from birth to PND 21. Gestation and lactation groups (GL-Ac, GL-AA, GL-Ac+AA) were administered acrylamide and/or ascorbic acid from GD 6 to PND 21. All dams and offspring were euthanized at PND 21. Each male offsprings were selected at PND 21 for further evaluation and the testes of offsprings were removed to use histological and biochemical investigations.

Gonadosomatik Index

Testis and body weights of eight offspring for each group were measured and gonadosomatic index was determined as ratio of the testis weight to the body weight.

Histopathology

Testes were fixed with Bouin's solution for 12 hours at 20°C and embedded in paraffin. Paraffin blocks were sectioned at 5 µm thickness, and then stained with hematoxylin-eosin (Humanson, 1972). All slides were examined by light microscope for the histological features.

Histometric Analysis

Samples of each group were counted using standard point counting method. Spermatogonia, primary and secondary spermatocyte, round spermatid, Sertoli,

Leydig and degenerative cells were counted under 100 xmagnificationfor every individual 30 areas which contain 100 point (Wreford, 1995).

Homogenization of Testis Tissue

Testes were removed and washed with saline solution, then homogenized at 1500 rpm in ice-cold 1M PBS buffer. The homogenate was centrifuged at $10000 \times g$ for 20 min at 4°C, and the supernatant was used to determine malondialdehyde (MDA) level and catalase (CAT) activity.

Determination of Lipid Peroxidation

The level of malondialdehyde (MDA) was determined usingthe method of Devasagayam and Tarachand (1987). The reaction mixture consisted of 1.0 ml of 0.15 M Tris–HCl buffer (pH 7.4), 0.3 ml of 10 mM KH_2PO_4, and 0.2 ml supernatant in a total volume of 2 ml. The mixture was incubated at 37°C for 20 min with constant shaking. The reaction was stopped by the addition of 1 ml 10 per cent trichloroacetic acid. The tubes were shaken well, 1.5 ml thiobarbituric acid (TBA) added to the solution and it was heated in the boiling water bath for 20 min. Finally the solution was centrifuged and the color developed was measured at 532 nm. MDAlevels were expressed as nmol per μg protein. Protein concentration in testes supernatants was assayed by the method of Lowry *et al.* (1951)using bovine serum albumin as a standard.

Determination of Catalase (EC 1.11.1.6, CAT)

Catalase activity was assayed by the method of Sinha (1972). In brief, the assay mixture contained 0.4 ml of 0.2 M$H_2O_2$and 0,5 ml sodium phosphate buffer and the reaction was started by addition of 0.1 ml supernatant. Then, 2 ml dichromate–acetic acid reagent was added to arrest the reaction. Catalase was added to the control tube. The tubes were placed in boiling water bath for 10 min then allowed to cool. The green color that developed was read at 590 nm against blank containing all components except for catalase in spectrophotometer. CAT activity was calculated as mMH_2O_2 consumed/min/μg of tissue protein.

Statistical Analysis

Data are presented as the mean ±S.E.M. Gonadosomatic index, spermatogenic cells count, the number of Leydig and Sertoli cells in the testis, spermatogenic cell ratio, the lipid peroxidation level and catalase enzyme activity were observed in the test groups versus control groups. Then, statistical analysis was performed using SPSS package programme and Kruskal Wallis test was used for comparisons among all groups. Mann-Whithey U test were applied in dual comparisons to calculate statistical analyses. $p < 0.01$ and $p < 0.05$ significance levels were referenced to interpret the results.

Results

Gonadosomatic Index Ratio

Gonadosomatic index of offsprings in experimental groups that treated with Ac and AA during gestation and lactation is presented in Figure 19.1. Gonadosomatic

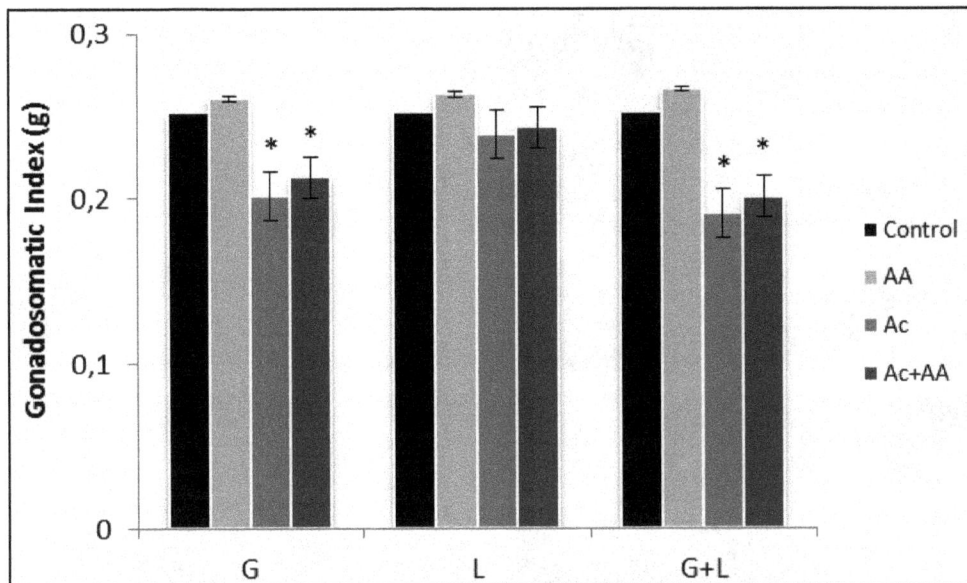

Figure 19.1: Gonadosomatic Index of Male Offspring at PND21 that were Exposed to Acrylamide and Ascorbic Acid Maternally.
Values are given as means ± SEM. *p<0.01.

index of offsprings that treated with acrylamideand acrylamide combined with ascorbic acid groups were lower compared to the control group, except for lactation group (Figure 19.1).

Histopathology

Multinuclear giant cells and degenerated cells in the lumen of seminiferous tubules, atrophic tubules and also structural changes in the connective tisue were observed in all groups that treated with acrylamide. Furthermore spermatogenesis was mostly arrested at the stage of primary spermatocytes in these groups (Figure 19.2).

Histometric Findings

Compared with the control group, histometric analysis showed a decrease in the number of spermatogonia, primary spermatocytes, secondary spermatocytes, early spermatids and Leydig cells while an increase in the number of degenerative cells in all acrylamide treated groups (Figures 19.3A, B, C and Figures 19.4B, C). Compared to the control group, Sertoli cell number was lower in the acrylamide groups except for acylamide treated groups during lactation (Figure 19.4A). Table 19.1 shows the spermatogenic cell rates (per cent) in the groups treated with acrylamide and both acrylamide and ascorbic acid. There was a significant increase in the rate of spermatogonia and primary spermatocytes while there was a reduction in the rate of secondary spermatocytes and early spermatids in all acrylamide treated groups.

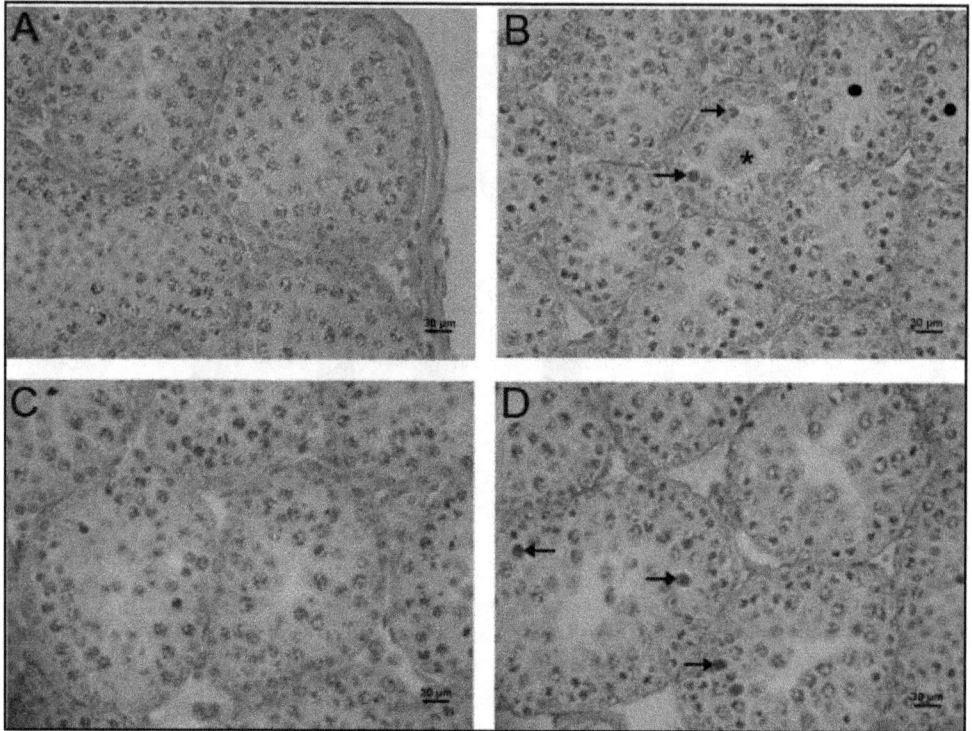

Figure 19.2: Histopathological Changes in Testis of Offsprings at PND21 that Exposed to 14 mg/kg/day Ac and/or1.5 mg/kg/day AA Maternally.
→: Degenerative cell, *: Multinuclear giant cell,
•: Atrofic Tubule. HE stain, Bars=30μm.

Table 19.1: Spermatogenic Cell Rates in Seminiferous Tubules of Male Offspring at PND21 that were Exposed to Acrylamide and Ascorbic Acid Maternally. Values are given as means ± SEM. *p<0.05.

	Per cent Spermatogonium	Per cent Primer Spermatocyte	Per cent Sekonder Spermatocyte	Per cent Round Spermatid
Control	28.02 ± 1.4	55.47 ± 2.6	12.38 ± 1.3	4.11 ± 0.3
G-Ac	30.52 ± 1.3*	61.41 ± 1.5*	5.38 ± 1.3*	2.67 ± 0.7*
L-Ac	30.12 ± 3.1*	61.46 ± 3.2*	5.71 ± 1.2*	2.68 ± 0.4*
GL-Ac	32.98 ± 3.4*	62.43 ± 3.2*	2.56 ± 0.6*	2.02 ± 0.5*
G-AA	27.99 ± 1.6	55.39± 2.1	12.42 ± 1.2	4.18 ± 0.2
L-AA	28.09 ± 1.3	55.14 ± 2.3	12.52 ± 1.2	4.22 ± 0.3
GL-AA	28.11 ± 1.5	54.95 ± 2.1	12.63 ± 0.9	4.29 ± 0.2
G-Ac+AA	31.19 ± 1.3*	60.29 ± 3.2*	5.71 ±1.2*	2.78 ± 0.6*
L-Ac+AA	30.35± 1.4*	61.04 ± 3.1*	5.72 ± 1.2*	2.87 ± 0.5*
GL-Ac+AA	33.95 ± 1.5*	60.92 ± 3.2*	2.91 ± 1.3*	2.20 ± 0.4*

**Figure 19.3: Effect of Maternal Exposure to Acrylamide and
Ascorbic Acid on Spermatogenic Cell Count.
A: The number of spermatogonia; B: The number of primer spermatocyte;
C: The number of secondary spermatocyte; D: The number of round spermatids.
Values are given as means ± SEM. *p<0.01.**

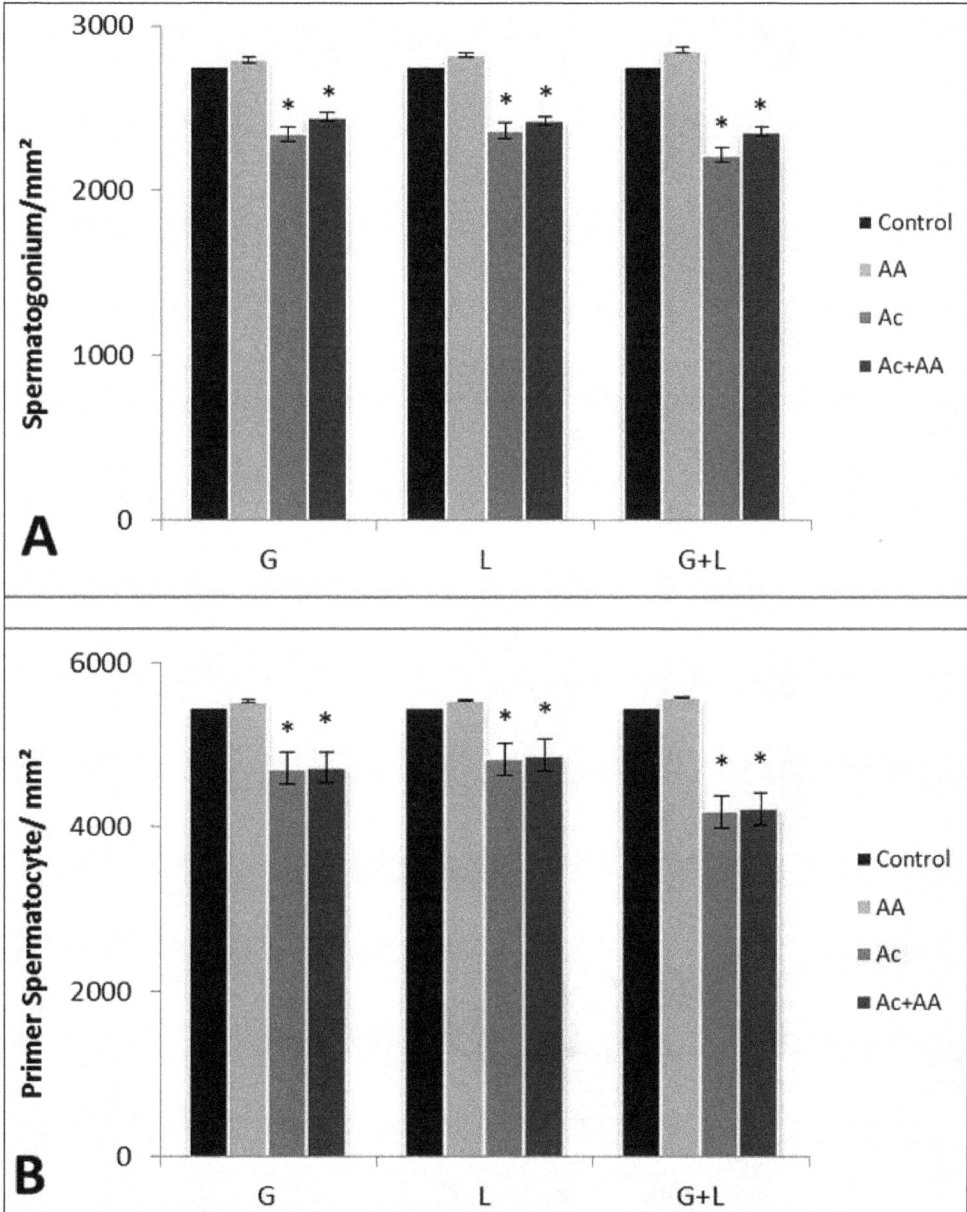

Contd...

Figure 19.3–*Contd...*

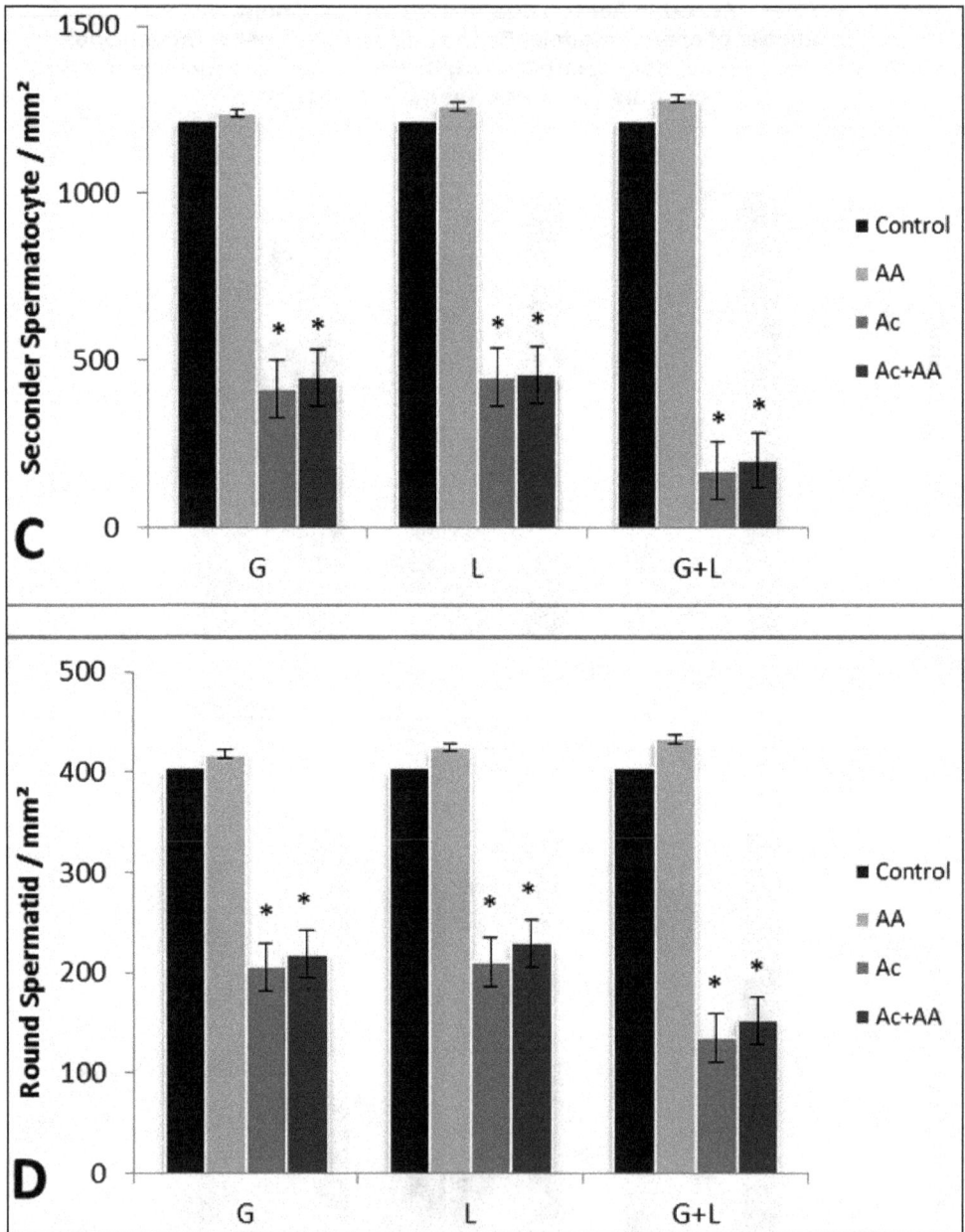

Lipid Peroxidation and Catalase Level

The levels of malondialdehyde (MDA), a lipid peroxidation product were increased in groups GL-Ac and GL-Ac+AA against the control group (Figure 19.5A).

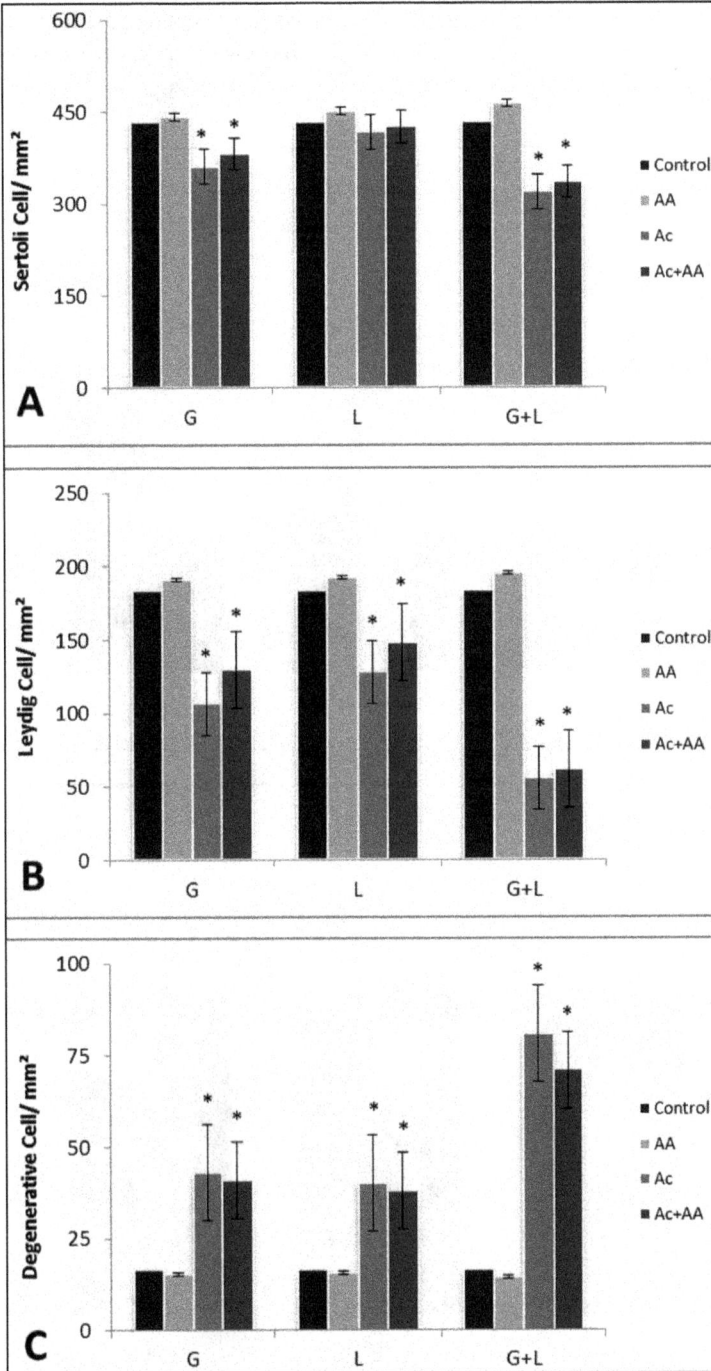

Figure 19.4: Effect of Maternal Exposure to Acrylamide and Ascorbic Acid on Sertoli (A), Leydig (B) and Degenerative (C) cell count. Values are given as means ± SEM. *p<0.01.

Catalase activity was not altered in the experimental groupsduring only gestation and only lactation periods, while the activity decreased in groups GL-Ac and GL-Ac+AA compared to the control (Figure 19.5B).

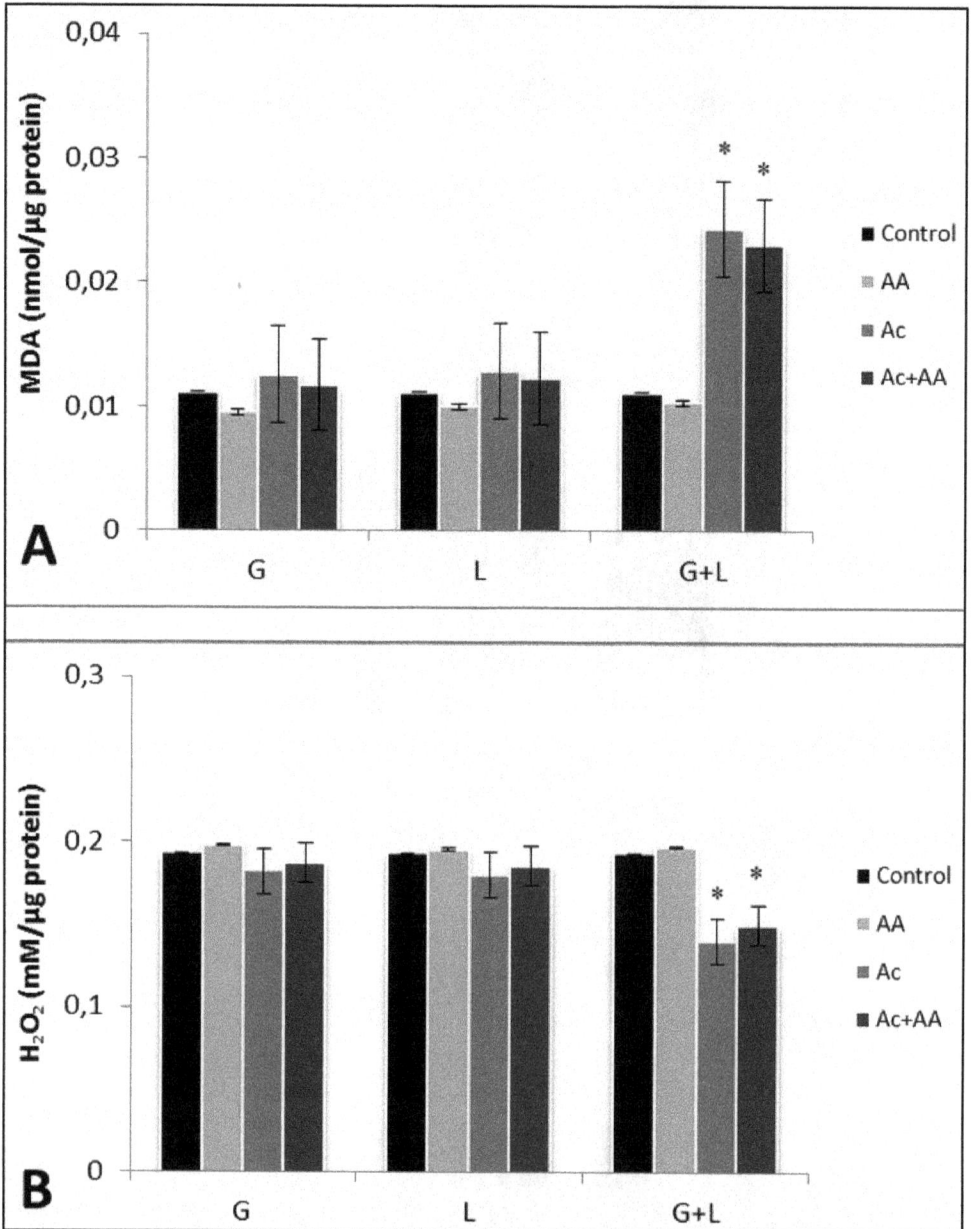

Figure 19.5: Effect of Maternal Exposure to Acrylamide and Ascorbic Acid on Lipid Peroxidation (A), Catalase activity (B) in the Testis tissue. Values are given as means ± SEM. *p<0.01.

Discussion

This study supports earlier studies, which indicate that gonadosomatic index is reduced in acrylamide and acrylamide combined with ascorbic acid treated groups except for lactation groups (Takahashi *et al.*, 2009; Allam *et al.*, 2011; Takahashi *et al.*, 2011). Different from the findings of Ji *et al.* (2012), but similar to the those of Hsu *et al.* (1998) and Sönmez *et al.* (2005), the findings of this study, reveal that ascorbic acid did not alter gonadosomatic index.

Previous reports showed that there were degenerated cells, multinuclear giant cells in the lumen of seminiferous tubules, atrophic tubules, maturation arrested tubules and also connective tisue deformation in acrylamide treatment (Sakamato *et al.*, 1986; Wang *et al.*, 2010; Yang *et al.*, 2005a; 2005b; Takami *et al.*, 2012).Takahashi *et al.* (2008 and 2009) observed that acrylamide treatment caused arrested tubules at the primary spermatocyte stage.Our histopathological findings in acrylamide treated groups were analogous with these studies. On the other hand, according to our findings ascorbic acid has no protective effect on acrylamide testicular toxicity similar with Uzunhisarcikli *et al.* (2007).

We observed a significant increase in the number of spermatogonia, primary spermatocytes, secondary spermatocytes, early spermatids and Leydig cells while a decrease in the number of degenerative cells in all acrylamide treated groups is similar to the study bywith the findings of Zhang *et al.* (2009).Wang *et al.* (2010) suggested an increase in the number of Leydig cell whereasYang *et al.* (2005a) and Camacho *et al.* (2012) reported a decrease in the number of Leydig cell along with our findings. Additionally, decreased Sertoli cell in the acrylamide groups except for acrylamide lactation group is different from the results of Takahashi *et al.* (2008). As shown by Yang *et al.* (2005a) there was a significant increase in the number of degenerative cells. Our results shows that ascorbic acid administration did not change cell numbers of acrylamide treated groups parallel with Ural *et al.* (2006).There was a significant difference in the cell rates of secondary spermatocytes and also early spermatids in acrylamide treated groups compared to control parallel with Gupta *et al.* (2007).These findings, in consistent with Sakamoto and Hashimoto (1986), Yang *et al.* (2005a), Takahashi *et al.* (2008, 2009, 2011) suggest that spermatogenesis was inhibited at the primary spermatocyte stage by acrylamide.

There are various studies investigating the effects of Ac on lipid peroxidation (Yousef *et al.*, 2006; Zhang *et al.*, 2009; Zhang *et al.*, 2010; Hamdy *et al.*, 2012). In recent years, some studies demostrated that adult exposure to acrylamide obviously increase lipid peroxidation in testis (Zhang *et al.*, 2010, Hamdy *et al.*, 2012). According to our study the level of lipid peroxidation increased in acrylamide treated groups during both gestation and lactation periods compared to the control group. However, we observed that ascorbic acid has no protective effect against this increase. These findings are parallel with Chang *et al.* (2007), however different from Gupta *et al.* (2004) and Sönmez *et al.* (2005).

Enzymatic and non-enzymatic antioxidants are the natural defense system against free radical-mediated tissue damage in several organs including testis (Aitken and Roman, 2008). Catalase is one of the enzymes that functions in the removal of hydrogen peroxide and it is generally located at the peroxisomes, and generates H_2O and O_2 by reacting to combining two hydrogen peroxide molecules (Cederbaum *et al.,* 2009). A recent study indicates that adult administration of acrylamide did not change catalase level in the testis (Zhang *et al.,* 2010). However, in the current study we found that catalase activity decreased in acrylamide treated groups during both gestation and lactation periods compared to the control group.

Conclusion

In conclusion, this study reveal that acrylamide not only decreases gonadosomatic index, spermatogenic cell counts and catalase activity but also increases lipid peroxidation level in testis, degenerative cell and multinuclear giant cell counts in seminiferous tubules and other histopathologic findings like vacuolization of the seminiferous epithelium, deformed connective tissue. However, ascorbic acid has not a protective effect against the harmful effects of acrylamide in testicular tissue.

Acknowledgements

This work was supported by Istanbul University Scientific Research Projects (Project Number:7502,3884).

References

Aitken, R.J. and Roman, S.D. 2008. Antioxidant systems and oxidative stress in the testes. *Oxidative Medicine and Cellular Longevity*, **1**: 15-24.

Allam, A., Gharee, A., Hamid, A., Baikry, A. andSabri, M.I. 2011. Prenatal and perinatal acrylamide disrupts the development of cerebellum in rat: Biochemical and morphological studies. *Toxicology and Industrial Health,* **27**: 291-306.

Camacho, L., Latendresse, J.R., Muskhelishvili, L., Patton, R., Bowyer, J.F., Thomas, M. and Doerge, D.R. 2012. Effects of acrylamide exposure on serum hormones, gene expression, cell proliferation, and histopathology in male reproductive tissues of Fischer 344 rats. *Toxicology Letters*, **211**: 135-143.

Cederbaum, A.I., Lu, Y., Wu, D. 2009. Role of oxidative stress in alcohol-induced liver injury. *Archives of Toxicology,* **83(6)**: 519-548.

Chang, S.M., Jin, B., Youn, P., Park, C., Park, J.D. and Ryu, D.Y. 2007. Arsenic-induced toxicity and the protective role of ascorbic acid in mouse testis. *Toxicology and Applied Pharmacology*, **218**: 196-203.

Dawson, E.B., Harris, W.A., and Powel, L.C. 1990. Relationship between ascorbic asid and male fertility, In"Aspects of Some Vitamins, Minerals and Enzymes in Health and Disease", (Ed by GH Bourne), World Review of Nutrition and Dietetic, **62**: 1-26.

Dearfield, K.L., Douglas, G.R., Ehling, U.H., Moore, M.M., Sega, G.A., and Brusick, D.J. 1995. Acrylamide: a review of its genotoxicity and an assessment of heritable genetic risk. *Mutation Research,* **330**: 71-99.

Devasagayam, T.P., and Tarachand, U. 1987. Decreased lipid peroxidation in the rat kidney during gestation. *Biochemical and Biophysical Research Communications,* **198**: 273-80.

Exon, J.H. 2006. A Review of the Toxicology of Acrylamide. *Journal of Toxicology and Environmental Health, Part B*, **9**: 397-412.

Friedman, M.A., Tyl, R.W., Marr, M.C., Myers, C.B., Gerling, F.S., and Ross, W.P. 1999. Effects of lactational administration of acrylamide on rat dams and offsprings. *Reprod. Toxicology*, **13**: 511-520.

Gupta, R.S., Kim, J., Gomes, C., Oh, S., Park, J., Im, W.B., Seong, J.Y., Ahn, R.S., Kwon, H.B., and Soh, J. 2004. Effect of ascorbic acid supplementation on testicular steroidogenesis and germ cell death in cadmium-treated male rats. *Molecular and Cellular Endocrinology*, **221**: 57-66.

Gupta, R.S., Khan, T.I., Agrawal, D., and Kachhawa, J.B.S. 2007. The toxic effects of sodium fluoride on the reproductive system of male rats. *Toxicology and Industrial Health*, **23**: 507-513.

Hamdy, S.M, Bakeer, H.M, Eskander, E.F, and Sayed, O.N, 2012. Effect of acrylamide on some hormones and endocrine tissues in male rats. *Human and Experimental Toxicology*, **31**: 483-491.

Hsu, P.C., Liu, M.Y., Hsu, C.C., Chen, L.Y., and Guo, L.G. 1998. Effects of vitamin E and: or C on reactive oxygen species-related lead toxicity in the rat sperm. *Toxicology*, **128**: 169-179.

Humanson, G.L., 1972. *Animal Tissue Techniques* 3rd Edn. W.H. Freeman and Company, San Francisco, 641.

Ji, Y.L., Wang, Z., Wang, H., Zhang, C., Zhang, H., Zhao, M., Chen, Y.H., Meng, X.H., and Xu, Z. 2012. Ascorbic acid protects against cadmium-induced endoplasmic reticulum stress and germ cell apoptosis in testes. *Reproductive Toxicology*, **34**: 357-363.

Lowry, O.H., Rosebrough, N.J., Farr, A.L., Randall, R.J. 1951. Protein measurement with folin phenol reagent. *Journal of Biological Chemistry*, **193**: 265-275.

Luck, M.R., Jeyaseelan, I., and Scholes, R.A. 1995. Ascorbic acid and fertility. *Biology of Reproduction*, **52**: 262-266.

Ma, Y, Shi, J, Zheng, M., Liu, J, Tian, S., He, Z., Zhang, H., Li, G., and Zhu, J. 2011. Toxicological effects of acrylamide on the reproductive system of weaning male rats. *Toxicology and Industrial Health*, **27**: 617-627.

May, J.M. 1999. Is ascorbic acid an antioxidant for the plasma membrane? *The FASEB Journal*, **13(9)**: 995-1006.

Mukhopadhyay, C.K., and Chatterjee, I.B. 1994. Free Metal Ion-independent Oxidative Damage of Collagen. *The Journal of Biological Chemistry*, **269(48-2)**: 30200-30205.

Norkus, E.P., and Rosso, P. 1981. Effects of maternal intake of ascorbic acid on the postnatal metabolism of this vitamin in the guinea pig. *The Journal of Nutrition*, **111(4)**: 624-630.

Parzefall, W., 2008. Minireview on the toxicity of dietary acrylamide. *Food and Chemical Toxicology*, **46(4)**: 1360-1364.

Sakamoto, J., and Hashimoto, K. 1986. Reproductive toxicity of acrylamide and related compounds in mice effects on fertility and sperm morphology. *Archives of Toxicology*, **59(4)**: 201-205.

Sega, G.A, Generoso, E.E., Brimer, P.A., and Malling, H.V. 1990. Acrylamide Exposure Induces a Delayed Unscheduled DNA Synthesis in Germ Cells of Male Mice That is Correlated with the Temporal Pattern of Adduct Formation in Testis DNA. *Environmental and Molecular Mutagenesis*, **16(3)**: 137-142.

Shipp, A., Lawrence, G., Gentry, R., Mcdonald, T., Bartow, H., Bounds, J., Macdonald, N., Clewell, H., Allen, B., and Landýngham, C.V. 2006. Acrylamide: Review of Toxicity Data and Dose-Response Analyses for Cancer and Noncancer Effects. *Critical Reviews in Toxicology*, **36**: 481-608.

Sinha, A.K. 1972. Colorimetric assay of catalase. *Analytical Biochemistry*, **47**: 389-394.

Sönmez, M, Türk, G., and Yüce, A.2005. The effect of ascorbic acid supplementation on sperm quality, lipid peroxidation and testosterone levels of male Wistar rats. *Theriogenology*, **63**: 2063-2072.

Stadler, RH, Blank I, Varga N, Robert F, Hau J, Guy PA, Robert MC, Riediker S. 2002. Acrylamide from Maillard Reaction Products. *Nature*, **419**: 449-450.

Sublet, V.H., Zenick, H., and Smith, M.K. 1989. Factors associated with reduced fertility and implantation rates in females mated to acrylamide-treated rats. *Toxicology*, **55(1-2)**: 53-67.

Takahashi, M., Shibutani, M., Nakahigashi, J., Sakaguchi, N., Inoue, K., Morikawa, T., Yoshida, M., and Nishikawa, A. 2009. Limited lactational transfer of acrylamide to rat offspring on maternal oral administration during the gestation and lactation periods. *Arch. Toxicol.*, **83**: 785-793.

Takahashi, M., Inoue, K., Koyama, N., Yoshida, M., Irine, K., Morikawa, A., Shibutani, M., Honma, M., and Nishikawa, A. 2011. Life stage-related differences in susceptibility to acrylamide-induced neural and testicular toxicity. *Arch Toxicol*, **85**: 1109-1120.

Takahashi, M., Shýbutaný, M., Inoue, K., Fujimoto, H., Hirose, M., and Nishikawa, A. 2008. Pathological assessment of the nervous and male reproductive systems of rat offspring exposed maternally to acrylamide during the gestation and lactation periods a preliminary. *The Journal of Toxicological Sciences*, **33(1)**: 11-24.

Takami, S., Imai, T., Cho, Y.M., Ogawa, K., Hirose, M., and Nishikawa, A. 2012. Juvenile rats do not exhibit elevated sensitivity to acrylamide toxicity after oral administration for 12 weeks. *J. Appl. Toxicol.*, **32**: 959-967.

Tyl, R.W., Friedman, M.A., Losco, P.E., Fisher, L.C., Johnson, K.A., Strotherd, D.E., and Wolf, C.E. 2000. Rat two-generation reproduction and dominant lethal study of acrylamide in drinking water. *Reproductive Toxicology* **14**: 385-401.

Ural, M., Özgüner, M., Büyükvanlý, B., Kuplay, H., and Köylü, H. 2006. Akut diazinon toksisitesinin testis dokusunda oluþturduðu histolojik deðiþiklikler ve bu deðiþikliklere C vitamini ve E vitaminin etkisi, S.D.Ü.. *Týp Fak. Derg.,* **13(2)**: 22-25.

Uzunhisarcikli, M., Kalender, Y., Dirican, K., Kalender, S., Ogutcu, A., and Buyukkomurcu, F. 2007. Acute, subacute and subchronic administration of methyl parathion-induced testicular damage in male rats and protective role of vitamins C and E. *Pesticide Biochemistry and Physiology*, **87**: 115-122.

Wang, H., Huanga, P., Lýe, T., Li, J., Hutzc, R.J., Li, K., and Shi, F. 2010. Reproductive toxicity of acrylamide-treated male rats. *Reproductive Toxicology*, **29**: 225-230.

Wise, L.D., Gordon, L.R., Soper, K.A., Duchai, D.M., and Morrissey, R.E. 1995. Developmental Neurotoxicty Evaluation of Acrylamide in Sprague-Dawley Rats. *Neurotoxicology and Teratology,* **17(2)**: 189-198.

Wreford, N.G. 1995. Theory and Practice of Stereological Techniques Applied tothe Estimation of Cell Number and Nuclear Volume inthe Testis. *Microscopy Research and Technique*, **32**: 423-436.

Yang, H.J., Lee, S.H., Jin, Y., Choi, J.H., Han, C.H., and Lee, M.H. 2005a. Genotoxicity and toxicological effects of acrylamide on reproductive systemin male rats. *Vet. Sci.*, **6(2)**: 103-109.

Yang, J.D., Lee, S.H., Jin, Y., Choi, J.H., Han, D.U., Chae, C., Lee, M.H., and Han, C.H. 2005b. Toxicological effects of acrylamide on rat testicular gene expression profile. *Reproductive Toxicology*, **19**: 527-534.

Young, I.S., and Woodside, V.J. 2001. Antioxidants in health and disease. *J. Clin. Pathol.*, **54**: 176-186.

Yousef, M.I., and Demerdash, F.M. 2006. Acrylamide-induced oxidative stress and biochemical perturbations in rats. *Toxicology*, **219**: 133-141.

Zhang, J. X., Yue, W.B., Ren, Y.S., and Zhang, C.X. 2010. Enhanced fat consumption potentiates acrylamideinduced oxidative stress in epididymis and epididymal sperm and effect spermatogenesis in mice. *Toxicology Mechanisms and Methods*, **20(2)**: 75-81.

Zhang, X., Chen, F., and Huang, Z. 2009. Apoptosis induced by acrylamide is suppressed in a 21.5 per cent fat diet through caspase-3-independent pathway in mice testis. *Toxicology Mechanisms and Methods*, **19(3)**: 219-224.

2015, Perspectives in Animal Ecology and Reproduction, Vol. 10   *Pages 331–335*
Editors: V.K. Gupta, Anil K. Verma and G.D. Singh
Published by: DAYA PUBLISHING HOUSE, NEW DELHI

Chapter 20

Mating Behaviour of Striped Hyena (*Hyaena hyaena*) in Captivity

Rajesh Kumar Mohapatra*,
Sarat Kumar Sahu and Sudarsan Panda
Nandankanan Zoological Park, Bhubaneswar, Odisha, India

ABSTRACT

Not much information is available on reproductive biology of striped hyena (*Hyaena hyaena*). The present study reports the observation on mating behaviour of striped hyena at Nandankanan Zoological Park, India. Observed mating behaviour continued for two consecutive days in stripped hyena. The mating sessions last for 32.5±12 minutes with an interval of 29.1±11.5 minutes between successive mating sessions. No pelvic thrust was observed.

Keywords: *Copulation, Hyaena hyaena; Mating behaviour, Mounting, Nandankanan Zoological Park, Striped hyena.*

Introduction

The striped hyenas (*Hyaena hyaena*) are carnivorous mammals belonging to the family Hyaenidae. They have a very wide range of distribution, extending from east and northeast Africa, through the Middle East, Caucasus region, Central Asia, and into the Indian subcontinent (Mills and Hofer, 1998; Prater, 2005). Not much information is available on the biology of striped hyena In wild, they are largely scavengers, but also feeds on insects, invertebrates, small vertebrates and actively hunt small mammals and ground nesting and ground-feeding birds (Gajera *et al.*, 2009). They live solitarily on in a polyandrous group *i.e.*, with multiple adult males

---

* *Corresponding author.* E-mail: rajesh.wildlife@gmail.com

sharing a territory with a single reproductively mature female (Wagner, 2006; Gajera *et al.,* 2009). They become sexually mature in 2-3 years (Rieger, 1979). Mating time is said to be in the cold weather for free living striped hyenas and the young are born during the hot season (Prater, 2005). But Reiger (1981) reported that female striped hyena can breed throughout the year in captivity. Little is known about the mating behaviour of striped hyenas mainly because mating is seldom observed in the wild. The present study reports the observation on mating behaviour of striped hyena at Nandankanan Zoological Park, Odisha, India (NKZP).

Material and Methods

A female striped hyena of an estimated age of one year and an adult male striped hyena were received at NKZP on 14.09.2001 and 25.04.2009, respectively. They were housed solitarily in dry moated, open air enclosures with an exhibit area of 50 sq.m. attached with retreating chamber of four sq.m. and a back kraal of 20 sq.m. dimensions. Each hyena was fed with 2kg of buffalo meat with bone, everyday except a weekly fasting day (Monday) and intermittent vitamin A and D supplements. The daily husbandry routine consisted of enclosure cleaning, water replacement and health monitoring. On 24th and 25th January, 2013 the male and female housed in the adjacent enclosure found interacting with each other, *i.e.,* standing straight on wire mesh, looking and smelling the other individual in the next enclosure and its scent marked spots. Assuming this as a sign of courtship and compatibility, the pair was housed together in the back kraal from 26th January, 2013 only during the day time. Focal behavioural observation (Altmann, 1974) was carried out to record their behaviour using data sheet and stop watch.

Results and Discussion

Observed mating behaviour in striped hyena last for two consecutive days, which generally involve a predictable sequence of events (Table 20.1: Figure 20.1). A total of 7 'complete' copulation (4 copulations in 1st day and 3 in the 2nd day), which include intromission by male hyena and an additional 18 pre-mounting events without intromission (13 pre-mounting events in the 1st day and 5 in the 2nd day) was observed. Temporal pattern of mating behaviour was given in Figure 20.2. The selected mating pair was not so comfortable initially, when they were housed together. But after an hour interval male started following the female and sniffing its genitalia. The male made a series of pre-mounts and repeatedly attempted to achieve intromission. But these pre-mount events only last for 1.7±0.4 minutes (N=18, range=1-3 minutes) and terminate before intromission was achieved. After several pre-mounts, the female become receptive and allow the male to mount her for longer duration. The act of mounting include initial bipedal stand by male hyena on the female placing its fore limb on the back of the female laterally and hind limb on ground. Thereafter the male move behind the female clasped her with its fore paws and arches his back bringing his genitalia in contact with that of female hyena. In response to the above act by the male hyena, the female hyena depressed her hind quarters by spreading its hind limbs apart and bent her tail to one side to facilitate intromission. 'Complete' mating sessions last for 32.5±12 minutes (N=7, range=18-66 minutes). The interval between mating is 29.1±11.5 minutes (N=5, range=14-47 minutes). The pelvic thrust with to

Figure 20.1: Mating Behaviour of Striped Hyena *(Hyaena hyaena).*
A: Male hyena sniffing female's genitalia; B: Male licking female's pelvic region;
C: Supportive bipedal stand; D: Intromission; E: Mild aggression on termination of
mating; F: Allo-genital licking by male hyena; G: Auto-genital licking male hyena;
H: Auto-genital licking female hyena.

Table 20.1: Ethogram of Mating Behaviour of Striped Hyena (*Hyaena hyaena*) in Captivity.

Behaviour	Description
Approaching	A hyena comes to vicinity of other and may followed by repetitive spotting and lowering of head
Following	A hyena walking closely behind the other
Sniffing	A hyena takes its nose down and inhale the scent of the other individual or a scent marked spot
Bipedal stand	Standing with fore limb on the back of other individual laterally and hind limb on ground
Pre-mount	The male attempts to mount the female but terminates midway without intermission
Intromission	Insertion of male genitalia in to the female's genitalia
Ejaculation	Sperm transfer in to female reproductive tract behaviourally assumed by mild aggressive display on termination of mating, followed by allo- and auto-genital licking
Aggression	Threat display or warning motions directed towards other hyena
Allo-genital licking	A hyena making tongue contact with genitalia of other hyena
Auto-genital licking	A hyena making tongue contact with genitalia of its own

and fro motion of male genitalia was not observed. But male hyena was found pressing its genitalia against that of female hyena intermittently duration the course of copulation. Ejaculation by male hyena was behaviourally assumed by mild aggressive display on termination of mating, followed by allo-genital licking by male hyena. Female hyena responded to the above act by raising its tail. Thereafter both male and female was found to exhibit auto-genital licking behaviour. The whole process of mating is repeated during each mating session with apparently similar sequence of events. On 2nd day afternoon and on 3rd day the male hyena was found avoiding the

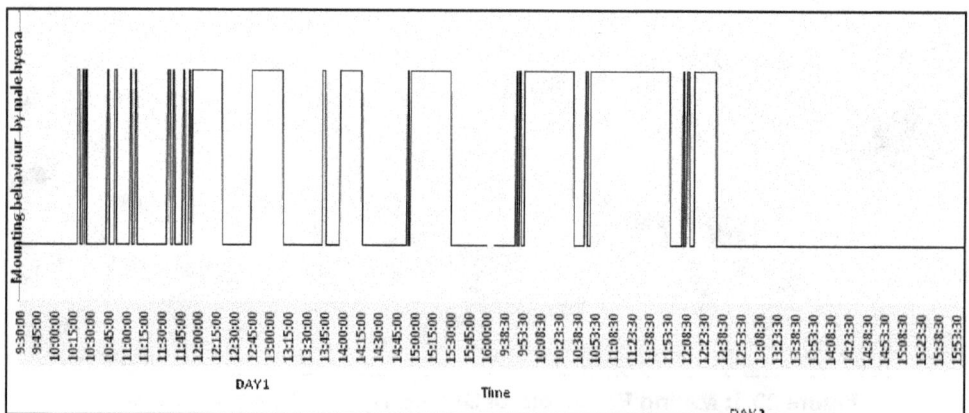

Figure 20.2: Temporal Pattern of Mating Behaviour including Successful Mating (Observed as long durations) and Pre-mounting Events (Observed as shorter duration) of striped hyena (*Hyaena hyaena*).

female followed by intermittent aggressive display. Therefore the pair was separated and housed solitarily on 29th January, 2013.

Reiger (1979) reported that copulation of striped hyenas has never been observed for longer than a single day at Zurich zoo suggestive of their brief oestrous period. The animals generally mate four or five times in the course of the day, each copulation lasting between 15-45 minutes. No pelvic thrusting has been noted. The interval between mating is about 20 minutes (Rieger, 1979).

References

Altmann, J. 1974. Observational study of behavior: sampling methods. *Behaviour*, **49**: 227-266.

Gajera, N., Dave, S.M., and Nishith, D. 2009. Feeding pattern and den ecology of striped hyena (*Hyaena hyaena*) in North Gujarat, India. *Tigerpaper,* **XXXVI(1)**: 13-17.

Mills, M.G.L., and Hofer, H. (1998): Hyaenas: Status Survey and Action Plan, IUCN/ SSC Hyaena Specialist Group. IUCN, Gland, Switzerland and Cambridge, UK. pp. 21-26.

Prater, S.H., 2005. *The Book of Indian Animals*, 3rd Edn. Bombay Natural History Society: Mumbai. pp. 103-110.

Rieger, I. 1979. Breeding the striped hyaena, *Hyaena hyaena* in captivity. *International Zoo Year Book,* **19**: 193-198.

Rieger, I. 1981. *Hyaena hyaena. Mammalian species,* **150**: 1-5.

Wagner, A.P. 2006. Behavioral ecology of the striped hyena (*Hyaena hyaena*). Ph.D. dissertation, Montana State University, Bozeman, MT. Available online: http:// etd.lib.montana.edu/etd/2006/wagner/WagnerA0506.pdf assessed on 09.02.2013.

2015, **Perspectives in Animal Ecology and Reproduction, Vol. 10**   *Pages 337–347*
Editors: **V.K. Gupta, Anil K. Verma and G.D. Singh**
Published by: **DAYA PUBLISHING HOUSE, NEW DELHI**

Chapter 21

Mobile Telephony, an Invisible Environmental Pollution and its Role in Rising Male Infertility: A Review

Mukesh Kumar[1]* and M.S. Srinivas[2]

[1]Reproductive Biotechnology Laboratory,
Department of Zoology, M.S.J. (Govt. P.G.) College,
Bharatpur – 328001, Rajasthan, India
[2]Caree Fertility Test Tube Centre, 324, RD Complex, 1st Cross,
8th Main, Basaveshwara Nagar, III- State,
Bangalore – 560 079, Karnataka, India

ABSTRACT

Mobile telephony is a recent phenomenon. It has revolutionized the tele-communication technology. Mobile phone has become an integral part of human life. Approximately 2 billion mobile users all over the world and their number is increasing rapidly and significantly everyday. Recently, serious concerns have been expressed about the safety of mobile telephony by some section of scientific community. As mobile phone uses radio frequency and emits electromagnetic radiation. It is attributed that the use of mobile phone may be contributory cause of rising cases of cancer, tumour (brain), heart diseases (strokes also), hypertension, asthma, sleeplessness, anxiety etc. Various research studies have also reported that the prolong use of mobile phone also causes the male infertility. Contrary to this, some research groups have claimed that mobile telephony is completely safe. Therefore; big confusion is prevailing regarding the safe use of mobile phone.

* *Corresponding author.* E-mail: prof.mukeshkumar@rediffmail.com

This article reviews the important researches, whether use mobile phone affects male fertility or not. If it affects then, what is the mechanism of action?

Keywords: *Mobile telephony, Electromagnetic radiation, Male fertility.*

Introduction

Electromagnetic fields are sourced from distribution and use of electricity (low frequency), mobile towers, mobile phones and wireless technology (high frequency, radio frequency phones or microwave). Serious concern have been expressed by scientific community world over regarding their possible untoward effect of Electromagnetic Radiation. Nowadays, major source of EMR are mobile towers and mobile hand-sets. Some researchers have claimed that mobile towers and mobile hand-sets do not affect the general and reproductive health. However, this is also a fact that there are rising percentage of cancer (various types), tumour, sleep disturbances, depression, blood pressure (hypertension) heart diseases (strokes also) anxiety, asthma, allergies, memory loss, infertility (male as well female) in recent years (Kunjaliwar and Behari, 1993, Frey, 1993; Sanford *et al.,* 1994; Hermann and Hossmann, 1997; Braune *et al.,* 1998; Michael *et al.,* 2000; Paulraj and Behari, 2002; Henderson, 2004; Aitken *et al.,* 2005; Behari and Kesari, 2006; Desai *et al.,* 2009; Agrawal *et al.,* 2009; Kesari *et al.,* 2010). However, we can not say that electromagnetic radiation is only the cause of this rising trend of above cited diseases due to inadequate research in this field. Electromagnetic Radiation (EMR) also affects the male fertility and various reports are coming up pertaining to the rising male infertility from all over world alongwith decline in general sperm count and motility (Fejes *et al.,* 2005; Erogul *et al.,* 2006; Wdowiak *et al.,* 2007; Agrawal *et al.,* 2008a; DeIuliis *et al.,* 2009; Falzon *et al.,* 2011). Various other environmental factors are responsible for rising male infertility. EMR may be one of them as million of mobile towers are mushrooming without any control and regulations all over the world in order to enhance the coverage area for mobile connectivity and almost 2.5 billion mobile users are in the world. This number (mobile users) is rising significantly every day. Now, the major issue is to ascertain whether mobile telephony affects (including mobile tower which are also major source of electromagnetic radiation) male fertility or male reproductive health or not. There are some problem in designing of human research experiments for assessing the effects on male fertility of EMR, because mobile hand set placing is not contact, talk-time over mobile phone/day vary person to person even in the same person and finding human volunteers and control group are also very difficult for research. Animal research studies do not provide enough base to develop guidelines for safer mobile telephony. *In vitro* studies reflect the effects of EMR on cellular level and human studies differ from animal experimentations being done in laboratories, where various factors remain constant and measurable. Another difficult aspect regarding the investigating untoward effects of EMR (from mobile towers and mobile telephony) as this is not very old technology (almost two decades old). Therefore, the correlation between EMR and its possible untoward effects seem to be inconclusive and needed extensive research (animal, *in vitro* and human studies as well). If this technology is proved to be harmful, then it is going to be major health hazardous technology as we can not live without this technology. This article reviews various studies based on EMR and its effect on male reproductive health as there is general decline in male fertility is being observed and it is generally attributed to various environmental pollutants (EMR may be one of them). Before reviewing various studies, we would discuss some fundamentals like radiation, its type and functioning of mobile technology.

What is Radiation?

Radiation is a combination of electrical and magnetic energy which travels through space at the speed of light. It is also called as electromagnetic radiation (EMR). Radiation is categorized into following groups.

(i) Ionizing Radiation (IR)

It is capable of causing characteristics changes in atoms or molecules in the body which can result in to tissue damage such as cancer, tumour etc. Example of IR are X-rays and gamma rays.

(ii) Non-Ionizing Radiation (NIR)

This kind of radiation does not cause above mentioned characteristics but it can cause molecule to vibrate. This can result into rise in temperature and other effects. Example of Non-Ionizing Radiation (NIR) are ultra-violet radiation in sun light, visible light, light bulb, infra-red radiation, microwave energy and radio-frequency energy. Radiation is divided in to four sub-section according to frequency range:

☆ Static – 0- Hz.

☆ Extremely Low Frequency (ELF)- O ↔ 300 Hz.

☆ Intermediate Frequency (IF)- 300 Hz ↔ 100 k Hz.

☆ Radio Frequency (RF)- It includes 100 k Hz ↔ 300 k Hz.

Table 21.1: Some Sources of Electromagnetic Fields

Sl.No.	Frequency Range	Frequencies	Some Example of Exposure
1	Static	0- Hz	Video Display Unit (VDU); MRI and other diagnostic scientific instruments, industrial electrolysis, welding devices.
2	ELF	0-300 Hz	Powerlines, domestic distribution lines, domestic electric appliances, electric engines in car, train and tramway.
3	IF	300-100 KHz	VDU anti theft devices in shops, hand free access control system, metal detector.
4.	RF	100 KHz – 300 GHz	Mobile telephony, broadcasting and TV, microwave oven, radar, portable and stationary radios, transistor and receivers, personal mobile, radios.

Functioning of Mobile Telephony

The mobile phones work like a two way radio and includes the individual hand set and the base station. Base station have their antennae mounted high off the ground (on a tower or on roof) to get widest coverage. A mobile phone has a radio-receiver and transmitter. When person makes a call, his phone uses radiofrequency (RF) radiation via its antenna to "talk" to near by station. Once the base station has received users signal, his call is routes through the land-line phone system.

Mobile phone base station emits constant level of RF radiation. The hand-set emits of RF radiation that vary depending on following three things :

(i) How long person uses the phone.

(ii) How close person hold the phone to his body.

(iii) Whereabouts person is in relation to the base station, if the link to the base station is weak, the hand set increases its radiation level to compensate it.

1ˢᵗ Generation Telecommunication Technology

Telecommunication technology has developed phenomenally in recent times. The earliest, fully automatic cellular phone system that were used they were called NORDIC mobile telephone, now classified as first generation cellular phone. They were introduced in the late 1970s and early 1980s. They were based on analogue technology.

2ⁿᵈ Generation Mobile Technology

The second generation cell phone that replaced the older analogue type are based on digital technology. These digital model have increased voice capacity, provided faster data transfer, speed, longer battery life, less power use and better signal quality than the 1ˢᵗ Generation cell phone. The cell phone technology used nowadays are Global System for Mobile communication (GSM) and Code Division Multiple Access System (CDMA). GSM technology uses narrow band Time Multiple Access (TDMA), whereas CDMA incorporates the wider band that allow more users without interference and better security by providing every user with unique code.

3ʳᵈ Generation Mobile Telephony

It is highly advanced technology. It is based on Universal Mobile Telecommunication System (UMTS) with Wide Band Code Division Multiple Access (W-CDMA). It will be a High Speed Down link Packet- Access (HSDPA) phones. It will have improved downlink speed that facilitates higher data transfer speed and capacity.

Frequency Band of Mobile Telephony

Generally two frequency band namely 850 MHz and 1900 MHz are used for mobile phone operation. In USA, frequency band 1900 MHz while 850 MHz is used in rest of the world.

Quad Band Mobile

Latest technological development in mobile offer a quad band feature that means mobile hand set can operate on the four different frequencies that are 850, 900, 1800, 1900 MHz with facility to switch automatically among these four frequencies.

Measurement of Radio-Frequency Electromagnetic Waves on Human Health

Impact of these EMR wave is measured via standard limit called Specific Absorption Rate (SAR) value. The SAR is measure of the rate of Radio Frequency energy absorption in the body and is calculated as watt/kg. Device specific SAR test are conducted with wireless device transmitting at its highest power level in all tested frequency band.

Although SAR is determined at a cell phones at maximum power level. The actual SAR values of operating wireless devices may be less than the reported maximum level. This value depends on multiple factors such as proximity to base station, proximity of the wireless device to the body while in use and mode of usage of device (talk versus standby mode).

Mobile phone have been in extensive use for relative short period and technology has progressively changed from analogue to digital system as mentioned above.

Mobile Phones and base stations emit radio frequency or microwave radiation. Exposure to such radiation could affect health directly.

Experimental research on the effect of electromagnetic radiation is very broad and heterogeneous. These research studies include laboratory animals (*in vivo* and *in vitro*) and people (volunteers) specifically human semen involving *in vitro* studies. These studies involved various health related aspects (like- brain tumour, cancer, cardiovascular diseases, asthma, blood pressure, hypertension, sleepless reproductive health etc.). Here, we will review only important research studies dealt with male reproductive health.

Effect of Electromagnetic Radiation on Male Reproduction: Animal Studies

Mobile Telephony has become an integral part of human life as there are approximately 2.5 billion mobile users all over world and their number is increasing every day but various doubts have been expressed by scientific community regarding safety of this technology and various kind of diseases like brain tumour, cancer, hypertension, sleeplessness, nausea, vomiting, restlessness, infertility (male as well female) increasing significantly (Kunjaliwar and Behari, 1993; Frey, 1993; Sanford *et al.*, 1994; Hermann and Hossmann, 1997; Braune *et al.*, 1998; Michael *et al.*, 2000; Paulraj and Behari, 2002; Henderson, 2004; Aitken *et al.*, 2005; Behari and Kesari, 2006; Desai *et al.*, 2009; Agrawal *et al.*, 2009; Kesari *et al.*, 2010). As mobile technology is not very old, therefore, we can not say that this electromagnetic radiation emitted by the mobile telephony is harmful to the human health specifically reproductive health. But it is very essential to review various studies being done in animal and human in relation to male reproductive health. Specifically analysis of animal studies based EMR of mobile telephony are very important because, animal experimentations are done in very controlled and well planned conditions as compared to human studies. Experiments being done by Yan *et al.* (2007) and Mailankot *et al.* (2009) showed that rats exposed to 6 hrs. of daily to cellular phone emission for 18-weeks resulted into reduction of sperm density, motility alongwith enhanced percentage of abnormal sperm. Research findings of Kesari *et al.* (2010) are also very important, they showed that the exposure of mobile, causes infertility in Wistar rats. To add further, results of Kumar *et al.,* 2011 revealed that exposure of microwave affects the male fertility. These animal studies were well planned and exposure of EMR was measurable and constant also. Therefore, it is very clear from above mentioned animal's studies that EMR causes decline in male fertility. To further invisage the mechanism of EMR induced male fertility, Aitken *et al.* (2011) showed EMR exposure causes the damage to the sperm mitochondrial DNA. Arguments given repeatedly by some section of scientific community that exposure of EMR from mobile phone does not affect the reproductive health. This agreement now does not hold the ground because various experimental studies done in animals reveal that EMR does affect the sperm and spermatogenesis process.

Animals experimentation in Wistar Albino rats that the exposure of mobile phone for 60 min/day for 3-months declined serum circulating level (Meo *et al.,* 2010) as we know that the testosterone is the key male reproductive hormone, which is essential for initiation and maintenance of spermatogenesis process and this hormone is secreted by Leydig cells. It is very interesting to take the notice of findings of Salama *et al.* (2010) where they showed that the EMR exposure greatly affected the ejaculatory frequencies and copulatory behaviour of rabbit used as an animal model in their studies, which gives the bases that copulatory behaviour and ejaculatory process is also controlled by testosterone hormones. Therefore, this can be attributed that the exposure of EMR may hit the functioning of Leydig cells.

Some research workers have also investigated the histological changes in testis of different animals like rats, mice, rabbit following the cell phone emitted EMR radiation. Testicular damage was dependent on duration of mobile phone exposure, SAR and energy of EMW (Sounder and Kowalczuk, 1981; Khillare and Behari, 1998; Dasdag *et al.,* 2003; Forgacs *et al.,* 2005; Ozguner *et al.,* 2005; Ribeiro *et al.,* 2007; Salama *et al.,* 2009).

Possible mechanism of action of EMR can be broadly attributed to continuous local heating due to use of mobile phone and non heating effects. Here, we will discuss first thermal effect, as we know that temperature regulation in testis through surface conduction rather than blood flow. Human testis is placed in scrotum *i.e.,* 2-inch below the human body which lowers at least 2°C temp than general temperature of human body in order to maintain the normal spermatogenesis process. Various research group have reported that the exposure EMR from mobile cell phones have direct effect on seminiferous tubules, which subsequently affect the semen parameters. However, it is worth mentioning that EMW energy used in these studies was much higher and greater than EMR emitted by cell phones (Verma and Trabaulay, 1975; Sounders and Kowalczuk, 1981; Kowalczuk *et al.,* 1983; Gautier *et al.,* 1988; Moon *et al.,* 2006). Another contributing factor which caused the male infertility is induction of oxidative stress in semen following the exposure of EMR (Grundler *et al.,* 1991; Oktem *et al.,* 2005, Ozguner *et al.,* 2006; Balci *et al.,* 2007; Meral *et al.,* 2007; Guney *et al.,* 2007; Kesari *et al.,* 2010). As it is well as established that sperm are susceptible to the damage from oxidative stress due to high content of polysaturated fatty acid in their membrane and it also affects the fertilizing ability of spermatozoa. Oxidative stress also damages many molecules like DNA, enzymes, lipids and many proteins (Jones *et al.,* 1979). Emission of EMR from mobile also affects the antioxidant enzyme activity (Dasdag *et al.,* 1999; Desai, 2009; Kumar *et al.,* 2010; Kumar *et al.,* 2011) which contributes in causing male fertility. From above discussion (based on animals), it is very clear that electromagnetic radiation certainly affects spermatogenesis process which results into the impaired semen parameters. Infertility factors include DNA damage, oxidative stress level, impairment of Leydig cells functioning. Now question arises here that findings of various animals studies can be extrapolated to human or not.

Effects of Electromagnetic Radiation on Male Reproduction: Human Studies

As discussed earlier that animals studies are done in laboratory set up which are having generally ideal, constant and measurable factors like EMR, exposure time, vicinity, access of dark : light duration, water, food alongwith parallel control group. Therefore; findings of animals experimentation more reliable and reproducible. Investigating the effect of EMR emitted by mobile phone in human is very difficult because above mentioned factors (mobile placing, talk time duration, exposure time, exposure vicinity) are not constant. Another important hurdle in human experimentation is finding control group (It is the group for research which include the human male volunteers who are not using mobile phone at all) very difficult. Despite all these difficulties in conducting human male research experimentation, many research group tried to do some studies. Here we are discussing some important research studies.

Research work done by Agrawal *et al.* (2008a) where they included 361 human male volunteers going for infertility evaluation, their semen sample were exposed to various period of mobile phones (2-4 hrs) and it was found that the exposure decreases sperm motility, viability in a duration of exposure dependent manner. Another study done by the same group (Agrawal *et al.,* 2009) showed that exposed semen sample to cellular phone (1- hr) in talk mode causes reduction in sperm motility, vitality and increase in ROS when compared with control group. Research findings of DeIuliis *et al.* (2009) substantiate various animals studies where they showed that mobile phone radiation induces reactive oxygen species (ROS) production and causes DNA damage in human spermatozoa with significant decrease in sperm motility and vitality. Similar effects have also been observed by Erogul *et al.* (2006).

Effects on sperm motility following exposure of EMR have also been observed by Fejes *et al.* (2005) and Wdowiak *et al.* (2007). Both these group included approximately 700- human male volunteers. It is very important to discuss research findings of Falzone *et al.* (2011), his group showed that the pulsed exposure of 900 M Hz of mobile phone caused significant reduction in sperm head area and acrosome percentage of head area. Sperm-zona binding was also decreased significantly following the exposure when compared with control group and this could affect the fertilization potential of spermatozoa. All these studies indicates some correlation between male infertility and mobile telephony. As mentioned earlier conducting well planned human experimentation based on mobile telephony is very difficult, more research are needed to derive conclusion whether mobile telephony is safe or not!

Conclusion

Although various studies (animals and human) suggest the possible correlation between mobile phone use and male infertility (by decreasing sperm count, motility, vitality and increasing sperm abnormalities). But mode of action is still not clear. Electromagnetic radiation (EMR) from mobile phone may affect male reproductive system via EMW specific effect and local heeting.

Well planned human studies are very difficult to conduct. Therefore, further studies with careful design are needed to assess the affect (specifically human) mobile phone on male fertility. Meanwhile, some guidelines/directives should be developed regarding installation of mobile towers and mobile hand sets in order to protect the human population from the possible untoward effect of mobile telephony.

References

Agrawal, A., Deepinder, F., Sharma, R.K., G. and Li J., 2008a. Effect of cell phone usage on semen analysis in men attending infertility clinic: an observational study. *Fertil. Steril.*, **89(1)**: 124-8.

Agrawal, A., Deepinder, F., Sharma, R.K., Ranga, G., and Lid J. 2008. Effect of cell phone usage on semen analysis in men attending infertility Clinic. An observational study. *Fertile. Steril.,* **84**: 124-8.

Agrawal, A., Desai, N.R., Makkar, K., Varghese, A., Mouradi, R. and Sabanegh, E., 2009. Effect of radio frequency electromagnetic wave (RF-EMW) from cellular phone on human ejaculated semen: An *in vitro* pilot study. *Fertile. Steril.*, **92**: 1318-25.

Aitken, R.J., Bennet, L.E., Sawyer, D., Wikkndt, A.M. and King, B.V., 2005. Impact of radio frequency electromagnet radiation of mDNA integrity in the male germ line. *Int. J. Androl.*, **28**: 171-9

Balci, M., Devrim, E. and Durak, I., 2007. Effect of mobile phone on oxidant/antioxidant balance in cornea and lens of rats. *Curr. Eye. Res.,* **32**: 21-5.

Behari, J, and Kesari, K.K. 2006. Effect of micro wave radiations on reproiductive system of male rats. *Embryo Talk 1, Supplement-***1:** 81-85.

Braune, S., Wreckages. C., Raczek, J., Gallus, T. and Lucking, C.H. 1998. Resting blood pressure increase during exposure to radio frequency electromagnetic field. *Lancet,* **351**: 1857-1858.

Dasdag, S., Ketani, M.A., Akdag, Z., Ersuy, A.R. Sari, I. and Demirtas O.C. 1999. Whole-body microwave exposure emitted by cellular phones and testicular function of rats. *Urol. Res.,* **27**: 219-23.

Dasdag, S., Zulkuf Akdag, M. and Aksen, F. 2003. Whole body exposure of rats to microwave emitted from a cell phone does not affect the testis. *Bioelectromagnetics,* **24**: 182-188.

de Seze, R., Fabbro Peray, P. and Miro, L. 1998. GSM radio cellular telephones do not disturb the secretion of antiputory hormones in human. *Bioelectromagnetics,* **19**: 271-8.

Deluliis, G.N., Newey, R.J., King, B.V. and Aitken, R.J. 2009. Mobile phone radiation induced reactive oxygen special production and DNA damage in human spermatozoa *in vitro. PLOS one,* **4**: e6446.

Desai, N.R., Kesari, K.K. and Agrawal, A.2009. Patho physiology of cell phone radiation oxidative stress car cinogenesis with focus on male reproductive system. *Respond. Bio. Endocrinol.,* **7**: 114.

Erogul, O., Oztas, E., Yieldirion, I., Kir, T., Aydur, E. and Komesli, G. 2006. Effect of electromagnetic radiation from a cellular phone on human sperm motility: in vitro study. *Arch. Med. Res.,* **37**: 840-3.

Falzone, N., Huyser, C., Becker, P., Leszczynski, D. and Franken, D.R., 2011. The effect of pulsed 900 MHz GSM mobile phone radiation on the acrosome reaction, head morphometry and Zona binding of human spermatozoa. *Int. J. Androl.,* **34 (1)**: 20-6.

Fejes. I., Zavazcki, Z., Szollosi, J., Kalozar, S., Daru, J. and Kovac, S. L.2005. Is there a relationship between cell phone use and semen quality? *Arch. Androl.,* 385-93.

Forgacs, Z., Kubinyi, G., Sinay, G., Bakos, J., Hudak, A. and Surjan, A. 2005. Effect of 1800 MHz GSM-like exposure on the gonadal function and hematological parameters of male mice. *Magy. Onkol.,* **49**: 149-51.

Frey, A.H. 1993. Electromagnetic field interactions with biological system. *FASEB;* **7**: 272-281.

Gautier, J., Norbury, C., Lohka, M., Nurse, P. and Maller, J. 1988. Purified maturation-promoting factors contain the product of a xenopus homolog of the fission cell cycle control gene, Cdc2y. *Cell.,* **54:** 433-9.

Grundler, W., Kaiser, F., Keilmann, F. and Walleczek, J. 1991. Mechanism of electromagnetic interaction with cellular system. *Naturewission Schaften,* **79**: 551-9.

Guney, M., Ozguner, F., Oral, B., Karahan, N. and Mungan, T., 2007. 900 MHz radio Frequency induced histopathological changes and oxidative stress in rat endometricom: Protection by Vitamin E and C. *Toxicol. Ind. Health,* **2.** 411-20.

Henderson, M. 2004. Mobile may decrease men's fertility. Science Editor times. In *New Orleans.*

Hermann, D.M. and Hossmann, K.A. 1997. Neurological effects of microwave exposure related to mobile communication, *J. Neurol. Sci.,* **152**: 1-14.

Hocking, B. 1998. Preliminary report: Symptoms associated with mobile phone use. *Occup. Med.,* **48**: 357-360.

Keasri, K.K. and Behari, J. 2010. Microwave exposure affecting reproductive system of male rats. *Appl. Biochem. Biotechnol.*, **162**: 416-28.

Kesari, K.K., Kumar, S. and Behari, J.2010: Mobile phone usage and male infertility in Wistar rats. *Ind. J. Exp. Biol.,* **48**: 987-92.

Khillare, B. and Bihari, J. 1998. Effect of amplitude-modulated radiofrequency radiation on reproduction pattern in rats. *Electromagn. Bio. Med.*, **1**: 55.

Kowelczuk, C.I., Sounders, R.D. and Stapleton, H.R. 1983. Sperm count and Sperm abnormality in male mice after exposure to 2.45 G Hz microwave radiation. *Metal Res.,* **122**: 155 -61.

Kumar, S., Kesari, K.K., and Behari, J. 2011. Influence of microwave exposure on fertility of male rats: *Fertil. Steril.,* **95**: 15002.

Kumar, S., Kesari, K.K. and Behari, J. 2010. Evaluation of geotaxis effect in male rats following microwave exposure. *Indian J. Exp. Biol.,* **48**: 586-92.

Kunjaliwar, K.K. and Behari, J. 1993. Effect of amplitude modulated radio frequency radiation on cholinergic system of developing rats. *Brain Research,* **601**: 321-324.

Mailankot, M., Kunnatt, A.P., Jayalekshmi, H., Koduru, B. and Valsalan, R., 2009. Radio-frequency electromagnetic radiation (RF-EMR) from GSM (0.9/1.8 GHz) mobile phones induces oxidative stress and reduces sperm motility in rats. *E.M.R.,* **64(6)**: 561-565.

Melachlan, R.I. and de Kretser, D.M. 2001. Male infertility: the case for continued research. *Med. J. Aust.,* **174**: 116-117.

Meo, S.A., Dress, A. I., Hussain, S.K.M.M. and Imran, M.B., 2010. Effect of mobile phone radiation on serum testosterone in Wistar Albino rats. *Saudi Med. J.,* **30 (8)**: 869-73.

Meral, I. Mert, H., Mert, N., Deger, Y., Yoruk, I. and Yetkin, A., 2007. Effect of 900 MHz electromagnetic field emitted from cellular phone on brain and some vitamin levels of guinea pig. *Brain Res.,* **1169**: 120-4.

Michael, M., Colins, B. and Mika, K. 2000: Electromagnetism and life. *Albany (NY) Suny Press.* 1982.

Moon, K., Shin, H.J., Ahn, H., Kim, J. Shin, S. and Yun, S. 2006. Long-term exposure of rats to 2.45 GHz electromagnetic Field: Effect on Reproductive Function. Magjarevic, R., Nagel, J.H. (Editors): World Congress on Medical Physics and Biomedical Engineering. *IFMBE Proceedings. Springer Berlin Heidelberg,* **14 (4)**: 2767-9.

Oktem, F., Ozguner, F., Mollaogle, H., Koyu, A. and Uz, E., 2005. Oxidative damage in Kidney induced by 900-MHz emitted mobile phone: Protection by melatonin. *Arch. Med. Res.,* **36**: 350-5.

Ozguner, F., Bardak, Y. and Comlekci, S. 2006. Protective effects of melatonin and acid phenethyl ester against retinal oxidative stress in long term use of mobile phone: A comparative study. *Mol. Cell Biochem.,* **282**: 83-8.

Ozguner, M., Koyu, A., Cesur, G., Ural, M., Ozguner, F. and Gokcimen, A. 2005. Biological and Morphological effects on the reproductive organs of rats after exposure to electromagnetic field. *Saudi Med. J.,* **26**: 405-10.

Paulraj. R. and Behari, J. 2002: The effect of low band continuous 2.45 GHz wave on brain enzymes of developing brain. *Electromagnetic Biology and Medicine,* **21**: 231-241.

Ribeiro, E.P., Rhoden, E.L., Horn, M.M., Rhoden, C., Lima, L.P. and Toniola, L. 2005. Effect of subchronic exposure to radio frequency from a conventional cellular telephone on testicular function in adult rats. *J. Urol.,* **177**: 395-9.

Salama, N., Kishimoto, T. and Kanagama, H.O.2010. Effect of exposure to a mobile phone on testicular tendies and structure in adult rabbit. *Int. J. Androl.,* **33**: 88-94.

Salama, N., Kishimoto, T., Kanayama, H.O. and Kagawa, S. 2009. The mobile phone decreases fructose but not citrate in rabbit semen, a longitudinal study. *Stud. Biol. Rerod. Med.,* **55**: 181-7.

Sanford, L.G., Brein, A., Sturesseon, K., Eberhasdt, J.L. and Persson, B.R. 1994. Permeability of blood – brain barrier induced by 915 MHz electromagnetic radiation, continuous wave and modulated at 8, 16, 50 and 200 Hz. *Micro. Sc. Res. Tech.,* **27**: 535-542.

Saunders, R.D. and Kowalczuk, C.I.1981: Effect of 2-45 G Hz microwave radiate heat on mouse spermatogenic epithelium. *J. Int. Radiat. Felut. Study Chen. Med.,* **40**: 623-32.

Verma, M.M. and Traoulay, E.A. Jr.1975. Biological Effect of microwave radiation on testis of Swiss Mice: *Experientia,* **31**: 31-3.

Wdowiak, A., Wdowiak, L. and Wiktor, H. 2007. Evaluation of the effect of using mobile phones on male fertility. *Ann. Agric. Environ. Med.,* **14 (1)**: 169-72.

Yan, J. G, Agresti, M, Bruce, T., Yan, Y.H., Granland, A. and Matlaub, H.S. 2007. Effect cellular phone emission on sperm motility in rats: *Fertil. Steril.,* **88**: 64.

2015, **Perspectives in Animal Ecology and Reproduction, Vol. 10**   *Pages 349–375*
Editors: **V.K. Gupta, Anil K. Verma and G.D. Singh**
Published by: **DAYA PUBLISHING HOUSE, NEW DELHI**

Chapter 22

Development of Female Birth Control Technologies: An Update

Mukesh Kumar

Reproductive Bio-technology Laboratory,
Department of Zoology, M.S.J. Government (P.G.) College,
Bharatpur – 328 001 Rajasthan, India

ABSTRACT

Continuously increasing population of the world (specially of India and China) posing serious threat to all developmental activities. This alarming population growth can be checked by effective implementing the family planning programme. Use of birth control methods by females has certainly contributed in restricting the family size. However, it is still far from satisfactory state. There are various approaches to control the female fertility like- Natural Family Planning Methods, Female Sterilization, Vaginal Contraception, Barrier Methods, Intrauterine Devices (IUD's), Hormonal Methods (pills, implants and injections), Birth Control Vaginal Rings, Post Ovulatory Approaches (RU 486) and Non-steroidal Synthetic Approach Centchroman (popular trade name Saheli). These birth control technologies function at different levels of female reproductive system with associated advantages and some disadvantages. Development and use of these female birth control approaches very significant in regulating their fertility and has contributed greatly in lowering the population growth rate. Researchers are trying to refine the existing birth control technologies and also making serious efforts to develop newer ideal female birth control approaches to widen the choice for users. This article reviews the current status of existing female birth control technology and also the future prospects of newer possible female contraceptives, which are under the process of development.

Keywords: *Population, Female fertility, Birth control approaches.*

---

* *Corresponding author.* E-mail: prof.mukeshkumar@rediffmail.com

Introduction

During the last few decades, the world has witnessed unprecedented population growth and this complex reality closely associated to untoward consequences such as environmental problems, retarded economic development, poverty and poor quality of life, primarily and specially in the developing countries. According to the latest medium-term U.N. projections-based on the optimistic assumption regarding continued use of birth control method by the eligible couples of reproductive potential-the world population will touch 85 billion by the year 2025. This means an average annual increase of 93 million or the growth rate of approximately 1.79/6, and is the result of imbalance in birth rate of 26 against the death rate of 9 per thousand population per year. This serious state is further compounded by the estimate that nearly 90 per cent of this population growth maybe witnessed in the developing world, Africa and South Asia contributing 50 per cent (Sadik, 1994). Not surprisingly therefore, this disturbing population scenario has been the issue of extensive and intensive discussion from political, economic, environmental, societal and development perspectives in a variety of world for at frequent intervals. Many complex issues have been directly associated to the evolution of human population reaching the current explosive state. These includes: (a) low education levels or illiteracy in women, restraining their control over sexual practice and birth control method's choice which is often fueled by culturally determined gender bias in traditional societies; (b) religious restriction pertaining to the control of undesired and unplanned pregnancies; (c) high child death rate with associated uncertainty regarding progeny survival for social security during old age. Additional population reasons, such as, greater proportion of younger generation of reproductive potential (due to earlier cycles of population increase) and declining death rate due to improved medical facilities and medical technology for disease diagnosis and treatment are also relevant in this context.

Despite this rather disappointing demographic state confronting the mankind, lessons learned during the last 25 years concerning population programmes in a wide variety of circumstances prevailing in underdeveloped countries, do instill a sense of guarded optimism. A revolution in reproductive behavior has taken place among adolescent and married women largely due to the success story of family planning ably aided by the development and appropriate use of variety of contraceptive technologies to regulate fertility. This has enabled women not only in controlling unplanned and unwanted pregnancies, but also in spacing them appropriately with consequent health benefit to herself and her progeny, thus having a major impact on the maternal and child health. Experience gained has shown that the Family Planning Programme can be successful in poor communities some times impressively and independent of socio-economic development. The dictum that "development is the best contraceptive" may be true in the long run, but the events of last few decades have clearly shown that lower fertility rate and, hence decreased population growth can precede other aspects of (relatively slower) developmental activities such as improvements in health education, economy and public hygiene. This dramatic change in reproductive behavior, despite being non-uniform globally due to a variety of reasons, has reduced the total fertility rate in developing countries

(Cross *et al.*, 2002). It has been estimated that to achieve 50 per cent reduction in level of fertility it took 58 year in U.S.A. compared to 27 years in Indonesia, 15 years in Columbia, 8 years in Thailand and 7 years in China (UNDP, 1994). This remarkable development in developing countries was propelled significantly by parallel enhanced prevalence of use of birth control methods (defined as the percentage of married couple of reproductive age using any form of contraception) which increased from 9 per cent in 1965-1970 to 51 per cent in 1985-1990; total number of contraceptive users shot up more than 10 fold, from 31 million to 381 million during this period (Fathalla, 1994).

Many factors have contributed to this remarkable achievement in fertility regulation which has given power to women to control their fertility, plan their lives and engage in more productive roles beyond child bearing and rearing. Political commitments of the governments in allocating requisite resources for promoting small family sizes besides providing information and services are a major contribution towards this end. The topmost among these is the technological advancement introducing a variety of effective and convenient birth control methodologies bringing them with more benefits to more people at less cost than any other single technology currently available to the human race (Grant, 1992).

Thus, these major and significant developments in contraceptive use have broadened the rather restricted view of small family norm to curtail population. It now includes the concept of reproductive health, which implies that the women should be able to engage in sex at their choice and be able to regulate their fertility without any side effects including prevention of STD and unwanted pregnancies which often results in unsafe abortion especially in under developed countries. Biologically, women are more vulnerable to reproductive ill-health than men since she carries the major load of child bearing and rearing; consequently the contraceptive use is three times more common among women than men. Reproductive biology of the female is totally different from that of the male, being more vulnerable to interference at multiple points. Therefore, greater research efforts in contraceptive development hitherto have been focused towards regulation of female fertility. It should also be emphasized that the spectacular break through in basic research in reproductive sciences has made tremendous contribution towards these technological innovations throughout these developments. In the discussion, an attempt has been made to give brief accounts of the various birth control technologies presently available for female fertility regulations with emphasis on their advantages and drawbacks. A lots of literature exists on each of these methods from basic discoveries, effectiveness, continuous product development, plus and minus points, long-term effects on reproductive health etc. in the form of symposia proceedings, monographs, conference/summit proceedings, WHO reports etc. Therefore, it is beyond the scope of this article to give detailed accounts of these methodologies except as brief overview of their current status.

United Nation approximate estimate of the percentage use of different methods of fertility regulating practiced world wide (Table 22.1) for the year 1987 have revealed that in general, methods with lowest failure rates predominate in developing countries, whereas in the developed nations, the methods with high failure rates were practiced more commonly (Table 22.2) (UNDIESA, 1991).

Table 22.1: Percentage Use of Contraceptive Method Worldwide

Currently, 390 million couples are using modern contraceptive, about 51 per cent of total

Tubectomy	29
IUDs	20
Oral Pills	14
Condoms	9
Vasectomy	8
Withdrawal Method	8
Rhythm	7
Injectables	2
Other Methods	2
Barrier Methods	1

Table 22.2: Prevalence of different Contraception Methods in Developing and Developed Regions of the World (per cent)

Method	More Developed Regions	Less Developed Regions	World
Tubectomy	10	33	26
Vasectomy	5	12	10
IUDs	8	24	19
Oral Contraception	20	12	15
Condoms	19	6	10
Injectable Method	–	2	1
Vaginal Barrier Approaches	3	1	2
Rhythm	13	5	7
Withdrawal Method	20	3	8
Other Methods	2	3	2

Current Methods

Natural Family Planning Methods (NFP)

According to definition of World Health Organization (WHO), Natural Family Planning methods for preventing pregnancies involve observation of naturally occurring signs and symptoms of fertile and non-fertile phases of the menstrual cycle with avoidance of intercourse during the fertile phase (or use of temporary method like condom during this phase) if pregnancy is to be avoided (Labbok and Queenan, 1989). This rhythm method involving periodic abstinence of intercourse is supported by several national and international interest groups and has the blessing and support of the Roman Catholic church. Its rate of success is dependent on the users ability to exactly assess the narrow window of fertility period during the menstrual cycle by either observing the changes in the viscosity of the cervical mucous or in the signs, symptoms and basal body temperature to predict ovulation. A multicentric and multinational efficacy study of this method carried out by WHO, 1981 did establish that if the rules and protocol are strictly adhered to, pregnancy rate ranged from 0 to 6.8. The largest proportions of unplanned pregnancies occur due to conscious departure from the rules. In terms of unplanned pregnancy rate, this method of intentional avoidance of sex on days believed to be fertile to prevent pregnancy compares favorably (16 per cent) with those of barrier methods like condoms (149/6), diaphragm (179/6) and is much better than spermicide (22 per cent). It is practiced reasonably widely by motivated and trained women of reproductive age in some Countries like U.S.A. (10 per cent), Mauritius and Peru (20 per cent), Bolivia and Kenya (30-50 per cent). However, the complex human behavioural factors greatly reduce in-use efficacy. Lack of education, motivation, inherent irregularity in menstrual cyclicity in individual women, (particularly in young adolescents) compounded by unpredictable nature of the sexual impulses of the partner makes this natural method of fertility regulation undependable as a method of contraception in many of the socio-cultural settings of the developing countries. Similar behavioural limitations leading to unacceptability high user-failure rates limit the practicality of the other natural method pregnancy prevention, *viz.,* coitus interruptus, particularly among young couples.

Female Contraceptives

Yet, another approach to contraception using the physiological method is based on the scientific finding that during breast feeding, women experience temporary infertility due to failure of ovulation. This phenomenon-based method, called lactation amenorrhoea method (LAM) delays the return of fertility due to ovulation suppression and luteal phase deficiency. It was developed after detail demographic, epidemiological and physiological research. Scientific basis of this short term infertility lies in the finding that suckling stimulus alters the periodicity of GnRH release at the hypothalamic level and consequently modifies the pattern of pituitary release of gonadotropins leading to inadequate follicle development and suppression of ovum release. It is now recognized that in traditional societies where mother's breast feed for 4-5years, child spacing of 2-3 years is the norm. Although, breast feeding is not a formal method of family planning, nevertheless LAM has been used as an interim postpartum method of delaying the time (at least six months) when complimentary mode of contraception such as IUD, barrier methods could be introduced to assure high efficacy. Another dimension of this method is its impact as a major child survival intervention since breast feeding has enormous benefits in terms of nutritional, immuno-protective, anti-diarrhoeal and psychological effects on infants.

Although the efficacy of LAM (less than 2 per cent risk of pregnancy) as a natural interim method of fertility intervention by postponing the return of menstrual cyclicity is now reasonably well established by prospective clinical case control intervention study at Santiago, Chile in 1991. The promotion of this approach on a mass scale especially among the uneducated masses of third world could be a monumental task at present especially in view of enormous efforts involved in the provision of information advice services and education, to make an impact. Nevertheless, several international agencies including Family Health International, Population Council and WHO are continuing their research efforts focusing on a scientific and behavioral understanding of the biological mechanisms and behavioural determinants of lactation and infertility increases. Only future will reveal the magnitude of contribution of such methods of fertility regulation in reducing the population problem worldwide. Success depends on the attitudinal changes and life style of working women in the developed world as well as economic, educational and social liberalization of women in the third world countries in adapting to breast feeding of infants to postpone postpartum return of fertility (Valdes *et al.,* 2000).

Female Sterilization

This is a permanent approach birth control involving tubectomy in women who have fulfilled their reproductive function in terms of having desired number of living children and do not wish to become pregnant again. This is possibly the most commonly used birth control approach according nearly 30 per cent of all contraceptive methods used worldwide. This method blockage of the fallopian tube where gamete fertilization takes place prior to the conceptus is transported to the uterus for implantation. It requires a skilled anaesthetist, a gynecologist trained in laparoscopy, the laparoscope and a hospital environment and is usually carried out after child birth or medical termination of pregnancy. Despite its being relatively expensive procedure with restricted availability, the fact that it is nevertheless most widely practiced in both developed and developing countries goes to prove the extent to which women would go to escape from the 'fear of pregnancy'. If the tubectomy can be simplified from a sophisticated surgical intervention to relatively simpler medical procedure capable of being handled by trained paramedics in a rural setting, this fertility regulating method should become even more attractive in terms of affordability and accessibility. The latter possibility appears bright in view of finding by Feldblum and collaborators that intrauterine application of the cheap antimalarial drug, quinacrine would cause occlusion and fibrosis of the fallopian tube (Feldblum *et al.,* 2000).

A large scale of field trial in Vietnam, involving 31,000 women volunteers has proved its efficacy; pregnancy rate per 100 women years of use was 2.6 per cent after one year of follow up with no death, no evidence of increased incidence of ectopic pregnancies and only 8 complications requiring treatment. This is in contrast to the situation with surgical sterilization in similar number of women wherein anticipated number of death is 30 and of medical complications, 1,800. This simplified procedure has not been approved by WHO for wide-spread field trials elsewhere so far and probably awaits further research and efficacy trials in other ethnic groups of women (Short, 1994: Pine and Pollak, 2000).

Indian Scenario

Since 1970, when laparoscopic operations including sterilization became operational in India, studies during the last two decades have shown that postpartum sterilization is the most widely employed female surgical sterilization. Two factors contributed to this popularity; (a) women is already in the hospital delivery room, and (b) the ease and safety of the procedure carried out under local anaesthesia. To improve the accessibility of sterilization to the rural masses, several laparoscopic "camps" for operations were organized in rural areas. This was done apparently without adequate environmental sanitation with consequent higher rates of complication and failure. Since then laparoscopic sterilization are confined to major medical institution where adequate facilities exist. After reaching performance peak around 1982, there was progressive fall in the number of such sterilization which now accounts no more than 30 per cent of all sterilization and is confined to medical institutes capable of providing adequate surgical and follow up facilities (Ramchandran, 1994; Mishra, 2002).

There have been several studies directed against assessing post-sterilization health risks in women with no clear cut pattern in term of their long-term impact. The existence and extent of physiological and psychological changes in women, commonly termed as "post-tubal ligation syndrome" characterized by subjective symptoms such as pelvic pain, change in sexual habits, changes in mental health and premenstrual cramps and cycle length duration do not generally exhibit identifiable pattern of symptoms and hence-require further examination. Sterilization as the terminal method of regulating fertility accounts for nearly one third of world fertility regulation and in most countries it has increased in prevalence more rapidly than other methods and is remarkably concentrated geographically due to uneven popularity. Twenty developing countries of the third world contain over 95 per cent of all users, China acounting for over 50 per cent and India 25 per cent. In general female sterilization outstrips its male counterpart, *viz.*, vasectomy and the gap between the two exhibits a widening trend (Pollack, 1994).

Vaginal Birth Control Methods: Physical and Chemical Methods

Despite the fact that birth control through practice of vaginal contraception has a long recorded past, it has received little research attention during the last 3 decades in terms of innovative methodology and technological improvement. This is probably due to various factors like inadequate contraceptive efficacy coupled with poor acceptability as well as availability of more user-friendly and efficacious hormonal and intrauterine contraceptives.

Female Contraceptives

However, if adequately improved, based on the knowledge of vaginal, cervical and sperm anatomy and physiology, pharmaceutics and device technology to lift them from the present antiquated state, vaginal contraceptive have potential to have greater impact on reproductive health particularly because of their potential to curtail the spread of STD including HIV, besides inhibiting conception. This, however, needs reorientation of support policies of the national and international agencies.

Barrier Approaches

Among physical barrier methods developed so far, disposal latex diaphragm containing spermicides, cervical cap, Leo's shield made of silicon rubber, sponges or tampon type devices containing spermicides have been tried to control fertility with varying degree of success. While, these approaches have shown high failure rate which is attributable to personal dislikes of the user, one should also remember the advantage that these are under woman's control. International agencies are currently evaluating a female condom with a focus on acceptability and contraceptive efficacy. As emphasised earlier, the whole field of barrier contraceptives has to be given an enhanced emphasis due to the STD/HIV pandemic (Andolsek and Schnare, 2000).

While, the conventional barrier methods like diaphragms and cervical caps have become outmoded due to high contraceptive failure rates and users bias due to discomfort during intercourse, the recent development of the "Today" sponge which is impregnated with the spermicide nonoxynol-9 initially held promise. But, later it fell out of favour due to unacceptably high failure rate. These studies have, however, proved that the chemical spermicide has bactericidal and viricidal properties to afford significant protection against gonorrhoea, chlamydia and HIV infection when placed in the vagina. If used repeatedly, however, high concentrations of this chemical causes significant vaginal irritation and ulceration which may augment STD/HIV transmission. An interesting recent development pertains to the basic discovery that during luteal phase of the normal menstrual cycle, the uterine fluid harbours high concentration of cholic acid which exhibits strong spermicidal and antiviral activity; this normal body constituent with profound detergent activity inhibits the reverse transcriptase of HIV and inhibits the ability of HIV to infect the human lymphocytes (Psychoyos *et al.*, 1993). A vaginal sponge, 'Protectaid' has been developed which is impregnated with a gel containing 0.5 per cent of sodium cholate, nonoxynol-9 and benzalkonium chloride. This device holds promise for the future, since in a small clinical study of one-year duration, the sponge prevented pregnancy in 20 women with no discernible side effects (Short, 1994). Praneem VILCI, a vaginal contraceptive cream containing a purified neem seed preparation developed by Prof. G.P. Taiwar has also shown similar promise in animal and human clinical trials to prevent fungal and bacterial infection. Besides, it has been claimed that the application of the cream mounts cell-mediated immunity locally against sperm antigens to prevent pregnancy in experiments with animals (Rai, 1995).

Therefore; people have at their disposal a range of spermicides, synthetic as well as natural (of plant origin) with potent bactericidal and viricidal activity. They may hold great promise for the future for use either postcoitally during coitus (as a lubricant for the condom! diaphragm). They may have substantial benefit to both the partners by serving as a means of prevention of pregnancy and of STD/HIV transmission. It is surely worthwhile to concentrate on research on chemical prophylaxis in future using more efficient and less toxic antibacterial and antiviral spermicides which could be used as female vaginal douches, pessaries and as lubricants for condoms for men to reduce the spread of these STDs of both microbial and viral nature.

Intra-uterine Devices (IUDs)

This birth control method is one of the most commonly used methods of fertility regulation particularly in developing countries. It is based on basic research finding that introduction of a foreign body into the uterus prevents pregnancy establishment due to creation of a rather hostile environment locally to prevent successful implantation of the early conceptus. Additionally IUD possibly interferes with sperm and ovum transport in the uterine lumen. Beginning with the Dalkon Shield in USA in early 1970s, the device has undergone various structural design and material modifications continuously during the last two decades. The contemporary devices such as copper T, copper T-380A with or without hormone (progestins) releasing mechanism have high continuation rates with low pregnancy rates. IUD's account for nearly 20 per cent of all the modern contraceptives and used among a total of around 390 million couples around the world. It is estimated that there are more than 80 million users with some 74 million in China alone, where however mostly stainless IUD rings have been used (instead of copper devices). This has unfortunately resulted in rather high rate (20 per cent) of method failure as well as expulsion rate of similar order (WHO, 1992). In China, the women born during 1960 baby boom are causing secondary baby boom in 1990s and it is anticipated that this will cause an IUD boom in the current decade. An UNFPA/WHO combined study of 1993 has predicted that if the Chinese authorities introduce a copper-T 380A device instead of stainless steel IUD, the following events could be averted: 55.6 million pregnancies resulting in 18.5 million live births and 35.6 million induced abortions, 16,300 maternal deaths, 3,56,000 infant deaths and 28,000 child deaths. The Chinese government has apparently decided to change over to copper-T-380 IUDs (WHO, 1993).

Chronologically, the Dalkon Shield of 1970s in USA was a very bitter experience in the minds of the users which resulted in a drastic decline in the number of users, from initial 2.5 million to 7,00,000. it had high pregnancy rate of over 5 per 100 women years, nearly two times higher than the modern IUDs. The high pregnancy rates with the Shield device coupled with low expulsion rate and its microbe laden multifilament tail have led to a number of serious consequences including pelvic inflammatory disease, septic abortions and death. This has finally led to the ban of Dalkon Shield by FDA which has cast a dark showdown on further technological innovations and user acceptability. Nevertheless, there have persistent innovative modifications later on starting with the arrival of Lippes loops (C and D) and then with the introduction of copper I series of IUD such as Nova T, MLcu250, copper T-220C, MLcu 375 and copper T-380 which progressively decreased the pregnancy rate to 0.8-1.2 per 100 women years (Sivin, 1994; Sivin and Moo-Young, 2002; Neuteboom *et al.,* 2003).

Besides high expulsion rate and occasional uterine perforation encountered, the major side effect of the IUD identified early in its history was the increase in menstrual blood loss at early times after, insertion. Although, this may not be a major problem with healthy women of higher strata of society with good nutritional status, it could be with women with anaemia or low nutritional status as in India. The Population Council of New York in collaboration with the Finnish Pharmaceutical Company, Leiras, has developed a new hormone releasing IUD-'Levonova' which releases continuously the synthetic progestin levonorgesterel to the uterine lumen from the silastic cylinder surrounding the stem of the device. This low progestin released locally creates a continuous progestational environment within the uterus which drastically curtails menstrual blood loss. Additionally, the presence of a viscous cervical mucous plug in women with hormone-releasing IUD was found to significantly reduce the incidence of ascending uterine infections which cause pelvic inflammatory disease. The contraceptive failure rate of 0.1 per 100 women years of use is the lowest of any method other than surgical sterilization. The device though relatively costly was found to be effective for over 7 years (Chaudhuri, 2001).

Indian Scenario

Coming to Indian Scenario, the Lippes loop was launched into the National Family Planning Programme during 1965 as a birth-control method initially through family planning services attached to major Hospitals with infrastructure for initial screening and follow up care to tackle the problems of expulsion, infection and excessive bleeding. Though, early response was encouraging, when the 'camp' approach was introduced to enhance the availability of this IUD to rural masses, expected success did not occur due to primarily lack of infrastructure for follow up care to alleviate the side effects when they arose. Copper IUD (T-200) was introduced into family planning programme during 1970s after Indian Council of Medical Research (ICMR) undertook a comparative evaluation of the IUDs. Under conditions prevailing in the country, copper T-200 was shown to be a safe and effective contraceptive with relatively few side effects and complications while the ICMR studies did show that IUD associated menstrual blood loss, other adverse effects decreased haemoglobin levels were not very serious even in anemic women and iron supplements improves the blood picture. On the basis of these findings, copper T-200 was introduced to the National Family Programme. Data collected during 1990-91 demonstrated that copper T-200 was the most widely used and effective birth spacing method in the national programme. With availability of such services in the primary health care centers capable of catering to 30-50 thousand people, the number of IUD used increased progressively and steadily. To ensure indigenous production, manufacturing plants for copper T-200 devices were set-up. During mid 1980's the newer long-acting copper 'r-380A with an efficacy of 10 years was taken up for comparative evaluation vis-à-vis copper-T-200. Early data did not reveal any major advantage of this over the other in terms of complications or side effects. Nevertheless, it may be more advantageous to use copper T-380 in lieu of copper 1-200 in view of its prolonged efficacy for over 10 years, particularly among younger women in their twenties/thirties, who have had already 1-2 children, but are unwilling for permanent sterilization because their children are still young and they are not sure of future course of action (Chaudhuri, 2001).

From the foregoing discussion it is clear that despite certain drawbacks such as expulsion, increased menstrual blood loss, tissue perforation and the possibility of pelvic infection and pain, the IUD developments and practices have made this contraceptive method a highly desirable reversible approach to family planning, particularly since it combines high efficacy, good continuation rates, low product cost and long life.

Hormonal Approaches of Female Fertility Regulation

The developmental history of hormonal methods of contraception is an interesting story which handsomely shows how development in basic research in reproductive sciences, chemistry and material technology can be profitably exploited to develop approaches that are successfully used today by millions of individuals world over for decades with great benefits in terms of fertility control and reproductive health (Diczfalusy, 1989). This is particularly true with the sequence of events in the development of oral contraceptives which were introduced into the global market during early 1960s. Prior to this, only methods available to women were sterilization and barrier methods. Exploiting the early knowledge that the female sex hormone, progesterone increases in the blood simultaneously with the implanting of a fertilized egg in the uterine wall and that premature exposure to this hormone could interfere with the developmental programme of the female egg in the follicle and hence with ovulation (by blocking the hypothalamus - pituitary - gonadal axis through feedback mechanism). Dr. G. Pincus and his team initiated, in 1956, a large clinical trial in Puerte Rico, to test the efficacy of this approach to prevent conception. This trial used a daily oral dose of 10 mg of the synthetic progestin (norethynodrel) combined with varying amounts of the estrogenic compound, menstranol, continuously for 21 days during the critical periods of menstrual cycle and successfully demonstrated a normal distribution of lengths of menstrual cycle. Since 1960, many different steroidal contraceptives, which contain progressively lower doses of both hormones have been developed (Diczfalusy, 1989).

Oral Contraceptives (OC) Approaches

During the last few decades, there has been exponential accumulation of knowledge on various aspects of steroidal contraceptives with focus on efficacy, epidemiological and toxicological aspects, long term metabolic side effects and cardiovascular risks etc. Oral contraceptives continue to be one of the most extensively researched type of drugs and attempts are still continuing to reduce the side effects and complications. 'The Pill' use by decreasing and modifying the steroidal contents of the pill without compromising the effectiveness. Currently, low dose pills and mini pills (progesterone only) are the most frequently prescribed oral contraceptives. Cardiovascular complications such as venous thrombosis, ischaemic heart disease and cerebrovascular diseases are still the safety concerns associated with combined OC which affect lipid-and carbohydrate metabolisms and the blood coagulation system thus predisposing women to cardiovascular problems. Continued risk-lowering research efforts have resulted in the very low-dose pills, the triphasic formulations and new progestin with reduced and even favourable metabolic effects without compromising efficacy. However, whether these modifications in formulation substantially improve their long term safety is yet to be decided by additional research efforts and continued post-marketing surveillance.

One of the most frequently debated issues in steroid contraceptive usage is the possible causal link between oral contraceptives and cancer of the reproductive organs. However, it is reassuring to note that a multinational study (25) conducted in 11 countries (both developed and developing) seems to largely dispel this fear. The conclusions backed up by clear cut epidemiological validity are (a) combined OC intake protect against ovarian and endometrial cancers in proportion to the duration of use, protection persists at least upto15 years after CC use is discontinued; (b) there is no evidence for overall increase in the risk of breast cancer associate with OC use;

and (c) the link between OC and cervical cancer is inconclusive and controversial, given the multiple factors (for example sexual activity characteristics) effecting the risk of this type of cancer (WHO, 1992).

However, it is relevant to emphasize some of the disappointing observations concerning OC use at a programmatic level in several social settings in the developing countries despite its undisrupted efficacy as a tool for safe, effective and reversible contraception (Rivera, 1994). These are: (a) failure rates are less than 0.1 per cent at one year in controlled clinical trials, while the same was more than 10 per cent at the programme level; and (b) compliance and continuation rates are very low, with less than 50 per cent of women continuing the pill use after one year. This dismal problem was multifactorial and varied from place to place. Limited contraceptive option, poor counselling, insufficient information and/or education, of both the user and health providers responsible for programme implementation combined with incorrect protocol use are some of the causative factors traced which resulted in unwanted pregnancies and frequent need to terminate such pregnancies. This has stirred up adverse reaction among the actual and potential user community often resulting in the contraceptive method falling to disrepute. Thus there is an urgent need to upgrade programme implementation strategies, particularly among the under privileged segments of the societies in developing countries.

Notwithstanding these limitations, in actual practice, the OC continue to be one of the most widely used means of reversible contraception representing nearly 14 per cent use of all the contraceptive methods prevalent globally, with the exception of China and Japan. This is despite the relatively higher cost of the method implementation and the tedium of having to take the pill by the women continuously for prolonged periods. Another, though not well- advertised advantage of the OCs is their use for "emergency contraception" for those who have had unprotected sexual intercourse or are victims of rape. The so called 'Yuzpe regime' of taking an increased dose of OCs immediately after the event is a very effective way of preventing pregnancy postcoitally and is practiced in Western societies frequently (Short, 1994).

In India, introduction of OC into National Family Planning Programme and subsequent implementation was delayed till mid 1970s for obvious reasons. Until then almost all the available data on hormonal contraceptives were derived from studies conducted among well nourished women in the developed countries and there was a need to obtain base-line information on the side effects and complications that might be encountered in a population like ours where, infection, under nutrition and anaemia are wide spread. Indian Council of Medical Research (ICMR) in collaboration with WHO undertook such an exploratory study among users of low dosage combination OC and women using progesterone only [minipill containing long-acting progestin *viz.*, depot medroxy progesterone acetate (DMPA)]. Studies clearly showed that these pills did not significantly aggravate the health parameters of these under-nourished women over and above what was to be anticipated in terms of the usual side effects encountered among the well-nourished women of the developed countries (Ramachandran, 1994).

In India, the majority of the women using hormonal contraceptives restrict their usage for 2-3 years only. On the basis of ICMR recommendation, the low-dosage combination pill was indeed introduced during 1976-77 and the pill was provided through nongovernmental agencies and social marketing channels at a subsidized price to improve its accessibility and availability. However, the pill was not accepted on a wide basis by the women in rural and semi urban areas; the majority of its users were urban middle-income women. As the consequence, proportion of OC users restricted to less than 1 per cent of all contraceptive users in India. This amply illustrates the point that the success of any contraceptive approach is dictated primarily by the strategic planning and efficiency of the health providers in popularizing the method by motivating the target population by developing infrastructure and delivery systems appropriate to the social setting and not on the scientific merit of the contraceptives *per se*.

Injectable Approaches

Continuous innovation and refinement are the indispensable attributes of any biomedical technology-oriented towards product development such as the contraceptives for wide spread use by the relatively healthy population worldwide. This is true with the steroid contraceptives. As mentioned earlier that the one of the major drawbacks of oral contraceptives is the regimented requirement of continuous daily intake of the pill which is generally difficult to implement among the target population anywhere, especially of the developing world with low education, poverty and tradition-bound social and cultural setting. To overcome this compliance drawback, researchers have discovered long-acting, slow-release synthetic steroid hormones (both progestins and estrogens) to be used as injectable/implants as an alternative approach to female fertility regulation. The extensive clinical experience accumulated over the last 20 years has indicated that available progestin only formulations, depot-medroxyprogesterone-(Depo-provera, DMPA) and norethisterone enanthate (NET-EN) are highly efficacious, safe and reversible contraceptives when given as injections, their effect lasting upto 3 months. However, in multicentric clinical trials, the continuation rates with these agents were unsatisfactory, the menstrual irregularities, especially bleeding problems being the most frequent reason for discontinuation among women using the injectable. With the inclusion of short/ medium-acting estrogen preparation, the endometrial bleeding pattern could be restored to normalcy, thus heralding the development of once-a-month injectable contraceptives. Further improvements could be achieved when combinations of either dihydroxyprogesterone acetophemide plus estradiol enanthate or 17 a-hydroxyprogesterone plus estradiol valerate were introduced as once-a month injectable contraceptives which have been tested extensively for their efficacy and safety for the last 20 years or so. During the last decade, two other progestogen-estrogen combination injectables have undergone multicentric clinical trials under the agesis of WHO: one is 'cyclofem' which contains low (25 mg) dose of DMPA and estradiol cypionate, and other is made up of NET-EN (50 mg) and estradiol valerate (5 mg). Both of them have proved their high efficacy in clinical trials and are in the process of being introduced into National Family Planning Programmes. It is anticipated that next decade will witness introduction of new aproaches to the sustained release of contraceptive steroids such as monolithic microspheres and novel synthetic steroid molecules with improved efficacy, safety, reversibility and above all, acceptability as superior injectable long term contraceptives (Garza-Flores *et al.,* 1994).

New Implant Systems

This revolutionary new long-term (5 years) approach to steroidal contraception for humans stems from the important observation that steroid hormones can be released at a constant rate for prolonged periods from silicon rubber tubing and forms the basis of subdermal implants like Norplant and Norplant II which have been approved for distribution in 24 countries. The New York - based Population Council provided research and developmental support for over 20 years for this new promising method of fertility regulation spending over US $ 20 million (Population Council, 1990).

The earlier version, Norplant consists of 6 capsules containing the synthetic progestin, levonorgestral, each capsule (3.4 cm long and 2.4 mm diameter) contains 36 mg of crystals of the synthetic steroid. The implants are inserted with a trocar through 2 mm incision in the upper arm of the woman to provide contraceptive protection upto 5 years. More recent Norplant II is a two implant system with a modified design to release the same amount of the progestin as its predecessor. The design consists of an inner rod containing 50 per cent levonorgestral and remaining 50 per cent silastic and is covered by a silastic membrane. This is easier to insert and quicker to remove and more amenable to automated large scale manufacture and hence of lower cost. Extensive pharmacokinetics-endocrine and serum biochemical parameters have been investigated to ensure their safety and efficacy and to detect metabolic side effects. One of the mechanisms of contraception is by inhibition of ovulation with the number of anovulatory cycles ranging from 82 per cent in the first year to 33 per cent during the 7th year which parallel release rates of the contraceptive steroid. Luteal function defect is suspected in view of diminished progesterone encountered in comparison to those during pre-treatment cycles. The second mechanism contributing to pregnancy prevention is the change in cervical mucus which is rendered thick, viscous and scanty impeding sperm motility and penetration. Extensive tests on function of various organs did not reveal clinically significant abnormalities nor major changes lipid and carbohydrate metabolism and hemoglobin levels during the study reflecting high degree of safety. Among more than one million of women using Norplant throughout the globe, the method has proved to be a very effective long term contraceptive approach with pregnancy rate per 100 women years being 0.2 during first 2 years and 1.1 by the 5th year. Body weight of the user is also contributing factor, women weighing>=70 kg have greater failure rate than their lower weight counterparts.

The major side effect and hence the most common reason for discontinuation during Norplant use is the disruption menstrual cyclicity. Analysis with different populations of women indicates that irregular uterine bleeding ranges, from frequent bleeding episodes through prolonged number of bleeding days, to amenorrhoea. The changes experienced occur primarily during the first 3-6 months after insertion of the implant and bleeding pattern improves subsequently. The gross annual discontinuation rate due to adverse effect of menstrual bleeding is 9.1 per 100 women during the 1st year to decrease to 2.9 by 5 year. The other non-menstrual medical reasons for discontinuation of the Norplant use are: headache, body weight change, mood change and depression. Other complications such as inflammation of the site of insertion, local tissue reactions and infection do occur albeit infrequently. Norplant is relatively costly to insert and remove, requires good clinical supervision and is principally used by those who have completed child bearing, but do not wish to take the irrevocable step of sterilization.

There are several other implants, effective for varying mid-term duration, currently in development stage under the auspices of various international and private sector agencies. These include those made of biodegradable material to avoid the otherwise inevitable visit to clinic to remove the implant at desired timing. Shorter duration (1-2 years) implants, used as a post-partum contraceptive should enable women to space child birth for 2-3 years apart thereby increasing infant survival rate. Progestin-only implant containing potent synthetic gestagen (3-ketodesogestrol) as a single biodegradable implant (Implanon) useful for lactating mothers being developed by Organon is currently under clinical trials. This should be useful since it has no adverse effect on lactation. The major advantage of long acting injectables and implants is that the contraception is under women control and she does not have to remember to do anything unlike OC.

Norplant and injectable have undergone limited clinical trials in India under the guidance of Indian Council of Medical Research and these trials have faced severe criticism from women's health advocates and some other social organization as to the way it was conducted without appropriate infrastructure for follow up. To the best of my knowledge, there has not been any major programme implementation in India so far as Norplant is concerned.

Birth Control Vaginal Rings

The hormone-releasing contraceptive vaginal rings are newer technological developments to disperse with the disadvantage of requirement for trained medical professionals for insertion and removal of long acting devices like copper releasing 380 A IUDs and Norplants; these are under user-control. The vaginal rings need no daily attention (like the pill) or coitus-related action (like diaphragm or shield). The development of hormone-releasing vaginal rings is based on the early knowledge that steroids are efficiently absorbed through the vaginal epithelium and silastic capsules can be used as a reservoir for constant release of steroids. During the past decade, two varieties of contraceptive rings, one containing a combination of a synthetic progestin (norethindrone acetate or levonorgestral) and ethynyl estradiol and the other the progestin only, have been developed and tested for efficacy and dose finding studies in multicentric trials under the guidance of WHO (20) and the Population Council at New York (The Population Council, 1992). Ring design varies in that the steroids may either be distributed evenly in the silastic matrix of the ring or concentrated in a silastic rod at the center. A combination ring containing potent progestin 3-keto-desogestrol and ethynyl estradiol is being developed by the pharmaceutical company "Oraganon". The method depends on the fact that slow release of small amounts of the steroid at constant rates results in thickening of cervical mucous making it impermeable to sperm; in some women, ovulation is also blocked. The method is reversible since removal of the ring promptly results in disappearance of circulating hormone levels. A world wide extensive trial in 19 centers confirmed its efficacy and acceptability. Side effects such as menstrual disturbance and bleeding, vaginal discomfort, repeated expulsion of the ring were encountered in a small proportion of women and the failure rate was approximately 4.5 per 100 women. There was positive correlation between the failure rate and body weight of the women using the ring. Extensive technology development and clinical trials under variety of social conditions in different ethnic groups may promote wide spread use of this new promising addition to contraceptive armamentarium particularly, since the ring use is both simple and entirely under the user's control.

Post Ovulatory Approaches of Birth Control

As the name implies, the post ovulatory methods of fertility regulation involve approaches that interfere with pregnancy establishment by curtailing either the fertilization of the gametes at the fallopian tube, subsequent transport of early conceptus, implantation on the uterine wall or subsequent establishment of pregnancy. In other words the postovulatory methods include those that prevent fertilization, *i.e.*, true contraceptives as well as those cause early abortion by interfering with the viability of the conceptus (contragestational agents). Most of these methods, currently available as well as those under development act around the time of implantation and represent important additions to available range of family planning methods since they offer efficient back-ups in case of contraceptive failure as well as an escape route from an unplanned/unwanted pregnancy from unprotected intercourse. The significance and overwhelming importance of these nonsurgical methods of menstrual induction/pregnancy termination can be gauzed from the yearly number of accidental pregnancies resulting from contraceptive failure which has been estimated at 8-30 millions per year globally (Segal and La Gaurdia, 1990). Induced abortion particularly accomplished clandestinely in an adverse legal, social setting under unhygienic conditions often by untrained personnel, is a major health problem in terms of high maternal mortality and morbidity particularly in developing countries. It has been estimated that every day, some 1,50,000 unwanted pregnancies (close to 53 million a year) are terminated globally by induced abortion under inappropriate conditions resulting in around 500 maternal death every day or 1,80,000 per year. This figure exceeds the annual total of all new cases of HIV infection in both men and women notified to WHO Global Programme on AIDS for every single year since 1980 (Von Look and Von Hertzen, 1994).

Earlier to 1984, options available for post ovulatory emergency/contraception were limited. For emergency contraception, the "Yuzpe" estrogen-progesterone combination (instead of high dose of estrogen alone practiced earlier) was a clear advantage in view of the fewer gastrointestinal upsets associated with the high dose of estrogen regimen. When steroid hormones are either contraindicated to the women or much time has lapsed after the act of unprotected intercourse, insertion of IUD was an alternate approach to prevent pregnancy establishment. However, the latter methods have not been popular with health providing professional despite their proven effectiveness. As far as early pregnancy termination is concerned the mechanical techniques such as suction-aspiration method was the only option available.

In India, the medical Termination of Pregnancy (MTP) Act of 1972 has legalized abortion which permits a woman to seek abortion if her unwanted pregnancy is injurious to her physical and mental health. At present most of these MTP services (suction method) are confined to urban areas and their outreach in rural areas are very poor. It is estimated that nearly 4-6 million illegal abortions are performed in the country, their number in the rural areas are nearly double (approximately 13.3 per 1000 pregnancies) compared to legal abortions. This has been attributed primarily to lack of awareness of legal MTP services as well as to poor assessibility and quality of MTP services available in the country (Maitra and Saxena, 1994).

Antiprogestin (RU 486)

Until mid 1980s, post ovulatory methods such as medically-induced abortion (in lieu of vacuum aspiration or suction method) received little thrust purely on political, ideological and religious reasons. Nevertheless, some of the most spectacular advances in fertility regulation in recent years were witnessed in this area, thanks to the discovery of a novel class of steroidal antihormones, *viz.*, synthetic antiprogestins. Among several hundreds of anti progestins synthesized during the last decade (which act by preventing progesterone binding to its cellular receptor in target tissues) only one of these, *viz.* RU 486 or mifepristone developed by Roussel-Uclaf of France, is registered for clinical use for first trimester pregnancy termination in France. After successful testing for safety and efficacy in a large clinical trial in France, it was used in Britain, Sweden and China along with uterotonic compounds like prostaglandin (taken orally) and in a defined time sequence (Hagenfeldt, 1994).

Most of the initial research on RU 486 and other antiprogestin were directed towards their application in medical termination of early pregnancies. These researches, spearheaded by the WHO Special Programme of Research Development and Research Training have shown that sequential administration of RU 486 and a low dose of synthetic prostaglandin analogue is a safe and effective medical alternative to suction-aspiration method. There are still, however, some unresolved issues, such as minimally-effective dose of RU 486, the most appropriate type and dose of prostaglandins and the maximum duration of pregnancy for which the treatment remains effective and safe. Other potential applications of RU 486 currently being investigated with very encouraging-results include emergency contraception, menses induction, once-a-month use during luteal phase, once-a-week and daily use.

In India, ICMR has carried out Phase II multi centre clinical trials in 1993 to examine dosage and efficacy of RU 486+ prostaglandin combination using 455 women as the study group. The success rate was 95 per cent in women within 7-14 days, 90 per cent within 15-28 days after the missed cycle. While, no serious complications were encountered about 15 per cent of women had complaints of vomiting, diarrhoea and abdominal pain (ICMR, 1993). Clearly expert medical care is required which is lacking mostly in rural India. Apart from biomedical issues, there is a need to add on questions relating to cost, user-acceptability and service facilities that should be made available to women seeking this non-surgical method of contragestation (ICMR, 2000).

Centchroman-Contribution of Indian Scientists to Antifertility Methods

Centchroman (ormeloxifene) is a non-steroidal synthetic drug with weak estrogenic and potent antiestrogenic activity developed by Scientists of Central Drug Research Institute (CDRI), Lucknow after >20 long years of research, development and clinical trials, which were often subjects of better controversy. The drug is thought to exert its post fertilization contraceptive activity by disrupting the delicate balance of estrogen and progesterone necessary for preparing the uterus for implantation. The drug was also shown to accelerate ovum transport to some extent in animal models. It is assumed that drug prevents implantation by precipitating 'asynchrony' between egg transport and uterine environment.

Centchrotnan has been claimed to be reasonable good postcoital contraceptive and has been released into the Indian market under the trade name 'Saheli'. A study was undertaken to simplify the dosing schedule without sacrificing the purpose of twice a week dosing regimen, using modeling measurement approaches. The drug was given to 60 female volunteers who were divided into seven groups: group I, 30 mg weekly; group II, 30 mg twice a week; group III, 30 mg twice a week for 12 weeks followed by 30 mg weekly; group IV, 30 mg twice a week for 6 weeks followed by 30 mg weekly; group V, 60 mg weekly; and groups VI and VII, single 60 mg loading dose followed by 30 mg weekly doses: The blood samples were collected and analyzed by high-performance liquid chromatography. In group 1, mean trough concentrations of centchroman and its active metabolites, 7-desmethyl centchroman, were comparable to the steady-state trough concentrations in groups III, IV, VI, and VII. The metabolite-to-parent-drug ratio remained constant in all the groups. The pharmacokinetic parameters in group VII were comparable to those reported after a single 30 mg dose. Dosage regimen VI was more convenient and provided better pregnancy protection (Pearl index 1.18) than regimen III, which is currently on the market and, thus, could be effectively used for contraception (Lal *et al., 2001*).

It is needless to emphasize that the development of a safe medical method of pregnancy curtailment and emergency contraception will have immense implications for the reproductive health of women in both the developed and developing countries. While, abortion is the most sensitive topic in the whole field of Family planning and is strongly opposed by the Roman Catholic Church, the need for abortion will always be there since no contraceptive method is 100 per cent effective. It is essential that our women have access to the safest and most effective means of terminating an unwanted pregnancy to avoid maternal mortality. It is hoped that these methods of medical termination of pregnancy become more popular among women and health providers in India in coming years since such approaches have been legalized since 1972.

Other Developing Methods

From the forgoing research, it is clear that the available contraceptive approaches to regulate human fertility are still inadequate particularly for developing countries in view of either lack of user compliance and/or incidence of birth that are unplanned and unwanted which in turn lead to abortion effecting the reproductive health of women. Considering varied needs of the user population in context or their education, religion and economic background additional methods are clearly required. For these and other reasons, the concept of a vaccine to prevent pregnancy is very fascinating. Feasibility of immunocontraception has been repeatedly demonstrated by 'natures' experiments on fertility due to immunological etiology of human infertility wherein all other causative factors-genetic, endocrine and anatomic- have been eliminated. For example, agglutinating and/or immobilizing antisperm antibodies in infertile men, otherwise perfectly normal, have long been documented. Several animal experiments have also demonstrated that infertility can be induced by immunization with selected antigens that are reproduction-specific, be it a reproductive protein or polypeptide hormone or an antigen that is specific to either one of the two gametes or to early embryo.

There are several attractive features associated with the vaccine approach to prevent pregnancy: (a) reasonably high acceptability of long-active injectable contraceptives indicates that administration of vaccine as injectables resulting in long testing infertility may be successful as a family planning programme; (b) as a vaccine is not coitally related unlike rings and condoms, it should be attractive; (c) since immunization would affect solely a vital stage of fertilization or implantation, few side effects can be excepted; (d) since contraceptive vaccines act by mobilizing internal process within the body, they do not entail the continuous pharmacological burden of synthetic drugs like steriods which, as we have been often have undesirable side effects such as cardiovascular complications and menstrual bleeding irregularities; and (e) depending upon the delivery system employed the use of vaccines must be easier than the most other methods, especially if administration is required infrequently with few anticipated method failures due to lack of compliance etc.

However, there are also some drawbacks to this approach to fertility regulation; (1) first of all, once a person is immunized the vaccine action cannot be stopped immediately unlike the case with oral contraceptives; (2) if there is 'do-it-yourself' diagnostic method to ascertain the antibody status, duration of infertility from the last inoculation will be unknown to the user without laboratory tests, and (3) periodic boosters are required unlike the case with the vaccine against infectious agents (Alexander, 1994).

Human Chorionic Gonadotropin (hCG) Vaccines

This human chorionic gonadotropin, secreted by the implanting human blastocyst acts by entering the maternal circulation to rescue the dying corpus luteum to continuously stimulate progesterone production to sustain the endomatrial environment for successful implantation. The principle behind hCG-based vaccine for female contraception relies on the fact that immunoneutralization of hCG would disrupt implantation and hence early pregnancy termination at pre-or peri-implantation stage. This concept of immunointerference with pregnancy was pioneered by Prof. C. P. Taiwar at National Institute of Immunology, New Delhi nearly 20 years ago (Talwar *et al.*, 1994). Preliminary researches were done with hormone-specific n-subunit of hCG coupled with tetanus toxoid as the carrier to overcome the T-cell tolerance of the self antigen to mount immune response. More recent modified version of the immunogen involves annealing of α-subunit of ovine

LH to 13-hCG before coupling the hybrid molecule to tetanus toxoid to elicit better immune response. These heterospecies dirners have resulted in generation of conformational antibodies with greater bioneutralizing activity. In phase II clinical trials, high hCG litres with higher affinity and enhanced bioneutralizing activity were elicited in 101 women. Only those women who achieved more than 50 mg/ml of serum antibody titres were admitted to efficacy phase of the trial. It was indeed remarkable that among sexually active women of proven fertility carefully observed over 750 menstrual cycles only one pregnancy occurred. There was no disruption of menstrual cycle in terms of duration of subsequent bleeding showing no apparent side effects. The contraceptive effect was reversible since women were able to conceive after antibody titres declined. While, this demonstration was a milestone achievement for Indian reproductive science, it has to be emphasized that about 20 per cent of the volunteers initially enrolled did not mount required level of immune response presumably due to inherent genetic factors, a problem faced with all vaccine formulation.

Another variant of anti-hCG vaccine developed by Prof. V. Stevens of U.S.A. under the aegis of WHO consists of a synthetic peptide of 36 amino acids corresponding to c- terminal 109-145 of -hCC which is coupled to di-diptheria toxoid and other components like synthetic rnuramyl peptides as the adjuvant. In phase I clinical studies conducted in Australia, healthy sterilized women recruited as volunteers for vaccination showed no adverse side effects, although histological observation did reveal some immunocrossreactivity in the cells of pancreas. However, immune responses observed were far less than in those participated in multi centre clinical trials with -hCG-1T conjugate. This peptide based vaccine has undergone a phase 2 clinical trials in Sweden (Griffin *et al.,* 1994).

One of the major constraints with hCG based vaccine is the prohibitive cost of the antigen from the natural sources. Several attempts are currently being made to produce-hCG through recombinant biotechnology with encouraging results. One approach was to produce a vaccinia virus construct that contains gene coding for J-hCG and a 48-residue membrane-anchor sequence. In preliminary experiments with monkeys immunized vaccine would be cheap to produce, simple to use and would be expected to require fewer doses. Another possibility is to use biocngineered baculovirus expression system for hCG production by recombinant technology. All these approaches are still in the experimental stages at present.

As stated earlier, one of the major hurdles to the immuno-contraception is the requirement for periodic boosting of women and in practice, this is difficult, if not impossible to achieve simply due to the fact that this requires frequent contact with the user population to motivate them to return for vaccine administration. To reduce this delivery burden of the health providers, various types of biodegradable microspheres with the vaccine entrapped therein to release the vaccine continuously or in pulsatile fashion are currently being tested for efficacy in terms of immune response (Steven, 1993; Chaudhuri, 2001).

Riboflavin Carrier Protein (RCP) Based Vaccines for Females

This approach to immuno contraception, pioneered by Professor P.R. Adiga of Indian Institute of Science, Bangalore is conceptually new and is based on a fundamental discovery that riboflavin carrier protein (RCP) is evolutionarilly highly conserved and is reproduction specific in mammals. This arose from the earlier finding that the chicken egg RCP, an estrogen-inducible reproduction-specific phosphoglycoprotein is obligatorily involved in yolk deposition of the vitamin for embryonic mutation in its structural gene, the maternal well-being including reproductive functions are unaffected except for the fact that there occurs 100 per cent embryonic mortality (Adiga and Karande, 1995). Its possibility to develop as an immuno contraceptive vaccine arose when it was found that this protein is evolutionarily highly conserved in mammals to human and immunization with RCP in female fertile rats and monkeys leads to very early loss of embryo at peri-implantation stage itself without any adverse effect, immunized animals returned to normalcy and animal were able to conceive normally on waning of antibody litre with time and delivered normal offspring (Adiga *et al.,* 1997). The conformation of the native protein was shown unnecessary to elicit neutralizing antibodies since denatured and linearized RCP was effective as the vaccine to interfere with pregnancy. Infact, two 21 amino acid peptidyl sequences, one at the N-terminus and the other at C- terminus were shown individually to elicit neutralizing antibodies to the maternal RCP, when they were administered as conjugate with diphtheria toxoid (for T cell help), to fertile female rats and monkeys. Protection from pregnancy could be achieved with efficiency extending over 15-16 consecutive cycles during 2-3 years of duration. Hopefully, the vaccine would be undergoing preclinical toxicological testing in the near future before Phase I clinical trials are initiated (Beena, 1994).

There are certain clear advantages associated with this novel approach to immunocontraception; (a) it is a cost-effective heteroantigen capable of mounting significant immune response without carrier conjugation for eliciting antibodies; (b) this antigen is a non-hormonal protein and hence no adverse effects due to endocrine imbalance; (c) it can be purified in relatively large quantities from a cheap source; infact it has been hyper- expressed as a recombinant protein in *E. coli* and is available in pure form; (d) it does not need refrigeration for storage since denatured protein is as efficient as the native protein; (c) its mode of pregnancy protection is through interference at implantation stage; (I) in rodent experiments, the protein elicits good immune response when immunized through oral route; and (g) RCP is also a sperm component, being localized on sperm head. Vaccination of fertile male rats and monkeys with chicken RCM-RCP elicits antisperm antibodies which coat the sperm and hence prevent fertilization. Thus, RCP vaccine can be used both in the male and female and this approach can be expected to further enhance the contraceptive efficacy. It is thought that this new vaccine approach will become a reality in terms of human contraception either by itself or in combination with other immunocontraception approaches (*e.g.,* Anti-hCG and ovine FSH) after careful planning and continuous trouble-shooting by technological development/mid-course correction as and when required. Author of this review article has been actively associated with this work.

Gamete Antigen Vaccines

Scientific and technological feasibility of developing immuno-contraceptive approaches to regulate female fertility based on sperm antigens stems from clinical observations that about 5-10 per cent of men and women attending infertility clinics are infertile due to antisperm antibodies in their systems; these people are otherwise perfectly healthy. Basic studies have revealed that the spermatozoa harbour a number of unique antigens that elicit antibodies capable of immobilizing, agglutinating and even killing the sperms. Thus, many of the naturally occurring sperm antibodies in the female reproductive tract interfere with sperm migration to sites of fertilization, others prevent fertilization either by blocking acrosome reaction and/or subsequent sperm binding to zona pellucida, which is a prerequisite for fertilization.

To be effective, a vaccine against sperm antigens must be directed against several components of sperm surface at most elicit high titre antibodies. Furthermore, the mammalian testis particularly, the seminiferous tubules which are sites of sperm production, are immunologically privileged sites to which antibodies are inaccessible. Therefore, sperm antibodies can bind to target antigens either at epididymal sites in the immunized male or in the reproductive tract of the immunized female. With this provision several potential antigens have been identified on the sperm surface for immunocontraceptive purposes. Among these are: (1) SP-10 located on inner acrosomal membrane which becomes exposed on acrosomal reaction (Frecmerman *et al.*, 1994), (2) A 90 KD protein, with tyrosine kinase activity, (3) PH-20, a 64 KD membrane protein localized on the plasma membrane and inner acrosomal region (Primakoff, 1994), and (4) a sperm-specific isoenzyme of lactic dehydrogenase, (LDH-C4) that have received much scientific attention for exploration as potential sperm antigens for immunocontraception (Goldberg and Shelton, 1926) All of these have been cloned and hyperexpressed by recombinant biotechnology and their molecular and immunological features have been thoroughly examined. One rather disappointing common observation is that while they are all capable of producing antisperm antibodies in the female sub-human primates, these antibodies are capable of interfering with pregnancy partially, maximum efficacy being confined to 50-60 per cent which is much less than that observed with RCP. While, underlying mechanism are awaiting elucidation, scientific and technological revolution taking place in this area together with mid-course correction contemplated should clearly bring about greater success in this line of endeavour for immuno-contraception.

Zona Pellucida Antigens

Like antisperm antibodies, antibodies directed against zona pellucida (a cellular amorphous glycoprotcin coat surrounding the mature egg in the mammals) are more commonly encountered in infertile women. This shows that zona protein could be a potential target for immuno-contraception. Obvious advantages of this aproach are: (1) zona proteins are synthesized transiently in the growing follicles during menstrual cycle only and laid on oocyte membrane just before maturation and subsequent ovulation, and (2) unlike sperms, the number of mature zona containing oocytes is small (mostly one in women during a menstrual cycle), zona proteins appear to be excellent target for fertilization inhibition by specific antibodies. Based on these considerations and on the fact that zona proteins are evolutionarily and immunologically highly conserved among mammals, there have been a large number of investigations on the feasibility of using vaccine approaches to zona proteins to curtail fertility in experimental animals including primates. This approach is claimed useful in halting the runway population increase of some of the wild animal species like wild horses and deers. Initial studies were very encouraging in terms of efficacy, but histopathological examination showed undesirable consequences such as ovarian damage due to immunological cross-reactivity with the other protein components of the normal follicles. More recent studies, particularly those involving purified zona components such as ZP3 (the sperm receptor) and its specific peptidyl fragments of defined sequences containing B-non T epitope gave aroused renewed great interest in this target protein for immunocontraception. It is hoped that ongoing and future investigations will lead to a better understanding of the molecular biology and immunology of these proteins. Some of these zona proteins including their human counterparts have been cloned and produced in large enough quantities by recombinant expression systems for further extensive investigations with the hope of success as a immuno-contraceptive target (Mahi-Brown and Tung, 1995).

From the foregoing investigations, it is clear that immunological approaches to fertility regulation offer several advantages over many other existing methods of fertility regulation. Enormous scientific interest generated during the last two decades in several laboratories world over are a testimony to this and researches in reproductive sciences in India occupy a pioneer status in this respect. As in any scientific endeavour, progress in initial stages has been rather slow, but recent upsurge in research activity with missionary zeal is dearly a welcome trend for future success. At this stage of scientific know how and technology development, some limitations of these approaches are perceptible, but they are not insurmountable. Some of the problems such as those concerning sustained maintenance of sufficient antibody titres to interrupt pregnancy have to be overcome by developing appropriate delivery systems to prolong the immune response. Genetic variability in immune response is a factor to be reckoned with and one should explore whether a 'cocktail' of potential antigens is an answer to this. Since secretory immune response appears to be important, it may be advantageous to adapt multiple (including oral and vaginal) routes of antigen administration. Some type of simple self-assessing method for antibody titres measurements would go a long way in determining whether a woman is immunologically protected. Above all, the safety and efficacy of the vaccine with minimum side effects have to be proved carefully and dispassionately.

Conclusion and Future Prospects

It has been very clear from foregoing discussion that there is no ideal contraceptive method suitable for all users, but there can be assortment of ideal birth control technology for the needs of different users. To develop any ideal birth control method, it takes 20-25 years (begins from basic research to-its application) if everything goes well. Contraceptive development research is also very expensive process. Despite all these problems, researchers are able to develop variety of birth control methods which are benefitting the users of different age groups. But, there is still greater need to develop newer birth control approaches and to refine the existing female contraceptives. It should be emphasized that Indian scientists significantly and greatly contributing in this area of contraceptive development research. However, it is absolutely essential that concerted research efforts backed by higher level of funding be made mandatory on priority basis to hasten progress towards reducing the population menace.

References

Adiga, P.R., and Karande A.A., 1995. Immunization with RCP for pregnancy curtailment. In: Taiwar CF. Raghupathy R, Eds. Birth Control Vaccines. *Austin, USA: RG Landes Company.* p. 103-114.

Adiga, P.R., Subramanium, S., Rao, J., and Kumar, M. 1997. Prospects of riboflavin carrier protein (RCP) as an antifertility vaccine in male and female mammals. *Human Reprod. Update* **3(4)**: 325-334.

Alexander, N.J. 1994. Contraceptive vaccine. In: van Look PFA, Perez-Palàcios C, Eds. Contraceptive Research and Development 1984-1994. The Road from Mexico City to Cairo and Beyond. New Delhi: *WHO, Oxford University Press.*

Andolsek, K.M. and Schnare, S.M. 2000. Update on barrier methods. *Dialog. Contracept.,* **7(6)**: 4-6.

Beena, T.K. 1994. Antigenic Determinants of Chicken Riboflavin Carrier Protein: Structural and Functional Aspects. *Ph.D. Thesis, Indian Institute of Science, Bangalore.*

Chaudhuri, S.K. 2001. Intrauterine devices. In: Chaudhuri, SK, Ed. Practice of Fertility Control: *A Comprehensive Textbook. Fifth edition, New Delhi: BI Churchill Livingstone.* p. 79-109, 2001.

Chaudhuri, S.K. 2001. Vaccines for fertility control. In: Chaudhuri SK, Ed. Practice of Fertility Control: *A Comprehensive Textbook. Fifth edition, New Delhi: BI Churchill Livingstone.* p. 312-317.

Cross, H., Hardee, K., and Ross, J. 2002. Completing the demographic transition in developing countries. *Washington DC: Futures Group International, POLICY Project.* p. 28.

Diczfalusy, E.1989. History of steriodal contraception. What is past and what is present keynote address. In: Safety Requirement for Contraceptive Steroids, Cambridge: Cambridge University Press, UK.

Fathalla, M.F.1994. Family planning and reproductive health- A global overview. In: Graham- Smith F, Ed. Population - The Complex Reality. *A Report of the Population Summit of the World's Scientific Academies. London: Royal Society*, 251-270.

Feldblum, P.J., Hays, M., Zipper, J., Guzman-Serani, R., and Sokal, D.C. 2000. Pregnancy rate among Chilean women who had non-surgical sterilization with quinacrine pellets between 1977 and 1989. *Contraception,* **61**: 379-38.

Frecmerman, A., Wright, R.M., Flickinger, C.J, and Herr, J.C. 1994. Tissue specificity of the acrosomal protein SP-10: a contraceptive vaccine candidate molecule. *Biol. Reprod.,* **50(3)**: 615-21.

Garza-Flores, J., and Cravioto, M.C.1994. Perez-Palacios G. Contraceptive research and development today: injectables. In: van Look PFA, Perez-Palacios G, Eds. Contraceptive Research and Development, 1984 to 1994. The Road from Mexico City to Cairo and Beyond. New Delhi: *WHO, Oxford University Press*, p. 53-68.

Goldberg, E., and Shelton, J.A.1986. Immunologic properties of LDH-C4 for contraceptive vaccine development. In: Zatuchni, C.l., Goldsmith, A, Spieler, J.M., Sciarra, J.J., Eds. *Male Contraception: Advances and Future Prospects. Philadelphia, Pennsylvania: Harper and Row,* p. 435-446.

Grant, J.P.1992. The State of world children, Oxford and New York: *Oxford University Press*. p58, .

Griffin, P.D., Jones, W.R., and Stevens, V.C.1994 Anti-fertility vaccines: current status and implications for family planning programmes. *Reprot Health Matt.*, **3**: 108-113.

Hagenfeldt, K.1994. Current status of contraceptive research and development. In: Graham- Smith F, Ed. Population - The Complex Reality. *A Report of the Population Summit of the World's Scientific Academies. London: Royal Society*, p. 271-285.

ICMR.1993. A multi centric clinical trial with RU 486 and prostaglandin contraception. *Contraception,* **49**: 87-98.

ICMR. 2000. A multicentre randomized comparative clinical trial of 200 mg RU 486 (mifepristone) single dose followed by either 5 mg 9-methylene PGE2 gel (meteneprost) or 600 micrograms oral PGEI (misoprostol) for termination of early pregnancy within 28 days of missed menstrual period. An ICMR Task Force study. India. Indian Council of Medical Research [ICMR]. Division of Reproductive Health and Nutrition. *Contraception,* **62(3)**: 125-130.

Labbok, M.H., and Queenan, J.T.1989. The use of periodic abstinence for family planning. *Clin. Obstet. Gynecol.*, **32(2)**: 387-402.

Lal, J., Nitynand, S., Asthana, O.P., Nagaraja, N.V., and Gupta, R.C. 2001. Optimization of contraceptive dosage regimen of Centchroman. *Contraception,* **63(1)**: 47-51.

Mahi-Brown, C.A., and Tung, K.S.K.1995. Vaccine against egg antigen and pathogenesis of ovarian autoimmune disease. In: Taiwar, G.P., Raghupathy, R., Eds., *Birth Control Vaccines. Austin, USA: RG Landes Company,* p. 41-62.

Maitra, K., and Saxena, B.N.1994. Current status of medical termination of pregnancy and usefulness of new technologies in India. In: Puri, C.P., van Look, P.F.A., Eds. *Current Concepts of Fertility Regulation and Reproduction. New Delhi: Wiley Eastern,* p. 257- 264.

Mishra, P. 2002.Concurrent evaluation of sterilisation beds scheme in Uttar Pradesh. *Series B, Survey Report No. 79, Population Research Centre, Lucknow; India: University of Lucknow*, p. 53.

Neuteboom, K., de Kroon, C.D., Dersjant-Roorda, M., and Jansen, F.W.2003. Follow-up visits after IUD-insertion: sense or nonsense? A technology assessment study to analyze the effectiveness of follow-up visits after IUD insertion. *Contraception,* **68**: 101-104.

Pine, R.N., and Pollack, A.E.2000. Putting an ear to the ground: where now with Quinacrine? *J. Gynecol. Obstet.*, **69(1)**: 55-65, .

Pollack, A.E.1994. One hundred and seventy million sterilization later. In: *Contraceptive Research and Development 1984-1994. New Delhi: WHO, Oxford University Press*. p. 215-232.

Primakoff, P.1994. Sperm proteins being studied for use in a contraceptive vaccine. *American J. Reprod. Immunol.*, **31(4)**: 208-210.

Psychoyos, A., Creatsas, C., Hassan, E., Georgoulias, V., and Gravanis, A.1993. Spermicidal and antiviral properties of cholic acid: contraceptive efficacy of a new vaginal sponge (Protectaid) containing sodium cholate. *Human. Reprod.*, **8(6)**: 866-869.

Rai, U.1995. India's vaccine inventor: Gursaran Talwar. *IDRC Reports*, **22(4)**: 10.

Ramachandran, P. 1994. Providing family planning: lessons learned and challenges ahead - India. In: Van Look, P.F.A., Perez-Palacios, C, Eds., *Contraceptive Research and Development, 1984 to 1994. The Road from Mexico City to Cairo and Beyond. New Delhi: Oxford University Press*. p. 291-303.

Rivera, R.1994. Oral contraceptive - The last decade. In: van Look PFA, Perez-Palacios G, Eds. *Contraceptive Research and Development. 1984-1994. The Road from Mexico City to Cairo and Beyond. New Delhi: WHO, Oxford University Press*. p. 23-25.

Sadik, N.1994. Population and development: preparing for the 21st century. A statement. In: Graham-Smith, F., Ed. *Population -The Complex Reality. A Report of the Population Summit of the World's Scientific Academies. London: Royal Society*. p. 77-81.

Segal, S.J., and La Gaurdia, K.D.1990. Termination of Pregnancy- A Global View. *Baillier's Clin. Obstet. Gynecol.*, **4**: 235-247.

Short, R.V.1994. Contraceptive Strategies for the future. In: Graham-Smith, F, Ed. *Population - The Complex Reality. A Report of the Population Summit of the World's Scientific Academies. London: Royal Society*. p. 323-339.

Sivin, I., and Moo-Young, A.2002. Recent developments in contraceptive implants at the Population Council. *Contraception*, **65**: 113-119.

Sivin, I.1994. IUD, A look to the future. In: *Contraceptive Research and Development 19M41994. New York: WHO, Oxford University Press*.

Stevens, V.C.1993. Vaccine delivery systems: potential methods for use in antifertility vaccines. *American.J. Reprod. Immunol.*, **29(3)**: 176-188.

Talwar, G.P., Pal, R., Dhawan, S., Singh, O., and Shaha, C.1994. Current status of future of immunological approaches to fertility control. In: Graham-Smith, F., Ed. *Population- The Complex Reality. A Report of the Population Summit of the World's Scientific Academies. London: Royal Society*. p. 287-302.

The Population Council.1992. Annual Report 1991, Centre for Biomedical Research. New York: *The Population Council*.

The Population Council.1990 Norplant levonorgestral implants. New York: *The Population Council*, p. 30.

United Nations Department of International Economic and Social Affair.1991. *World Contraceptive Use Data, New York: United Nation.*

United Nations Population Fund.1994. The State of World Population UNFPA, New York.

Valdes, V., Labbok, M.I.I., Pugin, E., and Perez, A. 2000. The efficacy of the lactational amenorrhea method (LAM) among working women. *Contraception,* **62**: 217-219.

Von Look, P.F.A., and von Hertzen, H.1994. Post ovulatory methods of fertility regulation, In: van Look, P.F.A., Perez-Palacios, C., Eds. *Contraceptive Research and Development 1984-1994. The Road from Mexico City to* Cairo and Beyond. New Delhi: WHO, Oxford *University Press.* p. 151-201.

WHO.1981. A prospective multi-centre trial of ovulation method of natural family planning II. *Fertil. Steril.,* **86**: 591-598.

WHO. Annual Technical Report 1992, Special Programme of Research, development and Research Training in Human Reproduction, Geneva: *World Health Organization.* 1993.

WHO.1992. Oral contraceptive and neoplasia, WHO Technical Report- Service no 817, Geneva: *World Health Organization.*

WHO.1992. Reproductive Health: A Key to a Brighter Future. Biennial Report, 1990-1991, Geneva: *World Health Organization.*

2015, **Perspectives in Animal Ecology and Reproduction, Vol. 10**   *Pages 377–388*
Editors: **V.K. Gupta, Anil K. Verma and G.D. Singh**
Published by: **DAYA PUBLISHING HOUSE, NEW DELHI**

Chapter 23

Tripterygium wilfordii: Possible Male Birth Control Pill? An Assessment Study in Langur Monkey

Mukesh Kumar

Reproductive Bio-technology Laboratory,
Department of Zoology, M.S.J. Government (P.G.) College,
Bharatpur – 328 001 Rajasthan, India

ABSTRACT

Present study was designed with higher dose level of *Tryptergium wilfordii* (15 mg/animal/day; oral of GTW) in male langur monkeys for 150 days based on the results we got from earlier study (6mg/animal/day; oral for 120 days) being done by our group in order to assess its potentiality to suppress fertility (spermatogenesis), mechanism of action, reversibility and possible side effects, if any. Semen examination, Semen biochemistry, Hematology and Blood biochemistry were done at monthly interval. Administration of GTW caused gradual decline in sperm density from 60-day onward and resulted into azoospermia (as 90-days semen examination revealed this state) alongwith decline in sperm motility and vitality significantly before reaching to azoospermic state. Sperm gradually started appearing in semen samples following 90-days of the cessation of GTW treatment and complete normalcy was reestablished during 150-days of discontinuation GTW treatment.

In conclusion, GTW treatment causes azoospermia without affecting the testosterone level and accessory sex glands functioning. GTW treatment effects were reversible, these results establish that the higher dose level of GTW

* *Corresponding author.* E-mail: prof.mukeshkumar@rediffmail.com

suppresses the spermatogenesis process and findings of this langur monkey's study could be extrapolated for human studies in order to develop GTW as male contraceptive.

Keywords: *GTW, Male contraceptive, Langur Monkey.*

Introduction and Rationale

Tryptergium wilfordii Hook F. a perennial Twining vine, member of the family – Celstraceae is grown widely in southern China. Since late 1970's active continuents *i.e.* so called multiglyconsides of *Tryptergium wilfordii* Hook F. (Here after it will be called GTW in this papers) are being used for treating the renal, liver, skin, collagenous diseases and arthritis and in few patients (male) the treatment caused necrospermia and reduction in fertility (Qin *et al.*, 1982; Li *et al.*, 1982: Jia, 1985).

Some studies done by Chinese researchers showed that GTW suppressed the fertility of male rats and mice (Qian *et al.*, 1986). Human male receiving GTW as medicine for above cited diseases showed reduction in fertility incidentally (Qian *et al.*, 1986; Qian, 1987). These studies were very encouraging but not systematically planned. Therefore, we planned research study in male langur monkeys as this animal model is very closer to human in its endo and exocrine profile and lack of seasonality provides opportunity to conduct long term experiments (Lohiya *et al.*, 1983) and findings could be extrapolated for future human studies. As our previous study with 6 mg dose level showed reduction in sperm density (near to oligospermic state) but azoospermia was not attained (Mukesh Kumar, personal communication). Therefore; present study was planned with relatively higher dose level *i.e.,* 15 mg of GTW in male langur monkeys in order to investigate complete suppression of fertility, possible mechanism of action, untoward effects of GTW treatments if any, process of reversibility.

Materials and Methods

Experimental Design

Adult male langur monkey (*Presbytis entellus entellus* Dufresne) were caught from the jungle around Jaipur, India and recruited for this study. The selection of adult male langur monkeys was based on following characters:

- ☆ The animal should have well developed musculature with pinkish oedomatous band on rump and large erupted canines.
- ☆ The animal should have good general health and should have the normal range of clinical laboratory tests in blood, urine and semen.

Housing and Feeding

Animals were housed in metallic cages under semi-natural conditions in our Primate Research Facility (PRF). Wheat chapaties, green leaves, seasonal fruits and roasted/soaked grams were given with free access to drinking water. Animals were quarantined for 3-4 months before commencing the experiment. Reproductive status of the animal was examined by semen analysis.

Experimental Design

One long term experiment with 15 mg dose level of GTW was done. Experiment composed of two groups and each group consisted of five animals.

Experiment Protocol

Group – A : Animal of this group received vehicle (banana) alone during treatment phase to serve as a control.

Group – B : Animals of this group received 15 mg GTW/animal/day; oral during treatment phase.

This study was divided into following three phases

- ☆ Pre-treatment phase   :   30 days
- ☆ Treatment phase   :   150 days
- ☆ Recovery phase   :   150 days

Body Weight

Body weight of the animals was recorded once a month during all the phases of study.

Semen Analysis

Semen samples were collected at monthly interval by penile electroejaculation (Mastroianni and Manson, 1963) and analysed for:

a) Seminal Characteristic

Semen weight, volume, colour, pH, sperm motility, density and sperm morphology (WHO, 1980).

b) Semen Biochemistry

Biochemical analysis of seminal plasma fructose (Man, 1964), acid phosphates (Gutman and Gutman, 1940) lactic dehydrogenase (Cabaud and Wroblewski, 1958), glycerylphosphorylcholene (White, 1959), magnesium (Neill and Neely, 1956) and citric acid (Lindner and Mann, 1960).

Circulating Level of Testosterone

To estimate the serum testosterone level (WHO, 1950) three blood samples at 15 minute intervals (08.00, 08.15, 08.30 am) were collected monthly. Animals were introduced to receptive female once every month for observing mounting and copulating behaviour (Lohiya and Sharma, 1983).

Clinical Investigation

Following laboratory clinical investigations were done during the entire course of study.

a) Haematology

Haemoglobin (Crosby *et al.,* 1954), haematocrit. (Natelson, 1951), RBC and WBC (Lynch *et al.,* 1969).

b) Serum biochemistry

SGOT and SGPT (Reitman and Frankel, 1957), LDH (Cabaund and Wrobelewski, 1958), CPK (Hughes, 1962) and blood glucose (Asatoor and King, 1954).

c) Electrolyte Metabolism

Na (John, 1960), K (John, 1960) Cl (Schales and Schales, 1941), PO_4 (Gomorri, 1942) and Mg (Neil and Neely, 1956).

Statistical Analysis

Data were expressed as mean ± S.E. and analysed for statistical difference using student "t" test.

Results

Body Weight

No significant changes were recorded in the body weight in all the treated animals during study period.

Semen Analysis

Physical examination of semen samples shows that GTW treatment did not cause any significant change in the semen weight, volume (Figures 23.1–23.2), colour and pH during entire course of study.

Treatment of GTW (with 15 mg dose level) caused gradual decline in sperm density from 60 days of dosing which result in azoospermia (this state was achieved

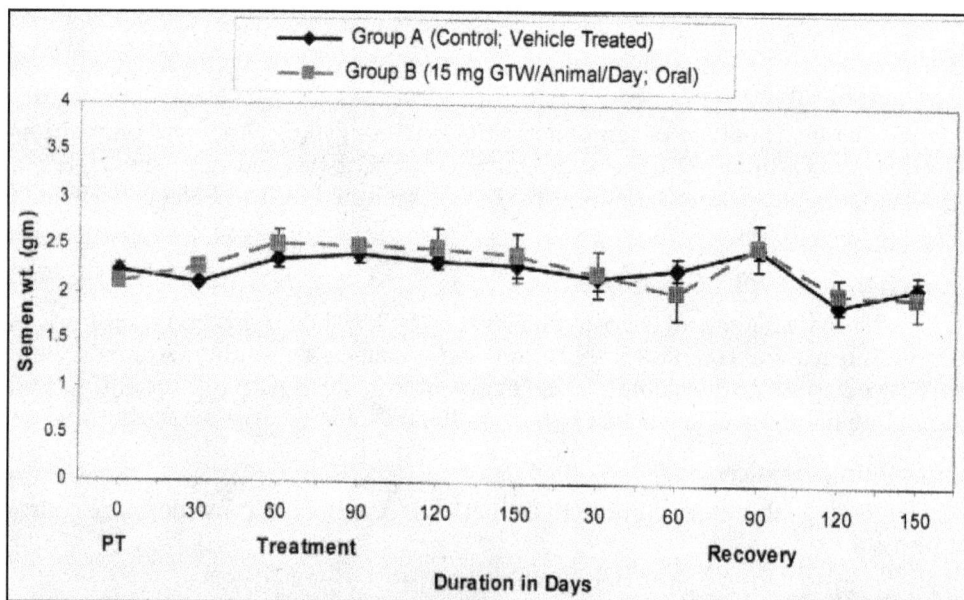

Figure 23.1: Showing Normal Semen Weight of Group-A (Control; vehicle treated) and Group-B (15 mg GTW/animal/day; Oral) during Pretreatment, Treatment and Recovery Phases. Each value represents mean±S.E. (Vertical bars).

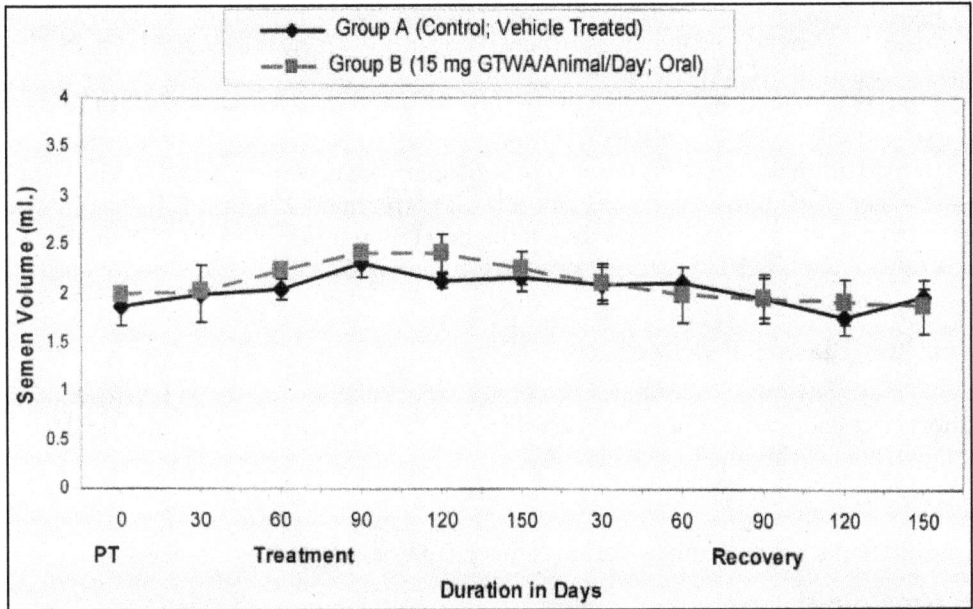

Figure 23.2: Showing Normal Semen Volume of Group-A (Control; vehicle treated) and Group-B (15 mg GTW/animal/day; Oral) during Pretreatment, Treatment and Recovery Phases. Each value represents mean±S.E. (Vertical bars).

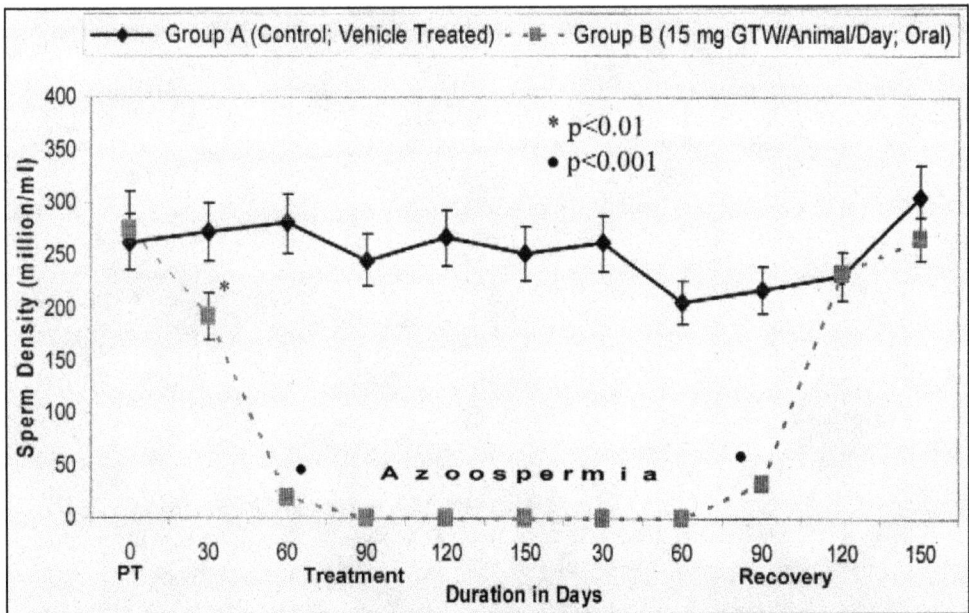

Figure 23.3: Showing the Significant Decrease in Sperm Density of GTW Recevied Animals from 30-days of Treatment and Azoospermic State was Observed during 90-days of GTW Treatment which Lasted upto 60-days of Recovery Period. The each value represents mean±S.E. (Vertical bars).

at the 90 days of treatment) in all treated animals uniformly (Figure 23.3) sperm gradually started appearing in semen samples following 90 days of the discontinuation of GTW treatment and normal sperm density was established at end of 150 days of recovering period (Figure 23.3) in all animals. Sperm motility also decreased following 30 days of GTW treatment. 5–10 per cent sperm were motile in all treated animals before attainment of azoospermic state and normal motility was established during 150 days of cessation of GTW treatment (Figure 23.4). Percentage of sperm vitality also showed gradual reduction following 30 days of GTW treatment and approximately 10 per cent of sperm were showing vitality before reaching to the azoospermic state (Figure 23.5).

Seminal Plasma Biochemistry

Treatment of GTW did not show any appreciable change in fructose, acid phosphateses, glyarylphosphorylcholine, lactic dehydrogenase, citric acid and magnesium during the course of study.

Libido

No remarkable changes were seen in ejaculatory responses latencies.

Haematology

Haematological parameters: haemoglobin, RBC, WBC and haemotocrit fell within normal range during experimentation period.

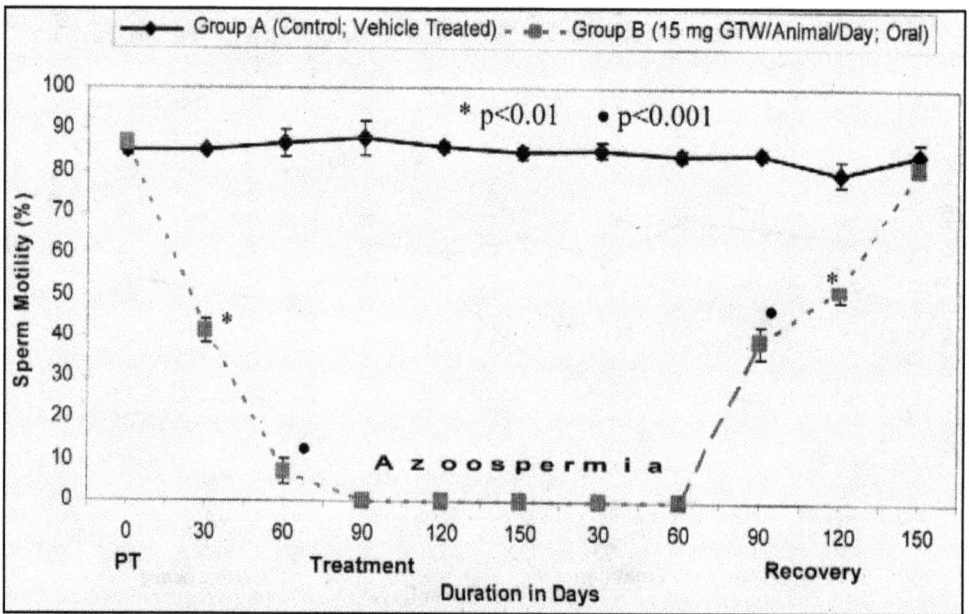

Figure 23.4: Showing the Significant Decrease in Sperm Motility of GTW Recevied Animals from 30-days of Treatment. The each value represents mean±S.E. (Vertical bars).

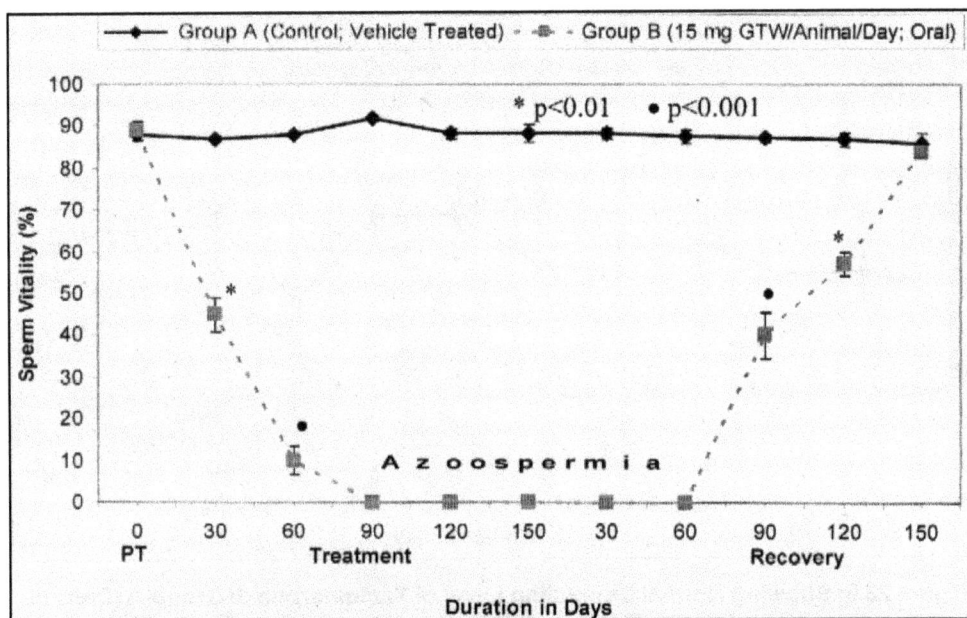

Figure 23.5: Showing the Significant Decrease in Sperm Vitality of GTW Recevied Animals from 30-days of Treatment. The each value represents mean±S.E. (Vertical bars).

Clinical Chemistry

Activities of serum transaminases (SGOT and SGPT), CPK, LDH were within the normal range when compared with control group during study period.

Electrolytes

Administration of GTW didn't affect the metabolism of serum electrolytes (Na, K, Mg, PO_4 and Cl) during this research investigation.

Circulating Level of Testosterone

No significant changes were found in the circulating level of testosterone (Figure 23.6).

Discussion

Our earlier study with 6 mg/animal/day for 120-days (Mukesh Kumar, personal communication) showed gradual decline in sperm density and reached to near oligospermic state at the end of 120-day of treatment in male langur monkeys. Present study is based on the our earlier findings with relatively higher dose (15 mg) level of GTW which caused gradual decline in sperm density from 60-onwards of the treatment and all animal were in azoospermic state during 90-days of treatment and this state lasted upto 60-days of the discontinuation of treatment. Sperm motility and vitality also gradually declined following GTW treatment. Effect of GTW on sperm density have also been shown in rats and mice by Chinese researchers (Qian, 1987;

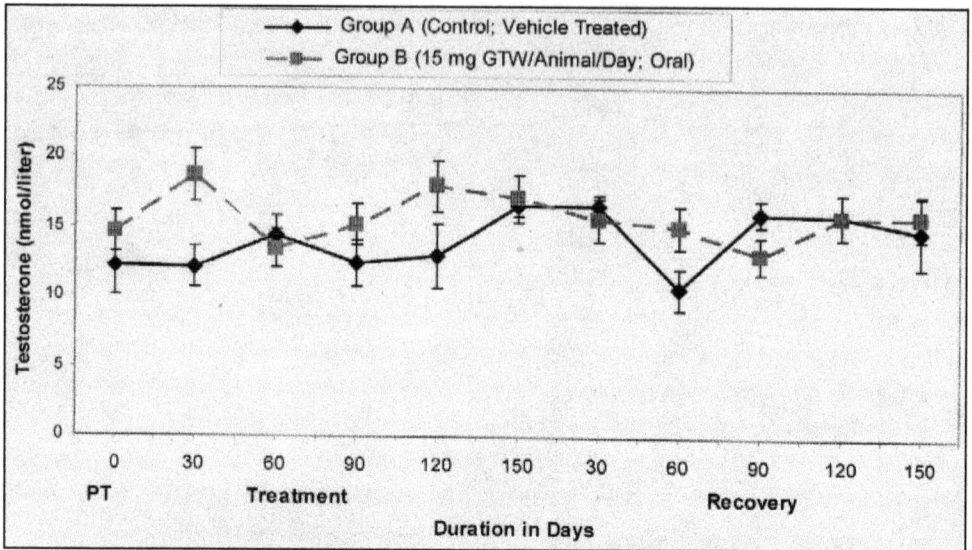

Figure 23.6: Showing Normal Circulating Level of Testosterone of Group-A (Control; vehicle treated) and Group-B (15 mg GTW/animal/day; Oral) during Pretreatment, Treatment and Recovery Phases. Each value represents mean±S.E. (Vertical bars).

Qian *et al.,* 1995). Chinese studies hypothesized that reduction in fertility is mainly due to the effect on epididymal spermatozoa (as sperm motility initially and severely got affected) and secondarily to the spermatogenic cells (Zhong, 1986). Our study revealed the reduction in sperm motility and complete disappearance/elimination of spermatozoa from semen samples, which reflects that GTW has got anti-motility and anti-spermatogenic effects but complete reversal of spermatogenesis was evident following 150-days of the cessation of treatment, which appears that terminal spermatogenic cells only got affected at this dose level being used. The very important finding of this study is that the normal circulating level of testosterone which reflects that GTW did not cause any damage to the Leydig cells which helped in the recovery of the spermatogenesis process and similar results have also been found in rat studies (Qian, 1987). Normal semen biochemical analysis shows that GTW did not affect accessory glands functioning which is also evident by the normal semen weight and volume being found in all treated animal when compared with our parallel control animals.

Normal haematological and serum biochemical parameters show that the GTW treatment did not affect the general well being of the male langur monkeys during entire period of the study.

In conclusion, GTW treatment has got anti-motility and anti-spermatogenic effects in male langur monkeys without affecting the circulating level of testosterone. These effects where dose and duration dependent. General health of treated animals did not get affected at this dose level being used and complete normalcy was restored following discontinuation of treatment. Findings of this langur monkey could be extrapolated to further human studies in order to develop this very important plant product (GTW) as an ideal male birth control pill.

Acknowledgements

This research study was sponsored by Indian Council Medical Research (ICMR), New Delhi and financial assistance to Dr. Mukesh Kumar in form of Senior Research Fellow (SRF) and Research Associateship (RA) from ICMR is gratefully acknowledged.

References

Asatoor, A.M. and King, E.J., 1954. Simplified colorimetric blood sugar method. *Biochem. T. J.*, **56**: XLIV

Cabaud, P.G. and Wroblewski, F., 1958. Colorimetric measurement of lactic dehydrogenase activity of body fluids. *Am. J.Clin.Pathol.*, **30**: 234 - 236

Crosby, W.H., Mann, J.I. and Furth, F.W., 1954. Standardizing a method for clinical hemo-globinometry. *U. S. Armed Forces Med. J.*, **5**: 693- 703.

Gomorri, C.J., 1942. Colorimetric determination of Phosphate. *Lab. Clin. Med.*, **27**: 955.

Gutman, E.B. and Gutman, A.B., 1940. Estimation of acid phosphates activity of blood serum. *J. Biol. Chem.*, **136**: 201–209.

Hughes, B.P., 1962. Colorimetric determination creatinine phosphokinase. *Clin. Chem. Acta.*, **7**: 597.

Jia, L., 1985 *Chin. Pharm. Bull.*, **20**: 101-105.

Jonh, A.D., 1960. Flame photometry McGraw-Hill Book Company Inc. 295.

Li, L.S., Zhang, Chen, H.P. Li, Y., Ji, DX, Zhan, J.H., How, F.F. and Chen, P.D. 1982. Effect of *Trypterygium wilfordii* in the treatment of nephritis. *Natl. Med., J. China*, **62**: 581–585.

Lindner, H.R. and Mann, T., 1960. Relationship between the content of androgenic steroids in the testes and the secretary activity of the seminal vesicles. *Bull. J. Endocr.*, **21**: 341 – 360.

Lohiya, N.K. and Sharma, O.P. 1983. Effect of cyproterone acetate with combination of testosterone enanthate on seminal characteristic, andragenecity and clinical chemistry in langur monkey. *Contraception*, **28**: 575 – 56.

Lohiya, N.K., David, G.F.X. and Anand Kumar, T.C. 1983. The longer monkeys as an experiment model for research in reproduction. *Contracept. Deliv. Syst.*, **4**: 89.

Lynch, M.J., Raphael, S.S., Mellore, L.D., Spare, P.D. and Inwood, M.J.H. 1969. *Medical laboratory: Technology* and *clinical pathology* WB Saunders Company, Philadelphia.

Mann, T. (Ed.) 1964. *The Biochemistry of Semen and of the male reproductive tract* Methuen, London.

Mastroianni, L., Jr. and Manson, W.A. Jr., 1963. Collection of monkey semen by electroejaculation. *Proc. Soc. Exp. Biol. Med.*, 1025 – 1027.

Mukesh Kumar. Personal communication.

Natelson, S., 1951. Routine use of ultramicro methods in the clinical laboratory. *Am. J. Clin. Pathol.,* **21**: 1153-1172.

Neill, D.W. and Neely, R.A., 1959. Estimation of magnesium in serum using titan yellow. *J. Clin. Pathol.,* **9**: 162 – 163.

Qian S Z (1987) *Trypterygium wilfordii,* a Chinese herb effective in male fertility regulation. *Contraception,* **36**: 335 – 345.

Qian, S.Z., Xu, Y. and Zhang, J.W., 1995. Recent progress is research on *Trypterygium*: A male antifertility agent. *Contraception,* **51**: 121 – 129.

Qian, S.Z., Zhang, C.Q. and Xu Y., 1986a. Effect of *Trypterygium Wilfordii* on the fertility of rats. *Contraception,* **33**: 105 – 110.

Qian, S.Z., Zhang, C.Q., Xu, N. and Xu, Y., 1986b. Anti-fertility effect of *Trypterygium wilfordii* in men. *Adv. Contracept.,* **2**: 253 – 254.

Qin, G.W., Yang, X.M., Gu, W.H., Wang, B.D., Chen, Z.X., Guo, R.X. and Shao, K.W., 1982. The structure of two new diterpene lactones from *Trypterygium wilfordii, wilforlide* A and B. *Acta. Chimica. Sinica.,* **40**: 637-647.

Reitman, S. and Frankel, S., 1957. A colorimetric method for determination of serum glutomic oxalo acetic and glutamic pyruvic transaminases. *Am. J. Clin. Pathol.,* **28**: 56-63.

Schales, O. and Schales, S., 1941. A simplified and accurate method for determination of chloride in biological fluids. *J. Biol.Chem.,* **140**: 879-884.

White, I.G., 1959. Studies on the estimation of glycerol, fructose and lactic acid with particular reference to semen. *Aust. J.Exp.Biol.,* **37**: 441 450

WHO, 1980.: Laboratory manual for the examination of Human Semen and Semen–Cervical Mucus interaction WHO special programme of Research, Development and Research Training in Human Reproduction. World Health Organization, Geneva, Switzerland.

Zhong, Q., Xu, Y. and Qian, S.Z. Unpublished Data.

Previous Volumes

— Volume 1 —

2002, xiii+308p., figs., tabls., ind., 23 cm   Rs. 995

ISBN 978-81-7035-272-3

Section I: Animal Ecology

Section II: Animal Reproduction

— Volume 2 —

2004, xi+387p., figs., tabls., ind., 23 cm   Rs. 1195
ISBN 978-81-7035-322-5

Section I: Animal Ecology

Section II: Animal Reproduction

— Volume 3 —

2006, xvii+443p., figs., tabls., ind., 23 cm   Rs. 1150

ISBN 978-81-7035-424-6

Section I: Animal Ecology

Section II: Animal Reproduction

— Volume 4 —

2007, xiv+550p., figs., tabls., ind., 23 cm   Rs. 1400

ISBN 978-81-7035-459-8

Section I: Animal Ecology

Section II: Animal Reproduction

— Volume 5 —

2008, xiv+476p., col. plts., figs., tabls., ind., 23 cm   Rs. 1600

ISBN 978-81-7035-563-2

Section I: Animal Ecology

Section II: Animal Reproduction

— Volume 6 —

2010, xiv+497p., col. plts., figs., tabls., ind., 23 cm   Rs. 1500

ISBN 978-81-7035-635-6

Section I: Animal Ecology

Section II: Animal Reproduction

— Volume 7 —

2011, xiv+494p., col. plts., figs., tabls., ind., 23 cm   Rs. 1700
ISBN 978-81-7035-746-9

Section I: Animal Ecology

Section II: Animal Reproduction

— Volume 8 —

2011, xviii+533p., col. plts., figs., tabls., ind., 23 cm   Rs. 1800

ISBN 978-81-7035-747-6

Section I: Animal Ecology

— Volume 9 —

2013, xvi+552p., col. plts., figs., tabls., ind., 23 cm   Rs. 2200

ISBN 978-81-7035-829-9

Index

* 9 7 8 9 3 5 1 3 0 6 6 2 7 *